Low Binder Concrete and Mortars

Low Binder Concrete and Mortars

Special Issue Editors

Jorge de Brito
Rawaz Kurda

MDPI • Basel • Beijing • Wuhan • Barcelona • Belgrade • Manchester • Tokyo • Cluj • Tianjin

Special Issue Editors

Jorge de Brito
University of Lisbon
Portugal

Rawaz Kurda
Erbil Polytechnic University
Iraq

Editorial Office
MDPI
St. Alban-Anlage 66
4052 Basel, Switzerland

This is a reprint of articles from the Special Issue published online in the open access journal *Applied Sciences* (ISSN 2076-3417) (available at: https://www.mdpi.com/journal/applsci/special_issues/concrete_mortars).

For citation purposes, cite each article independently as indicated on the article page online and as indicated below:

LastName, A.A.; LastName, B.B.; LastName, C.C. Article Title. *Journal Name* **Year**, *Article Number*, Page Range.

ISBN 978-3-03936-583-8 (Hbk)
ISBN 978-3-03936-584-5 (PDF)

Contents

About the Special Issue Editors

Jorge de Brito is a Full Professor of Civil Engineering in the Department of Civil Engineering, Architecture and Georesources; the former head of the CERIS Research Centre (2017–2018), and the director of the Eco-Construction and Rehabilitation Doctoral Programme at the Instituto Superior Técnico (IST), University of Lisbon, Portugal, from which he graduated and obtained his MSc and Ph.D. degrees. Although his research covers bridge management systems and construction technology, his main research area is sustainable construction, with an emphasis on the use of recycled aggregates in concrete and mortar. He has participated in 25 competitively financed research projects and supervised 40 Ph.D. and 180 MSc theses. He is the author of 7 books, 28 book chapters, and 440 papers in peer-reviewed international journals and has two patents. He is the Editor-in-Chief of the *Journal of Building Engineering*, an Associate Editor of the *European Journal of Environmental and Civil Engineering*, and a member of the Editorial Board of 44 international journals and of the following scientific/professional organizations: CIB, FIB, RILEM, IABMAS, and IABSE.

Rawaz Kurda received his Ph.D. in Civil Engineering from Instituto Superior Técnico (IST), University of Lisbon, Portugal. Since then, he has been working for the same university as a post-doctorate researcher. Additionally, he is also working as a researcher at Erbil Polytechnic University. He has published more than 23 ISI papers in scientific journals related to materials science, life cycle assessment, costs, multi-criteria analysis and optimization, sustainable concrete, supplementary cementitious materials, and construction and demolition waste.

Editorial

Special Issue Low Binder Concrete and Mortars

Jorge de Brito [1,*] **and Rawaz Kurda** [1,2,*]

[1] Instituto Superior Técnico, Universidade de Lisboa, Av. Rovisco Pais, 1049-001 Lisbon, Portugal
[2] Department of Civil Engineering, Technical Engineering College, Erbil Polytechnic University, Erbil 44001, Kurdistan-Region, Iraq
[*] Correspondence: jb@civil.ist.utl.pt (J.d.B.); rawaz.kurda@tecnico.ulisboa.pt (R.K.)

Received: 28 April 2020; Accepted: 28 April 2020; Published: 2 June 2020

1. Introduction

It is well known that, after water, concrete and mortars are the most demanded materials worldwide. Therefore, they have a significant influence on environmental impacts, namely because of their cement content. Thus, many alternative materials, such as supplementary cementitious materials, have been proposed to be used in order to decrease the environmental impacts of mortar and concrete. However, studies regarding extremely low binder content, namely cement, are still very scarce.

This Special Issue of Applied Sciences provides a forum for original studies and comprehensive reviews on the technical performance (e.g., mechanics and durability), economics and environmental impacts (e.g., global warming and energy consumption) of concrete and/or mortar containing low binder content. In other words, any attempts or techniques (i.e., using any type of supplementary cementitious materials, alkali activation (e.g., geopolymers), strengthening systems (e.g., fibers), by-product nano materials, additives, etc.) that help to decrease the cement content of concrete and mortars are welcome.

So far, 18 papers have been published in the Special Issue, of a total of 25 submitted. The next sections provide a brief summary of each of the papers published.

2. Developing Zero-Cement Binder

One way towards zero cement binder can be through the alkali activation technique. D'Elia et al. [1] studied the potential feasibility of the use of thermally pre-treated carbonate-high illite clay (LCR) as a precursor to produce alkali-activated paste. For that purpose, different activating solution concentrations (4, 6 and 8 M NaOH), with an alkaline solution/clay ratio of 0.55, were used. In terms of cost and technical performance, the paste with 6 M NaOH solution can be considered an optimum concentration. Additionally, an aluminium-enriched C-S-H (C-A-S-H) gel of paste with a higher concentration (e.g., 6 M NaOH) is less carbonated than that of the lower concentration. Furthermore, the output of this study showed that the LCR can be considered a potential material to be used in alkali-activated material with acceptable mechanical strength (20 and 30 MPa at 2 and 28 days, respectively).

Šimonová et al. [2] enhanced the fracture properties of alkali-activated mortar made with slag by using steel microfibers. The effective crack model, work-of-fracture method and double-K fracture model were considered to evaluate the fracture properties of the mortar. This study concluded that the addition of 10–15% of the precursor's mass in microfibers is an optimum percentage to enhance the fracture properties of the alkali-activated mortar. For example, the fracture toughness and fracture energy values of the mortar increased by up to 50% and 250% with the addition of the steel fiber of 20%, respectively. In addition, the compressive strength increased by up to 40% with the incorporation of the steel fiber of 10%.

3. Ecotoxicological and Chemical Characteristics of the Non-Conventional Materials Used to Replace Cement and Natural Aggregates

Concrete can be produced with low environmental impact and cost by incorporating some non-conventional binders and/or aggregates. Nevertheless, it may still not be considered a sustainable product due to its ecotoxicological issues. Thus, a study [3] evaluated the ecotoxicological potential of using fly ash (FA) and recycled concrete aggregates in concrete. In order to consider all aspects that relate to the toxicity, chemistry (non-metallic and metallic parameters) and ecotoxicology (battery of tests with the luminescent bacterium *Vibrio fischeri*, the freshwater crustacean *Daphnia magna*, and the yeast *Saccharomyces cerevisiae*) of eluates obtained from leaching, tests of raw materials and concrete containing FA and/or recycled concrete aggregates (RCA) were evaluated, and then compared to the limit values given by standards (non-hazardous or non-hazardous). The output of the results shows that, in terms of chemical characterization, although FA as a raw material can be classified as hazardous to landfill, concrete containing FA is classified as a non-hazardous material due to the solidification effect. In terms of ecotoxicity, although FA and RCA presented no or low levels of potential ecotoxicity, concrete with a high volume of FA and RA was classified as clearly ecotoxic, mainly due to *Daphnia magna* mobility.

4. Reduce the Environmental Impacts and Resources Use of Binders

As mentioned by Baghban and Mahjoub [4], reducing the environmental impacts and resource-use of binders are the key drivers for today's technology uptake. The mentioned study made a global review to show the feasibility of using non-conventional materials to reduce cement content and improve the durability of concrete. They show that limestone-calcined clay cement (LC3) is one of the promising binders that can be used in the concrete industry to improve mechanical performance, environmental impacts and cost. The authors then focused on another path to improve the performance of concrete containing LC3. This is due to the fact that concrete with high-performance characteristics leads to long service life, which is the main contributor to sustainability. For that purpose, they made an extensive literature review to see the effect of natural fibers, specifically "kenaf", on LC3 concrete, as a step toward green cementitious composite. They concluded that most mechanical (e.g., toughness, tensile strength and impact resistance) and durability (e.g., carbonation, sulphate and chloride resistance) properties can be enhanced compared to ordinary Portland cement concrete, specifically when the natural fiber is about 1% of binder's mass, with a length of about 50 mm. The authors suggested using a combination of LC3 with natural fibers such as kenaf fiber due to the fact that the kenaf plant absorbs a significant amount of CO_2 from the air.

Hu et al. [5] showed that cement content can be reduced by decreasing the porosity of the paste matrix. For that purpose, cement was replaced with different components and different grade tailings of waste limestone powder. Then, the specimens were characterized via nuclear magnetic resonance and scanning electron microscopy to understand their microstructure and pore properties. Based on the results, with a given grade tailing of the waste limestone powder, 20% of cement content can be replaced with limestone powder without jeopardizing the mechanical properties. Further, porosity, macropore proportion and the average pore radius decreased by 7%, 46% and 16%, respectively.

As mentioned before, the durability of concrete can be an effective key factor to obtain sustainable concrete. Accordingly, Rehman et al. [6] studied the influence of the thermo-mechanical activation of bentonite on the mechanical and durability performance of concrete. For that purpose, they studied the mechanical (compressive and tensile strength) and durability (water absorption, sorptivity and acid attack resistance) properties of concrete containing different incorporation ratios (10%, 15%, 20%, 25%, 30% and 35% mass replacements of ordinary Portland cement) of bentonite that was thermally and mechanically activated. Apart from the fact that the environmental impacts of bentonite are way lower than those of ordinary Portland cement, specifically when mechanically activated, both the durability and the mechanical properties of the concrete mixes improved with the incorporation of bentonite in both activation techniques. The results also show that the incorporation ratio of bentonite of 15–25% is the optimum in terms of technical performance of concrete, and specifically for durability.

Hafez et al. [7] highlighted, through a systematic review of more than 300 peer reviewed articles, that the current methodology for environmental impact assessment of concrete is not conclusive. The reviewed values for the environmental impact of green concrete seemed to spread widely, with a mean quite close to that of ordinary Portland cement concrete mixes. Hence, the authors established five golden rules as a cornerstone for a standard concrete life cycle assessment methodology: (i) Include the whole life cycle, including the CO_2 uptake throughout the service life; (ii) Develop a functional unit based on performance-based specifications of the concrete product; (iii) Rely on primary inventory data whenever possible; (iv) Combine several mid-point indicators; (v) Perform sensitivity and scenario analyses to tackle the intrinsic uncertainty of the data. The same authors [8] made another study to validate the optimization process of green concrete, based on the cost, environmental impacts and technical performance.

Narloch et al. [9] used an artificial neural network to establish a model for predicting the compressive strength of cement-stabilized rammed earth, using only scanning electron microscope images. The model showed reliable prediction results relative to other conventional methods, achieving 84% prediction accuracy without using a significant amount of materials to test the samples.

5. Modify the Characteristics of the Cement-Based Materials

Yeon [10] focused on the way to enhance the deformability of three-dimensional printable mortar containing FA and silica fume, using ethylene-vinyl acetate (EVA). Thus, they mainly focused on the effect of the EVA/binders ratio on the modulus of elasticity, drying shrinkage, and the thermal expansion of the mortar. The results show that as the EVA/binders ratio increased, the elastic modulus and compressive strength tended to reduce, and the opposite occurred for drying shrinkage and thermal expansion. Nevertheless, the maximum compressive strain seemed to increase. Accordingly, this path, specifically using a high EVA/binders ratio, is not advisable to enhance the deformability of three-dimensional printable mortar.

In another study, Fan et al. [11] also intended to improve the mechanical properties of mortar using approximately 0–2.0% (by cement's mass) polyvinyl alcohol (PVA). To understand the effect of PVA on the mechanical properties, hydration products and microstructure of the mortar, differential scanning calorimetry (DSC), Fourier transform infrared spectroscopy (FTIR) and scanning electron microscopy (SEM) results were obtained. They show that the optimum PVA content was 0.6% and 1.0% (by cement's mass) for the tensile and compressive strength of mortar. The authors explained that this is due to the fact that PVA formed evenly dispersed, network-like thin films within the cement matrix, and subsequently can act as bridges across microcracks and small voids to arrest cracking in the cement matrix.

Additionally, Park et al. [12] made a study under the title "experimental study on flexural behaviour of TRM-strengthened reinforced concrete (RC) beam: various types of textile-reinforced mortar with non-impregnated textile", and they showed the feasibility of strengthening RC beams using textile-reinforced mortar (TRM) with non-impregnated textile. The results show that the bearing capacity, strengthening efficiency and crack uniformity of beams can be improved with the use of TRM, especially when the mesh size of the textile was reduced and mechanical end anchorage implemented, and also when the textile lamination and textile reinforcement ratios increased.

Niu et al. [13] also attempted to modify the performance of the cement-based materials, namely concrete, using different types and combinations of fibers. For that purpose, the basalt fiber (BF) and polypropylene fiber (PF) were jointly and individually incorporated in concrete. The results show that the fibers can be beneficial to the compressive and tensile strength of concrete when a small percentage (each 0.05% of concrete's volume) of both fibers are jointly incorporated into concrete. The authors recommended using a high percentage of the fibers, but accurate control is necessary to disperse the fibers in the cement matrix. Similar to the performance of most fibers, PF and BF are more effective in enhancing tensile strength than compressive strength.

Yousuf et al. [14] made an extensive literature review to find a way to solve the issue of the crossover effect phenomenon in cement-based materials when they are subjected to accelerated initial curing. The higher curing temperature (e.g., 40–60 °C) may promote the cement matrix to rapidly create non-uniform hydration products, and subsequently generate a great porosity at later ages. The authors proposed using different types and incorporation ratios of cement-based materials to overcome this issue.

6. Low Binder Concrete On-Site Application

Apart from the production of the raw materials, the total environmental impacts and cost of cement-based materials also depend on the execution process. One way to promote this path can be with the three-dimensional printable technique. For example, Chen et al. [15] focused on the way to use non-conventional materials in the mentioned technique. Thus, they used high, medium and low grades (ratio of metakaolin) of calcined clay. They recommended the use of a high content of metakaolin in calcined clay because it considerably shortens the initial setting time, and increases the extrusion pressures and compressive strength at early and later ages. The same attempts can be seen in study [10] (Section 5).

7. Sustainable Cement-Based Materials in Road Engineering

Another way to reduce the consumption of natural aggregates is by reusing the existing pavement as aggregate. Gonzalo-Orden et al. [16] showed that this path has some uncertainty due to the fact that there is no specific test to determine the behavior of recycled pavements stabilized (full-depth reclamation—FDR) with cement. Thus, FDR with cement has been treated as cement-bound granular material or soil-cement. Since the bituminous material may be attached to the recycled pavement, the stiffness of the FDR with cement may be reduced. Therefore, this study intended to find long-term correlations between indirect flexural and tensile strength, and unconfined compressive strength in the FDR with cement, and to compare the outputs with the same relationships obtaining between soil-cement and cement-bound granular materials, in order to confirm whether their behaviors are similar or different; the latter was the case (the similar behavior hypothesis was not entirely accurate).

As in the previous study, the next ones [17,18] also focused on sustainable road engineering. For that purpose, Liang et al. [17] recommended the use of a large amount of crushed construction waste, namely clay brick, to replace the fine aggregates of a cement stabilized macadam (CSM) subbase. The results showed that using 50% fine recycled clay aggregates is optimum for the CSM subbase. Du et al. [18] also studied the application of CSM. The authors confirmed the advantage of using sandstone in CSM, and that it can be used for road base construction.

Funding: This research received no external funding.

Acknowledgments: Thanks are due to all the authors and peer reviewers for their valuable contributions to this special issue. The MDPI management and staff are also to be congratulated for their untiring editorial support for the success of this project.

Conflicts of Interest: The authors declare no conflict of interest.

References

1. D'Elia, A.; Pinto, D.; Eramo, G.; Laviano, R.; Palomo, A.; Fernández-Jiménez, A. Effect of Alkali Concentration on the Activation of Carbonate-High Illite Clay. *Appl. Sci.* **2020**, *10*, 2203. [CrossRef]
2. Šimonová, H.; Frantík, P.; Keršner, Z.; Schmid, P.; Rovnaník, P. Components of the Fracture Response of Alkali-Activated Slag Composites with Steel Microfibers. *Appl. Sci.* **2019**, *9*, 1754. [CrossRef]

3. Rodrigues, P.; Silvestre, J.D.; Flores-Colen, I.; Viegas, C.A.; Ahmed, H.H.; Kurda, R.; de Brito, J. Evaluation of the Ecotoxicological Potential of Fly Ash and Recycled Concrete Aggregates Use in Concrete. *Appl. Sci.* **2020**, *10*, 351. [CrossRef]

4. Baghban, M.H.; Mahjoub, R. Natural Kenaf Fiber and LC3 Binder for Sustainable Fiber-Reinforced Cementitious Composite: A Review. *Appl. Sci.* **2020**, *10*, 357. [CrossRef]

5. Hu, J.; Ding, X.; Ren, Q.; Luo, Z.; Jiang, Q. Effect of incorporating waste limestone powder into solid waste cemented paste backfill material. *Appl. Sci.* **2019**, *9*, 2076. [CrossRef]

6. Rehman, S.U.; Yaqub, M.; Noman, M.; Ali, B.; Khan, A.; Nasir, M.; Fahad, M.; Muneeb Abid, M.; Gul, A. The Influence of Thermo-Mechanical Activation of Bentonite on the Mechanical and Durability Performance of Concrete. *Appl. Sci.* **2019**, *9*, 5549. [CrossRef]

7. Hafez, H.; Kurda, R.; Cheung, W.M.; Nagaratnam, B. A Systematic Review of the Discrepancies in Life Cycle Assessments of Green Concrete. *Appl. Sci.* **2019**, *9*, 4803. [CrossRef]

8. Hafez, H.; Kurda, R.; Kurda, R.; Al-Hadad, B.; Mustafa, R.; Ali, B. A Critical Review on the Influence of Fine Recycled Aggregates on Technical Performance, Environmental Impact and Cost of Concrete. *Appl. Sci.* **2020**, *10*, 1018. [CrossRef]

9. Narloch, P.; Hassanat, A.; Tarawneh, A.S.; Anysz, H.; Kotowski, J.; Almohammadi, K. Predicting Compressive Strength of Cement-Stabilized Rammed Earth Based on SEM Images Using Computer Vision and Deep Learning. *Appl. Sci.* **2019**, *9*, 5131. [CrossRef]

10. Yeon, J. Short-Term Deformability of Three-Dimensional Printable EVA-Modified Cementitious Mortars. *Appl. Sci.* **2019**, *9*, 4184. [CrossRef]

11. Fan, J.; Li, G.; Deng, S.; Wang, Z. Mechanical Properties and Microstructure of Polyvinyl Alcohol (PVA) Modified Cement Mortar. *Appl. Sci.* **2019**, *9*, 2178. [CrossRef]

12. Park, J.; Hong, S.; Park, S.-K. Experimental Study on Flexural Behavior of TRM-Strengthened RC Beam: Various Types of Textile-Reinforced Mortar with Non-Impregnated Textile. *Appl. Sci.* **2019**, *9*, 1981. [CrossRef]

13. Niu, D.; Huang, D.; Zheng, H.; Su, L.; Fu, Q.; Luo, D. Experimental Study on Mechanical Properties and Fractal Dimension of Pore Structure of Basalt–Polypropylene Fiber-Reinforced Concrete. *Appl. Sci.* **2019**, *9*, 1602. [CrossRef]

14. Yousuf, S.; Shafigh, P.; Ibrahim, Z.; Hashim, H.; Panjehpour, M. Crossover Effect in Cement-Based Materials: A Review. *Appl. Sci.* **2019**, *9*, 2776. [CrossRef]

15. Chen, Y.; Li, Z.; Chaves Figueiredo, S.; Çopuroğlu, O.; Veer, F.; Schlangen, E. Limestone and Calcined Clay-Based Sustainable Cementitious Materials for 3D Concrete Printing: A Fundamental Study of Extrudability and Early-Age Strength Development. *Appl. Sci.* **2019**, *9*, 1809. [CrossRef]

16. Gonzalo-Orden, H.; Linares-Unamunzaga, A.; Pérez-Acebo, H.; Díaz-Minguela, J. Advances in the study of the behavior of Full-Depth Reclamation (FDR) with cement. *Appl. Sci.* **2019**, *9*, 3055. [CrossRef]

17. Liang, C.; Wang, Y.; Song, W.; Tan, G.; Li, Y.; Guo, Y. Potential Activity of Recycled Clay Brick in Cement Stabilized Subbase. *Appl. Sci.* **2019**, *9*, 5208. [CrossRef]

18. Du, Q.; Pan, T.; Lv, J.; Zhou, J.; Ma, Q.; Sun, Q. Mechanical Properties of Sandstone Cement-Stabilized Macadam. *Appl. Sci.* **2019**, *9*, 3460. [CrossRef]

Article

Effect of Alkali Concentration on the Activation of Carbonate-High Illite Clay

Angela D'Elia [1], Daniela Pinto [1,*], Giacomo Eramo [1], Rocco Laviano [1], Angel Palomo [2] and Ana Fernández-Jiménez [2,*]

[1] Dipartimento di Scienze della Terra e Geoambientali, Università degli Studi di Bari Aldo Moro, 70121 Bari, Italy; angela.delia@uniba.it (A.D.); giacomo.eramo@uniba.it (G.E.); rocco.laviano@uniba.it (R.L.)
[2] Instituto Eduardo Torroja (CSIC), 28033 Madrid, Spain; palomo@ietcc.csic.es
* Correspondence: daniela.pinto@uniba.it (D.P.); anafj@ietcc.csic.es (A.F.-J.)

Received: 25 February 2020; Accepted: 16 March 2020; Published: 25 March 2020

Abstract: The present study explores the effect of activating solution concentration (4, 6 and 8 M NaOH) on mechanically and thermally pre-treated carbonate-high illite clay (LCR). Pastes were prepared with an alkaline solution/clay (S/B) ratio of 0.55, which were cured at room temperature and relative humidity > 90% in a climatic chamber. At two and 28 days, compressive mechanical strength was determined, and the reaction products were characterised by X-ray Powder Diffraction analysis (XRPD), Fourier-transform infrared spectroscopy (FTIR) and Scanning Electron Microscopy - Energy Dispersive X-ray spectroscopy (SEM/EDX). Results obtained showed that the presence of reactive calcium in the starting clay induces co-precipitation of a mix of gels: An aluminium-enriched C-S-H gel (C-A-S-H) and a N-A-S-H gel, in which sodium is partially replaced by calcium (N,C)-A-S-H. Pastes prepared with higher (6 or 8 M) activator concentrations exhibit a more compact matrix than the ones prepared with 4 M NaOH. The findings show that the use of a 6 M NaOH solution yields a binder with two days compressive strength >20 MPa and 28 days strength of over 30 MPa.

Keywords: carbonate-high illite clay; alkali activation; NaOH; activator concentration

---

1. Introduction

Alkali-activated materials (AAMs) are binders obtained when aluminosilicate materials ("precursors") react chemically with an alkaline activator (normally an alkaline hydroxide or alkaline silicate solution) [1–6]. As materials with a low environmental impact, AAMs are attracting considerable research as cements for the future, along with limestone calcined clay (LC3) [7,8] and belite–ye'elimite–ferrite (BYF) cements [9,10].

One of the major technological challenges to be faced in alkaline activation is to find precursors that are both widely available and technically suitable. The aluminosilicates most commonly used include burnt kaolinite clay and industrial by-products such as fly ash (FA) and ground granulated blast furnace slag (GGBFS) [11–13]. Output of FA and GGBFS is very high at this time and both will continue to be readily available during the twenty-first century. Nonetheless, the environmental challenges facing humanity herald a decline in future production, and the universal geographic availability for industrial scale AAM manufacture is not guaranteed. Pure kaolinite clay (metakaolin) deposits, in turn, are characterised by limited market availability and competition from other industries (paper, fired clay ceramics), which raise the cost of this raw material. Moreover, metakaolin-based AAMs are subject to workability [14] and drying shrinkage [15] problems.

In contrast, clay minerals, outside of very pure kaolinite, constitute an abundant source of aluminosilicates, available worldwide. Studies carried out in recent years on ways to enhance the reactivity of "low purity" clay minerals, with a view to AAM [14,16] and LC3 [7,8] cement manufacture, have shown that their mineralogy (crystallinity), texture, and particle size [17–21] affect its reactivity.

As a rule, lower crystallinity, greater surface amorphism and smaller particle size lead to higher initial reactivity in precursors and hence to binders with greater initial mechanical strength. Most of the studies published described clays with a low calcium oxide content such as kaolinite [6,22–24], illite [25–27], bentonite [28,29] or even sedimentary [5,20] materials.

The main post-alkali-activation reaction product in precursors with a low CaO content (<10%) is a three-dimensional alkaline aluminosilicate hydrate gel (Na_2O-Al_2O_3-SiO_2-H_2O) commonly named as, N-A-S-H. Conversely, if the CaO content in the precursor is fairly high (30% to 40%, such as in blast furnace slag), the main reaction product is a two-dimensional calcium aluminosilicate hydrate gel (CaO-Al_2O_3-SiO_2-H_2O), commonly named as, C-A-S-H [11,12]. These two gels differ both chemically and structurally from the calcium silicate hydrate (CaO-SiO_2-H_2O), commonly named as C-S-H, that precipitates as the main hydration product in ordinary Portland cement (OPC) [7].

Activator CaO content and pH, which also play a significant role in the chemical composition and structure of the cementitious gels forming in AAMs, constitute a promising field of research at this time [11,12]. Many studies have shown [30,31] that mixing a precursor with moderate (20% to 30%) amounts of calcium in 4 to 5 M caustic solutions favours the solution/precipitation of a C-A-S-H gel at ambient temperature. Higher alkalinity may raise the degree of reaction and Si and Al (although not Ca) solubility, favouring the precipitation of a N-A-S-H gel or a mix of N-A-S-H/C-A-S-H gels that may interact over time [32] to make (N,C)-A-S-H (calcium-bearing N-A-S-H) gel or C-(N)-A-S-H (sodium bearing C-A-S-H) gel. The composition and structure of the aforementioned gel/mix of gels and the parameters affecting their nature and stability (pH, temperature, moisture …) are studied in the so-called "hybrid alkaline cements" obtained in the alkaline activation of blends bearing approximately 20 wt.% of CaO, SiO_2 and Al_2O_3 [33,34].

Against that backdrop, the present study forms part of broader research that explores the potential of carbonate-bearing (~30% CaO) illite clay in alkaline cement formulation. In earlier experiments [21] focused on optimising milling, temperature and thermal treatment to raise clay reactivity, the authors observed that intense clay grinding lowered clay mineral decomposition temperatures. Milling for 15 min in a high energy mixer mill followed by heating at 700 °C for 1 h was found to be the most effective pre-activation treatment for calcium carbonate-high illite clays for the intents and purposes of geopolymer manufacture.

The present study explores the effect of activating solution concentration (4, 6 and 8 M NaOH) on carbonate-bearing, mechanically and thermally pre-treated illite clays. More specifically, the material used was a carbonate-high illite clay largely occurring in southern Italy (Apulia and Basilicata regions), which may constitute an innovative, widely available and suitable precursor for alkaline cement production. The findings can be extrapolated to other carbonate-bearing clays as well as to other possible precursors with a similar composition. The authors note that this study also affords the opportunity for in-depth research into the role of calcium and the degree of alkalinity on alkaline activation.

2. Materials and Methods

The material used in this study was a carbonate-high illite clay (LCR) mined in the region of Apulia, Lucera, FG, Italy. Based on earlier findings [21], this clay was pre-treated mechanically (15 min at 1800 rpm in a Retsch MM 400 mixer mill) and thermally (700 °C for 1 h) to yield a clay here labelled MA157. Table 1 gives the chemical compositions of the original and pre-treated clays as found on a Panalytical AXIOS-Advanced XRF spectrometer fitted with an X SST-mAX (Rh anode) X-ray source. The table also lists the XRD (Rietveld method)-determined mineralogical composition for the clays [21].

Table 1. Bulk chemical composition and mineralogical content of the original untreated clay and pre-treated MA157.

Chemical Composition (wt%) XRF			Mineralogical Composition (wt%, 3σ)		
	Untreated Clay	**MA157** [1]		**Untreated Clay**	**MA157** [1]
SiO_2	35.67	43.05	**Quartz**	16.3 (0.7)	20.6 (1.4)
Al_2O_3	11.47	13.14	**Calcite**	31.2 (1.3)	5.4 (1.1)
CaO	21.15	29.97	**Illite/Muscovite**	3.8 (0.5)	3.5 (1.0)
K_2O	1.73	2.28	*R0* illite(0.3)/smectite mixed layers [2]	24.5 (2.9)	-
Na_2O	0.34	0.36	**Kaolinite**	5.9 (1.0)	
Fe_2O_3	6.67	9.34	**K-feldspar**	4.2 (1.0)	10.6 (1.7)
MgO	1.91	2.22	**Albite**	3.2 (0.5)	3.4 (0.8)
MnO	0.15	0.21	**Goethite**	1.7 (0.2)	-
TiO_2	0.53	0.73	**Hematite**	-	2.8 (0.5)
P_2O_5	0.07	0.09	**Dolomite**	1.0 (0.2)	-
LoI	20.32	0.61	**Chlorite**	1.0 (0.3)	-
			Amorphous matter	7.1 (6.3)	53.9 (3.6)
			Tot C.M.	35.2 (3.7)	3.5 (1.0)

[1] Clay ground in a mixer mill and subsequently heated to 700 °C; [2] illite/smectite interstratifications of *R0* (Reichweite) ordering type and with composition of 30 mol% illite and 70 mol% smectite according to the nomenclature from Moore and Reynolds [35]. Tot C. M. stands for total clay minerals.

Figure 1 shows the particle size distribution for clay MA157 dispersed via ultrasound in a DOLAPIX CE64 1/100 ethanol and dispersion medium, as determined on a SYMPATEC laser sizer with a measuring range of 0.05 to 875 μm.

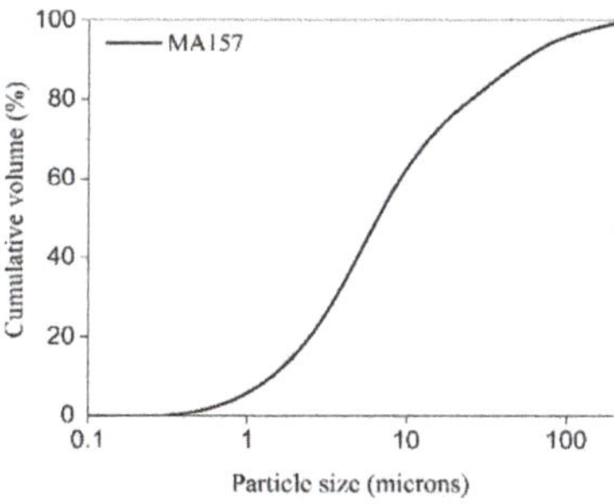

Figure 1. MA157 particle size distribution.

MA157 clay pastes were prepared with 4, 6 and 8 M solutions of NaOH at an alkaline solution/clay (S/B) ratio of 0.55, previously found to be the most suitable to ensure good paste workability. The pastes were subsequently poured into prismatic ($1 \times 1 \times 6$ cm^3) moulds and cured at ambient temperature and relative humidity > 90% in a climate chamber. They were removed from the moulds 20 h later and stored in the climate chamber until the 2 and 28 days test ages, when six samples at each age were tested for compressive strength on an IBERTEST® test frame (six specimens per test age and material were tested).

The reaction products were analysed by X-ray Powder Diffraction analysis (XRPD) with a Bruker AXS D8 Advance Cu-Kα radiation diffractometer operating at 30 mA and 40 kV across a 2θ range of 5° to 60° and scanning at 0.0197° (2θ) steps. They were likewise characterised Fourier-transform infrared spectroscopy (FTIR), on KBr pellets (1 mg of sample in 200 mg of KBr) with a Nicolet Thermo FTIR spectrometer, recording 64 scans per sample at 4000 to 400 cm^{-1}. Lastly, the products were analysed along the sputter-coated fracture surfaces of the samples under Scanning Electron Microscopy - Energy Dispersive X-ray spectroscopy (SEM/EDX) a JEOL JSM 5400 SEM equipment fitted with an Oxford

Link-ISIS-EDX spectrometer, collecting the images with secondary electrons (SE) at a working distance of 15 mm and a 20 kV accelerating voltage. The EDX (point) analyses were conducted with a 20 μm beam, also at a 20 kV accelerating voltage and a working distance of 15 mm. Acquisition time was 30 s per point analysis.

3. Results and Discussion

3.1. Mechanical Strength

Figure 2 shows the two and 28 days compressive strengths for the MA157 pastes prepared with 4, 6 and 8 M NaOH solutions. In the two days materials, the highest strength, at >25 MPa, was recorded for the pastes prepared with 6 M NaOH. The 28 days pastes exhibited strength of 32 MPa at 4 M NaOH, 33 MPa at 6 M NaOH and 36 MPa at 8 M NaOH. The increase in mechanical strength with time was attributed to the greater degree of reaction, although strength was also observed to depend on reaction product chemical, mineralogical and microstructural composition, as discussed in a later section.

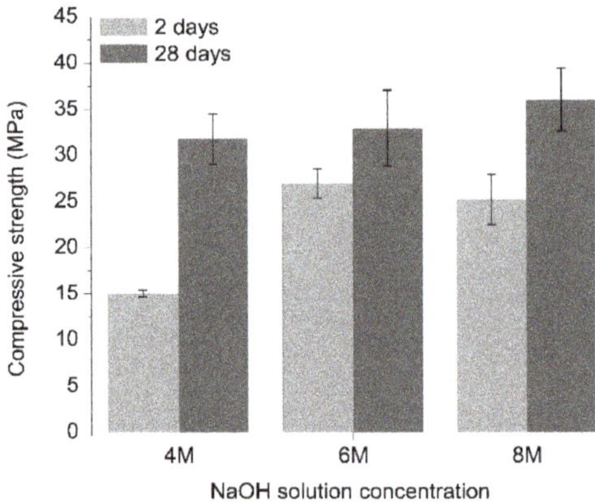

Figure 2. Two and 28 days compressive strength in 4, 6 and 8 M NaOH-activated pastes (six specimens per test age and material were tested).

The compressive strength values recorded here were higher than reported by other authors, who worked with thermally treated alkali-activated illite/smectite clay with CaO contents of <5%. Buchwald et al. [25] found compressive strength values of around 13 MPa with 6 M NaOH, (solution/solid ratio = 0.41, curing 20 h at 20 °C in sealed moulds avoiding drying out, after demoulding; seven days at 60 °C in 100% relative humidity (RH), 3 h at 80 °C in 100% RH and drying over 24 h at 40 °C). Essaidi et al. [26], using a potassium silicate solution, observed strength of 25 MPa (initial curing in an opened polystyrene sealed mould, 48 h at 70 °C, subsequently, the mould was closed into oven at 70 °C for 2 h; after demoulding ambient air (25 °C, 40% RH) during 21 days). Seiffarth et al. [19] obtained values of <10 MPa with sodium silicate molar ratio of $SiO_2/Na_2O = 0.31$ (initial curing in wet conditions, seven days, 25 °C and 85% RH, followed by curing in dry conditions: Three days, 40 °C, 55% RH).

The higher strength values found in this study were attributed to the high degree of clay reaction, the outcome of the initial mechanical-thermal treatment. As the benefits of that treatment, which induces phyllosilicate amorphism and carbonate decomposition, were described in detail in an earlier paper [21], they are not addressed hereunder. Another factor deemed to have contributed significantly to the good initial mechanical performance of this alkali-activated, ambient temperature-cured clay, was the presence of a fairly high (~30%) proportion of calcium. Like silicon and aluminium, calcium solubility is heavily dependent upon system alkalinity. Given that solution alkalinity played a substantial role in these complex systems (~43% SiO_2, 13% Al_2O_3 and 30% CaO, see Table 1), it was defined as the primary object of the present study.

3.2. Reaction Products

The XRD patterns for clay MA157 and the three (4, 6 and 8 M) pastes are reproduced in Figure 3. The precursor exhibited a significant amorphous component at 2θ values of 20° to 40°. The crystalline phases detected in MA157 included quartz, calcite, illite/muscovite, K-feldspar (orthoclase, albite and anorthite) and plagioclase. With the exception of calcite, the intensity of these diffraction lines barely varied after alkaline activation, an indication that the respective phases were inert. The patterns for all the pastes showed a slight shift in the hump located in the precursor at 2θ = 20° to 40° to around 2θ = 30° to 40° (the hump is a characteristic of amorphous and/or glassy phases) denoting a reaction between the amorphous phases in the clay and the formation of new, low structural order phases, primarily cementitious gels [4]. The lines attributed to the presence of calcite were more intense in the paste activated with 4 M NaOH. Although less calcite was formed at higher alkali concentrations (6 and 8 M), sodium carbonates or calcites such as natron ($Na_2CO_3 \cdot 10H_2O$ (PDF # 00-015-0800)) and gaylussite ($Na_2Ca(CO_3)_2 \cdot 5H_2O$) (PDF # 00-021-0343)) were detected on their diffractograms.

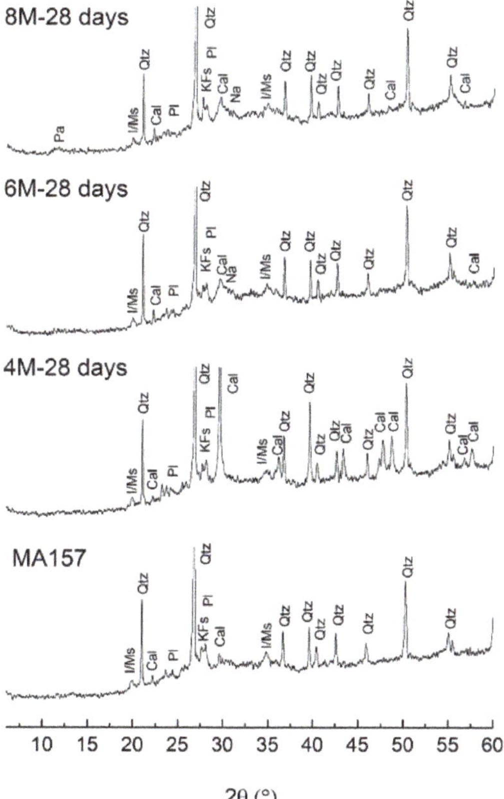

Figure 3. XRD patterns for anhydrous MA157 clay and 28 days, 4, 6 and 8 M NaOH-activated pastes Legend: I/Ms = illite/muscovite; Qtz = quartz; Pl = plagioclase, Kfs = K-feldspar, Cal = calcite; Na = natron; Pa = para-alumohydrocalcite/hemicarboaluminate.

Unlike other alkali-activated binders made with dehydroxylated clay as precursors [25–27], the pastes prepared here contained no zeolite crystals. That absence was attributed in part to the fairly low (13.4%) aluminium content in MA157 and/or to ambient temperature curing which would lower Al solubility. A number of authors have reported that zeolite formation as a secondary product in alkaline cements depends on both the reactive aluminium content and the synthesis parameters, primarily activator type and concentration [11,12,36], and curing conditions (time, temperature, humidity) [4,37].

A small hump was observed at $2\theta = 10°$ to $11°$, possibly associated with the formation of para-alumohydrocalcite ($CaAl_2(CO_3)_2(OH)_4\cdot6H_2O$ (PDF# 00-030-0222)), a hydrotalcite group mineral, or hemicarboaluminate ($Ca_4Al_2O_7CO_2\cdot11H_2O$ (PDF# 00-036-0377)). These phases are often present as secondary reaction products in calcium-high alkaline cements such as alkali-activated blast furnace slag [38] or limestone-metakaolin blends [39].

The FTIR spectra for clay MA157 and the two and 28 days pastes are reproduced in Figure 4. The spectrum for the starting material exhibited a wide band at approximately 1079 to 1086 cm^{-1}, associated with the asymmetric stretching vibrations generated by the Si-O-T (T = Si, Al) bonds possibly present in the amorphous component of the material [40]. It also contained vibration bands characteristics of quartz at 1084, 796, 694 and 463 cm^{-1} [41], a low intensity band at 1438 cm^{-1} attributed to the presence of calcium carbonate and another at 1635 cm^{-1} to the bending vibrations generated by the OH groups in water [41].

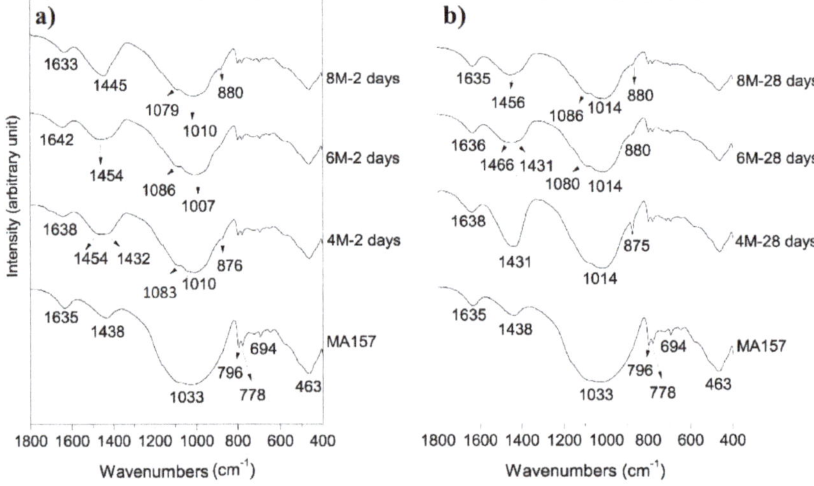

Figure 4. FTIR spectra for anhydrous MA157 clay and (**a**) two days and (**b**) 28 days pastes alkali-activated with 4, 6 and 8 M NaOH.

The Si-O-T (T = Si, Al) stretching band in the pastes was shifted to lower frequencies and its signal grew narrower at longer reaction times. That band appeared at 1010 cm^{-1} in the two days and at 1014 cm^{-1} in the 28 days pastes. Further to the literature [4,9,40,41], these variations stem from the initial alkaline-mediated dissolution of the amorphous/reactive component in the clay and the ensuing precipitation of a cementitious gel with a fairly high aluminium content (intermediate reaction product, band at 1010 cm^{-1}). The formation of that intermediate product would be explained by the synchronised dissolution of the aluminium and silicon ions in the alkaline medium in the initial stages. As the aluminium content was fairly low, upon its depletion the silicon would continue to react, affording the gel a higher proportion of silicon and shifting the FTIR band to higher values (1014 cm^{-1}).

The carbonate band on the spectra for the activated pastes were both shifted to slightly higher or lower wave numbers (1400 to 1500 cm^{-1}) and somewhat more intense. A shoulder appearing at 875 cm^{-1} confirmed the formation of the carbonates detected with XRD in the alkali-activated MA157 clay pastes. More specifically, vibration bands characteristic of calcite were observed at approximately 1432, 876 and 710 cm^{-1} [42,43] in the 4, 6 and 8 M NaOH-activated pastes, whereas the signal at approximately 1450 cm^{-1} might be attributed to sodium carbonate [43]. The bands associated with the presence of calcite were most intense on the spectrum for the 28 days sample activated with 4 M NaOH (Figure 4b), a finding consistent with the XPRD results (Figure 3).

Figure 5 depicts the two days and Figure 6 the 28 days micrographs for the 4, 6 and 8 M NaOH-activated pastes. A cementitious matrix consisting primarily in Si, Al and Ca was identified in all. The EDX analyses showed that the proportions of these elements, calcium in particular, varied depending on activator concentration and curing time. That finding was associated with the possible formation of a mix of (N,C)-A-S-H/N-(C)-A-S-H gels. The formation of that mix of gels and the high early mechanical strength at ambient temperature were primarily the outcome of the presence of calcium in these dehydroxylated clays [32,44,45]. Activator alkalinity also affected the type and proportion of the C-A-S-H (Al-enriched C-S-H) and/or (N,C)-A-S-H (calcium-bearing N-A-S-H) gels formed, as well as the amount and composition of the secondary reaction products (such as carbonates or AFm phases) (Figure 3).

As a rule the microstructure of the 2 days, 4 M NaOH-activated paste was less uniform than in the other materials (Figure 5a). The phases in the form of Ca-high plates in the Si, Al and Ca-containing cementitious matrix suggested the formation of an Al-tobermorite-like compound. In the 28 days 4 M-activated paste, the matrix was less compact and more granular (Figure 6a). That observation was associated with the significant carbonation in the 4 M NaOH-activated pastes (as revealed by the XRD and FTIR results) that apparently induced decalcification of the C-A-S-H-like cementitious gel. The EDX analyses denoted a lower calcium content in the 28 days than in the two days gels. Similar decalcification was described by Puertas et al. [46] in a study of C-A-S-H gel carbonation in a study of the alkaline activation of blast furnace slag.

At the two higher alkalinities, the microstructure was more uniform and compact. Nonetheless, the excess sodium from the higher concentration of the activating solution (especially in the 8 M NaOH-activated samples) induced calcium and sodium carbonate salt precipitation, as confirmed by the EDX analyses (Figure 5(c2)). The resulting calcite crystals were primarily located in, and filling, the matrix pores.

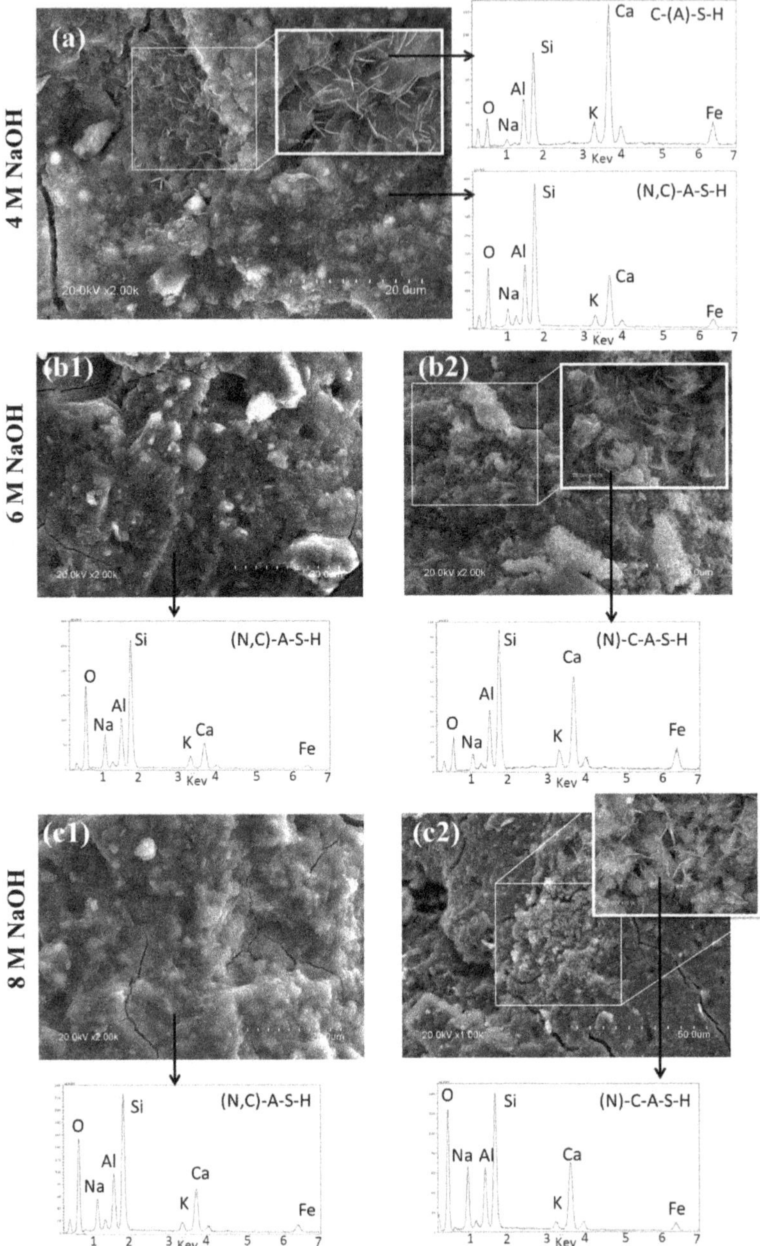

Figure 5. Two day SEM micrographs and EDX spectra for (**a**) 4 M, (**b**) 6 M (**b1,b2**) and (**c**) 8 M (**c1,c2**) NaOH-activated pastes.

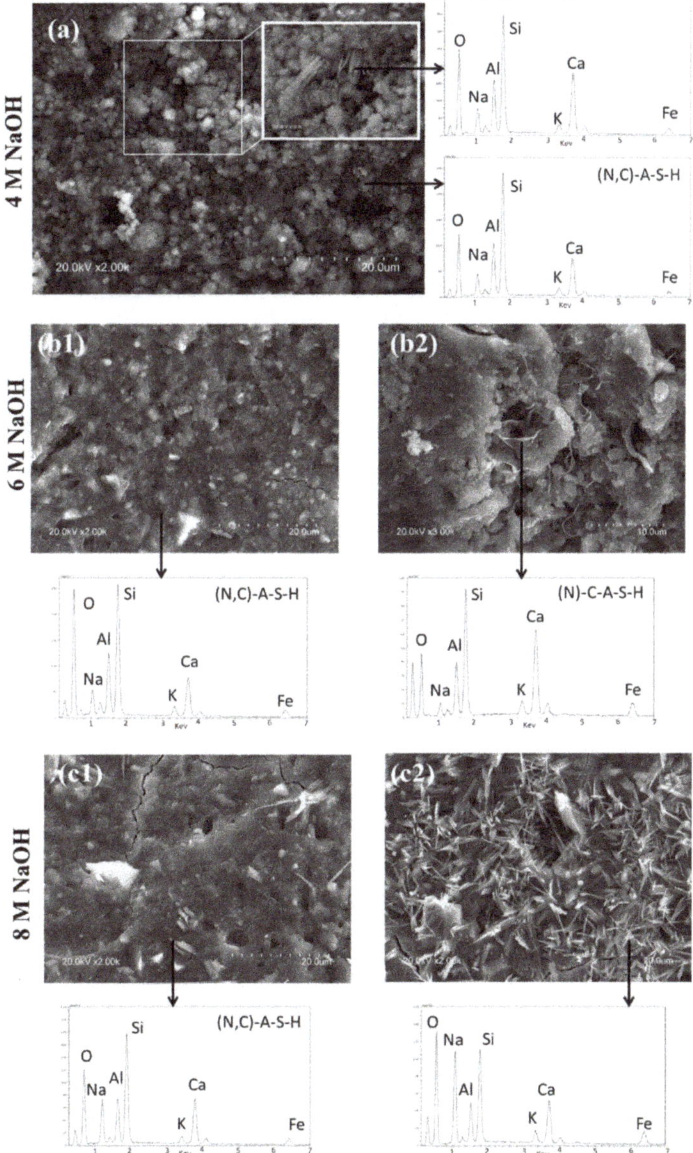

Figure 6. Twenty-eight day SEM micrographs and EDX spectra for (**a**) 4 M, (**b**) 6 M (**b1,b2**) and (**c**) 8 M (**c1,c2**) NaOH-activated pastes.

The EDX-determined variability in gel composition was graphed on CaO-Al_2O_3-SiO_2 and Na_2O-Al_2O_3-SiO_2 ternary diagrams (Figure 7) to study the effect of activator concentration on the type or types of gels formed. The findings showed that in all the cases studied, the alkaline activation of carbonate-high clays with NaOH solutions entailed the co-precipitation of a mix of C-A-S-H- and (N,C)-A-S-H-like gels, which had a higher SiO_2 and a lower Al_2O_3 content (i.e., a higher SiO_2/Al_2O_3 ratio) than other gels described in the literature [32]. Chemically speaking, these gels are more like the ones forming in blast furnace slag [47,48] than in dehydroxylated kaolinite alkaline activation [14,27].

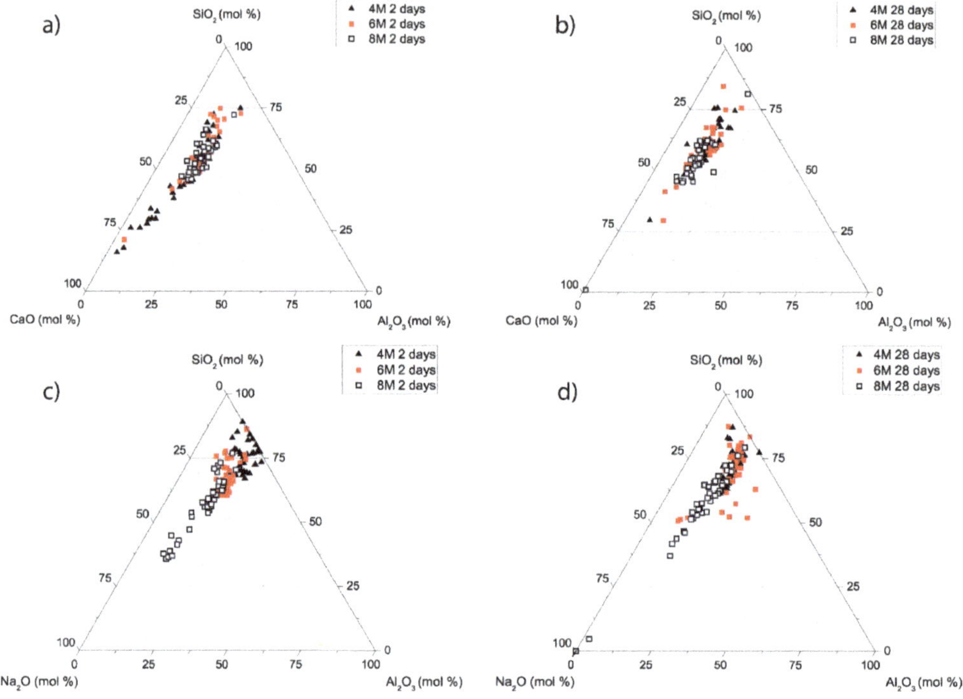

Figure 7. EDX data for the two and 28 days, 4, 6 and 8 M NaOH-activated pastes graphed on (**a,b**) CaO-Al$_2$O$_3$-SiO$_2$; and (**c,d**) Na$_2$O-Al$_2$O$_3$-SiO$_2$ ternary diagrams.

When the activator was 4 M NaOH and the time two days, although scattered, calcium clearly prevailed. The points with >60% CaO were associated with the Ca-high, plate-like particles visible in Figure 5a. Given their chemical composition and shape, those phases would appear to be consistent with an aluminium-enriched C-S-H such as Al-tobermorite [32,49,50].

The clusters observed on the micrographs for the two days, 6 and 8 M NaOH-activated pastes (Figure 7a) were characterised by a non-uniform distribution of CaO, ranging from 25% to 60% and associated with the precipitation of a mix of an Al-enriched C-S-H gel (C-A-S-H) and a calcium-bearing N-A-S-H gel (N,C)-A-S-H). With time, higher activator alkalinity (especially visible for 8 M NaOH) induced (N,C)-A-S-H gel evolution toward a C-A-S-H gel. Earlier studies [32] showed that N-A-S-H gel may persist at lower pH (<12), whereas in the presence of calcium, ionic exchange prompts the formation of (N,C)-A-S-H gel, in which calcium may be taken up until it ultimately replaces all the sodium in the cementitious gel's three-dimensional structure.

The 28 days, 8 M NaOH-activated MA157 paste exhibited a chemical composition primarily attributable to a C-A-S-H gel, although at 0.24 ± 0.07 the Al$_2$O$_3$/SiO$_2$ molar ratio was lower than the 0.3 to 0.5 range normally observed for such gels [32]. That was due the higher SiO$_2$ than Al$_2$O$_3$ content in the starting clay. In contrast, the matrix for that same paste exhibited a CaO/SiO$_2$ molar ratio of 0.64 ± 0.17, consistent with the range suggested in the literature (0.6 < CaO/SiO$_2$ < 1) for C-A-S-H gels [32,48]. The chemical composition of the 28 days, 6 M NaOH-activated pastes also moved toward higher CaO contents in the CaO-Al$_2$O$_3$-SiO$_2$ ternary diagrams (Figure 7b) than the two days pastes (Figure 7a).

In contrast, the matrices for the 28 days, 4 M NaOH-activated pastes were characterised by a lower CaO/SiO$_2$ ratio than the two days materials (Figure 7a,b) and a higher Na$_2$O uptake in the gel as the reaction progressed (Figure 7d). Those findings may be attributed to the higher carbonation observed in the latter pastes. As the XRPD and FTIR data showed, the 28 days, 4 M NaOH-activated pastes had a high calcite content, a result of the interaction between atmospheric CO_2 and the calcium ions in the material. Inasmuch as these pastes had no portlandite, the calcium was necessarily sourced from the C-S-H/C-A-S-H gel initially formed [46].

The mechanisms governing alkali-activated materials are presently under study by a number of authors, for many factors are involved, including type of starting materials [51], activator type [47], relative humidity [52] and CO_2 concentration [53]. The reasons underlying the higher carbonation in the 4 M than in the 6 M or 8 M NaOH-activated pastes are fairly complex. Bernal et al. [54] observed the carbonation rate to decline with rising activator concentration in calcium-high alkali-activated systems (such as metakaolin + slag) due to the more intense carbonate salt precipitation generated by the interaction between the highly alkaline solution and the CO_2 in the air. Carbonate salt precipitation would lower the pore volume in the matrix, hindering additional CO_2 ingress in the sample and carbonation progression [54]. As noted earlier, further to the XRD and SEM data, the 6 and 8 M NaOH-activated pastes, rather than developing high calcite contents, precipitated sodium and sodium-calcium carbonates such as natron and para-alumohydrocalcite/hemicarboaluminate.

Briefly, the rise in the degree of reaction with higher activator concentration, proven by the appearance of progressively more compact matrices in the pastes studied (Figure 5), would explain the higher compressive strength values in the 6 and 8 M than in the 4 M NaOH-activated pastes. The higher two and 28 days strength in the pastes studied here, than by other authors, was associated with a higher degree of reaction over time and the formation of gels with greater silicon and calcium contents.

There are different factors that affect the development of mechanical strength such as: total porosity, size and porous distribution, degree of reaction, type and morphology of the reaction product, structure and composition of cementitious gels, etc. All these aspects are important and depend on the reactivity of the precursor and the alkaline solution used and the curing procedure. In the literature, there are already works that use molecular dynamics simulations to evaluate the fracture toughness of C-S-H [55] techniques. In this work, we have focused on the chemical aspects (instead of the physical characteristics). In addition, the high SiO_2/Al_2O_3 ratio and the co-precipitation of C-A-S-H and (N,C)-A-S-H gels observed in all the pastes studied might explain the high strength of our materials (higher than the ones reported for studies using common low-calcium clays) [5,22–25].

4. Conclusions

The most prominent conclusions to be drawn from this study are listed below.

- NaOH (at concentrations of 4, 6 and 8 M) activation of a carbonate-high illite clay contributes to paste hardening at ambient temperature. The presence of reactive calcium in the starting clay induces co-precipitation of a mix of gels: an aluminium-enriched C-S-H gel (C-A-S-H) and a N-A-S-H gel, in which sodium is partially replaced by calcium (N,C)-A-S-H.
- When the activator is a 4 M NaOH solution, the C-A-S-H gel is more intensely carbonated than with 6 or 8 M activators.
- Pastes prepared with higher (6 or 8 M) activator concentrations exhibit a more compact matrix than the ones prepared with 4 M NaOH. Given the comparability of the findings for the 6 and 8 M NaOH-activated pastes and the cost effectiveness of using a lower concentration activator, the 6 M NaOH solution may be deemed to be the most suitable of those studied here. Activating a mechanically and thermally pre-treated carbonate-high illite clay with a 6 M NaOH solution can deliver a binder with two days compressive strength of >20 MPa and 28 days strength of >30 MPa.

Author Contributions: A.D., Formal analysis; writing—original draft. D.P. and A.F.-J., Methodology; Investigation; Supervision; Resources; Writing—review & editing; Funding acquisition. G.E., R.L. and A.P., Writing—review & editing. All authors have read and agreed to the published version of the manuscript.

Funding: The research was also partially funded by the Spanish Ministry of the Economy, Industry and Competitiveness and ERDF under project BIA2016-76466-R.

Acknowledgments: The authors gratefully acknowledge the Italian Ministry of Education, University and Research (MIUR Funds—PhD in Geosciences, University of Bari) and the University of Bari for ordinary financial support (ex 60% Founds).

Conflicts of Interest: The authors declare no conflict of interest.

References

1. Xu, H.; van Deventer, J. The geopolymerisation of alumino-silicate minerals. *Int. J. Miner. Process.* **2000**, *59*, 247–266. [CrossRef]
2. Alonso, S.; Palomo, A. Alkaline activation of metakaolin and calcium hydroxide mixtures: Influence of temperature, activator concentration and solids ratio. *Mater. Lett.* **2001**, *47*, 55–62. [CrossRef]
3. Granizo, M.L.; Alonso, S.; Blanco-Varela, M.T.; Palomo, A. Alkaline Activation of Metakaolin: Effect of Calcium Hydroxide in the Products of Reaction. *J. Am. Ceram. Soc.* **2002**, *85*, 225–231. [CrossRef]
4. Fernández-Jiménez, A.; Monzo, M.; Vincent, M.B.A.; Palomo, A. Alkaline activation of metakaolin-fly ash mixtures: Obtain of Zeoceramics and Zeocements. *Microporous Mesoporous Mater.* **2008**, *108*, 41–49. [CrossRef]
5. Ferone, C.; Colangelo, F.; Cioffi, R.; Montagnaro, F.; Santoro, L. Use of reservoir clay sediments as raw materials for geopolymer binders. *Adv. Appl. Ceram.* **2013**, *112*, 184–189. [CrossRef]
6. Davidovits, J. Geopolymers: Inorganic Polymeric New Materials. *J. Therm. Anal.* **1991**, *37*, 1633–1656. [CrossRef]
7. Scrivener, K.; Martirena, F.; Bishnoi, S.; Maity, S. Calcined clay limestone cements (LC3). *Cem. Concr. Res.* **2018**, *114*, 49–56. [CrossRef]
8. Dhandapani, Y.; Sakthivel, T.; Santhanam, M.; Gettu, R. Pillai Mechanical properties and durability performance of concretes with limestone calcined clay cement (LC^3). *Cem. Concr. Res.* **2018**, *107*, 136–151. [CrossRef]
9. Scrivener, K.L.; John, V.M.; Gartner, E.M. Eco-efficient cements: Potential economically viable solutions for a low-CO_2 cement-based materials industry. *Cem. Concr. Res.* **2018**, *114*, 2–26. [CrossRef]
10. Alvarez-Pinazo, G.; Santacruz, I.; Aranda, M.A.; De la Torre, A.G. Hydration of belite–ye'elimite–ferrite cements with different calcium sulfate sources. *Adv. Cem. Res.* **2016**, *28*, 1–15. [CrossRef]
11. Palomo, A.; Krivenko, P.; García-Lodeiro, I.; Kavalerova, E.; Maltseva, O.; Fernández-Jiménez, A. A review on alkaline activation: New analytical perspectives. *Materiales Construcción* **2014**, *64*, e022. [CrossRef]
12. Provis, J.L.; van Deventer, J.S.J. (Eds.) *Alkali-Activated Materials: State-of-the-Art Report, RILEM TC 224-AAM*; Springer: Dordrecht, The Netherlands, 2014.
13. Provis, J.L. Alkali-activated materials. *Cem. Concr. Res.* **2018**, *114*, 40–48. [CrossRef]
14. Zhang, Z.H.; Zhu, H.J.; Zhou, C.H.; Wang, H. Geopolymer from kaolin in China: An overview. *Appl. Clay Sci.* **2016**, *119*, 31–41. [CrossRef]
15. Kuenzel, C.; Vandeperre, L.J.; Donatello, S.; Boccaccini, A.R.; Cheeseman, C. Ambient Temperature Drying Shrinkage and Cracking in Metakaolin-Based Geopolymers. *J. Am. Ceram. Soc.* **2012**, *95*, 3270–3277. [CrossRef]
16. Liew, Y.-M.; Heah, C.-Y.; Mustafa, A.B.M.; Kamarudin, H. Structure and properties of clay-based geopolymer cements: A review. *Prog. Mater. Sci.* **2016**, *83*, 595–629. [CrossRef]
17. Brigatti, M.F.; Galán, E.; Theng, B.K.G. Structure and Mineralogy of Clay Minerals. In *Handbook of Clay Science*; Bergaya, F., Lagaly, G., Eds.; Elsevier: Oxford, UK, 2013; pp. 21–81.1148. [CrossRef]
18. Murray, H.H. *Applied Clay Mineralogy: Occurrences, Processing and Application of Kaolins, Bentonites, Palygorskite-Sepiolite, and Common Clays*; Elsevier: Amsterdam, The Netherlands, 2006.
19. Seiffarth, T.; Hohmann, M.; Posern, K.; Kaps, C. Effect of thermal pre-treatment conditions of common clays on the performance of clay-based geopolymeric binders. *Appl. Clay Sci.* **2013**, *7*, 35–41. [CrossRef]
20. Ferone, C.; Liguori, B.; Capasso, I.; Colangelo, F.; Cioffi, R.; Cappelletto, E.; di Maggio, R. Thermally treated clay sediments as geopolymer source material. *Appl. Clay Sci.* **2015**, *107*, 195–204. [CrossRef]
21. D'Elia, A.; Pinto, D.; Eramo, G.; Giannossa, L.C.; Ventruti, G.; Laviano, R. Effects of processing on the mineralogy and solubility of carbonate-rich clays for alkaline activation purpose: Mechanical, thermal activation in red/ox atmosphere and their combination. *Appl. Clay Sci.* **2018**, *152*, 9–21. [CrossRef]

22. Duxson, P.; Lukey, G.C.; van Deventer, J.S.J. Physical evolution of Na-geopolymer derived from metakaolin up to 1000 °C. *J. Mater. Sci.* **2007**, *42*, 3044–3054. [CrossRef]

23. Granizo, M.L.; Blanco-Varela, M.T.; Palomo, A. Influence of the starting kaolin on alkali-activated materials based on metakaolin. Study of the reaction parameters by isothermal conduction calorimetry. *J. Mater. Sci.* **2000**, *35*, 6309–6315. [CrossRef]

24. Palomo, A.; Blanco-Varela, M.T.; Granizo, M.L.; Puertas, F.; Vazquez, T.; Grutzeck, M.W. Chemical stability of cementitious materials based on metakaolin. *Cem. Concr. Res.* **1999**, *29*, 997–1004. [CrossRef]

25. Buchwald, A.; Hohmann, M.; Posern, K.; Brendler, E. The suitability of thermally activated illite/smectite clay as raw material for geopolymer binders. *Appl. Clay Sci.* **2009**, *46*, 300–304. [CrossRef]

26. Essaidi, N.; Samet, B.; Baklouti, S.; Rossignol, S. Effect of calcination temperature of Tunisian clay on the properties of geopolymers. *Ceramics Silikáty* **2013**, *57*, 251–257.

27. Ruiz-Santaquiteria, C.; Fernández-Jiménez, A.; Skibsted, J.; Palomo, A. Clay reactivity: Production of alkali activated cements. *Appl. Clay Sci.* **2013**, *73*, 11–16. [CrossRef]

28. García-Lodeiro, I.; Cherfa, N.; Zibouche, F.; Fernández-Jimenez, A.; Palomo, A. The role of aluminium in alkali-activated bentonites. *Mater. Struct.* **2015**, *48*, 585–597. [CrossRef]

29. Garcia-Lodeiro, A.; Fernandez-Jimenez, A. Palomo Hybrid Alkaline Cements: Bentonite-Opc Binders. *Minerals* **2018**, *8*, 137. [CrossRef]

30. Khater, H.M. Effect of Calcium on Geopolymerization of Aluminosilicate Wastes. *J. Mater. Civ. Eng.* **2012**, *24*, 92–101. [CrossRef]

31. Temuujin, J.; van Riessen, A.; Williams, R. Influence of calcium compounds on the mechanical properties of fly ash geopolymer pastes. *J. Hazard. Mater.* **2009**, *167*, 82–88. [CrossRef]

32. García-Lodeiro, I.; Palomo, A.; Fernández-Jiménez, A.; Macphee, D. Compatibility studies between N-A-S-H and C-A-S-H gels. Study in the ternary diagram Na_2O-CaO-Al_2O_3-SiO_2-H_2O. *Cem. Concr. Res.* **2011**, *41*, 923–931. [CrossRef]

33. Alahrache, S.; Winnefeld, F.; Champenois, J.B.; Hesselbarth, F.; Lothenbach, B. Chemical activation of hybrid binders based on siliceous fly ash and Portland cement. *Cem. Concr. Compos.* **2006**, *66*, 10–23. [CrossRef]

34. García-Lodeiro, I.; Donatello, S.; Fernández-Jiménez, A.; Palomo, A. Hydration of Hybrid Alkaline Cement Containing a Very Large Proportion of Fly Ash: A Descriptive Model. *Materials* **2016**, *9*, 605. [CrossRef] [PubMed]

35. Moore, D.M.; Reynolds, R.C., Jr. *X-ray Diffraction and the Identification and Analysis of Clay Minerals*, 2nd ed.; Oxford University Press: New York, NY, USA, 1997.

36. Komljenović, M.; Bascarević, Z.; Bradić, V. Mechanical and microstructural properties of alkali-activated fly ash geopolymers. *J. Hazard. Mater.* **2010**, *181*, 35–42. [CrossRef]

37. Criado, M.; Palomo, A.; Fernández-Jiménez, A. Alkali activation of fly ashes. Part 1: Effect of curing conditions on the carbonation of the reaction products. *Fuel* **2005**, *84*, 2048–2054. [CrossRef]

38. Myers, R.J.; Bernal, S.A.; Provis, J.L. Phase diagrams for alkali-activated slag binders. *Cem. Concr. Res.* **2017**, *95*, 30–38. [CrossRef]

39. Cwirzen, A.; Provis, J.L.; Penttala, V.; Habermehl-Cwirzen, K. The effect of limestone on sodium hydroxide-activated metakaolin-based geopolymers. *Constr. Build. Mater.* **2014**, *66*, 53–62. [CrossRef]

40. Rees, C.; Provis, J.L.; Luckey, G.C.; van Deventer, J.S.J. In Situ ATR-FTIR Study of the Early Stages of Fly Ash Geopolymer Gel Formation. *Langmuir* **2007**, *23*, 9076–9082. [CrossRef] [PubMed]

41. Criado, M.; Fernández-Jiménez, A.; Palomo, A. Alkali activation of fly ash: Effect of the SiO_2/Na_2O ratio Part I: FTIR study. *Microporous Mesoporous Mater.* **2007**, *106*, 180–191. [CrossRef]

42. Yu, P.; Kirkpatrick, R.J.; Poe, B.; McMillan, P.F.; Cong, X. Structure of Calcium Silicate Hydrate (C-S-H): Near-, Mid-, and Far-Infrared Spectroscopy. *J. Am. Ceram. Soc.* **1999**, *82*, 742–748. [CrossRef]

43. Joshi, S.; Kalyanasundaram, S.; Balasubramanian, V. Quantitative Analysis of Sodium Carbonate and Sodium Bicarbonate in Solid Mixtures Using Fourier Transform Infrared Spectroscopy (FT-IR). *Appl. Spectrosc.* **2013**, *67*, 841–845. [CrossRef]

44. Yip, C.; Lukey, G.; Provis, J.; van Deventer, J. Effect of calcium silicate sources on geopolymerisation. *Cem. Concr. Res.* **2008**, *38*, 554–564. [CrossRef]

45. Yip, C.; Provis, J.; Lukey, G.; van Deventer, J. Carbonate mineral addition to metakaolin-based geopolymers. *Cem. Concr. Compos.* **2008**, *30*, 979–985. [CrossRef]

46. Puertas, F.; Palacios, M.; Vazquez, T. Carbonation process of alkali-activated slag mortars. *J. Mater. Sci.* **2006**, *41*, 3071–3082. [CrossRef]
47. Puertas, F.; Palacios, M.; Manzano, H.; Dolado, J.S.; Rico, A.; Rodríguez, J. A model for the C-A-S-H gel formed in alkali-activated slag cements. *J. Eur. Ceram. Soc.* **2011**, *31*, 2043–2056. [CrossRef]
48. Fernández-Jiménez, A.; Zibouche, F.; Boudissa, N.; García-Lodeiro, I.; Abadlia, M.T.; Palomo, A. Metakaolin-Slag-Clinker Blends. The role of Na$^+$ or K$^+$ as Alkaline Activators of Theses Ternary Blends. *J. Am. Ceram. Soc.* **2013**, *96*, 1–8. [CrossRef]
49. García-Lodeiro, I.; Fernández-Jiménez, A.; Palomo, A. Hydration kinetics in hybrid binders: Early reaction stages. *Cem. Concr. Compos.* **2013**, *39*, 82–92. [CrossRef]
50. Liu, B.; Ray, A.S.; Thomas, P.S. Strength development in autoclaved aluminosilicate rich industrial waste cement systems containing reactive magnesia. *J. Aust. Ceram. Soc.* **2007**, *43*, 82–87.
51. Bernal, S.A.; Provis, J.L.; Walkley, B.; Nicolas, R.S.; Gehman, J.D.; Brice, D.G.; Kilcullen, A.; Duxson, P.; van Deventer, J.S.J. Gel nanostructure in alkali-activated binders based on slag and fly ash, and effects of accelerated carbonation. *Cem. Concr. Res.* **2013**, *53*, 127–144. [CrossRef]
52. Bernal, S.A.; Provis, J.L.; de Gutiérrez, R.M.; van Deventer, J.S.J. Accelerated carbonation testing of alkali-activated slag/metakaolin blended concretes: Effect of exposure conditions. *Mater. Struct.* **2015**, *48*, 653–669. [CrossRef]
53. Bernal, S.A.; Provis, J.L.; Brice, D.G.; Kilcullen, A.; Duxson, P.; van Deventer, J.S.J. Accelerated carbonation testing of alkali-activated binders significantly underestimate the real service life: The role of the pore solution. *Cem. Concr. Res.* **2012**, *42*, 1317–1326. [CrossRef]
54. Bernal, S.A. Effect of the activator dose on the compressive strength and accelerated carbonation resistance of alkali silicate-activated slag/metakaolin blended materials. *Constr. Build. Mater.* **2015**, *98*, 217–226. [CrossRef]
55. Bauchy, M.; Laubie, H.; Qomi, M.A.; Hoover, C.G.; Ulm, F.J.; Pellenq, R.M. Fracture toughness of calcium–silicate–hydrate from molecular dynamics simulations. *J. Non-Cryst. Solids* **2015**, *419*, 58–64. [CrossRef]

Article

Components of the Fracture Response of Alkali-Activated Slag Composites with Steel Microfibers

Hana Šimonová *, Petr Frantík, Zbyněk Keršner, Pavel Schmid and Pavel Rovnaník

Brno University of Technology, Faculty of Civil Engineering, Veveří 331/95, 602 00 Brno, Czech Republic; kitnarf@centrum.cz (P.F.); kersner.z@fce.vutbr.cz (Z.K.); pavel.schmid@vutbr.cz (P.S.); rovnanik.p@fce.vutbr.cz (P.R.)
* Correspondence: simonova.h@vutbr.cz; Tel.: +420-541-147-381

Received: 4 April 2019; Accepted: 24 April 2019; Published: 27 April 2019

Abstract: Knowledge of the mechanical and primarily fracture parameters of composites with a brittle matrix is essential for the quantification of their resistance to crack initiation and growth, and also for the specification of material model parameters employed for the simulation of the quasi-brittle behavior of structures made from this type of composite. Therefore, the main target of this paper is to quantify the mechanical fracture parameters of alkali-activated slag composites with steel microfibers and the contribution of the matrix to their fracture response. The first alkali-activated slag composite was a reference version without fibers; the others incorporated steel microfibers amounting to 5, 10, 15 and 20% by weight of the slag. Prism specimens with an initial central edge notch were used to perform the three-point bending fracture tests. Load vs. displacement (deflection at midspan) and load vs. crack mouth opening displacement diagrams were recorded during the fracture tests. The obtained diagrams were employed as inputs for parameter identification, the aim of which was to transfer the fracture test response data to the desired material parameters. Values were also determined for fracture parameters using the effective crack model, work-of-fracture method and double-*K* fracture model. All investigated mechanical fracture parameters were improved by the addition of steel microfibers to the alkali-activated matrix. Based on the obtained results, the addition of 10 to 15% of microfibers by weight is optimal from the point of view of the enhancement of the fracture parameters of alkali-activated slag composite.

Keywords: alkali-activated slag; steel microfibers; fracture test; identification; work-of-fracture method; double-*K* model; crack propagation

---

1. Introduction

The global production of cement in 2018 was about 4.1 billion tons [1]. That immense volume of cement production is related to a very substantial impact on the environment: the carbon dioxide emissions produced by the cement industry contribute up to 8% of worldwide CO_2 emissions [2].

The majority of Portland cement (PC) is employed to produce concrete, mortars and plasters in the building industry. To decrease cement consumption, supplementary cementitious materials with good hydraulic cementitious properties (especially fly ash and ground granulated blast furnace slag) are often used as partial substitutes for PC in specific applications. They are also employed as constituent parts of blended cements [1]. The other possibility is to use alternative types of binder. Alkali-activated materials (AAM) are one example of the relatively new binders now being produced via the alkaline activation of different materials of geological origin or by-product materials that are rich in silicon and aluminum. The utilization of secondary raw materials (fly ashes, slags, etc.) or other aluminosilicate materials during alkali activation leads to a decrease in cement consumption, resulting in the more efficient reduction of CO_2 emissions and energy consumption. Although the production of alkaline activators is connected with CO_2 emissions it is assumed that the global warming potential of alkali-activated composites is approximately 40–70% lower than that of ordinary PC-based composites [3,4]. AAMs also show good durability compared to Portland cement [5,6].

Just like materials based on Portland cement, AAMs are quasi-brittle materials that show what is known as tensile softening. The different types of steel, synthetic or natural fibers which are used in Portland cement-based materials [7–10] are added to improve the material's resistance to crack propagation. Knowledge of the mechanical and primarily fracture parameters of composites with a brittle matrix is essential for the quantification of their resistance to crack initiation and growth, as well as for the specification of material model parameters employed for the simulation of the quasi-brittle behavior of structures or their parts made from this type of composite. Studying the mechanical response of specimens made of such composites under static and dynamic/fatigue loading is complicated due to their highly nonlinear nature. Numerical tools for modeling both elastic (elastic-plastic) behavior, and also the fracture process, are commonly used to predict or assess this response. Such tools—often based on the finite element method [11] or physical discretization of the continuum [12]—usually exploit a type of nonlinear fracture model that simulates the cohesive nature of the cracking of quasi-brittle material [13–15]. The parameters of this fracture model are determined from records of fracture tests; this is carried out either using evaluation methods built on the principle of the used non-linear fracture model, e.g., the work-of-fracture method [16] or the size effect method [17], or using inverse analysis with the possible application of advanced identification methods [18–20]. The fracture models for quasi-brittle composites are most often based on the standardized geometry of specimens with stress concentrators; the three-point bending test [14] or wedge splitting test [21,22] are typically used.

As mentioned above, in order to perform the realistic numerical modeling of the response of quasi-brittle composite structures it is essential to determine the parameters of the used material models from experimental measurements. Unfortunately, the literature on the fracture properties of alkali-activated mortars which can be used as suitable inputs for material models is still fairly limited. Most of the published articles are only concerned with the determination of basic mechanical parameters, i.e., compressive and flexural strengths, or in some cases modulus of elasticity [23–30]. Only a few researchers' findings connected with the fracture behavior of this kind of material have been published. Goncalves et al. [31] presented a study about the crack growth resistance of fiber-reinforced alkali-activated fly ash concrete exposed to extreme conditions. Alomayri [32] investigated the effects of glass microfiber content on the mechanical properties of fly ash-based geopolymer. It was found that the optimal amount of glass microfibers is 2 mass% from the point of view of the enhancement of fracture toughness, compressive strength, Young's modulus and hardness. Ding et al. [33,34] examined the fracture properties of alkali-activated slag (AAS) and ordinary Portland cement (OPC) concrete and mortar. It was observed that the fracture energy value was lower in the case of AAS mortar, as compared to OPC mortar with the same compressive strength. Sarker et al. [35] investigated the effect of geopolymer binder on the fracture characteristics of concrete. The fracture energy of geopolymer determined by the work-of-fracture method was similar to that of the investigated OPC concrete. The critical stress intensity factor was higher in the case of geopolymer compared to OPC concrete with the same compressive strength. Ngyuen et al. [36] ascertained that the addition of polypropylene fibers to AAS mortar leads to an increase in fracture energy and fracture toughness compared to mortar without fibers.

Because of the lack of information about the fracture properties of composites with alkali-activated matrix which could be used as relevant inputs for the material model, the main aim of the present work is to determine the fracture parameters of alkali-activated slag composites with steel microfibers and quantify the contribution of the matrix of AAS composites to their fracture response. Five AAS composites were investigated. The first was a reference version without fibers; the others contain steel microfibers amounting to 5, 10, 15 and 20% by weight of the slag. The AAS mixtures were cast into molds with dimensions of 40 × 40 × 160 mm so as to prepare prismatic specimens for use in fracture testing. The fracture characteristics were determined based on the results of three-point bending tests conducted on specimens which were provided with an initial central edge notch before testing. Load vs. displacement (deflection at midspan) and load vs. crack mouth opening displacement diagrams were recording during the fracture tests. Each diagram was processed in order to obtain the component that corresponds to the structural response of the matrix of the composite, which consists of AAS matrix and the steel microfibers reinforcing that matrix. The obtained diagrams were employed as inputs for parameter identification, the aim of which was to transfer the fracture test response data to the required material parameters. Values were also determined for the fracture parameters using the effective crack model [14], work-of-fracture method [16] and double-K fracture model [37].

2. Materials and Methods

2.1. Mixtures

The first alkali-activated slag composite was a reference version without fibers. Granulated blast furnace slag provided by Kotouč, s.r.o. (CZ) was chosen as a binder. The specific surface and mean grain size of the slag were 383 m^2/kg and 15.5 μm, respectively. Solid sodium silicate (Susil MP 2.0) fabricated by Vodní sklo, a.s. (CZ) was used to achieve the alkali activation of the slag. The alkaline activator has a molar SiO_2/Na_2O ratio equal to 2.0, and a SiO_2 content of 52.4%. The fine-grained AAS composites were produced using quartz sand with a maximum grain size of 2.5 mm. Brass coated steel microfibers with an average length of 6 mm and a diameter of 0.175 mm supplied by KrampeHarex CZ s.r.o. were used as reinforcement (see Figure 1). The added steel microfibers amounted to 5, 10, 15 and 20% of the weight of the slag.

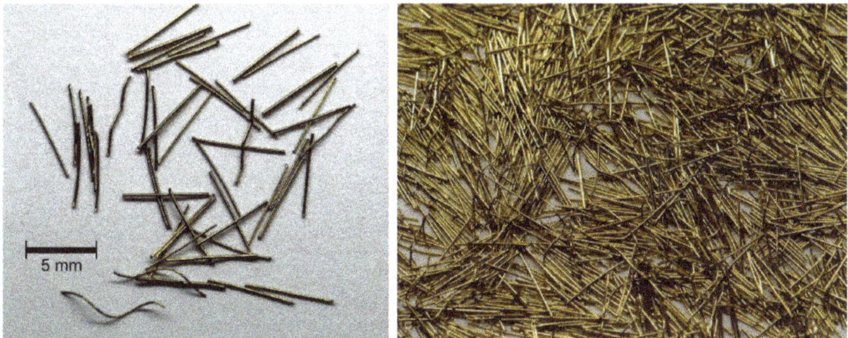

Figure 1. The used steel microfibers with an average length of 6 mm and a diameter of 0.175 mm.

The AAS mixtures used for casting the prismatic specimens were prepared according to the following previously optimized procedure. At first, solid alkaline activator was suspended in water in which it dissolved partially. Then, the slag and quartz sand were added to the activator and the mixture was stirred in the planetary mixer for about 5 min to provide a fresh slurry. The aggregate to slag ratio was equal to 3.0. Finally, steel microfibers were added to the mixture and further mixed for another 3 min so as to disperse them properly. The AAS mixture composition is presented in Table 1. The designation of individual mixtures is based on their steel microfiber content: SF00, SF05, SF10, SF15, and SF20. Mixture SF00 is a reference mix without steel microfibers.

The prepared fresh mixtures were cast into prismatic moulds of 40 × 40 × 160 mm in size. Three specimens were made from each mixture. After 24 h the hardened specimens were immersed in a water bath at 20 °C for further 27 days. Before the fracture tests were performed, all specimens were pulled out of the water and allowed to dry spontaneously under ambient conditions for 24 h.

Table 1. Composition of alkali-activated slag mixtures.

Component	Unit	SF00	SF05	SF10	SF15	SF20
Slag	g	450	450	450	450	450
Sodium silicate	g	90	90	90	90	90
Aggregate	g	1350	1350	1350	1350	1350
Steel microfibers	g	–	22.5	45.0	67.5	90.0
Water	mL	190	190	190	195	195

2.2. Fracture Test Configuration

The determination of the mechanical fracture parameters of composites with brittle matrix is most often based on fracture tests conducted on specimens of standardized geometry with stress concentrators; the three-point bending test [14] or wedge splitting test [21,22] are typically used. In this case, the three-point bending configuration was chosen because of the availability of testing equipment. Standardized prism specimens with a nominal size of 40 × 40 × 160 mm, which are typically used for the determination of basic mechanical properties of mortars, were used for the fracture tests. An initial notch was cut by a diamond blade saw in the center of the prisms. The nominal depth was about 13 mm. The span length was 120 mm. The fracture tests were performed at the age of 29 days.

A very stiff mechanical testing machine (LabTest 6-1000.1.10, LaborTech s.r.o., Opava, Czech Republic) was used to perform the fracture tests. The stiffness of the testing machine is required to be adequate in comparison to the specimen's stiffness so as to enable stable fracture tests to be conducted without any interruption in the post-peak branch. The loading was conducted so that displacement occurred in constant increments, which were equal to 0.02 mm/min. This loading procedure is slow enough for the whole post-peak behavior of test specimens to be recorded.

The dependence between loading force and the deflection of the center of the prism specimen (*F-d* diagram), as well as crack mouth opening displacement (*F-CMOD* diagram), constituted the outputs of the performed fracture tests. The deflection and *CMOD* values were gauged using an inductive sensor placed above the support and by extensometer placed between blades fixed close to the initial notch, respectively (see Figures 2 and 3). The mentioned parameters together with time were continuously recorded by an HBM Quantum X data logger (HBM, Darmstadt, Germany); the frequency was 5 Hz.

Figure 2. The selected reference specimen after the fracture test was performed with crack propagation from initial notch; detail of blades fixed close to the initial notch used for measurement of *CMOD*.

(a)   (b)

Figure 3. Fracture test configuration: (**a**) Overall view of the testing machine; (**b**) Detail of the positions of measuring sensors (1—deflection sensor fixed in the measurement frame, 2—extensometer).

2.3. Adjustment of Measured F−d and F−CMOD Diagrams

At the beginning of the specimen loading, small-sized deviations in the measured values of monitored parameters are often recorded. This effect is caused by small projections on the specimen's surface being crushed due to the pressure at the support and loading points. These phenomena usually occur over a short period at the beginning of the loading test, after which the measured diagram proceeds with a linear part. It follows that it is appropriate to adjust the beginning part of the diagram in order to obtain the correct input values for the subsequent evaluation of diagrams using the selected fracture model. The first step is to construct a straight best-fit line for the linear part of the diagram, after which the intersection of the line with the horizontal axis must be pinpointed. The second step consists in the shifting of all points of the diagram equidistantly, thus making the intersection the new origin of the coordinate system.

The adjustment of the recorded diagrams was performed in GTDiPS software (v3.01, developed by Petr Frantík and Jan Mašek, Brno University of Technology, Czech Republic) [38], which is based on advanced transformation methods used for the processing of extensive point sequences. The adjustment of diagrams in this case incorporated the erasing of duplicate points, the moving of the origin of the coordinate system, the smoothing of the diagram and the reduction of the number of points.

Thereafter, each diagram was processed to obtain the component that corresponds to the structural response of the matrix of the composite and the steel microfibers reinforcing that matrix. The individual steps of the decomposition procedure are as follows: first, the measured diagram of the steel microfiber-reinforced AAS composite specimen is plotted; then, the last part of this diagram (after the substantial drop in the curve) is assumed to be the result of the contribution of the steel microfibers only and the composite matrix is not expected to have an effect here. That last part is subjected to a straightforward linear regression analysis so that an approximation of the initial part of the diagram can also be obtained (a polynomial function is used here with extrapolation to the origin of the diagram space). Finally, the approximation is subtracted from the recorded diagram, which results in a simulated diagram corresponding to the plain AAS matrix for the next evaluation. A detailed illustration of the used procedure can be found in [39].

The above-described procedure was applied to all measured *F–d* and *F–CMOD* diagrams. For the purpose of illustration, Figure 4 shows corrected *F–CMOD* diagrams of AAS composites with various amounts of microfibers. The use of the *F–d* diagrams after the decomposition procedure is described in the following section.

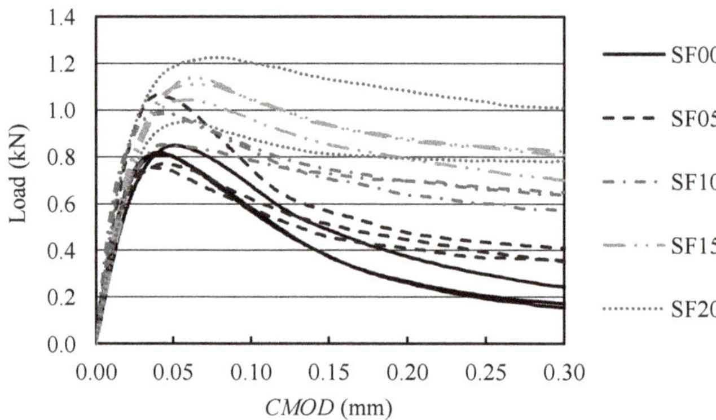

Figure 4. *F–CMOD* diagrams of alkali-activated slag (AAS) composites with various amounts of microfibers.

2.4. Identification of Material Parameters

After the previously mentioned adjustment, the *F–d* diagrams were utilized as input data for parameter identification with the aim of transferring the fracture test response data to the desired material parameters. The FiCubS application [40] developed by co-author Petr Frantík was used for this purpose. The FiCubS application is used to simulate the performance of a fracture test in the three-point bending configuration on a fiber-reinforced composite prism with a notch.

Using symmetry, half of the specimen is modeled as an elastic body connected to the plane of symmetry by boundary conditions, which can be released and replaced by cohesive forces. The release of a particular condition occurs after the tensile strength of the material is overstepped. Cohesive forces are dependent on displacements of the released ligament area and on the identified cohesive function. The used cohesive function consists of two components: cohesion provided by the matrix and by fiber resistance. The matrix cohesive function is modeled by the Hordijk function [41]:

$$\sigma = f_t\left[\left(1 + \left(c_1\frac{w}{w_c}\right)^3\right)e^{\left(-c_2\frac{w}{w_c}\right)} - \frac{w}{w_c}\left(1 + c_1{}^3\right)e^{-c_2}\right] \tag{1}$$

where σ is cohesive stress, f_t is tensile strength, w is crack opening displacement, w_c is critical crack opening displacement, and c_1 and c_2 are material constants.

The fiber resistance is modeled by the proposed function:

$$\sigma = \sigma_{max}\left(1 - \frac{1}{w_{cf}}\left(we^{p(w-w_{cf})} + \left(w_{cf} - w\right)e^{-kw}\right)\right) \tag{2}$$

where σ is the cohesive stress applied for displacement w, σ_{max} is the approximate maximum cohesive stress, w_{cf} is the displacement limit where cohesion disappears and k, p are parameters determining the initial and finite slope. Initial and finite slopes are given by:

$$\frac{\partial}{\partial w}\sigma\bigg|^{w=0} = \frac{\sigma_{max}}{w_{cf}}\left(1 + kw_{cf} - e^{-pw_{cf}}\right) \tag{3}$$

$$\frac{\partial}{\partial w}\sigma\bigg|^{w=w_{cf}} = \frac{\sigma_{\max}}{w_{cf}}\left(1 - \left(pw_{cf} + 1\right)e^{kw_{cf}}\right)e^{-kw_{cf}} \tag{4}$$

The total fracture energy G represented by this function is given by the relation:

$$G = \sigma_{\max}\left(\frac{1}{w_{cf}}\left(\frac{1}{k^2}\left(1 - e^{-w_{cf}k}\right) + \frac{1}{p^2}\left(1 - e^{-w_{cf}p}\right)\right) - \frac{1}{k} - \frac{1}{p} + w_{cf}\right) \tag{5}$$

From the identified parameters of the model (seven independent values) it was necessary to determine the effective modulus of elasticity of the composite, the tensile strength of the composite, the fracture energy and the coefficient of transverse contraction.

To illustrate, Figure 5 shows F–d diagrams for selected specimens of AAS composite with various amounts of steel microfibers. The following diagrams are plotted in the graphs: a measured diagram of composite reinforced by microfibers (COMP); a simulated diagram corresponding to the plain AAS matrix (MTX); an identified diagram of AAS composite reinforced by microfibers (ID).

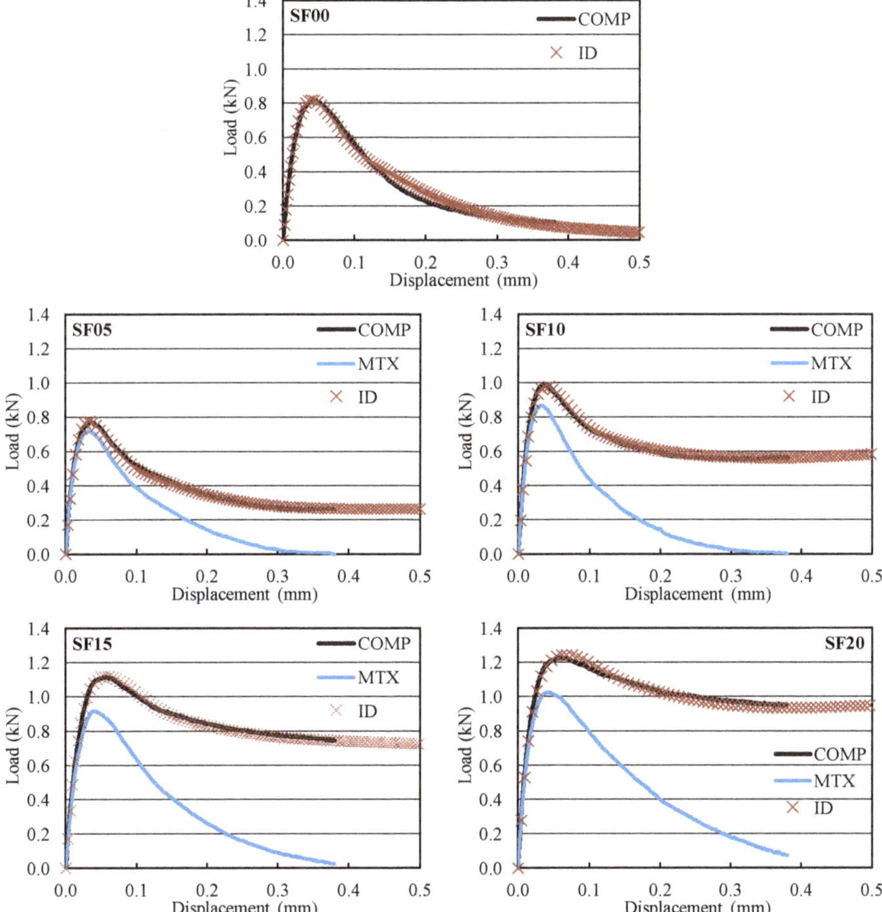

Figure 5. F–d diagrams for selected specimens of individual AAS composites with various amounts of steel microfibers: COMP—a measured diagram of composite reinforced by microfibers; MTX—a simulated diagram corresponding to the plain AAS matrix; ID—an identified diagram of composite reinforced by microfibers.

2.5. Evaluation of F–d Diagrams

After the previously mentioned diagram adjustment, the ascending linear parts of the *F-d* diagrams were utilized to estimate the modulus of elasticity E_c values according to [14]:

$$
E_c = \frac{F_i}{4Bd_i}\left(\frac{S}{D}\right)^3\left[1 + \frac{5qS}{8F_i} + \left(\frac{D}{S}\right)^2\left\{2.70 + 1.35\frac{qS}{F_i}\right\} - 0.84\left(\frac{D}{S}\right)^3\right]
$$
$$
+ \frac{9}{2}\frac{F_i}{Bd_i}\left(1 + \frac{qS}{2F_i}\right)\left(\frac{S}{D}\right)^2 F_1(\alpha_0)
$$

$$(6)$$

where F_i is load in the ascending linear part of the diagram; d_i is deflection at midspan corresponding with load F_i; B and D are the breadth and depth of the specimen, respectively; q is the self-weight of the specimen per unit length; S is length of the span and with $\alpha_0 = a_0/D$, and $Y(x)$ is the geometry function for the three-point bending configuration [14] (a_0 is the depth of the initial notch).

The effective fracture toughness K_{Ice} was determined based on the *F–d* diagrams using the effective crack model [14]. First, the effective crack length a_e corresponding with the maximum load F_{max} and matching deflection at midspan d_{Fmax} was calculated. From the effective crack concept, it follows that the a_e can be calculated from rearranged Equation (6) using F_{max} and d_{Fmax} instead of F_i and d_i. Subsequently, the effective fracture toughness values were calculated using a linear elastic fracture mechanics formula according to [14,42]:

$$
K_{Ice} = \frac{3F_{max}S}{2BD^2}Y(\alpha_e)\sqrt{a_e}
$$

$$(7)$$

where $Y(\alpha_e)$ is the geometry function with $\alpha_e = a_e/D$ [14].

The complete *F-d* diagrams, including their post-peak parts, were employed to determine the work of fracture W_F^*, which is given by the area under the *F-d* diagram. After that, the specific fracture energy values were determined according to the RILEM method [16,43]:

$$
G_F^* = \frac{W_F^*}{B(D - a_0)}
$$

$$(8)$$

2.6. Evaluation of F–CMOD Diagrams

After the previously mentioned diagram adjustment, the double-*K* fracture model was employed for subsequent evaluation of the *F-CMOD* diagrams. The benefit of the double-*K* fracture model lies in its ability to predict the different phases that occur during crack propagation in quasi-brittle material: crack initiation and both stable and unstable crack propagation. The different phases of the fracture process in quasi-brittle material can be connected with two size-independent parameters: the initiation fracture toughness K_{Ic}^{ini} and unstable fracture toughness K_{Ic}^{un}. The determination of double-*K* model parameters is based on an approach involving the action of cohesive forces on the faces of the fictitious (effective) crack increment combined with the stress intensity factor criterion (for details refer to (for example) Kumar and Barrai [37]).

In the instance of the present research, as the first step the unstable fracture toughness K_{Ic}^{un} was determined. In the second step, the cohesive fracture toughness K_{Ic}^{c} was determined. As the last step, the following formula based on the formerly obtained parameters was utilized to calculate the initiation fracture toughness K_{Ic}^{ini}:

$$
K_{Ic}^{ini} = K_{Ic}^{un} - K_{Ic}^{c}
$$

$$(9)$$

The exact procedure concerning the determination of cohesive and unstable fracture toughness values can be found in many published works, e.g., [37,44].

In general, the relation between the cohesive stress and the effective crack opening displacement is given by the cohesive stress function in the cohesive crack model. The cohesive stress at the tip of an initial notch at the critical state can be gained from the softening function. In this study, a non-linear softening function (1) (as stipulated in Hordijk [41]) was used. The parameters of the softening function were considered to be as follows: the tensile strength f_t was considered based on identification described in Section 2.4, the fracture energy was determined according to Equation (8), and the material constants were considered according to [41] as being $c_1 = 3$ and $c_2 = 6.93$.

Finally, the load F_{ini}, which expresses the load at the outset of stable crack propagation from the initial crack/notch, was determined according to this relation:

$$F_{ini} = \frac{4 \cdot W \cdot K_{Ic}^{ini}}{S \cdot F_1(\alpha_0) \cdot \sqrt{a_0}} \tag{10}$$

where W is the section modulus (calculated as $W = 1/6 \cdot B \cdot D^2$), S is span length, $F_1(\alpha_0)$ is the geometry function for a three-point bending configuration [14] and α_0 is the a_0/D ratio.

3. Results and Discussion

The average values (determined based on 3 independent measurements) and sample standard deviations (given by the error bars) of selected mechanical fracture parameters of AAS composites with different amounts of steel microfibers obtained from F–d and F–$CMOD$ diagrams are summarized in the following figures. The values obtained for the monitored parameters using the above-described non-linear fracture models were determined for composite reinforced by microfibers (COMP) and for the plain AAS matrix (MTX) which results when the decomposition procedure is applied to the measured diagrams. Selected parameters were determined via the inverse analysis of F–d diagrams of AAS composites reinforced by microfibers (ID).

Compressive and tensile strength values gained for AAS composite with different amounts of steel microfibers are shown in Figure 6. The compressive strength was determined according ČSN EN 196-1 [45] from two parts of the prismatic specimens obtained after the fracture tests were finished. The reference AAS composite achieved a compressive strength equal to 65 MPa, which is comparable with values gained for alkali-activated slag [33] and fly ash-based composite [46] that have been published in the literature. The addition of steel microfibers did have a reinforcing effect: compressive strength gradually increased with the addition of microfibers amounting to 5 and 10% by weight. The highest mean value of 93 MPa was obtained for the composite with a 15% microfiber content. However, there is no significant difference between microfiber contents of 10 to 20%.

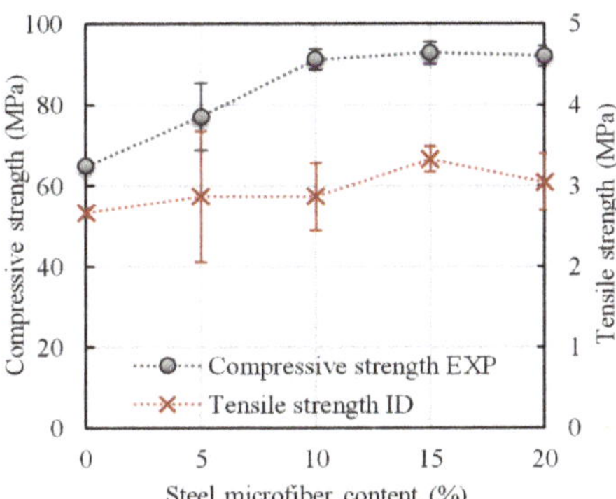

Figure 6. Compressive and tensile strength of AAS mortar with different amounts of steel microfibers.

The tensile strength value was obtained by identification from measured F–d diagrams. The tensile strength of the reference AAS composite was 2.7 MPa. As in the case of compressive strength, the highest mean value was obtained for the composite with a 15% microfiber content, the increase being about 25% in comparison with composite without fibers.

The modulus of elasticity of the reference AAS composite was 15.1 GPa (see Figure 7), which is comparable with values for alkali-activated slag-based composite published in the literature [33]. The modulus of elasticity increased with the addition of microfibers by about 30–40%. The highest mean value of 21.1 GPa was reached for the composite with a 5% microfiber content. However, the standard deviation for higher amounts of microfibers is so high that the modulus of elasticity can be considered to be almost the same. The higher standard deviation especially for 20% microfiber content is caused by heterogeneity of material when the uniform dispersion of fibers became more difficult to achieve. The same trend seen for the modulus of elasticity with the addition of microfibers was observed for the values obtained by identification. The identified values are about 5% higher. If only the contribution of plain AAS matrix is taken into consideration, then the modulus of elasticity is up to 5% lower in comparison with AAS composite containing steel microfibers.

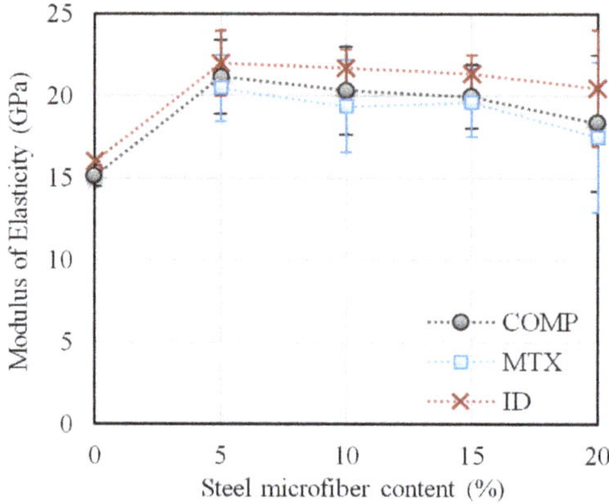

Figure 7. Modulus of elasticity of AAS mortar with different amounts of steel microfibers.

The fracture toughness values for AAS mortar with different amounts of steel microfibers determined by two different non-linear fracture models are presented in Figure 8a. Fracture toughness gradually increased with the addition of steel microfibers in both cases. The fracture toughness values obtained by both models are the same in the case of microfiber contents of 5–10%. The fracture toughness obtained by the double-K model is about 15% lower in the case of the reference composite and composites with a microfiber content of 15% and more. The fracture toughness of the reference AAS composite is comparable with values gained for alkali-activated slag-based composite with similar compressive strength that have been published in the literature [36].

If only the contribution of plain AAS matrix is taken into consideration, then the fracture toughness is about 95, 85, 80 and 70% of the fracture toughness of AAS composites containing steel microfibers SF05, SF10, SF15 and SF20, respectively (see Figure 8b). The same trend was also observed for fracture toughness determined by the double-K fracture model.

The fracture energy values gained for AAS mortar with various amounts of steel microfibers determined by the work-of-fracture method are presented in Figure 9. The specific fracture energy of the reference composite is 113 J/m^2, and this gradually increases as the amount of steel microfibers rises. The fracture energy of the reference AAS composite is comparable with that of an AAS composite with similar compressive strength published in the literature [33]. The highest mean value was obtained for the composite with a 20% microfiber content. This is more than 2.5 times higher than the reference composite value. However, the standard deviation is so high that the value can be considered to be almost the same as for the composite with a 15% microfiber content. The same trend in fracture energy values with the addition of microfibers was observed for values gained via identification. The values obtained in this way are about 15–25% lower in comparison to those obtained via the work-of-fracture method. The sample standard deviation is significantly higher in the case of values determined by identification. If only the contribution of plain AAS matrix is taken into consideration, the fracture energy ranges between 40–50% of that of AAS composite containing steel microfibers.

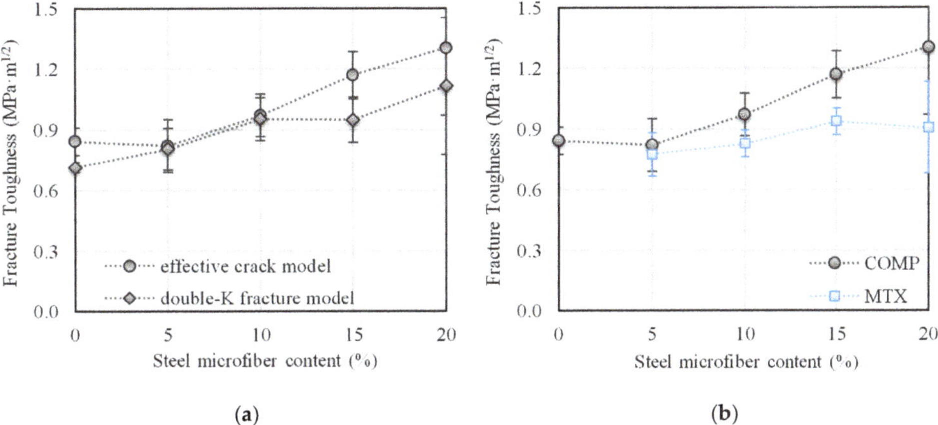

(a)   (b)

Figure 8. Fracture toughness of AAS mortar with different amounts of steel microfibers determined by: (a) Two different fracture models; (b) The effective crack model—contribution of plain AAS matrix.

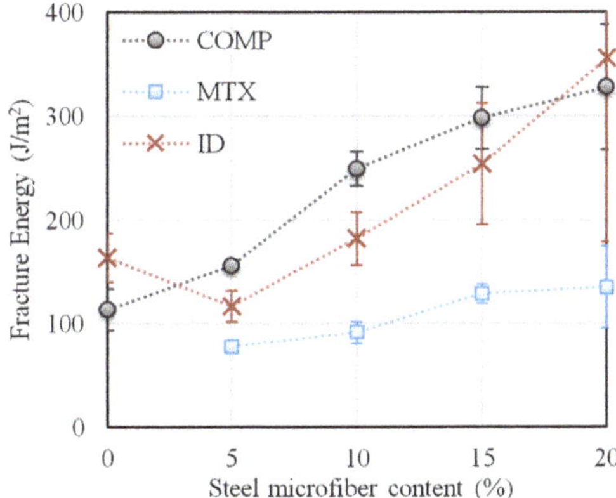

Figure 9. Fracture energy of AAS mortar with different amounts of steel microfibers.

The K_{Ic}^{ini}/K_{Ic}^{un} ratio (fracture toughness ratio, see Figure 10a), which expresses resistance to stable crack propagation, decreased by about 15% in the case of composite with a steel microfiber content of up to 10%. For composite with a higher steel microfiber content, the fracture toughness ratio is comparable with that of the reference composite.

The ratio between the load at the outset of stable crack propagation and the maximum load obtained during the test (load ratio, see Figure 10b) shows a trend analogous to that of the fracture toughness ratio. The addition of steel microfibers (5% by weight) caused a slight decrease in the load ratio. The load ratio for higher amounts of steel microfibers is comparable with that of the reference composite when standard deviation is taken into consideration.

From the obtained results it is obvious that the addition of steel microfibers has a positive effect on resistance to unstable crack propagation and the post-peak behavior of composite. However, resistance to stable crack propagation is rather negatively affected by the addition of steel microfibers, which can be attributed to the heterogeneity of the specimens which occurs when fibers are added. If only the contribution of plain AAS matrix is taken into consideration, the resistance to stable crack propagation is about 10–15% higher than in the case of AAS composite with a steel microfiber content of more than 10%.

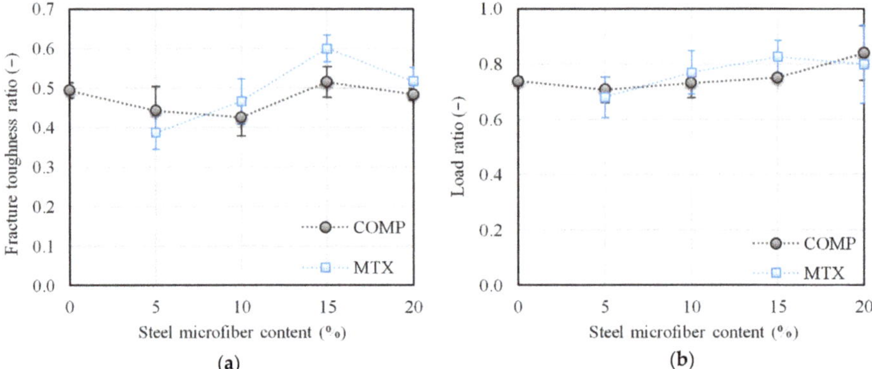

Figure 10. Fracture toughness ratio (**a**) and load ratio (**b**) of AAS mortar with different amounts of steel microfibers.

4. Conclusions

Because of the lack of information concerning the fracture properties of composites with alkali-activated matrix which could be used as inputs for material models, the main aim of the present research was to assess the fracture parameters of alkali-activated slag composites with steel microfibers and quantify the contribution of the matrix of AAS composites to their fracture response. The following conclusions can be drawn based on the obtained experimental research results:

- The compressive strength value increased by up to 40% in the case of composite with a steel microfiber content of more than 10%;
- The modulus of elasticity increased with the addition of steel microfibers by about 30–40%. The highest mean value was obtained for the composite with a 5% content of steel microfibers by weight. If a comparison is made between AAS matrix alone and the composite including steel microfibers, it can be seen that the values are almost the same.

Generally, the addition of steel microfibers to AAS matrix should contribute to a lowering in the tendency to crack, and to an enhancement in the tensile properties of AAS based materials. The presented research results are in line with these suppositions:

- The addition of steel microfibers to the AAS matrix caused the resistance to unstable crack propagation expressed here by the fracture toughness to gradually increase by up to 50% for the composite with a 20% content of steel microfibers;
- The steel microfiber-reinforced AAS composites proved to have a much better load carrying capacity after the maximum load is reached in comparison with the reference composite;
- The addition of steel microfibers to the AAS matrix caused the fracture energy value to gradually increase by up to 2.5 times the reference composite value for the composite with a 20% content of steel microfibers; the energy absorption mechanism is related to the de-bonding and pull-out of microfibers that bridge cracks.

Based on the performed study, the addition of 10 to 15% by weight of microfibers to AAS composite is optimal from the point of view of the enhancement of the fracture properties of this composite. The obtained results can be used when designing alternative binders for Portland cement, as well as relevant input data for material models for the realistic numerical modeling of the response of structures or their parts which are made from this type of composite.

Author Contributions: Conceptualization, P.R.; methodology, H.S. and P.F.; software, P.F.; validation, P.R., P.F. and Z.K.; formal analysis, H.S. and P.F.; investigation, P.R., P.S. and H.S.; resources, P.R. and P.S.; data curation, H.S., P.F. and P.R.; writing—original draft preparation, H.S. and P.F.; writing—review and editing, P.R. and Z.K.; visualization, H.S.; supervision, P.R. and Z.K.; project administration, P.R.; funding acquisition, P.R.

Funding: This researcher was funded by the Czech Science Foundation, project No. 16-00567S, and the Ministry of Education, Youth and Sports of the Czech Republic under "National Sustainability Programme I" (project No. LO1408 AdMaS UP).

Conflicts of Interest: The authors declare no conflict of interest.

References

1. U.S. Geological Survey. *Mineral Commodity Summaries 2019*; U.S. Geological Survey: Reston, VA, USA, 2019; 200p. [CrossRef]

2. Andrew, R.M. Global CO_2 emissions from cement production. *Earth Syst. Sci. Data* **2018**, *10*, 195–217. [CrossRef]

3. McLellan, B.C.; Williams, R.S.; Lay, J.; van Reissen, A.; Corder, G.D. Costs and carbon emissions for geopolymer pastes in comparison to ordinary portland cement. *J. Clean. Prod.* **2011**, *19*, 1080–1090. [CrossRef]

4. Keun-Hyeok, Y.; Jin-Kyu, S.; Keum-Il, S. Assessment of CO_2 reduction of alkali-activated concrete. *J. Clean. Prod.* **2013**, *39*, 265–272. [CrossRef]

5. Provis, J.L.; van Deventer, J.S. (Eds.) *Alkali Activated Materials: State-of-the-Art Report RILEM TC 224-AAM*, 1st ed.; Springer: Dordrecht, The Netherlands, 2014; 388p. [CrossRef]

6. Shi, C.; Krivenko, P.V.; Roy, D. *Alkali-Activated Cements and Concretes*, 1st ed.; CRC Press: London, UK, 2003; 392p. [CrossRef]

7. Johnston, C.D. *Fiber-Reinforced Cements and Concretes*, 1st ed.; Taylor & Francis Group: London, UK, 2010; 261p.

8. Zanichelli, A.; Carpinteri, A.; Fortese, G.; Ronchei, C.; Scorza, D.; Vantadori, S. Contribution of date-palm fibres reinforcement to mortar fracture toughness. *Procedia Struct. Integr.* **2018**, *13*, 542–547. [CrossRef]

9. Smarzewski, P. Influence of basalt-polypropylene fibres on fracture properties of high performance concrete. *Compos. Struct.* **2019**, *209*, 23–33. [CrossRef]

10. Kizilkanat, A.B.; Kabay, N.; Akyüncü, V.; Chowdhury, S.; Akça, A.H. Mechanical properties and fracture behavior of basalt and glass fiber reinforced concrete: An experimental study. *Constr. Build. Mater.* **2015**, *100*, 218–224. [CrossRef]

11. Červenka, V.; Jendele, L.; Červenka, J. *ATENA Program Documentation—Part 1: Theory*; Červenka Consulting: Prague, Czech Republic, 2016.

12. Frantík, P. FyDiK Application. 2015. Available online: http://fydik.kitnarf.cz/ (accessed on 1 April 2019).

13. Bažant, Z.P.; Planas, J. *Fracture and Size Effect in Concrete and other Quasibrittle Materials*, 1st ed.; CRC Press: Boca Raton, FL, USA, 1998; 640p.

14. Karihaloo, B.L. *Fracture Mechanics and Structural Concrete*, 1st ed.; Longman Scientific & Technical: Harlow, UK, 1995; 330p.

15. Shah, S.P.; Swartz, S.E.; Ouyang, C. *Fracture Mechanics of Structural Concrete: Applications of Fracture Mechanics to Concrete, Rock, and Other Quasi-Brittle Materials*, 1st ed.; John Wiley& Sons, Inc.: New York, NY, USA, 1995; 592p.

16. RILEM TC—50 FMC (Recommendation). Determination of the fracture energy of mortar and concrete by means of three-point bend tests on notched beams. *Mater. Struct.* **1985**, *18*, 287–290. [CrossRef]

17. RILEM TC—89 FMT (Recommendation). Size-effect method for determining fracture energy and process zone size of concrete. *Mater. Struct.* **1990**, *23*, 461–465. [CrossRef]

18. Frantík, P. CheCyId Application. 2017. Available online: http://checyid.kitnarf.cz (accessed on 1 April 2019).

19. Novák, D.; Lehký, D. ANN inverse analysis based on stochastic small-sample training set simulation. *Eng. Appl. Artif. Intell.* **2006**, *19*, 731–740. [CrossRef]

20. Lehký, D.; Novák, D.; Keršner, Z. FraMePID-3PB software for material parameter identification using fracture tests and inverse analysis. *Adv. Eng. Softw.* **2014**, *72*, 147–154. [CrossRef]

21. Brühwiler, E.; Wittmann, F.H. The wedge splitting test, a new method of performing stable fracture mechanics tests. *Eng. Fract. Mech.* **1990**, *35*, 117–125. [CrossRef]

22. Tschegg, E.K. New equipments for fracture tests on concrete. *Mater. Test.* **1991**, *33*, 338–342.

23. Bagheri, A.; Nazari, A.; Sanjayan, J.G. Fibre-reinforced boroaluminosilicate geopolymers: A comparative study. *Ceram. Int.* **2018**, *44*, 16599–16605. [CrossRef]

Appl. Sci. **2019**, *9*, 1754

24. Nath, P.; Sarker, P.K. Flexural strength and elastic modulus of ambient-cured blended low-calcium fly ash geopolymer concrete. *Constr. Build. Mater.* **2017**, *130*, 22–31. [CrossRef]
25. Marjanović, N.; Komljenović, M.; Baščarević, Z.; Nikolić, V.; Petrović, R. Physical–mechanical and microstructural properties of alkali-activated fly ash–blast furnace slag blends. *Ceram. Int.* **2015**, *41*, 1421–1435. [CrossRef]
26. Aydin, S.; Baradan, B. Effect of activator type and content on properties of alkali-activated slag mortars. *Compos. B Eng.* **2014**, *57*, 166–172. [CrossRef]
27. Rashad, A.M. A comprehensive overview about the influence of different additives on the properties of alkali-activated slag—A guide for Civil Engineer. *Constr. Build. Mater.* **2013**, *47*, 29–55. [CrossRef]
28. Aydin, S.; Baradan, B. The effect of fiber properties on high performance alkali-activated slag/silica fume mortars. *Compos. B Eng.* **2013**, *45*, 63–69. [CrossRef]
29. Natali, A.; Manzi, S.; Bignozzi, M.C. Novel fiber-reinforced composite materials based on sustainable geopolymer matrix. *Procedia Eng.* **2011**, *21*, 1124–1131. [CrossRef]
30. Puertas, F.; Amat, T.; Fernández-Jiménez, A.; Vázquez, T. Mechanical and durable behaviour of alkaline cement mortars reinforced with polypropylene fibres. *Cem. Concr. Res.* **2003**, *33*, 2031–2036. [CrossRef]
31. Goncalves, J.R.A.; Boluk, Y.; Bindiganavile, V. Crack growth resistance in fibre reinforced alkali-activated fly ash concrete exposed to extreme temperatures. *Mater. Struct.* **2018**, *51*, 42. [CrossRef]
32. Alomayri, T. The microstructural and mechanical properties of geopolymer composites containing glass microfibers. *Ceram. Int.* **2017**, *43*, 4576–4582. [CrossRef]
33. Ding, Y.; Dai, J.G.; Shi, C.J. Fracture properties of alkali-activated slag and ordinary Portland cement concrete and mortar. *Constr. Build. Mater.* **2018**, *165*, 310–320. [CrossRef]
34. Ding, Y.; Dai, J.G.; Shi, C.J. Mechanical properties of alkali-activated concrete: A state-of-the-art review. *Constr. Build. Mater.* **2016**, *127*, 68–79. [CrossRef]
35. Sarker, P.K.; Haque, R.; Ramgolam, K.V. Fracture behaviour of heat cured fly ash based geopolymer concrete. *Mater. Des.* **2013**, *44*, 580–586. [CrossRef]
36. Nguyen, H.; Carvelli, V.; Adesanya, E.; Kinnunen, P.; Illikajnen, M. High performance cementitious composite from alkali-activated ladle slag reinforced with polypropylene fibers. *Cem. Concr. Compos.* **2018**, *90*, 150–160. [CrossRef]
37. Kumar, S.; Barai, S.V. *Concrete Fracture Models and Applications*, 1st ed.; Springer: Berlin, Germany, 2011; 262p. [CrossRef]
38. Frantík, P.; Mašek, J. GTDiPS Software. 2015. Available online: http://gtdips.kitnarf.cz/ (accessed on 24 August 2018).
39. Havlíková, I.; Frantík, P.; Schmid, P.; Šimonová, H.; Veselý, V.; Abdulrahman, A.; Keršner, Z. Components of Fracture Response of Steel Fibre Reinforced Concrete Specimens. *Appl. Mech. Mater.* **2016**, *827*, 287–291. [CrossRef]
40. Frantík, P. FiCubS Application. 2017. Available online: http://ficubs.kitnarf.cz (accessed on 2 October 2018).
41. Hordijk, D.A. Local Approach to Fatigue of Concrete. Ph.D. Thesis, Technische Universiteit Delft, Delft, The Nederlands, 1991.
42. Nallathambi, P.; Karihaloo, B.L. Determination of specimen-size independent fracture toughness of plain concrete. *Mag. Concr. Res.* **1986**, *38*, 67–76. [CrossRef]
43. Karihaloo, B.L.; Abdalla, H.M.; Imjai, T. A simple method for determining the true specific fracture energy of concrete. *Mag. Concr. Res.* **2003**, *55*, 471–481. [CrossRef]
44. Zhang, X.; Xu, S. A comparative study on five approaches to evaluate double-K fracture toughness parameters of concrete and size effect analysis. *Eng. Fract. Mech.* **2011**, *78*, 2115–2138. [CrossRef]
45. *ČSN EN 196-1 Methods of Testing Cement—Prat 1: Determination of Strength*; ÚNMZ: Prague, Czech Republic, 2016.
46. Al-mashhadani, M.M.; Canpolat, O.; Aygörmez, Y.; Uysal, M.; Erdem, S. Mechanical and microstructural characterization of fiber reinforced fly ash based geopolymer composites. *Constr. Build. Mater.* **2018**, *167*, 505–513. [CrossRef]

Article

Evaluation of the Ecotoxicological Potential of Fly Ash and Recycled Concrete Aggregates Use in Concrete

Patrícia Rodrigues [1], José D. Silvestre [1,*], Inês Flores-Colen [1], Cristina A. Viegas [2], Hawreen H. Ahmed [1,3], Rawaz Kurda [1,3] and Jorge de Brito [1]

[1] CERIS, Instituto Superior Técnico, Universidade de Lisboa, Av. Rovisco Pais, 1049-001 Lisbon, Portugal; patricia.rodrigues@tecnico.ulisboa.pt (P.R.); ines.flores.colen@tecnico.ulisboa.pt (I.F.-C.); hawreen.a@gmail.com (H.H.A.); rawaz.kurda@tecnico.ulisboa.pt (R.K.); jb@civil.ist.utl.pt (J.d.B.)

[2] Bioengineering Department and iBB-Institute for Bioengineering and Biosciences, Instituto Superior Técnico, Universidade de Lisboa, Av. Rovisco Pais 1, 1049-001 Lisboa, Portugal; cristina.viegas@tecnico.ulisboa.pt

[3] Department of Civil Engineering, Technical Engineering College, Erbil Polytechnic University, Kirkuk Road, Erbil 44001, Kurdistan Region, Iraq

* Correspondence: jose.silvestre@tecnico.ulisboa.pt; Tel.: +351-21-841-9709

Received: 29 November 2019; Accepted: 30 December 2019; Published: 3 January 2020

Abstract: This study applies a methodology to evaluate the ecotoxicological potential of raw materials and cement-based construction materials. In this study, natural aggregates and Portland cement were replaced with non-conventional recycled concrete aggregates (RA) and fly ash (FA), respectively, in the production of two concrete products alternative to conventional concrete (used as reference). The experimental program involved assessing both the chemical properties (non-metallic and metallic parameters) and ecotoxicity data (battery of tests with the luminescent bacterium *Vibrio fischeri*, the freshwater crustacean *Daphnia magna*, and the yeast *Saccharomyces cerevisiae*) of eluates obtained from leaching tests of RA, FA, and the three concrete mixes. Even though the results indicated that RA and FA have the ability to release some chemicals into the water and induce its alkalinisation, the respective eluate samples presented no or low levels of potential ecotoxicity. However, eluates from concrete mixes produced with a replacement ratio of Portland cement with 60% of FA and 100% of natural aggregates and produced with 60% of FA and 100% of RA were classified as clearly ecotoxic mainly towards *Daphnia magna* mobility. Therefore, raw materials with weak evidences of ecotoxicity could lead to the production of concrete products with high ecotoxicological potential. Overall, the results obtained highlight the importance of integrating data from the chemical and ecotoxicological characterization of materials' eluate samples aiming to assess the possible environmental risk of the construction materials, namely of incorporating non-conventional raw materials in concrete, and contributing to achieve construction sustainability.

Keywords: chemical characterization; concrete; construction material; ecotoxicology; fly ash; leachate; raw material; recycled aggregates; sustainability

---

1. Introduction

The interest in using building construction materials with better mechanical, durability, and environmental characteristics is continuously growing [1–6]. However, to obtain such properties, particularly in cement-based materials, some non-conventional raw materials (such as fly ash (FA) and recycled aggregates (RA)), with unknown hazard levels, are being incorporated. Therefore, it is important to determine the ecotoxicological potential of these innovative construction materials and of their raw materials, so that their use and handling are consciously made and contribute to the sustainability of buildings.

Toxicology is a multidisciplinary science that studies the harmful effects resulting from the interaction between a toxic agent and a biological system in order to predict the advent of these effects. This science encompasses several areas that, depending on the field, have different names, namely ecotoxicology (or environmental toxicology), which studies the toxic effects of xenobiotics on organisms living in ecosystems [7], and occupational toxicology (or human toxicology), which focuses on the study of the harmful effects caused by chemicals present in the workplace [8]. Toxicology can contribute to technological advancement through the implementation of safe conditions of exposure to given substances [9]. To improve construction sustainability, ecotoxicology may have an important contribution in the assessment of the environmental risk associated with the products to be used in the construction sector. For that purpose, it is necessary to evaluate the products' potential ecotoxicity using leaching tests, ecotoxicity tests, and chemical analyses. The assessment of the environmental contamination risk triggered by the contact between surface and groundwater and the materials is done through leaching tests, which allow assessing the mobility of hazardous substances in aqueous media and their environmental impact, although, by themselves, these tests do not reproduce a real situation. These tests should be supplemented with chemical analyses and ecotoxicity testing using biological parameters to understand the influence of the aqueous extracts (e.g., leachates, eluates) on living organisms.

Ecotoxicity tests using the bioluminescent marine bacterium *Vibrio fischeri* are often used because they are simple, fast, and reproducible. *Vibrio fischeri* cells emit light when they are healthy, and this luminescence can decrease when the bacterium cells are brought into contact with a toxic agent that causes inhibition of metabolism. This test can be used in the assessment of the potential ecotoxicity of, for instance, leachates or eluates from solid samples like construction materials or ashes [10]. The study by Tsiridis et al. [10] sought to analyse the toxicity of FA and of its elution products to determine the most sensitive bio test in the toxicological assessment and the most effective leaching process. These authors concluded that the *Vibrio fischeri* test (Microtox test) is suitable for the preliminary screening of solid waste eluate toxicity. However, comparison of data from tests performed with the former bacterium, a freshwater planktonic crustacean *Daphnia magna* and a freshwater planktonic rotifer *Brachionus calyciflorus* indicated *Daphnia magna* as the most sensitive test organism to assess the ecotoxicity of FA eluates [10]. The study also revealed that FA samples obtained from various coal-fired power plants contain heavy metals, namely: Cr, Cu, Mn, Ni, Pb and Zn. The results of the eluates' chemical analysis in accordance with EN 12457-2 [11] and with the toxicity characteristic leaching procedure (TCLP) [12] indicated that the chemical compounds present in FA can be transferred to the liquid phase depending on the pH of the leaching medium. Thus, the pH of the eluate was also considered an important factor in determining environment contamination and ecotoxicity of FA [10]. In general, the chemical analysis showed the elements Al, Ca, Fe, Mg, and Cr are the most frequently found in both solid samples and eluates of FA [10].

Choi et al. [13] proved *Daphnia magna* could be suitable to evaluate the ecotoxicity of leachate from concrete. The leachates of cement samples in acute toxicity test showed 100% *Daphnia magna* immobilization upon 24 h exposure. The authors pointed out the potential negative impact of alkali leaching from concrete into the aquatic environment [13].

The yeast species *Saccharomyces cerevisiae* is a model eukaryotic microorganism and has been used as a test organism in the assessment of the potential toxicity of pesticides and other environmental contaminants [14]. A study developed by Gil et al. [14] showed that this model is suitable for the development of simple low-cost toxicity tests, which can be useful to provide quick preliminary screening of environmental samples' potential toxicity level, before advancing further to ecotoxicity testing with more complex higher eukaryotes.

Kanare and West [15] studied the composition of leachates obtained from concrete mixes produced with two types of aggregates and four types of Portland cement. The authors used two leachants: acetic acid solution and deionised water. The results with acetic acid showed that Hg, Cr, Pb, and Ni were leached, whereas with the deionised water, only partial leaching of Hg, Cr, and Ni occurred.

Rankers and Hohberg [16] investigated various leaching tests on cementitious mortars with FA. The experiment included four types of tests: glass/polyethylene column test (pH 4 water with nitric acid); DEV-S4 batch test (demineralized water); bottle test (pH 4 water with nitric acid); and tank-leaching test (pH 4 water with nitric acid). The first two tests were considered not adequate to construction materials. In the third test, maximum toxic element concentrations were estimated and the results were considered reproducible. The diffusion leaching rates were calculated according to the fourth test, which was also considered reproducible.

Hillier et al. [17] investigated the long-term (up to 256 days) leaching of toxic metals into deionized water from cementitious mortars. The tested samples were analysed according to Directive 80/778/EEC [18] for different metals by using atomic absorption spectroscopy. Of the metals analysed, only vanadium leaching was found in poorly cured mixes [17].

Gwenzi and Mupatsi [19] studied the leachability of heavy metals from cement, sand, and coal ash into water. Total concentrations of Zn, Pb, and Mn in coal ash were lower or similar to those in the other materials, whereas total Fe and Cu were higher in coal ash. In general, the leached concentrations of Fe, Zn, Cu, Mn, and Pb from coal ash and from concrete made with it were found low and comparable to that of conventional concrete, suggesting leaching risk. The authors suggested that coal ash can be incorporated in concrete since it is not anticipated to have harmful impact on environmental and public health related to this constituent [19].

In conclusion, the literature review shows that the incorporation of alternatives (FA and RA) to replace conventional components (cement and natural aggregates (NA)) is insufficiently studied in terms of ecotoxicological potential [20]. Moreover, studies concerning concrete mixes with the simultaneous and high incorporation ratio of FA and RA to produce sustainable RA concrete are missing. In this context, this study focuses on the application of an innovative methodology developed in a previous study by the authors [21] that reveals the environmental profile and the potential ecotoxicological risk of raw materials and building materials. In the present study, the developed methodology was applied to three concrete mixes and to raw materials incorporated in them, namely RA and FA. This was intended to assess their environmental risk when considered unusable and sent to landfills. The ecotoxicity results obtained by applying the proposed methodology can be used independently or as part of life cycle assessment (LCA) studies. Generally, these risks are not considered in LCA studies, although being of high relevance to researchers developing innovative and sustainable materials.

2. Methodology

A global methodology was proposed in a previous study by the authors [21] for assessing the ecotoxicological potential of materials based on chemical and ecotoxicological tests [22–26]. The proposed global methodology assumes the division of materials into two distinct groups: raw materials (e.g., cement, NA, RA, FA) and construction materials (e.g., concrete). The raw materials group includes materials that are incorporated in cement-based products and is divided into four subgroups: virgin raw materials (VRM); processed raw materials (PRM); recycled and by-product raw materials (RBPRM); and nanomaterials.

A simplified flowchart explaining the methodology adopted to perform the chemical characterization (CC) and ecotoxicological characterization (EC) is presented in Figure 1. Detailed information regarding this methodology can be found in the previous study of the authors [21]. The procedure started with crushing a previously produced source concrete and collecting the remaining RA. Then, the collected RA and other raw materials (FA, cement, NA) were used to produce three new concrete mixes (NAC, C40FA60NA100, C40FA60RA100). The materials' characterization and concrete production are further described in Section 3. The three concrete mixes were then crushed and three types of recycled aggregates (A1, A2, and A3) were produced. After that, the eluates of RA (from the source concrete), FA (raw material), A1, A2, and A3 (from the three concrete mixes) were collected by a leaching procedure and tested for CC and EC (Figure 1) as described in Section 4.

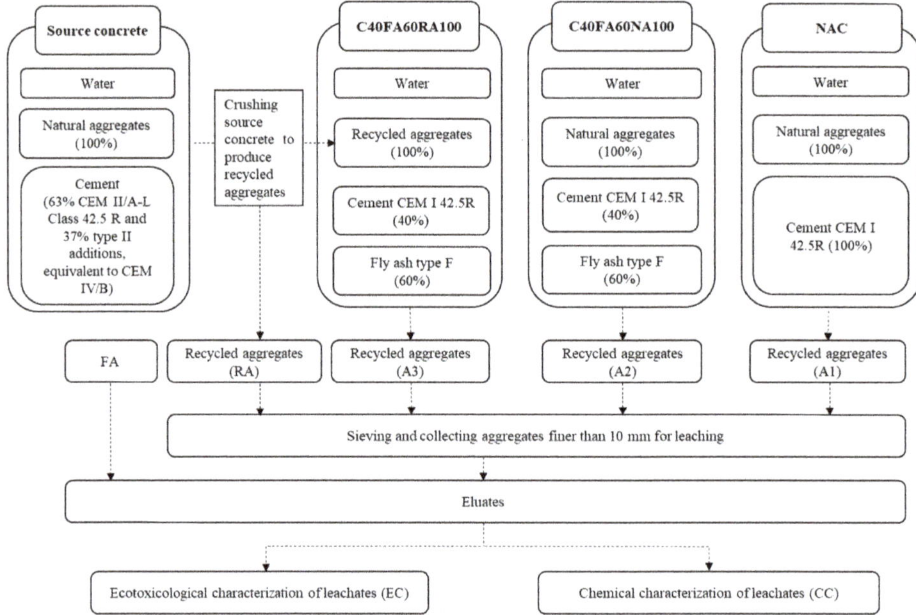

Figure 1. Flowchart of the methodology to perform chemical and ecotoxicological characterizations.

3. Material Characterization and Concrete Compositions

The raw materials chosen to produce concrete mixes include: fine natural aggregates (FNA); coarse natural aggregates (CNA); Portland cement (type CEM I 42.5 R); fine recycled aggregates (FRA); coarse recycled aggregates (CRA); and industrial by-products (FA type F from a Portuguese coal power plant). The materials' characteristics are presented in Tables 1 and 2. FRA and CRA were initially recycled from a source concrete provided by Unibetão Company (Secil, Lisboa, Portugal) with 63% CEM II/A-L Class 42.5 R and 37% type II additions, equivalent to CEM IV/B. Prior to crushing, the source concrete was cured and kept outside (exposed to air) for 28 days.

Regarding the construction materials to be studied, three concrete mixes (NAC, C40FA60NA100 and C40FA60RA100) were produced, with the incorporation of the mentioned raw materials (Table 3 and Figure 1). NAC was produced with 100% NA and 100% Portland cement, C40FA60NA100 with 40% cement, 60% FA and 100% NA, and C40FA60RA100 with 40% cement, 60% FA and 100% RA in weight.

Table 1. Characterization of cement and FA.

Property		Cement (CEM 42.5 R)	FA (Type F)
f_{ctm} at 28 days (MPa)		10.1	-
f_{cm} at 28 days (MPa)		57.7	-
Specific gravity (g/cm^3)		3.1	-
Residue on 45 µm sieve (%)		6.2	14.42
Residue on 32 µm sieve (%)			22.48
Final setting (min)		231.7	
Initial setting (min)		161.1	
Chemical composition (%)	Al_2O_3	5.0	20.9
	CaO	63.5	3.6
	CaO (L)	1.3	
	Cl^-	0.0	
	Fe_2O_3	3.3	7.4
	K_2O	0.6	1.7
	MgO	1.3	1.0
	Na_2O	0.2	1.0
	SiO_2	19.5	57.8
	SO_3	3.3	0.6
	LOI	2.4	
	IR	1.2	3.8

Legend: f_{ctm} = Flexural strength; f_{cm} = Compressive strength; LOI = Loss on ignition; IR = Insoluble residue; FA = Fly ash.

Table 2. Characterization of natural and recycled aggregates.

Property	FNA		CNA			FRA	CRA
	Coarse Sand	Fine Sand	Coarse Gravel	Fine Gravel	"Rice Grain"		
Bulk density (kg/m^3) [27]	1684	1626	1385	1391	1449	1385	1280
Oven dried density (kg/m^3) [27]	2600	2594	2625	2742	2681	2218	2389
Water absorption (%) [27]	0.5	0.4	1.4	1.2	1.0	8	5
LA abrasion mass loss (%) [28]	-	-	28.6	27.2		-	43
Shape index [29]	-	-	15.6	18.0	17.1	-	25.6

Legend: LA = Los Angeles; FNA = Fine natural aggregates; CNA = Coarse natural aggregates; FRA = Fine recycled aggregates; CRA = Coarse recycled aggregates.

Table 3. Concrete mix design.

Concrete	Binder (%)		Natural Aggregates (%)		Recycled Aggregates (%)	
	Cement	FA	FNA	CNA	FRA	CRA
NAC	100	0	100	100	0	0
C40FA60NA100	40	60	100	100	0	0
C40FA60RA100	40	60	0	0	100	100

Legend: FA = Fly ash; FNA = Fine natural aggregates; CNA = Coarse natural aggregates; FRA = Fine recycled aggregates; CRA = Coarse recycled aggregates.

The produced concrete specimens were cured for 28 days in a dry chamber with relative humidity of 50 ± 5% and 22 ± 2 °C air temperature. Then, the concrete specimens were fragmented through a jaw crusher in conjunction with a vertical shaft impact crusher. After that, the aggregates were sieved with 10 mm mesh aperture [30]. The three types of aggregates produced from crushing NAC, C40FA60NA100, and C40FA60RA100 were named A1, A2, and A3, respectively. The aggregates recycled from source concrete (named RA) and NAC (named A1) were almost identical, since they were both made with similar NA and cement content. The only differences were in the cement type and curing process of concrete mixes.

4. Testing Methods

The tests performed are summarized in Table 4. The leaching test and further CC and EC of the obtained eluate samples were carried out in the Analytical Laboratory of Instituto Superior Técnico (LAIST, Tecnico-Lisboa, Portugal) following recommended guidelines as indicated (Table 4), except for the yeast growth inhibition test, which used an in-house microplate susceptibility test (Table 4).

Table 4. Testing methods.

	Tests	Methodology
Leaching	Production of eluates	EN 12457-4 [31]
	As, Hg, Sb, Se	
	Ba, Cd, Cr, Cu, Mo, Ni, Pb, Zn	ISO 11885 [32]
	Chloride, Fluoride, Sulphate	SMEWW 4110 [33]
Chemical characterization (CC)	Total dissolved solids (TDS)	SMEWW 2540 [34]
	Dissolved organic carbon (DOC)	SMEWW 5310 [35]
	pH	SMEWW 4500 [36,37]
	Electrical conductivity (ELC)	EN 27888 [38]
Ecotoxicological characterization (EC)	Inhibition of *Vibrio fischeri* luminescence	15 to 30 min exposure in static test (ISO 11348-3 [39])
	Inhibition of the mobility of *Daphnia magna*	24 and 48 h of exposure in static test (ISO 6341 [40])
	Inhibition of yeast growth	16 h exposure ([14,21]; microplate susceptibility test)

The leaching test was intended to simulate the behaviour of materials when they come into contact with water as leaching agent under given conditions. This test allows predicting the behaviour of the waste if it is deposited in a landfill or in contact with an ecosystem, assuming this process may occur similarly in the environment. In this study, eluate samples were obtained in accordance with the requirements of EN 12457-4 [31]. The test consists of leaching a portion of the solid raw materials with a liquid/solid ratio of 10 over a period of 24 h at 20 ± 2 °C. Deionized water (ultrapure) through the Millipore system (Merck KGaA, Darmstadt, Germany) was used as a leaching agent, with practically zero electrical conductivity. The pH of this water was close to that of rainwater (pH 5–7). The leaching test according to EN 12457-4 [31] required the materials to be fragmented and sieved to obtain particles finer than 10 mm. The mixture of the obtained solid material with the deionized water was placed in a vessel and mechanically agitated for a period of 24 ± 1 h in a shaking equipment, with orbital motion. Subsequently, the samples were subjected to vacuum filtration through 0.45 μm porosity filter paper and placed in a sterile container at 4 °C until further analysis (CC and EC).

The CC of the eluate samples aimed at classifying the materials in relation to their potential danger. Therefore, the parameters analysed for CC (Table 4) were selected from the ones established in Directive 1999/31/EC [23] and in EC Decision 2003/33/EC [24] that regulate the acceptance of solid waste in landfills; most of them are also present in the French proposal on Criteria on the Evaluation Methods for Waste Ecotoxicity (CEMWE) [25].

The EC of the eluates allows detecting the effect of chemical components that may be present in solution, translating this information into an effect on a test organism used as a bioindicator. In this study, the EC of the samples was made using a battery of three short-term acute toxicity tests with the following test organisms: *Vibrio fischeri*, *Daphnia magna*, and *Saccharomyces cerevisiae* (Table 4), as described before [21]. Briefly, the degree of toxicity was determined from the measurement of the harmful effects caused by various concentrations of each sample, comprising the undiluted eluate sample (100% sample) and a dilution series of it (nominal concentrations: 75%, 50%, 25%, 12.5%, 6.25%, and 3.125%), on the test organisms. For each undiluted or diluted sample, the endpoint measured was bioluminescence upon 30-min exposure for *Vibrio fischeri*, mobility after 24- and 48-h exposure for *Daphnia magna*, or growth after 16-h exposure for the yeast (Table 4). The toxicity index EC_{50}, namely the median-effective concentration of each eluate sample (an eluate concentration that causes a response decrease in the exposed organism population to 50% of that measured in a non-toxic control sample), was determined [21]. These toxicity indexes were further processed following the Toxicity

Classification System (TCS) of Persoone et al. [41] in order to facilitate sorting the samples according to their measured ecotoxicity level. Therefore, the EC_{50} values obtained for each test and eluate sample were converted into toxicity units (TU) as per Equation (1) [41]:

$$TU = (1/EC_{50}) \times 100 \tag{1}$$

The calculated TU values allowed classifying each eluate sample into one of the five ecotoxicity classes defined by Persoone et al. [41], as seen in Table 5. This classification is based on the highest TU value determined for each sample in the battery of tests performed battery (i.e., based on the most sensitive ecotoxicity test).

Table 5. Ecotoxicity classes defined according to the TCS system [41].

Classes	I	II	III	IV	V
Toxicity level	No acute toxicity	Low acute toxicity	Acute toxicity	High acute toxicity	Very high acute toxicity
TU	≤0.4	0.4 < TU ≤ 1.0	1.0 < TU ≤ 10	10 < TU < 100	TU ≥ 100

5. Results and Discussion

The application of the proposed methodology to classify the materials under study in terms of hazard regarding the acceptance of solid waste at landfills and ecotoxicological potential allowed obtaining the results that are presented in Table 6. NA are considered VRM, which were previously classified as inert raw materials with no evidence of ecotoxicity (NOE) (Table 6 and [21]); this result can be supported by the study of Barbudo et al. [42].

Table 6. Classification of the raw materials and of the concrete mixes according to the EC Decision 2003/33/EC (pursuant to Directive 1999/31/CE), the document of the CEMWE French proposal and the TCS system.

Materials		Chemical Characterization		Ecotoxicological Characterization		
		Directive 1999/31/EC [23] EC Decision 2003/33/EC [24]		CEMWE [25]		TCS (Persoone et al. [41])
		Class	Parameter	Class	Parameter	
Raw materials	NA	Inert	-	NOE	n/a	n/a
	Cement	-	-	NOE	n/a	n/a
	RA	Non-dangerous	TDS	NOE	n/a	Class III
	FA	Hazardous	Se, Cr, Mo	NOE	n/a	Class III
Construction materials	A1	Non-dangerous	TDS	NOE	n/a	Class III
	A2	Non-dangerous	TDS, Cr, Mo	Ecotoxic	*D. magna*	Class IV
	A3	Non-dangerous	TDS, Cr, Mo	Ecotoxic	*D. magna*	Class IV

Legend: NA = Natural aggregates; RA = Recycled aggregates; FA = Fly ash; A1 = Aggregates recycled from NAC; A2 = Aggregates recycled from C40FA60NA100; A3 = Aggregates recycled from C40FA60RA100; TDS = Total dissolved solids; NOE = No evidence of ecotoxicity; n/a = Not applicable; *D. magna* = *Daphnia magna*; Class III = Acute Toxicity; Class IV = High Acute Toxicity.

According to the European regulation for Registration, Evaluation, Authorization and Restriction of Chemicals (REACH) [43], Portland cement is considered a well-defined mixture (Table 1); it is in the RBPRM group of raw materials. The existence of a safety data sheet (SDS) for Portland cement shows that this mixture meets the criteria for the classification of hazardous substances set out in the CLP regulation [44]. According to the information provided in the SDS, this material could be classified as non-dangerous for the aquatic environment [44]. Thus, it was not considered necessary to perform the CC and EC of cement, and it may be classified as NOE raw material (Table 6). This result is consistent with reports of Hillier et al. [17] and Gwenzi and Mupatsi [19].

RA prepared from source concrete also belong to the group of RBPRM. In this study, the CC and EC of eluate sample obtained from RA were performed (results described in Sections 5.1 and 5.2, respectively). The original RA constituents were the following raw materials: water, NA and cement (CEM II/AL 42.5R), which already have been classified [2]. Therefore, in the particular case in which RA were obtained from the crushing of factory-produced concrete with the desired characteristics (Figure 1), and its composition is fully known (Table 2), it can be assumed that the classification of this raw material can be made from the results obtained for the eluates of the aggregates produced from the reference concrete. The compositions are identical, and both the CEM II/AL 42.5R and the CEM I 42.5R cements (Figure 1) are regulated by the same SDS.

A SDS is not available for the FA used in this study; this type of raw material can be considered as potentially non-hazardous (unknown or variable composition substances) and expected to cause no or low ecotoxicological risk to the aquatic environment [21]. However, a REACH product information sheet is available for "coal ashes (residues)" under the REACH Registration number 01-2119491179-27-0012, which reports measurable no-observed-effect-concentration (NOEC) values for coal ashes towards representative organisms of freshwater, marine and soil ecosystem. Therefore it was found important to examine eluate samples obtained from the FA used in the present study in terms of CC and EC; the results were reported for the first time in [21] and are herein compared with the CC and EC data for the concrete mixes under study (in Sections 5.1 and 5.2, respectively).

The incorporation of RA and FA in cementitious building materials, respectively replacing NA and cement, may have environmental benefits because they avoid landfilling of waste considered hazardous such as FA.

5.1. Chemical Characterization (CC) of RA, FA, A1, A2 and A3 Eluate Samples

The results for the non-metallic and metallic parameters from the CC of the eluate samples obtained from of the raw materials (RA and FA) and concrete mixes (A1, A2, and A3) under study are presented in Table 7.

Table 7. Data of the chemical characterization of the eluate samples.

Parameters		RA	FA	A1	A2	A3
Non-metallic parameters	pH at 22 °C	12.5	11.8	12.5	12.4	12.1
	ELC (µS/cm)	6950	1672	6950	4300	3560
	DOC (mg/kg)	88	23	88	11	100
	TDS (mg/kg)	17,000	6450	17,000	12,000	9200
	Chloride (mg/kg)	15	<30	15	19	34
	Fluoride (mg/kg)	<10	<10	<10	<10	<10
	Sulphate (mg/kg)	<30	4400	<30	<30	37
Metallic parameters (mg/kg)	As	<0.4	<0.4	<0.4	<0.4	<0.4
	Ba	4.0	4.6	4.0	9.0	8.0
	Cd	<0.1	<0.1	<0.1	<0.1	<0.1
	Cr	<0.5	2.5	<0.5	0.5	0.7
	Cu	<1.0	<0.5	<1.0	<1.0	<1.0
	Hg	<0.2	<0.2	<0.2	<0.2	<0.2
	Mo	<0.3	10.0	<0.3	0.6	1.00
	Ni	<0.4	<0.4	<0.4	<0.4	<0.4
	Pb	<0.5	<0.5	<0.5	<0.5	<0.5
	Sb	<0.4	<0.4	<0.4	<0.4	<0.4
	Se	<0.2	4.0	<0.2	<0.2	<0.2
	Zn	<0.5	<0.5	<0.5	<0.5	<0.5

Legend: ELC = Electrical conductivity; DOC = Dissolved organic carbon; TDS = Total dissolved solids; RA = Recycled aggregates; FA = Fly ash; A1 = Aggregates recycled from NAC; A2 = Aggregates recycled from C40FA60NA100; A3= Aggregates recycled from C40FA60RA100.

Regarding the non-metallic parameters, it was found that all the eluate samples were highly alkaline, with pH values varying in the following order: FA (pH = 11.8) < A3 (pH = 12.1) < A2 (pH = 12.4) < A1 = RA (pH = 12.5) (Table 7). The lower pH value of FA was compared with the results of Moreno et al. [45] and Tsiridis et al. [10]. The value from our study was closer to the mean value of Tsiridis et al. [10], possibly because the leachates were produced based on the same standard procedure [31]. The pH values of RA, A1, A2 and A3 eluates are common in this type of construction materials; they are close to concrete's pH (approximately 12.5) and are slightly higher than that of FA (Table 7). These high pH values may be due to the presence of carbonates, oxides and hydroxides formed during combustion processes [26]. On what concerns the concrete mixes, it is worthy to note that the eluate pH slightly decreased (about 0.4 pH units) with the incorporation of 60% of FA and 100% of RA (comparing A3 to A1; Table 7). However, as the pH value of FA depends on the sulphur content of the source coal, it can vary between 4.5 and 12.0 [46,47]. Therefore, this study suggests that the incorporation of FA in concrete may lead to a decrease of the pH of the respective eluates, possibly contributing to a lower possible negative impact of alkaline eluate runoff (for instance, due to heavy raining) into aquatic environments, in landfills and/or during building service.

Regarding acceptability as solid waste in landfills, the raw materials (FA and RA) and concrete mixes (A1, A2 and A3) cannot be considered inert because the TDS concentration values measured in the respective eluate samples (Table 7) are higher than the limit value of 4000 mg/kg defined in the EC Decision 2003/33/EC [24] for inert waste. However, based solely on TDS measurements, they can be considered as non-hazardous waste (Table 6) since the TDS values are lower than the limit values (60×10^3 mg/kg) defined for non-hazardous waste in the same EC Decision [24].

Regarding the concentration of sulphate (SO_4^{2-}), a sulphur (VI) compound, FA showed the highest value among all the materials (4400 mg/kg), much higher than the limit values defined in the EC Decision 2003/33/EC [24]. Therefore, in terms of the sulphate concentration, it is advisable to use FA in concrete because the solidification process caused by the hydration of this binder is effective on retaining heavy metals, namely sulphate. The sulphate concentration in the remaining samples was close to 30 mg/kg, much lower than the limit values [24].

On what concerns the metallic parameters analysed in the eluate samples (Table 7), of the studied elements, Cd, Hg, Ni, and Pb belong to the list of priority substances in the field of water policy and Cd and Hg are considered priority hazardous substances [48]. From this set of metals evaluated in the RA sample, only Ba presented a concentration value above the limit of quantification (4.0 mg/kg) (Table 7). RA can be considered a non-hazardous waste because the concentrations of all the elements in the respective eluate (Table 7) were lower than the limit values established for non-hazardous waste the EC Decision 2003/33/EC [24] (Table 6). On the contrary, in the FA eluate sample, Ba (4.6 mg/kg), Cr (2.5 mg/kg), Mo (10 mg/kg) and Se (4.0 mg/kg) were detected at concentrations above the quantification limits (Table 7). The concentration values of metals leached from the FA in this study were close to the average values obtained by Moreno et al. [45] (determined based on 23 different samples of FA). In the present work, among these metal concentrations, only Se is above the limit value (0.5 mg/kg) defined for non-hazardous waste in EC Decision 2003/33/EC [24] and thus FA is classified as potentially hazardous for the environment (Table 6 and [21]).

For the eluate samples from A1, A2, and A3, only Ba, Cr, and Mo showed concentrations above the minimum detection limit (Table 7). A1 eluate presented a Ba concentration (4.0 mg/kg) lower than that of FA (4.6 mg/kg); the remaining metal parameters showed concentration values lower than the detection limit (Table 7). For the A2 sample, the Ba concentration (9.0 mg/kg) was approximately twice the one recorded in the FA sample, while the concentrations of Cr and Mo were 80% and 94% lower in A2 compared with FA, respectively (Table 7). A similar trend was observed for the A3 sample (Table 7). It is worthy of note that the concentration of Se for all three concrete samples was zero or below the detection limit (Table 7). Importantly, the results show that all the metals concentrations in the A1, A2 and A3 eluates were lower than the limit values defined for non-hazardous waste in the landfill EC Decision [24].

Overall, based on the results obtained from the CC, the samples RA, A1, A2, and A3 were classified as non-dangerous and the sample FA as potentially hazardous regarding possible contamination of aquatic environments associated to possible leaching of metals from waste disposal in landfills (Table 6).

5.2. Ecotoxicological Characterization (EC) of RA, FA, A1, A2 and A3 Eluate Samples

The results of the EC are presented in Table 8. Despite the alkaline nature and slightly different pH of the tested eluates samples (Table 7), no corrections were made to their pH values in toxicity testing, since it was intended to simulate environmental exposure conditions as close as possible to the real ones. Therefore, the ecotoxicity indexes and TU values presented in Table 8 combine the contributions of possible harmful effects associated not only with the chemical components present in the eluate samples but also with their alkaline pH values.

Table 8. Ecotoxicological characterization results.

Materials	pH	Bacterium *Vibrio fischeri*		Crustacean *Daphnia magna*				Yeast *Saccharomyces cerevisiae*	
		EC_{50} (%) [30 min]	TU	EC_{50} (%) [24 h]	TU [24 h]	EC_{50} (%) [48 h]	TU [48 h]	EC_{50} (%) [16 h]	TU
RA	12.5	>100	<1	18.8	5.3	14.6	6.8	19.8	5.1
FA	11.8	49.3	2.0	30.8	3.2	30.8	3.2	>100	<1
A1	12.5	>100	<1	18.8	5.3	14.6	6.8	19.8	5.1
A2	12.4	>100	<1	6.8	15	5.5	18.8	30.2	3.3
A3	12.1	>100	<1	13.6	7.4	7.7	12.9	40.7	2.5

Legend: RA = Recycled aggregates; FA = Fly ash; A1 = Aggregates recycled from NAC; A2 = Aggregates recycled from C40FA60NA100; A3 = Aggregates recycled from C40FA60RA100; TU = Toxicity Units; EC_{50} = median-effective concentration for the endpoint in each test.

The eluate sample of RA (equal to A1) did not cause inhibition of *Vibrio fischeri* luminescence. Comparatively, the toxicity of the same samples to *Daphnia magna* and *Saccharomyces cerevisiae* test organisms was higher (Table 8). These results suggest that the bioluminescent bacterium was not sensitive to exposure to the RA (or A1) sample. On the contrary, the crustacean and the yeast tests presented similar results between them, with TU values between 5.3 and 6.8 in the *Daphnia magna* toxicity test (24 h and 48 h exposure, respectively) and equal to 5.1 in the yeast toxicity test (16 h exposure) (Table 8). These TU values higher than 1 but considerably lower than 10 (TCS system [41]; Table 5) in both toxicity tests with RA (or A1) indicate potentially low acute ecotoxicity (class III) of these materials (Table 6).

Regarding the FA eluate sample, *Vibrio fischeri* bioluminescence and *Daphnia magna* mobility were both moderately affected by this sample, contrary to the yeast growth which was not inhibited upon exposure to the FA sample (Table 8 and [21]). The results reported in the study of Tsiridis et al. [10,49] indicated *Daphnia magna* among the diverse test organisms they used (bacteria, microalga, crustacean, and rotifer), to be the most sensitive test organism in the evaluation of the ecotoxicity of coal FA leachates [10,49]. The authors attributed this result to the lower tolerance of this invertebrate to the presence of high Cr concentration in the samples [49]. This study also revealed that the presence of Cu, Ni, and Zn in the leachate samples resulted in a significant bioluminescence inhibition of *Vibrio fischeri*. In our study, the concentrations of these metals in all eluate samples are very low (Table 7), so the resulting effect is expected to be low. The TU values much lower than 10 (TCS system [41]; Table 5) in both *Daphnia magna* and *Vibrio fischeri* toxicity tests (TU = 3.2 and 2, respectively; Table 8), suggest very low acute ecotoxicity (class III) of FA eluate (Table 6).

The eluate samples of A2 and A3 did not cause inhibition of *Vibrio fischeri* bioluminescence. However, both the crustacean mobility and the yeast growth were considerably affected by the A2 and the A3 samples (Table 8). In both cases, the comparison of the respective TU values indicated that the *Daphnia magna* toxicity test (TU [24 h] = 7.4 and 15 for A3 and A2, respectively) was much more sensitive than the yeast one (TU [16 h] = 2.5 and 3.3, respectively) (Table 8). Based on the TU values > 10 for the most sensitive ecotoxicity test used, i.e., 48 h mobility for *Daphnia magna* (Table 8), the samples A2 and A3 were classified as moderate-highly toxic (class IV) (Table 6). Comparatively, both the A3 and A2 samples were found to exert higher levels of ecotoxicity towards the freshwater model crustacean than the A1 sample, and the order of potential ecotoxicity is: A2 > A3 > A1 (Tables 6 and 8).

Given the results obtained, there was also no evidence to classify the eluate samples of RA, FA and A1 as ecotoxic (Table 6), since the results of CC and EC for these materials (Tables 7 and 8) comply with the limit values established in the French proposal CEMWE [25]. However, the results obtained for samples A2 and A3 allow classifying these materials as potentially ecotoxic based on the fact that the limit value of 10% for 48 h mobility EC_{50} in *Daphnia magna* (corresponding to TU = 10 in Table 5) defined in the French proposal CEMWE [25] was exceeded (Table 6).

There are reports by other authors of a considerable sensitivity of *Daphnia magna* invertebrates when suspended in alkaline media (pH values as high as 11.3) [26]. In our work, despite the high pH values of all the eluate samples tested (Table 8), pH is not anticipated to be the only relevant factor contributing to the negative effects of the samples of this model invertebrate. This is because eluates from RA (A1) with a pH value of 12.5 were found much less toxic to the crustacean than the A2 and A3 eluates with pH values equal to 12.4 and 12.1, respectively (Table 8). Therefore, other factors, for instance the release of chemical compounds, including metals, from these materials into water due to leaching, may be influencing the biological response of the test-organism to the eluates' samples, as reported by others [13,19].

Of the three short-term ecotoxicity tests used (with *Vibrio fischeri*, *Daphnia magna*, and *Saccharomyces cerevisiae*), the most sensitive one in analysing the ecotoxicity of cement-based raw and construction materials was the standard acute mobility test with the freshwater crustacean *Daphnia magna*. On the contrary, the bioluminescent bacterium test was the least sensitive to all eluates' samples, while the microplate susceptibility test with *Saccharomyces cerevisiae* showed an intermediate response (except for the FA eluate sample). In this study, the yeast-based test is proposed as a suitable test for preliminary screening of the potential ecotoxicity of raw and construction materials. Even though a relatively low sensitivity of the yeast test, compared with the *Daphnia magna* one, should be recognized, the yeast test provides some relevant advantages as test system, namely: sparing of higher eukaryotes in toxicity testing (required by regulators, e.g., REACH); meaningfulness of data for eukaryotic microorganisms that provide relevant services in ecosystems; simplicity; low-cost; small-scale (96-well microplate format) and reproducibility [21].

The potential leaching of metals and other chemicals from FA, A2 and A3 was very low for almost all analysed metals, including the ones in the European Union list of priority substances in the field of water policy (e.g., Cd, Hg, Pb, and Ni) [44]. Nevertheless, regarding landfill deposition and based on the CC alone, FA eluate was classified as hazardous (due to the Se level) while the eluates of A2 and A3 were considered non-hazardous (due to Mo and TDS levels) (Table 6). On the contrary, based on EC, the former was NOE or class III (according to CEMWE or TCS, respectively) while the latter were classified as class IV based on ecotoxicity towards *Daphnia magna* (Table 6). These changes can be due to possible differences in the chemical composition of the eluates and/or in other parameters not analysed in this work. Based on the data presented in this work and on other studies [19,26], it can also be suggested that the alkaline pH of eluates from FA (pH 11.8) and concrete (pH 12.5 and 12.4 for A1 and A2, respectively) may be relevant in this respect and may contribute to possible environmental risks. Such risks can be especially important if leachates or eluates from concrete or FA are produced in buildings in service (e.g., due to rain events) and/or in landfills and can reach freshwater ecosystems leading to water alkalinisation. Therefore, results obtained in this study highlight the importance of integrating data from CC and EC of materials' eluate samples aiming at assessing the possible environmental risk of the construction materials, as suggested by other authors [26,50].

The incorporation of raw materials that have a low ecotoxicity level (e.g., FA, RA) in construction materials can however generate construction materials leading to eluates with a significantly higher toxicity level. For example, A1, coming from a concrete with 100% NA and 100% cement, has a toxicity level lower than that of A2 and A3 where natural aggregates and Portland cement were partially replaced with non-conventional RA and FA. In fact, it seems relevant that although FA has a low ecotoxicological potential (Table 6), when this raw material was incorporated in concrete mixes in a given quantity (60% in the present study) it did not contribute to diminish the potential ecotoxicity of the respective eluates.

The literature states that the cement-based construction materials that raise the most concerns in terms of ecotoxicological risk are those that incorporate RA. However, the incorporation of by-products (FA) in such materials may also contribute to some non-negligible environmental risk. These raw materials have a variable chemical composition that is often unknown.

6. Conclusions

From the application of a methodology proposed in a previous study [21] to three concrete mixes (A1, A2, A3), in order to compare the ecotoxicological potential of raw materials (fly ash—FA, recycled aggregates—RA) and cement-based construction materials, the following conclusions were drawn:

- Taking into account only the chemical characterization (CC) of the eluates and the comparison of the results obtained with limit values established in the landfill waste EC Decision 2003/33/EC [24], the concrete mixes formulated with raw materials classified as hazardous (FA) may be classified as non-hazardous materials (A2 and A3).
- Materials with a high ecotoxicological potential, namely samples A2 and A3, can be formulated from raw materials with evidence of lower ecotoxicity (FA and RA). This result is corroborated by TCS, which showed that the ecotoxicity degree of construction materials formulated with ecotoxic raw materials was greater than that of their raw materials; an increase in toxicity occurs from the raw materials (Class III) to the construction materials (Class IV). Thus, the incorporation of 100% RA and 60% FA in concrete may result in an adverse effect on human health and on the environment.
- The incorporation of high quantities of FA with an extremely high sulphate concentration (4400 mg/kg) in concrete reduced the concentration to 30 mg/kg for the samples A1, A2, and A3. The concentration of chlorides was low in FA, but increased slightly when incorporated in concrete.

In this study, the importance of assessing the possible environmental risks of the incorporation of non-conventional raw materials in cement-based construction materials was demonstrated. The chemical and ecotoxicological characterization of materials' eluate samples can complement environmental life cycle assessment studies and contribute to achieving construction sustainability.

Author Contributions: P.R. developed and applied the methodology to concrete mixes. J.D.S. and I.F.-C. supervised the research work based on a master thesis and reviewed the paper. C.A.V. carried out ecotoxicity tests, helped in the interpretation of the results and reviewed the paper. H.H.A. wrote the paper. R.K. formulated and produced the concrete mixes. J.d.B. reviewed and edited the paper. All authors have read and agreed to the published version of the manuscript.

Funding: This research was funded by FCT—Foundation for Science and Technology, Portugal, via the Project FCT PTDC/ECM/118372/2010—EXCELlentSUStainableCONcrete.

Acknowledgments: The authors wish to thank CERIS research centre and iBB-Institute for Bioengineering and Biosciences from IST, FCT—Foundation for Science and Technology (UID/BIO/04565/2013 for iBB), Programa Operacional Regional de Lisboa 2010 (Lisboa-01-0145-FEDER-007317 for iBB) and to WGB Shield research project (FCT-PTDC/ECI-EGC/30681/2017).

Conflicts of Interest: The authors declare no conflict of interests.

References

1. Sunayana, S.; Barai, S.V. Performance of fly ash incorporated recycled aggregates concrete column under axial compression: Experimental and numerical study. *Eng. Struct.* **2019**, *196*, 109258. [CrossRef]
2. Kurad, R.; Silvestre, J.D.; de Brito, J.; Hawreen, A. Effect of incorporation of high volume of recycled concrete aggregates and fly ash on the strength and global warming potential of concrete. *J. Clean. Prod.* **2017**, *166*, 485–502. [CrossRef]
3. Kurda, R.; de Brito, J.; Silvestre, J.D. Water absorption and electrical resistivity of concrete with recycled concrete aggregates and fly ash. *Cem. Concr. Compos.* **2019**, *95*, 169–182. [CrossRef]
4. Arezoumandi, M.; Drury, J.; Volz, J. Effect of recycled concrete aggregate replacement level on the fracture behavior of concrete. *ACI Mater. J.* **2015**, *3*, 1–8.
5. Kurda, R.; de Brito, J.; Silvestre, J.D. Carbonation of concrete made with high amount of fly ash and recycled concrete aggregates for utilization of CO_2. *J. CO_2 Util.* **2019**, *29*, 12–19. [CrossRef]
6. Mocová, K.A.; Sackey, L.N.A.; Renkerová, P. Environmental impact of concrete and concrete-based construction waste leachates. *IOP Conf. Ser. Earth Environ. Sci.* **2019**, *290*, 012023. [CrossRef]
7. APA. (Portuguese Environment Agency). 2015. Available online: http://apambiente.pt/_zdata/LRA/Ecotoxicologia.pdf (accessed on 2 April 2015). (In Portuguese).
8. Santos, A. *Occupational Toxicology—Toxicity of Organic Solvents*; Portuguese Chemical Society: Lisboa, Portugal, 2014; Available online: https://www.spq.pt/magazines/BSPQuimica/567/article/3000492/pdf (accessed on 2 April 2015). (In Portuguese)
9. Nassis, C. *Toxicology Fundamentals*. 2015. Available online: http://pt.scribd.com/doc/128059485/Fundamentos-de-Toxicologia (accessed on 2 April 2015). (In Portuguese).
10. Tsiridis, V.; Samaras, P.; Kunglos, A.; Sakellaropoulos, G. Application of leaching tests for toxicity evaluation of coal fly ash. *Environ. Toxicol.* **2006**, *21*, 409–416. [CrossRef]
11. *Characterisation of Waste. Leaching. Compliance Test for Leaching of Granular Waste Materials and Sludges. Part 2: One Stage Batch Test at a Liquid to Solid Ratio of 10 L/kg for Materials with Particle Size below 4 mm (without or with Size Reduction)*; EN 12457-2:2002; Comité Européen de Normalisation: Brussels, Belgium, 2002.
12. *Toxicity Characteristic Leaching Procedure*; TCLP 1311:1992; United States Environmental Protection Agency (USEPA): Washington, DC, USA, 1992.
13. Choi, J.; Bae, S.; Shin, T.; Ahn, K.; Woo, S. Evaluation of daphnia magna for the ecotoxicity assessment of alkali leachate from concrete. *Int. J. Indust. Entomol.* **2013**, *26*, 41–46. [CrossRef]
14. Gil, F.; Santos, M.; Chelinho, S.; Pereira, C.; Feliciano, J.; Sousa, L.J.; Ribeiro, R.; Viegas, C. Suitability of a *Saccharomyces cerevisiae*-based assay to assess the toxicity of pyrimethanil sprayed soils via surface runoff: Comparison with standard aquatic and soil toxicity assays. *Sci. Total Environ.* **2015**, *505*, 161–171. [CrossRef]
15. Kanare, H.M.; West, B.W. Leachability of selected chemical elements from concrete. In *Emerging Technologies Symposium on Cement and Concrete in the Global Environment*; Portland Cement Association: Chicago, IL, USA, 1993.
16. Rankers, R.H.; Hohberg, I. Leaching tests for concrete containing fly ash-Evaluation and mechanism. *Stud. Environ. Sci.* **1991**, *48*, 275–282.
17. Hillier, S.; Sangha, C.; Plunkett, B.; Walden, P. Long-term leaching of toxic trace metals from Portland cement concrete. *Cement Concrete Res.* **1991**, *29*, 515–521. [CrossRef]
18. *Council Directive 80/778/EEC of 15 July 1980 Relating to the Quality of Water Intended for Human Consumption—Repealed by Council Directive 98/83/EC on the Quality of Water Intended for Human Consumption*; Council of the European Union: Brussels, Belgium, 1980.
19. Gwenzi, W.; Mupatsi, N. Evaluation of heavy metal leaching from coal ash-versus conventional concrete monoliths and debris. *J. Waste Manag.* **2006**, *49*, 114–123. [CrossRef] [PubMed]
20. Kurda, R.; Silvestre, J.D.; de Brito, J. Toxicity and environmental and economic performance of fly ash and recycled concrete aggregates use in concrete: A review. *Heliyon* **2018**, *4*, e00611. [CrossRef] [PubMed]
21. Rodrigues, P.; Silvestre, J.D.; Flores-Colen, I.; Viegas, C.A.; de Brito, J.; Kurad, R.; Demertzi, M. Methodology for the assessment of the ecotoxicological potential of construction materials. *Materials* **2017**, *10*, 649. [CrossRef]

22. *Directive 2008/98/EC of the European Parliament and of the Council of 19 November 2008 on Waste and Repealing Certain Directives (Text with EEA Relevance)*; European Parliament and Council of the European Union: Brussels, Belgium, 2008.
23. Directive 1999/31/EC of 26 April 1999 on the landfill of waste. *Off. J. Eur. Communities* **1999**, *L182*, 1–19.
24. Council decision 2003/33/EC of the European Community of 19 December 2002 Establishing Criteria and Procedures for the Acceptance of Waste and Landfills Pursuant to Article 16 of and Annex II to Directive 1999/31/EC. *Off. J. Eur. Communities* **2003**, *L11*, 27–49.
25. *Criteria on the Evaluation Methods of Waste Ecotoxicity: Proposal—CEMWE*; ADEME 1998; ADEME: Paris, France, 1998.
26. Lapa, N.; Barbosa, R.; Lopes, M.; Mendes, B.; Abelha, P.; Gulyurtlu, I.; Oliveira, J. Chemical and ecotoxicological characterization of ashes obtained from sewage sludge combustion in a fluidised-bed reactor. *J. Hazard Mater.* **2007**, *147*, 175–183. [CrossRef]
27. *Tests of the Mechanical and Physical Properties of the Aggregates. Part 6: Determination of Density and Water Absorption*; EN 1097-6:2013; Comité Européen de Normalisation: Brussels, Belgium, 2013.
28. Tests for mechanical and physical properties of aggregates. In *Methods for the Determination of Resistance to Fragmentation*; EN 1097-2:1998; Comité Européen de Normalisation: Brussels, Belgium, 1998.
29. *Tests to Determine the Geometric Characteristics of the Aggregates. Part 4: Determination of the Shape of the Particles*; EN 933-4:2002; Form index; Comité Européen de Normalisation: Brussels, Belgium, 2002.
30. *Aggregates for Concrete*; IPQ:2002; EN 12620:2002+A1:2010; Instituto Português da Qualidade: Lisboa, Portugal, 2002.
31. *Characterization of Waste. Leaching—Compliance Test for Leaching of Granular Waste Materials and Sludges. Part 4: One Stage Batch Test at a Liquid to Solid Ratio of 10 L/kg for Materials with Particle Size below 10 mm (without or with Size Reduction)*; EN 12457-4:2002; Comité Européen de Normalisation: Brussels, Belgium, 2002.
32. *Water Quality—Determination of Selected Elements by Inductively Coupled Plasma Optical Emission Spectrometry (ICP-OES)*; ISO 11885:2007; International Organization for Standardization: Geneva, Switzerland, 2007.
33. Determination of Anions by Ion Chromatograph. In *Standard Methods for the Examination of Water and Wastewater*; SMEWW 4110:2017; American Public Health Association, American Water Works Association and Water Environment Federation: Washington, DC, USA, 2017.
34. Solids. In *Standard Methods for the Examination of Water and Wastewater*; SMEWW 2540:2017; American Public Health Association, American Water Works Association and Water Environment Federation: Washington, DC, USA, 2017.
35. Total Organic Carbon (TOC). In *Standard Methods for the Examination of Water and Wastewater*; SMEWW 5310:2017; American Public Health Association, American Water Works Association and Water Environment Federation: Washington, DC, USA, 2017.
36. H+ pH value. In *Standard Methods for the Examination of Water and Wastewater*; SMEWW 4500:2017; American Public Health Association, American Water Works Association and Water Environment Federation: Washington, DC, USA, 2017.
37. Boron. In *Standard Methods for the Examination of Water and Wastewater*; SMEWW 4500:2017; American Public Health Association, American Water Works Association and Water Environment Federation: Washington, DC, USA, 2017.
38. *Determination of Conductivity—Condutimetry*; IPQ:1996. EN 27888:1996; Instituto Português da Qualidade: Lisboa, Portugal, 1996. (In Portuguese)
39. *Water Quality—Determination of the Inhibitory Effect of Water Samples on the Light Emission of Vibrio Fischeri (Luminescent Bacteria Test). Part 3: Method Using Freeze Dried Bacteria*; ISO 11348-3:2007; International Organization for Standardization: Geneva, Switzerland, 2007.
40. *Determination of the Inhibition of the Mobility of Daphnia Magna Straus (Cladocera, Crustacea)—Acute Toxicity Test*; ISO 6341:2012; International Organization for Standardization: Geneva, Switzerland, 2012.
41. Persoone, G.; Marsalek, B.; Blinova, I.; Torokne, A.; Zarina, D.; Manusadzianas, L.; Nalecz-Jawecki, G.; Tofan, L.; Stepanova, N.; Tothova, L.; et al. A practical and user-friendly toxicity classification system with microbiotests for natural waters and wastewaters. *Environ. Toxicol.* **2003**, *18*, 395–402. [CrossRef]
42. Barbudo, A.; Galvín, A.; Agrela, F.; Ayuso, J.; Jiménez, J. Correlation analysis between sulphate content and leaching of sulphates in recycled aggregates from construction and demolition wastes. *J. Waste Manag.* **2012**, *32*, 1229–1235. [CrossRef]

43. Regulation (EC) 1907/2006 of 18 December 2006, of the European Parliament concerning the Registration, evaluation, authorization and restriction of chemicals (REACH). *Off. J. Eur. Union* **2006**, *L396*, 1–849.

44. *Regulation (EC) 1272/2008 of the European Parliament and of the Council of 16 December 2008 on Classification, Labelling and Packaging of Substances and Mixtures*; Europa EU: Brussels, Belgium, 2008.

45. Moreno, N.; Querol, X.; Andrés, J.; Stanton, K.; Towler, M.; Nugteren, H.; Janssen-Jurkovicova, M.; Jones, R. Physico-chemical characteristics of European pulverized coal combustion fly ashes. *Fuel* **2005**, *84*, 1351–1363. [CrossRef]

46. Plank, C.; Martens, D. Boron availability as influenced by application of fly ash to soil. *Soil Sci. Soc. Am. J.* **1974**, *38*, 974–977. [CrossRef]

47. Page, A.; Elseewi, A.; Straughan, I. Physical and chemical properties of fly ash from coal-fired power plants with special reference to environmental impacts. *Residue Rev.* **1979**, *71*, 83–120.

48. *Directive 2008/105/EC of the European Parliament and of the Council of 16 December 2008 on Environmental Quality Standards in the Field of Water Policy*; Europa EU: Brussels, Belgium, 2008.

49. Tsiridis, V.; Petala, M.; Samara, P.; Kungolos, A.; Sakellaropoulos, G. Environmental hazard assessment of coal fly ashes using leaching and ecotoxicity tests. *Ecotoxicol. Environ. Saf.* **2012**, *84*, 212–220. [CrossRef]

50. *Directive 2000/60/EC of the European Parliament and of the Council Establishing a Framework for the Community Action in the Field of Water Policy*; Europa EU: Brussels, Belgium, 2008.

 *applied sciences*

Review

Natural Kenaf Fiber and LC^3 Binder for Sustainable Fiber-Reinforced Cementitious Composite: A Review

Mohammad Hajmohammadian Baghban [1,*] and Reza Mahjoub [2,*]

1 Department of Manufacturing and Civil Engineering, Norwegian University of Science and Technology (NTNU), 2815 Gjøvik, Norway
2 Civil Engineering Department, Faculty of Engineering, Khorramabad Branch, Islamic Azad University, Khorramabad 6815139417, Iran
* Correspondence: mohammad.baghban@ntnu.no (M.H.B.); mahjoub@khoiau.ac.ir or r_mahjoob@yahoo.com (R.M.); Tel.: +47-48-351-726 (M.H.B.); +98-91-2372-4133 (R.M.)

Received: 28 October 2019; Accepted: 30 December 2019; Published: 3 January 2020

Abstract: Low impact on the environment and low cost are the key drivers for today's technology uptake. There are many concerns for cement production in terms of negative environmental impact due to greenhouse gas (GHG) emission, deficiency of raw materials, as well as high energy consumption. Replacement of the cement by appropriate additives known as supplementary cementitious materials (SCMs) could result in reduction in GHG emission. Limestone-calcined clay cement (LC^3) is a promising binder in the concrete sector for its improvements to environmental impact, durability, and mechanical properties. On the other hand, the advantages of fiber-reinforced concrete such as improved ductility, versatility, and durability have resulted in increasing demand for this type of concrete and introduction of new standards for considering the mechanical properties of fibers in structural design. Thus, using natural fibers instead of synthetic fibers can be another step toward the sustainability of the concrete industry, which is facing increasing demand for cement-based materials. This review studies the potential of natural Kenaf fiber-reinforced concrete containing LC^3 binder as a step toward green cementitious composite. While studies show that energy consumption and GHG emission can be reduced and there is a significant potential to enhance mechanical and durability properties of concrete using this composition, adjustment of the mix design, assessing the long-term performance and standardization, are the next steps for the use of the material in practice.

Keywords: supplementary cementitious materials; natural fiber; LC^3; kenaf fiber; mechanical properties; durability; calcined clay; limestone

---

1. Introduction

Introducing environmentally friendly materials is one of the most fascinating research fields in engineering. In civil engineering and related disciplines, concrete is a high-demand material for the building and construction sector. Nearly one ton of CO_2 is released in the production of every ton of ordinary Portland cement (OPC) [1–3]. In addition, SO_2 and NO_x which have a role in greenhouse effects and acidic rains are also released during OPC manufacturing [4]. Therefore, different countries and communities should have plans to reduce air pollution. For example, the European Commission has a plan to reduce CO_2 emissions under the EU's emissions trading system, to speed the transition to a low-carbon economy in four phases by 2031. Scientists are continuously trying to improve concrete mixes not only to enhance different properties but also to reduce the material impact on the environment. Fiber-reinforced concrete and supplementary cementitious materials (SCMs) have also emerged in alignment with such goals. While fiber-reinforced concrete is becoming more commonly used and fibers are becoming an alternative to steel reinforcing bars, using natural fibers would be interesting from a sustainability point of view [5]. On the other hand, the cementitious binder needs to be compatible and in alignment with sustainability goals, which has sparked the interest in using natural fibers. There is still an important issue regarding the use of natural fibers in the cementitious binder as natural fiber-reinforced concrete (NFRC). The main drawback of the NFRC is the deterioration of fiber in the alkaline surrounding of OPC concrete [6,7]. Moreover, less greenhouse gas (GHG) emission, less energy consumption, and avoiding deficiency in binder quality makes SCMs an attractive option [8–12]. The use of by-products (slag, fly ash (FA), silica fume (SF), and other waste materials) as SCMs is an effective solution for mitigating air pollution, but the appropriate SCM for NFRC should cause a decrement in pore solution alkalinity in the binder. Since the resources for by-products as SCM are limited and may be challenging to approach in the near future, limestone-calcined clay cement (LC^3) is an alternative for a compatible binder for natural fibers. A review of the latest developments in SCMs (especially LC^3) and NFRC (specifically in combination with SCMs) are the goals of this paper.

2. Supplementary Cementitious Material (SCM)

Green buildings are essential elements for acquiring sustainability, and the concrete industry can move toward sustainability by introducing SCMs [13,14], which can also lead to improved durability and mechanical properties of concrete [15,16]. Industrial residues, such as SF and FA, are extensively employed as SCMs due to the high pozzolanic reactivity [16–19]. Furthermore, continuous effort is made to introduce agriculturally sourced pozzolanic substances, such as rice husk ash (RHA) [20,21], corn cob ash [22,23], wood ash [24], natural zeolite [4,25], bamboo leaf ash [26], and palm oil fuel ash [27–31]. As an example, when cement is replaced by micro-palm oil fuel ash (mPOFA) at certain levels, the compressive strength will be increased [32,33]. Replacing 10% weight of cement with mPOFA increases compressive strength up to 33%. This could be caused by mPOFA occupying the space between the particles of cement and enhancing the creation of calcium-silicate-hydrate (C-S-H) gel [34]. Table 1 expresses the composition and properties of typical OPC and some common SCMs. The higher value of SiO_2 and Al_2O_3 and Fe_2O_3 means the higher pozzolanic feature of SCM. According to ASTM C618, the value of all these three oxide compositions together should be more than 70%, as the requirement of material to be considered to be class N or F natural pozzolan.

FA is known as one of the most common SCMs which can reduce GHG emission of concrete and enhance durability, fracture toughness, and compressive strength of this material [13,42,43]. On the other hand, it is noted in previous studies that although enhancement in durability and compressive strength was observed by using FA, extensive usage may lead to challenges such as higher carbonation and delayed hydration [43]. Moreover, SF is the other commonly used pozzolanic substance that can also enhance the strength and durability significantly due to the high purity of silica content with fine particle size increasing its reactivity [35,41,44]. However, the main challenge for SCMs which are by-products of other industries (such as SF and FA) is the limitation in global production [45]. Furthermore, the quality of by-products is the other issue. For example, over 66% of the accessible FA, which has one of the highest quantities between these by-products, is not appropriate for mixing with cement [46].

Calcined clay is a kind of artificial pozzolan and its pozzolanic activity is affected by parameters such as the quantity of calcined minerals, impurity measure, activation technique, and post-calcination. [47]. Calcined clays appear as a confident source of SCM, able to offer a considerable replacement of the Portland cement clinker in mixed cement [48]. Various types of clay minerals include illite, kaolinite, palygorskite, and montmorillonite [49,50]. It is documented that among the different type of clay minerals, kaolinite has the highest pozzolanic activity [47,51]. After calcining the kaolinite-containing clay, metakaolin is created that is an amorphous alumino-silicate ($Al_2Si_2O_7$), which may make a reaction with calcium hydroxide to provide calcium-aluminate-silicate-hydrate (C-A-S-H) and aluminate hydrates [45]. Furthermore, carbo-aluminate hydrates could be produced as the reaction between the alumina and limestone [52]. The metakaolin (clay) is an abundant material and also its quality is further stable compared to FA and slag [53–56]. Mayo and Hassan reported that by the presence of 20% metakaolin in the self-compacting concrete mixture, the tensile and compressive strengths (28 days) could be increased to 25% and 30%, respectively [57]. This indicates that calcined clay has the potential to show higher pozzolanic reactivity than FA.

Table 1. Typical composition and properties of ordinary Portland cement (OPC), SF, FA, limestone, metakaolin and calcined clay [6,13,35–41].

Chemical Composition and Physical Properties	Ordinary Portland Cement	SF	FA	Limestone	Metakaolin	Calcined Clay (50.3 wt.% Kaolinite Content)
SiO_2 (%)	19.2–21.63	90–95	46.44–50.96	0.1–0.8	51.8–57.37	44.9
Al_2O_3 (%)	4.18–4.27	0.25–1.2	25.88–38.01	0.3	38.63–42.4	32.3
Fe_2O_3 (%)	3.32–3.45	0.15–1.3	3.12–8.25	0.3	0.77–4.15	15.4
TiO_2 (%)	—	—	1.36	—	1.07	2.4
CaO (%)	63.25–64.93	0.36–1.5	2.15–7.5	55–58	0.03–0.071	1.3
MgO (%)	1.61–2.77	0.47–2	0.23–2.60	1.8–0.2	0.07	0.8
Na_2O (%)	0.09	0.13–0.5	0.33–1.26	0.01–0.1	0.39	0.4
MnO (%)	—	0.02–0.07	—	—	—	0.1
K_2O (%)	0.78	0.2–0.84	0.88–2.65	0.01	0.218–0.49	0.2
SO_3 (%)	2.02–3.35	0.69	0.65–0.69	0.05	0.105–0.15	0.1
P_2O_5 (%)	0.09	0.04–0.17	0.06–0.35		0.61	0.4
Loss of ignition (%)	1.24–2.49	2.29–3	2.79–3.2	42.6–43.4	1.04	1.7
Specific gravity (g/cm^3)	3.2	1.9–2.15	2.14	—	2.59	—
Specific surface (cm^2/g)	3280–9000	2730	3640	18,000	—	45,700
Bulk density (kg/m^3)	—	300–660	—	—	—	—

Increase in compressive strength of mortar blending at an early age was also observed by using metakaolin as SCM [48,58,59]. Furthermore, combination of limestone and metakaolin resulted in higher compressive strength compared to using typical OPC [58]. Avet et al. stated that compressive strength of mortars containing different types of calcined clay appeared mainly dependent on the calcined kaolinite content irrespective to the other parameters [59]. Sulfate resistance is also reported to be significantly good for the investigated mortars with calcined clays (either calcined montmorillonite or metakaolin), non-dependent to the pore structures and compressive strength [48].

2.1. Limestone-Calcined Clay Cement (LC^3)

Among various SCMs available for substituting Portland cement clinker, the features of a ternary blend identified as LC^3 is evaluated broadly in terms of its benefits over OPC [45,52,59,60]. Calcination at temperatures between 600 and 800 °C results in the pozzolanic activity of kaolinite [50]. Limestone and kaolinitic clay are present in the earth crust abundantly, and much lower heating temperature compared to Portland cement clinker is required to produce calcined clay. Only 0.3 tons of CO_2 may be emitted for producing 1 ton of calcined clay [45,61], which is much less than the production of the same mass of OPC [62] (which is typically 1 ton of CO_2). There are many types of clay with different mineral composition depending on the region. Usually, most of clay types have about 40% kaolinite content or higher, which means they are suitable for calcination to produce highly reactive pozzolan. There are three common methods for calcination including rotary kilns, flash calcination and fluidized bed [56,63,64]. Therefore, LC^3 mixes have considerable variations in performance and color, based on the material source as well as calcination and use method. It is noteworthy that by substitution of clinker with limestone in LC^3 blends, both cost and the environmental impacts are reduced [45]. Optimal mechanical characteristics and enhancement in durability tests are observed with the replacement of 50% of clinker [45,52]. The viability of any technology depends on four key elements, including economic viability, technical feasibility, easy accessibility of raw materials and low capital investment. The developed LC^3 technology meets all the mentioned criteria [45]. One of the challenges with using limestone and calcined clay as SCM is the reduction in workability compared to the OPC binder. This issue could be managed by using the appropriate dosage of superplasticizer (SP) and also viscosity-modifying admixture (VMA) [56,64]. The relation between the dosage of SP and VMA with the content of calcined clay and limestone is still unformulated properly. There are many parameters related to the physical and chemical features of materials (e.g., particle size distribution, calcination temperature and chemical adsorption) and application of binder for getting the proper correlation and formula. Developing a chemical admixture designed for LC^3 mixes is still under demand.

2.1.1. Mechanical Properties of LC^3 Cement Binder

Antoni et al. and Avet et al. reported the highest compressive strength was found by the weight proportion of 1:2 for the limestone to metakaolin [52,59]. The most significant parameter of the calcined clay is kaolinitic content leading the mechanical properties of calcined clay-based binder systems. The comparable pozzolanic reactivity is also obtained by the lower levels of metakaolin in calcined clay (metakaolin content: 40–50%) which was confirmed by Avet et al. [59]. Moreover, Chen et al. showed an increase in the metakaolin content of calcined clay, increases the compressive strength (Figure 1). MIX-R in Figure 1 is the reference mixture with no calcined clay and MIX-L, MIX-M, and MIX-H contain low (40–50%), medium (62.5%), and high (75%) amounts of metakaolin content of calcined clay, respectively [65]. Moreover, recent studies indicate that improvement in mechanical and durability properties is significant even by introducing low or medium kaolinite content to the mix [59,66]. The LC^3 mortars with different kaolinite content (41.9%, 50.3%, 79.4% and 95.0%) indicated 9%, 9%, 27%, and 34% greater compressive strength compared to OPC mortar, respectively [66]. The results of previous studies introduce the LC^3 mixture as a promising ternary blend for improving the mechanical properties of concrete or mortar. Moreover, using calcine clay with low kaolinite content which is widely accessible seems to be an economical choice for the concrete industry. These statements are verified by other studies. [65,67].

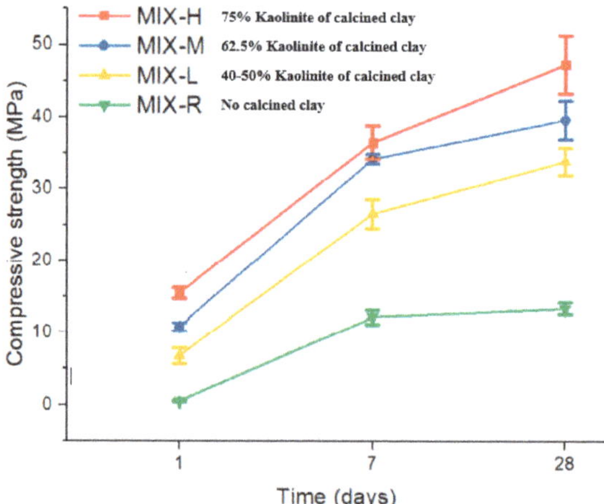

Figure 1. Compressive strength of concrete containing calcined clay with different metakaolin content [65].

2.1.2. Durability and Microstructure of LC^3 Cement Binder

In LC^3 technology, not only can the larger amount of OPC be replaced by SCMs to enhance the mechanical properties and reduce GHG emission, but also carbo-aluminate hydrates are generated, which can occupy the capillary pores [45]. Considerable incorporation of aluminum is discovered for

LC^3 blends compared to the OPC and the C-A-S-H gel of LC^3 system shows significant variations in the composition. The kaolinite content of calcined clay has the main role of increasing the aluminum incorporation and its integration [39]. Furthermore, the chemical composition of calcium carbonate and alumina creates supplemental aluminate ferrite monosulfate (AFm) phases and stabilizes ettringite [52].

LC^3 has shown considerable improvement in chloride resistance of concrete compared to the other mixes with similar compressive strength [67,68]. This is mainly due to the creation of a large quantity of C-A-S-H and a synergetic impact between limestone and calcined clay [52,69]. Resistance against chloride ingress is tested by using high to intermediate grades of kaolinite content in clay and where it was found that chloride resistance of clay containing intermediate amount of kaolinite was in a similar range to high kaolinite content (which is an expensive choice) [66]. Furthermore, studies indicated that the chloride penetration in OPC mortar after two years of exposure was four times higher than the LC^3 mixtures (of 50% and higher kaolinite content) [66]. The main reason for this feature was ascribed to the denser structure of the LC^3 compared to OPC. The action against capillary water absorption and gas permeability indicates that LC^3 can provide significant performance in comparison to OPC. Moreover, using hydrophobic agents can lead to a material with considerably higher resistance to moisture ingress [70–74] meaning that LC^3 concrete has the potential to be exploited at environmental conditions where there is a risk of chloride ingress, including marine environments [67].

Furthermore, sulfate attack and Alkali–Silica Reaction (ASR) could both be mitigated by the LC^3 binder. A minimum quantity of limestone and calcined clay (around 30%) is reported to mitigate the sulfate attack [75]. Shi et al. suggested that LC^3 could be involved in standards as an innovative kind of sulfate-resisting Portland pozzolan cement [48].

3. Fiber-Reinforced Concrete (FRC)

Regardless of numerous benefits, concrete has some weaknesses including low energy-absorption capacity and low tensile resistance resulting in spalling, cracking, and lower lifespan of the structures [76,77]. Recently, using macro fibers as concrete reinforcement has become prevalent to present a solution for enhancing the mechanical properties of the OPC concrete [5,78–80]. The use of fiber in concrete matrix has less or no impact on concrete pre-cracking behavior. However, fibers improve post-cracking response, control the brittle fracture process, provide strength, and offer post-cracking toughness through the advantage of reliable deformation behavior and post-cracking strength [81–83]. Furthermore, incorporating fibers in cementitious substances has become important to some extent, because of reducing the shrinkage cracking, which is able to enhance the material durability. The preventing of shrinkage cracks may contribute to diminishing the material permeability [84]. When considering the structural behavior of a material, it is important to consider both strength and toughness together. The capacity of the material to absorb energy in the plastic range is considered to be the toughness index [83]. The propagation of the crack is inhibited by adding fibers to the concrete matrix, resulting in improvement of energy-absorption capacity [85–87]. The range of the enhancement in the mechanical properties and toughness index of concrete mixture depends on the fiber length and its amount [88–91].

Toughness index of FRC is increased by reduction in fiber cross-sectional area. Higher specific surface area is achieved by reduction in the diameter of the fibers, leading to higher contact areas between fibers and the matrix. Moreover, it causes a significant increment in FRC energy absorption compared to plain concrete [92]. In the mechanisms of energy-absorption in FRC, de-bonding and fiber pull-out are features related to the fiber surface area. Therefore, fiber-specific surface (FSS), the fiber content (FC), and the reinforcement area are relevant parameters to be investigated to find their effect on FRC properties. The implication is that the length of individual fibers influences computing the reinforcement area. To calculate these parameters, many analytical expressions are mentioned in reference [81].

Hasan et al. stated that the splitting tensile, compressive, and flexural strengths reached their maximum with 0.36% fiber volume fraction in comparison to plain concrete. The concrete strengths started to decrease due to high-volume fiber interface with the cohesiveness of the concrete matrix causing difficulty in concrete compaction lowering its workability [93].

According to past scientific findings, short fiber reinforcement in concrete could enhance the properties of plain concrete in the appropriate fiber volume fraction. Presently, regarding financial problems and environmental concerns, natural fibers are fascinating for industrial applications. Therefore, the natural (bio) fiber-reinforced concrete (NFRC) is an attractive subject for further research.

3.1. Natural-Fiber-Reinforced Concrete (NFRC)

Recently, many studies have considered bio-fibers for reinforcement of Portland cement-based concrete structures to increase the tensile strength, flexural strength, tensile ductility, and flexural toughness, reducing the drying shrinkage and density of concrete [94–99]. The benefits of NFRC, such as incremented toughness, improved cracking behavior, greater durability, and enhanced impact resistance and fatigue were well demonstrated formerly [91,97,100–102]. Also, research outcome has shown NFRC possesses the potential for repairing, retrofitting, and rehabilitation of reinforced concrete structures and as a new construction material [103].

3.1.1. Mechanical Properties of NFRC

In terms of impact resistance, Zhou et al. [104] and Wang et al. [105] reported the positive effects of the impact resistance of concrete reinforced by jute fiber and hybrid bamboo-steel fiber, respectively. More results on impact energy absorption of NFRC slabs were reported by Ramakrishna and Sundararajan [101]. Investigations of jute FRC show that compressive strength is not considerably influenced by adding fibers; however, flexural, tensile strengths and toughness are all considerably incremented [91]. Moreover, it is stated that the modulus of elasticity, compressive strength, and repetitive impact resistance of coconut FRC (CFRC) were reduced by increasing the length of the fiber [106]. Also, a similar finding reported that the short flax fiber (12 mm) had the most effects on the flexural strength of flax FRC [107]. Al-Oraimi and Seibi tested many FRC samples by using the different FC of glass and palm trees. They stated that adding fibers, in general, improves the toughness of concrete and enhances its impact resistance; also, bio-fibers are comparable to the synthetic fibers in improving the toughness and impact resistance [83]. Moreover, from another study, hemp fibers improve the concrete fracture energy for 70% in NFRC. By bridging the cracks, fibers provide a post-cracking ductility, leading to significant improvement of toughness [102]. Furthermore, higher flexural strength is exhibited by alkaline treatment of hemp fibers compared to their non-treated equivalents [7].

3.1.2. Durability of Bio Fiber-Reinforced Concrete

Degradation of natural fibers are investigated by treatment in aging environments [108–114]. Pretreatment of natural fibers is a well-documented method to enhance the degradation resistance. Silane coating [115,116], out-of-autoclave method [117], hornification [118], sodium silicate [119], potassium silicate [119], alkaline treatment [115,120], and coating fibers with bacterial nano-cellulose [121] were used to enhance the durability and mechanical characteristics of NFRC by creating protective layers on the fiber surface or enhancing cellulose structure of natural fiber.

The other method for increasing the durability of natural fibers is use of appropriate SCMs in the concrete matrix. The lignin quantity of natural fiber has a main role in the sensitivity of NFRC to natural weathering. This is caused by the more susceptibility of hemicelluloses and lignin to chemical deterioration and alkaline environment of cement. The findings show that by combining the calcined clay minerals, alkalinity of pore solution is decreased which can lead to mitigating fiber deterioration. Both alkali hydrolysis and mineralization of natural fiber can be alleviated significantly by this technique [6]. Different studies agree that decreasing the alkalinity of the matrix using SCMs as cement replacement can prevent the chemical attack to lignocellulosic fibers in the matrix [6,7,110,111,122–125].

The rate of natural fiber deterioration in cementitious materials could be reduced by using some SCMs such as nano-calcined clay [7]. Hakamy et al. stated that the substitution of cement with 1 wt.% nano-calcined clay results in not only enhancing the microstructure but also facilitating the pozzolanic activity which results in stronger bonds between the matrix and the surface of treated hemp fibers [7]. Furthermore, initial flexural strength as well as durability of NFRC are improved by the coupled replacement of metakaolin and montmorillonite, due to modifying the mineralization and alkaline degradation of the fibers. For example, the degradation of sisal fibers was moderated most considerably at high cement substitution level (about 50%) [6]. Using short-length natural fibers and the effects of cement and SCMs on fiber degradation is well documented [6,7,75,111]. Moreover, incorporating a pozzolanic substance into the matrix results in a considerable reduction in capillary absorption and chloride penetration. Overall, the use of appropriate amount and quality of SCMs can enhance the mechanical and durability properties of FRC [84]. Table 2 shows the effect of SCMs on the mechanical performance of natural fiber cementitious composites. Future studies need to be conducted on the adjustment of the mix design, assessing the long-term performance and standardization of the natural-fiber-reinforced LC^3 concrete.

Table 2. The effect of SCMs on the natural fiber performance in cementitious binder.

Name of Natural Fiber + SCM	Deteriorating Environment	Duration	Type of Mechanical Properties	Percentage of Changes Compared to Conventional Concrete	Notes	Reference
Sisal fiber	Alkaline solution	28 days of immersion	Impact strength	About−2%	The impact strength of 2% sisal NFRC was about 2 times more than plain mortar.	[114]
Sisal fiber + 30% SF	Water bath (Alkalinity of binder)	730 days of aging	Flexural strength	+28%	The flexural strength increased after aging period due to using 30% SF.	[111]
Sisal fiber	Outdoors	322 days	Flexural strength	−70%	The first crack strength increased by about 53% due to the use of sisal fiber.	[108]
Sisal fiber + 45% MK + 5% montmorillonite	Wetting and drawing cycle	30 cycles	Tensile strength of fiber embedded	+500%	The positive effects of MK on the mitigation of alkalinity was proved.	[6]
Hemp fiber + 1% calcined Nano clay	N/A	N/A	Flexural strength	+38%	——	[7]
Coir fiber	Sulfate attack	2 years of immersion	Compressive strength	−14%	The deterioration value for conventional concrete was 54%.	[112]
Sugarcane fiber	Sulfate attack	2 years of immersion	Compressive strength	−20%	The deterioration value for conventional concrete was 54%.	[112]
Coir fiber	Freezing and thawing	300 cycles	Modulus of elasticity	−10%	The deterioration value for conventional concrete was 8%.	[112]
Sugarcane fiber	Freezing and thawing	300 cycles	Modulus of elasticity	−14%	The deterioration value for conventional concrete was 8%.	[112]

3.2. Kenaf Fiber-Reinforced Concrete (KFRC)

The kenaf plant is able to grow to heights of 3.5–4.5 m within 4–5 months [126]. Studies indicate that the kenaf plant had the optimal CO_2 absorption among the investigated plants. Kenaf plant can absorb 1.5 times the carbon dioxide by its weight [127]. The findings show that the tensile strength of kenaf fibers vary between 223 MPa and 1191 MPa and the elastic modulus and final tensile strain of the kenaf fiber vary within 2860 MPa to 60,000 MPa and 0.012 to 0.1, respectively [120]. Kenaf fiber shows a linear stress–strain diagram [120,128]. Currently, kenaf fiber is used in bio-materials with a wide application area [94,129–132]. Table 3 presents the mechanical properties of some natural fibers. The elastic modulus of some natural fibers such as hemp, kenaf, and flax are comparable to glass fibers, while the density of these natural fibers is one half the density of glass fiber. According to the nature of bio-fibers, the properties may differ in different origins, so the range of properties are reported in the following table.

Table 3. Mechanical properties of natural fiber.

Fiber	Elastic Modulus (GPa)	Tensile Strength (MPa)	Elongation at Break (%)	Density (g/cm^3)	Reference (s)
Kenaf	40	731.64	1.8	1.2	[120,129]
Jute	26.5	393–773	1.5–1.8	1.3	[133]
Sisal	9–22	400–700	2.0–2.5	1.43–1.5	[134,135]
Flax	27.6–65.5	345–1500	1.86–3.2	1.5	[107,136]
Hemp	70	690	1.6–4.0	1.47	[136–138]
Pineapple	34.5–82.5	170–1627	1–3	1.44–1.56	[139–141]
Cotton	5.5–12.6	400	7.0–8.0	1.5–1.6	[140]
Oil Palm	0.48–9	24.9–550	4–18	0.7–1.55	[142,143]
E–glass	70–71	2000–3500	0.5–3.4	2.5–2.55	[120,137,144]
Carbon	224–240	2650–4000	1.4–1.8	1.4–1.75	[145,146]

Surface treatment of kenaf fibers by sodium hydroxide (NaOH) can reduce its hydrophilic properties [120]. This reduction in the fiber water sorption characteristic causes an improvement of fiber durability and reduces its biodegradability [147]; however, it may affect the bond strength with the cementitious matrix.

According to the literature, higher toughness is exhibited by NFRC (such as KFRC) generally compared to the normal concrete [94,104,124]. Also, microstructural analysis by scanning electron microscopy (SEM) shows a good bonding between the kenaf fibers and concrete matrix [124]. Lam and Yatim conducted research on KFRC by changing the fiber volume content and the fiber length. They stated the indirect tensile and flexural strength increased by an increment of FC and fiber length [94]. This statement seems to be in contrast with the previous statement about coconut FRC in terms of the effects of fiber length [106] but both studies suggested that the 50 mm fiber length was suitable. Moreover, the ductile failure mode was observed compared to plain concrete, which resulted in an enhancement in cracking behavior and ductility. Moreover, another study reported a toughness index in KFRC almost 3 times higher than the OPC concrete control samples [124]. Use of natural fibers such as kenaf fiber to cast NFRC (specifically KFRC) can result in not only economic profit in terms of production cost and material weight, but also in terms of health benefits for society when compared to synthetic fibers [104,147]. The green concrete developed has an environmental benefit which is of immense importance in the present context of the sustainability of natural resources [127,147]. Moreover, the quantity of CO_2 would be reduced in the atmosphere by using kenaf fibers in concrete. It may decline the high CO_2 released within the manufacturing of Portland cement. Thus, kenaf fiber-reinforced concrete (KFRC) is a potential green material for various construction purposes [94,124].

4. Discussion and Conclusions

Using natural fibers as an alternative for concrete reinforcement is of interest not only due to increasing ductility and versatility of the material but also from an environmental perspective. On the other hand, the binder needs to be compatible with the fibers and be environmentally friendly to make a favorable composition. SCMs including SF, FA, slag, and LC^3 can enhance mechanical and durability properties, reduce the environmental impacts, and adjust the alkaline environment and pore structure of the matrix. The latter can be of interest when dealing with the durability of nature-based materials into concrete. While resources for SCM materials which are industrial by/products are limited and may be challenging to approach in the near future, LC^3 can be an available choice in most parts of the world. Furthermore, the clay can be calcined by using renewable energy which can lead to zero emissions for the calcination process. The weight proportion of calcined clay to limestone as 2:1 and the cement replacement ratio of 50% are reported to be optimal for normal uses in different studies.

The kaolinite content of the clay, which is reported to play an important role in cementitious functionality of calcined clay, varies significantly in different types of clay. However, studies have shown that calcined clay with low or medium kaolinite content can also be used in LC^3 achieving acceptable mechanical properties for common applications. This means the LC^3 is not sensitive to kaolinite concentration and different types of clay available with minimum transport can be suitable for concrete production leading to reduction in cost and environmental impacts. Furthermore, improvement in durability properties in terms of chloride resistance, ASR, and sulfate attack are reported for LC^3 concrete with medium kaolinite content in the clay.

On the other hand, the performance of kenaf fibers as short-length natural fiber concrete mixture is investigated in different studies mainly using OPC. Mechanical properties of concrete such as toughness, tensile strength, and impact resistance can be improved using this type of fiber. Moreover, durability properties such as carbonation, sulfate, and chloride resistance were reported to be enhanced compared to OPC concrete. Natural fiber volume content under 1% and fiber length of about 50 mm are reported to be a proper performance in the concrete mix.

Combination of LC^3 with natural fibers such as kenaf fiber can be a promising composition to get green concrete with low GHG emission and energy consumption due to the replacement of cement by LC^3 as well as significant properties of kenaf plant in absorbing the CO_2 from the air and introducing proper fibers for concrete mix. Current studies on this composition are limited and need to be taken into account for further investigation. Furthermore, adjustment of the mix design, assessing the long-term performance, as well as standardization, are the next steps for use of the kenaf fiber-reinforced LC^3 concrete in practice.

Funding: The APC was funded by Department of Manufacturing and Civil Engineering, Norwegian University of Science and Technology (NTNU).

Conflicts of Interest: The authors declare no conflict of interest.

References

1. Naqi, A.; Jang, J.; Naqi, A.; Jang, J.G. Recent Progress in Green Cement Technology Utilizing Low-Carbon Emission Fuels and Raw Materials: A Review. *Sustainability* **2019**, *11*, 537. [CrossRef]
2. Hasanbeigi, A.; Price, L.; Lu, H.; Lan, W. Analysis of energy-efficiency opportunities for the cement industry in Shandong Province, China: A case study of 16 cement plants. *Energy* **2010**, *35*, 3461–3473. [CrossRef]
3. Meyer, C. The greening of the concrete industry. *Cem. Concr. Compos.* **2009**, *31*, 601–605. [CrossRef]
4. Valipour, M.; Yekkalar, M.; Shekarchi, M.; Panahi, S. Environmental assessment of green concrete containing natural zeolite on the global warming index in marine environments. *J. Clean. Prod.* **2014**, *65*, 418–423. [CrossRef]
5. Afroughsabet, V.; Biolzi, L.; Ozbakkaloglu, T. High-performance fiber-reinforced concrete: A review. *J. Mater. Sci.* **2016**, *51*, 6517–6551. [CrossRef]
6. Wei, J.; Meyer, C. Degradation of natural fiber in ternary blended cement composites containing metakaolin and montmorillonite. *Corros. Sci.* **2017**, *120*, 42–60. [CrossRef]
7. Hakamy, A.; Shaikh, F.U.A.; Low, I.M. Effect of calcined nanoclay on microstructural and mechanical properties of chemically treated hemp fabric-reinforced cement nanocomposites. *Constr. Build. Mater.* **2015**, *95*, 882–891. [CrossRef]
8. Yang, K.-H.; Jung, Y.-B.; Cho, M.-S.; Tae, S.-H. Chapter 5—Effect of Supplementary Cementitious Materials on Reduction of CO_2 Emissions from Concrete. In *Handbook of Low Carbon Concrete*; Nazari, A., Sanjayan, J., Eds.; Butterworth-Heinemann: Oxford, UK, 2017; pp. 89–110. ISBN 978-0-12-804524-4.
9. Crossin, E. The greenhouse gas implications of using ground granulated blast furnace slag as a cement substitute. *J. Clean. Prod.* **2015**, *95*, 101–108. [CrossRef]
10. Li, C.; Sun, H.; Li, L. A review: The comparison between alkali-activated slag (Si + Ca) and metakaolin (Si + Al) cements. *Cem. Concr. Res.* **2010**, *40*, 1341–1349. [CrossRef]

11. Pacheco-Torgal, F.; Castro-Gomes, J.; Jalali, S. Alkali-activated binders: A review: Part 1. Historical background, terminology, reaction mechanisms and hydration products. *Constr. Build. Mater.* **2008**, *22*, 1305–1314. [CrossRef]

12. Scrivener, K.L.; Gartner, E.M. Eco-efficient cements: Potential economically viable solutions for a low-CO_2 cement-based materials industry. *Cem. Concr. Res.* **2018**, *114*, 2–26. [CrossRef]

13. Golewski, G.L. Green concrete composite incorporating fly ash with high strength and fracture toughness. *J. Clean. Prod.* **2018**, *172*, 218–226. [CrossRef]

14. Tahanpour Javadabadi, M.; De Lima Kristiansen, D.; Bayene Redie, M.; Baghban, M.H. Sustainable Concrete: A Review. *Int. J. Struct. Civ. Eng. Res.* **2019**, *8*, 126–132.

15. Fallah, S.; Nematzadeh, M. Mechanical properties and durability of high-strength concrete containing macro-polymeric and polypropylene fibers with nano-silica and silica fume. *Constr. Build. Mater.* **2017**, *132*, 170–187. [CrossRef]

16. Chatterji, S.; Thaulow, N.; Christensen, P. Puzzolanic activity of byproduct silica-fume from ferro-silicom production. *Cem. Concr. Res.* **1982**, *12*, 781–784. [CrossRef]

17. Mehta, P.K.; Gjørv, O.E. Properties of portland cement concrete containing fly ash and condensed silica-fume. *Cem. Concr. Res.* **1982**, *12*, 587–595. [CrossRef]

18. Diamond, S. The utilization of flyash. *Cem. Concr. Res.* **1984**, *14*, 455–462. [CrossRef]

19. Chindaprasirt, P.; Jaturapitakkul, C.; Sinsiri, T. Effect of fly ash fineness on microstructure of blended cement paste. *Constr. Build. Mater.* **2007**, *21*, 1534–1541. [CrossRef]

20. Rodríguez de Sensale, G. Effect of rice-husk ash on durability of cementitious materials. *Cem. Concr. Compos.* **2010**, *32*, 718–725. [CrossRef]

21. Zain, M.F.M.; Islam, M.N.; Mahmud, F.; Jamil, M. Production of rice husk ash for use in concrete as a supplementary cementitious material. *Constr. Build. Mater.* **2011**, *25*, 798–805. [CrossRef]

22. Adesanya, D.A.; Raheem, A.A. Development of corn cob ash blended cement. *Constr. Build. Mater.* **2009**, *23*, 347–352. [CrossRef]

23. Adesanya, D.A.; Raheem, A.A. A study of the permeability and acid attack of corn cob ash blended cements. *Constr. Build. Mater.* **2010**, *24*, 403–409. [CrossRef]

24. Sigvardsen, N.M.; Kirkelund, G.M.; Jensen, P.E.; Geiker, M.R.; Ottosen, L.M. Impact of production parameters on physiochemical characteristics of wood ash for possible utilisation in cement-based materials. *Resour. Conserv. Recycl.* **2019**, *145*, 230–240. [CrossRef]

25. Vejmelková, E.; Koňáková, D.; Kulovaná, T.; Keppert, M.; Žumár, J.; Rovnaníková, P.; Keršner, Z.; Sedlmajer, M.; Černý, R. Engineering properties of concrete containing natural zeolite as supplementary cementitious material: Strength, toughness, durability, and hygrothermal performance. *Cem. Concr. Compos.* **2015**, *55*, 259–267. [CrossRef]

26. Villar-Cociña, E.; Morales, E.V.; Santos, S.F.; Savastano, H.; Frías, M. Pozzolanic behavior of bamboo leaf ash: Characterization and determination of the kinetic parameters. *Cem. Concr. Compos.* **2011**, *33*, 68–73. [CrossRef]

27. Aprianti, E.; Shafigh, P.; Bahri, S.; Farahani, J.N. Supplementary cementitious materials origin from agricultural wastes—A review. *Constr. Build. Mater.* **2015**, *74*, 176–187. [CrossRef]

28. Kroehong, W.; Sinsiri, T.; Jaturapitakkul, C. Effect of Palm Oil Fuel Ash Fineness on Packing Effect and Pozzolanic Reaction of Blended Cement Paste. *Procedia Eng.* **2011**, *14*, 361–369. [CrossRef]

29. Lim, N.H.A.S.; Ismail, M.A.; Lee, H.S.; Hussin, M.W.; Sam, A.R.M.; Samadi, M. The effects of high volume nano palm oil fuel ash on microstructure properties and hydration temperature of mortar. *Constr. Build. Mater.* **2015**, *93*, 29–34. [CrossRef]

30. Rajak, M.A.A.; Majid, Z.A.; Ismail, M. Morphological Characteristics of Hardened Cement Pastes Incorporating Nano-palm Oil Fuel Ash. *Procedia Manuf.* **2015**, *2*, 512–518. [CrossRef]

31. Islam, M.M.U.; Mo, K.H.; Alengaram, U.J.; Jumaat, M.Z. Mechanical and fresh properties of sustainable oil palm shell lightweight concrete incorporating palm oil fuel ash. *J. Clean. Prod.* **2016**, *115*, 307–314. [CrossRef]

32. Jaturapitakkul, C.; Tangpagasit, J.; Songmue, S.; Kiattikomol, K. Filler effect and pozzolanic reaction of ground palm oil fuel ash. *Constr. Build. Mater.* **2011**, *25*, 4287–4293. [CrossRef]

33. Jaturapitakkul, C.; Kiattikomol, K.; Tangchirapat, W.; Saeting, T. Evaluation of the sulfate resistance of concrete containing palm oil fuel ash. *Constr. Build. Mater.* **2007**, *21*, 1399–1405. [CrossRef]

34. Wi, K.; Lee, H.-S.; Lim, S.; Ismail, M.A.; Hussin, M.W. Effect of Using Micropalm Oil Fuel Ash as Partial Replacement of Cement on the Properties of Cement Mortar. *Adv. Mater. Sci. Eng.* **2018**, *2018*, 1–8. [CrossRef]

35. Khodabakhshian, A.; Ghalehnovi, M.; de Brito, J.; Asadi Shamsabadi, E. Durability performance of structural concrete containing silica fume and marble industry waste powder. *J. Clean. Prod.* **2018**, *170*, 42–60. [CrossRef]

36. Lothenbach, B.; Le Saout, G.; Gallucci, E.; Scrivener, K. Influence of limestone on the hydration of Portland cements. *Cem. Concr. Res.* **2008**, *38*, 848–860. [CrossRef]

37. Song, Q.; Yu, R.; Wang, X.; Rao, S.; Shui, Z. A novel Self-Compacting Ultra-High Performance Fibre Reinforced Concrete (SCUHPFRC) derived from compounded high-active powders. *Constr. Build. Mater.* **2018**, *158*, 883–893. [CrossRef]

38. Avet, F.; Scrivener, K. Hydration Study of Limestone Calcined Clay Cement (LC3) Using Various Grades of Calcined Kaolinitic Clays. In *Calcined Clays for Sustainable Concrete*; Springer: Dordrecht, The Netherlands, 2018; pp. 35–40.

39. Avet, F.; Boehm-Courjault, E.; Scrivener, K. Investigation of C-A-S-H composition, morphology and density in Limestone Calcined Clay Cement (LC3). *Cem. Concr. Res.* **2019**, *115*, 70–79. [CrossRef]

40. Sharaky, I.A.; Megahed, F.A.; Seleem, M.H.; Badawy, A.M. The influence of silica fume, nano silica and mixing method on the strength and durability of concrete. *SN Appl. Sci.* **2019**, *1*, 575. [CrossRef]

41. Meddah, M.S.; Ismail, M.A.; El-Gamal, S.; Fitriani, H. Performances evaluation of binary concrete designed with silica fume and metakaolin. *Constr. Build. Mater.* **2018**, *166*, 400–412. [CrossRef]

42. Saha, A.K. Effect of class F fly ash on the durability properties of concrete. *Sustain. Environ. Res.* **2018**, *28*, 25–31. [CrossRef]

43. Uthaman, S.; Vishwakarma, V.; George, R.P.; Ramachandran, D.; Kumari, K.; Preetha, R.; Premila, M.; Rajaraman, R.; Mudali, U.K.; Amarendra, G. Enhancement of strength and durability of fly ash concrete in seawater environments: Synergistic effect of nanoparticles. *Constr. Build. Mater.* **2018**, *187*, 448–459. [CrossRef]

44. King, D. The effect of silica fume on the properties of concrete as defined in concrete society report 74, cementitious materials. In Proceedings of the 37th Conference on Our World in Concrete and Structures, Singapore, 29–31 August 2012; pp. 29–31.

45. Scrivener, K.; Martirena, F.; Bishnoi, S.; Maity, S. Calcined clay limestone cements (LC3). *Cem. Concr. Res.* **2018**, *114*, 49–56. [CrossRef]

46. Snellings, R. Assessing, Understanding and Unlocking Supplementary Cementitious Materials. *Rilem Tech. Lett.* **2016**, *1*, 50. [CrossRef]

47. Tironi, A.; Trezza, M.A.; Scian, A.N.; Irassar, E.F. Assessment of pozzolanic activity of different calcined clays. *Cem. Concr. Compos.* **2013**, *37*, 319–327. [CrossRef]

48. Shi, Z.; Ferreiro, S.; Lothenbach, B.; Geiker, M.R.; Kunther, W.; Kaufmann, J.; Herfort, D.; Skibsted, J. Sulfate resistance of calcined clay—Limestone—Portland cements. *Cem. Concr. Res.* **2019**, *116*, 238–251. [CrossRef]

49. Taylor-Lange, S.C.; Lamon, E.L.; Riding, K.A.; Juenger, M.C.G. Calcined kaolinite–bentonite clay blends as supplementary cementitious materials. *Appl. Clay Sci.* **2015**, *108*, 84–93. [CrossRef]

50. Fernandez, R.; Martirena, F.; Scrivener, K.L. The origin of the pozzolanic activity of calcined clay minerals: A comparison between kaolinite, illite and montmorillonite. *Cem. Concr. Res.* **2011**, *41*, 113–122. [CrossRef]

51. Garg, N.; Skibsted, J. Thermal Activation of a Pure Montmorillonite Clay and Its Reactivity in Cementitious Systems. *J. Phys. Chem. C* **2014**, *118*, 11464–11477. [CrossRef]

52. Antoni, M.; Rossen, J.; Martirena, F.; Scrivener, K. Cement substitution by a combination of metakaolin and limestone. *Cem. Concr. Res.* **2012**, *42*, 1579–1589. [CrossRef]

53. Tafraoui, A.; Escadeillas, G.; Lebaili, S.; Vidal, T. Metakaolin in the formulation of UHPC. *Constr. Build. Mater.* **2009**, *23*, 669–674. [CrossRef]

54. Souza, P.S.L.; Dal Molin, D.C.C. Viability of using calcined clays, from industrial by-products, as pozzolans of high reactivity. *Cem. Concr. Res.* **2005**, *35*, 1993–1998. [CrossRef]

55. Ding, J.-T.; Li, Z. Effects of metakaolin and silica fume on properties of concrete. *Mater. J.* **2002**, *99*, 393–398.

56. Chen, Y.; Chaves Figueiredo, S.; Yalçinkaya, Ç.; Çopuroğlu, O.; Veer, F.; Schlangen, E. The Effect of Viscosity-Modifying Admixture on the Extrudability of Limestone and Calcined Clay-Based Cementitious Material for Extrusion-Based 3D Concrete Printing. *Materials* **2019**, *12*, 1374. [CrossRef] [PubMed]

57. Hassan, A.A.A.; Mayo, J.R. Influence of mixture composition on the properties of SCC incorporating metakaolin. *Mag. Concr. Res.* **2014**, *66*, 1036–1050. [CrossRef]

58. Alvarez, G.L.; Nazari, A.; Bagheri, A.; Sanjayan, J.G.; De Lange, C. Microstructure, electrical and mechanical properties of steel fibres reinforced cement mortars with partial metakaolin and limestone addition. *Constr. Build. Mater.* **2017**, *135*, 8–20. [CrossRef]

59. Avet, F.; Snellings, R.; Alujas Diaz, A.; Ben Haha, M.; Scrivener, K. Development of a new rapid, relevant and reliable (R3) test method to evaluate the pozzolanic reactivity of calcined kaolinitic clays. *Cem. Concr. Res.* **2016**, *85*, 1–11. [CrossRef]

60. Sánchez Berriel, S.; Favier, A.; Rosa Domínguez, E.; Sánchez Machado, I.R.; Heierli, U.; Scrivener, K.; Martirena Hernández, F.; Habert, G. Assessing the environmental and economic potential of Limestone Calcined Clay Cement in Cuba. *J. Clean. Prod.* **2016**, *124*, 361–369. [CrossRef]

61. Huang, W.; Kazemi-Kamyab, H.; Sun, W.; Scrivener, K. Effect of replacement of silica fume with calcined clay on the hydration and microstructural development of eco-UHPFRC. *Mater. Des.* **2017**, *121*, 36–46. [CrossRef]

62. Arbi, K.; Nedeljković, M.; Zuo, Y.; Ye, G. A Review on the Durability of Alkali-Activated Fly Ash/Slag Systems: Advances, Issues, and Perspectives. *Ind. Eng. Chem. Res.* **2016**, *55*, 5439–5453. [CrossRef]

63. Almenares, R.S.; Vizcaíno, L.M.; Damas, S.; Mathieu, A.; Alujas, A.; Martirena, F. Industrial calcination of kaolinitic clays to make reactive pozzolans. *Case Stud. Constr. Mater.* **2017**, *6*, 225–232. [CrossRef]

64. Ferreiro, S.; Herfort, D.; Damtoft, J.S. Effect of raw clay type, fineness, water-to-cement ratio and fly ash addition on workability and strength performance of calcined clay—Limestone Portland cements. *Cem. Concr. Res.* **2017**, *101*, 1–12. [CrossRef]

65. Chen, Y.; Li, Z.; Chaves Figueiredo, S.C.; Çopuroğlu, O.; Veer, F.; Schlangen, E.; Chen, Y.; Li, Z.; Chaves Figueiredo, S.; Çopuroğlu, O.; et al. Limestone and Calcined Clay-Based Sustainable Cementitious Materials for 3D Concrete Printing: A Fundamental Study of Extrudability and Early-Age Strength Development. *Appl. Sci.* **2019**, *9*, 1809. [CrossRef]

66. Maraghechi, H.; Avet, F.; Wong, H.; Kamyab, H.; Scrivener, K. Performance of Limestone Calcined Clay Cement (LC3) with various kaolinite contents with respect to chloride transport. *Mater. Struct.* **2018**, *51*, 125. [CrossRef]

67. Dhandapani, Y.; Sakthivel, T.; Santhanam, M.; Gettu, R.; Pillai, R.G. Mechanical properties and durability performance of concretes with Limestone Calcined Clay Cement (LC3). *Cem. Concr. Res.* **2018**, *107*, 136–151. [CrossRef]

68. Pillai, R.G.; Gettu, R.; Santhanam, M.; Rengaraju, S.; Dhandapani, Y.; Rathnarajan, S.; Basavaraj, A.S. Service life and life cycle assessment of reinforced concrete systems with limestone calcined clay cement (LC3). *Cem. Concr. Res.* **2019**, *118*, 111–119. [CrossRef]

69. Steenberg, M.; Herfort, D.; Poulsen, S.L.; Skibsted, J.; Damtoft, J.S. Composite cement based on Portland cement clinker, limestone and calcined clay. In Proceedings of the Xiii International Congress on the Chemistry of Cement, Madrid, Spain, 3–8 July 2011.

70. Baghban, M.H.; Hovde, P.J.; Jacobsen, S. Effect of internal hydrophobation, silica fume and w/c on water sorption of hardened cement pastes. In Proceedings of the International Conference on Durability of Building Materials and Components, Porto, Portugal, 12–15 April 2011.

71. Baghban, M.H. Water Sorption of Hardened Cement Pastes. *Cem. Based Mater.* **2018**. [CrossRef]

72. Justnes, H.; Østnor, T.; Barnils Vila, N. Vegetable oils as water repellents for mortars. In Proceedings of the 1st International Conference of Asian Concrete Federation, Chiang Mai, Thailand, 28–29 October 2004; pp. 28–29.

73. Baghban, M.; Hovde, P.; Jacobsen, S. Effect of internal hydrophobation, silica fume and w/c on compressive strength of hardened cement pastes. *World J. Eng.* **2012**, *9*, 7–12. [CrossRef]

74. Baghban, M.H.; Holvik, O.K.; Hesselberg, E.; Javadabadi, M.T. Cementitious Composites with Low Water Permeability through Internal Hydrophobicity. *Key Engineering Materials.* **2018**, *779*, 37–42. [CrossRef]

75. Favier, A.; Scrivener, K. Alkali Silica Reaction and Sulfate Attack: Expansion of Limestone Calcined Clay Cement. In *Calcined Clays for Sustainable Concrete*; Springer: Dordrecht, The Netherlands, 2018; pp. 165–169.

76. Carpinteri, A.; Cadamuro, E.; Ventura, G. Fiber-reinforced concrete in flexure: A cohesive/overlapping crack model application. *Mater. Struct.* **2015**, *48*, 235–247. [CrossRef]

77. Oh, B.H.; Kim, J.C.; Choi, Y.C. Fracture behavior of concrete members reinforced with structural synthetic fibers. *Eng. Fract. Mech.* **2007**, *74*, 243–257. [CrossRef]

78. Yin, S.; Tuladhar, R.; Shi, F.; Combe, M.; Collister, T.; Sivakugan, N. Use of macro plastic fibres in concrete: A review. *Constr. Build. Mater.* **2015**, *93*, 180–188. [CrossRef]

79. Mobasher, B. *Mechanics of Fiber and Textile Reinforced Cement Composites*; CRC Press: New York, NY, USA, 2011; ISBN 9780429131387.

80. Choe, G.; Kim, G.; Kim, H.; Hwang, E.; Lee, S.; Nam, J. Effect of amorphous metallic fiber on mechanical properties of high-strength concrete exposed to high-temperature. *Constr. Build. Mater.* **2019**, *218*, 448–456. [CrossRef]

81. Zollo, R.F. Fiber-reinforced concrete: An overview after 30 years of development. *Cem. Concr. Compos.* **1997**, *19*, 107–122. [CrossRef]

82. Brandt, A.M. Fibre reinforced cement-based (FRC) composites after over 40 years of development in building and civil engineering. *Compos. Struct.* **2008**, *86*, 3–9. [CrossRef]

83. Al-Oraimi, S.K.; Seibi, A.C. Mechanical characterisation and impact behaviour of concrete reinforced with natural fibres. *Compos. Struct.* **1995**, *32*, 165–171. [CrossRef]

84. De Gutiérrez, R.M.; Díaz, L.N.; Delvasto, S. Effect of pozzolans on the performance of fiber-reinforced mortars. *Cem. Concr. Compos.* **2005**, *27*, 593–598. [CrossRef]

85. Mobasher, B.; Li, C.Y. Effect of interfacial properties on the crack propagation in cementitious composites. *Adv. Cem. Based Mater.* **1996**, *4*, 93–105. [CrossRef]

86. Alberti, M.G.; Enfedaque, A.; Gálvez, J.C. On the mechanical properties and fracture behavior of polyolefin fiber-reinforced self-compacting concrete. *Constr. Build. Mater.* **2014**, *55*, 274–288. [CrossRef]

87. Fujikake, K. Impact Performance of Ultra-High Performance Fiber Reinforced Concrete Beam and its Analytical Evaluation. *Int. J. Prot. Struct.* **2014**, *5*, 167–186. [CrossRef]

88. Soroushian, P.; Marikunte, S. Moisture effects on flexural performance of wood fiber-cement composites. *J. Mater. Civ. Eng.* **1992**, *4*, 275–291. [CrossRef]

89. Soroushian, P.; Marikunte, S. Moisture Sensitivity of Cellulose Fiber Reinforced Cement. In Proceedings of the Durability of Concrete: Second International Conference, Montreal, QC, Canada, 4 August 1991; Volume II, pp. 821–835.

90. Soroushian, P.; Marikunte, S. Reinforcement of cement-based materials with cellulose fibers. *Spec. Publ.* **1990**, *124*, 99–124.

91. Mansur, M.A.; Aziz, M.A. A study of jute fibre reinforced cement composites. *Int. J. Cem. Compos. Lightweight Concr.* **1982**, *4*, 75–82. [CrossRef]

92. Rostami, R.; Zarrebini, M.; Sanginabadi, K.; Mostofinejad, D.; Abtahi, S.M.; Fashandi, H. The effect of specific surface area of macro fibers on energy absorption capacity of concrete. *J. Text. Inst.* **2019**, *110*, 707–714. [CrossRef]

93. Hasan, A.; Maroof, N.; Ibrahim, Y. Effects of Polypropylene Fiber Content on Strength and Workability Properties of Concrete. *Polytech. J.* **2019**, *9*, 7–12. [CrossRef]

94. Lam, T.F.; Yatim, J.M. Mechanical properties of kenaf fiber reinforced concrete with different fiber content and fiber length. *J. Asian Concr. Fed.* **2015**, *1*, 11. [CrossRef]

95. Ramaswamy, H.S.; Ahuja, B.M.; Krishnamoorthy, S. Behaviour of concrete reinforced with jute, coir and bamboo fibres. *Int. J. Cem. Compos. Lightweight Concr.* **1983**, *5*, 3–13. [CrossRef]

96. Hasan, N.M.S.; Sobuz, H.R.; Sayed, M.S.; Islam, M.S. The Use of Coconut Fibre in the Production of Structural Lightweight Concrete. *J. Appl. Sci.* **2012**, *12*, 831–839.

97. Ali, M.; Liu, A.; Sou, H.; Chouw, N. Mechanical and dynamic properties of coconut fibre reinforced concrete. *Constr. Build. Mater.* **2012**, *30*, 814–825. [CrossRef]

98. Vajje, S.; Murthy, N.R.K. Study On Addition Of The Natural Fibers Into Concrete. *Int. J. Sci. Technol. Res.* **2013**, *2*, 213–218.

99. Güneyisi, E.; Gesoğlu, M.; Akoi, A.O.M.; Mermerdaş, K. Combined effect of steel fiber and metakaolin incorporation on mechanical properties of concrete. *Compos. Part B Eng.* **2014**, *56*, 83–91. [CrossRef]

100. Uzomaka, O.J. Characteristics of akwara as a reinforcing fibre. *Mag. Concr. Res.* **1976**, *28*, 162–167. [CrossRef]

101. Ramakrishna, G.; Sundararajan, T. Impact strength of a few natural fibre reinforced cement mortar slabs: A comparative study. *Cem. Concr. Compos.* **2005**, *27*, 547–553. [CrossRef]

102. Merta, I.; Tschegg, E.K. Fracture energy of natural fibre reinforced concrete. *Constr. Build. Mater.* **2013**, *40*, 991–997. [CrossRef]

103. Banthia, N.; Zanotti, C.; Sappakittipakorn, M. Sustainable fiber reinforced concrete for repair applications. *Constr. Build. Mater.* **2014**, *67*, 405–412. [CrossRef]

104. Zhou, X.; Ghaffar, S.H.; Dong, W.; Oladiran, O.; Fan, M. Fracture and impact properties of short discrete jute fibre-reinforced cementitious composites. *Mater. Des.* **2013**, *49*, 35–47. [CrossRef]

105. Wang, X.D.; Zhang, C.; Huang, Z.; Chen, G.W. Impact Experimental Research on Hybrid Bamboo Fiber and Steel Fiber Reinforced Concrete. *Appl. Mech. Mater.* **2013**, *357–360*, 1049–1052. [CrossRef]

106. Wang, W.; Chouw, N. The behaviour of coconut fibre reinforced concrete (CFRC) under impact loading. *Constr. Build. Mater.* **2017**, *134*, 452–461. [CrossRef]

107. Page, J.; Khadraoui, F.; Boutouil, M.; Gomina, M. Multi-physical properties of a structural concrete incorporating short flax fibers. *Constr. Build. Mater.* **2017**, *140*, 344–353. [CrossRef]

108. Toledo Filho, R.D.; Scrivener, K.; England, G.L.; Ghavami, K. Durability of alkali-sensitive sisal and coconut fibres in cement mortar composites. *Cem. Concr. Compos.* **2000**, *22*, 127–143. [CrossRef]

109. Mohr, B.J.; Biernacki, J.J.; Kurtis, K.E. Supplementary cementitious materials for mitigating degradation of kraft pulp fiber-cement composites. *Cem. Concr. Res.* **2007**, *37*, 1531–1543. [CrossRef]

110. Harper, S. Developing asbestos-free calcium silicate building boards. *Composites* **1982**, *13*, 123–128. [CrossRef]

111. Bergström, S.G.; Gram, H.-E. Durability of alkali-sensitive fibres in concrete. *Int. J. Cem. Compos. Lightweight Concr.* **1984**, *6*, 75–80. [CrossRef]

112. Sivaraja, M.; Velmani, N.; Pillai, M.S. Study on durability of natural fibre concrete composites using mechanical strength and microstructural properties. *Bull. Mater. Sci.* **2010**, *33*, 719–729. [CrossRef]

113. Omoniyi, T.; Akinyemi, B. Durability based suitability of bagasse cement composite for roofing sheets. *J. Civ. Eng. Const. Tech.* **2012**. [CrossRef]

114. Ramakrishna, G.; Sundararajan, T.; Kothandaraman, S. Evaluation of durability of natural fibre reinforced cement mortar composite-a new approach. *ARPN J. Eng. Appl. Sci.* **2010**, *5*, 44–51.

115. Javadi, A.; Srithep, Y.; Pilla, S.; Lee, J.; Gong, S.; Turng, L.S. Processing and characterization of solid and microcellular PHBV/coir fiber composites. *Mater. Sci. Eng. C* **2010**, *30*, 749–757. [CrossRef]

116. Bilba, K.; Arsene, M.-A. Silane treatment of bagasse fiber for reinforcement of cementitious composites. *Compos. Part Appl. Sci. Manuf.* **2008**, *39*, 1488–1495. [CrossRef]

117. Cooke, A.M. Durability of Autoclaved Cellulose Fiber Cement Composites. In Proceedings of the 7th Inorganic-Bonded Wood and Fibre Conference, Sun Valley, ID, USA, 25–27 September 2000; Available online: www.fibrecementconsulting.com/publications/990925.durabilitypaper.pdf (accessed on 15 October 2019).

118. Claramunt, J.; Ardanuy, M.; García-Hortal, J.A.; Filho, R.D.T. The hornification of vegetable fibers to improve the durability of cement mortar composites. *Cem. Concr. Compos.* **2011**, *33*, 586–595. [CrossRef]

119. Pehanich, J.L.; Blankenhorn, P.R.; Silsbee, M.R. Wood fiber surface treatment level effects on selected mechanical properties of wood fiber–cement composites. *Cem. Concr. Res.* **2004**, *34*, 59–65. [CrossRef]

120. Mahjoub, R.; Yatim, J.M.; Mohd Sam, A.R.; Hashemi, S.H. Tensile properties of kenaf fiber due to various conditions of chemical fiber surface modifications. *Constr. Build. Mater.* **2014**, *55*, 103–113. [CrossRef]

121. Mohammadkazemi, F.; Doosthoseini, K.; Ganjian, E.; Azin, M. Manufacturing of bacterial nano-cellulose reinforced fiber−cement composites. *Constr. Build. Mater.* **2015**, *101*, 958–964. [CrossRef]

122. Mohammadhosseini, H.; Yatim, J.M.; Sam, A.R.M.; Awal, A.S.M.A. Durability performance of green concrete composites containing waste carpet fibers and palm oil fuel ash. *J. Clean. Prod.* **2017**, *144*, 448–458. [CrossRef]

123. Sharman, W.R.; Vautier, B.P. *Durability Studies on Wood Fibre Reinforced Cement Sheet*; Building Research Association of New Zealand: Rochdale, UK, 1986.

124. Elsaid, A.; Dawood, M.; Seracino, R.; Bobko, C. Mechanical properties of kenaf fiber reinforced concrete. *Constr. Build. Mater.* **2011**, *25*, 1991–2001. [CrossRef]

125. Marikunte, S.; Soroushian, P. Statistical evaluation of long-term durability characteristics of cellulose fiber reinforced cement composites. *Mater. J.* **1995**, *91*, 607–616.

126. Zaveri, M.D. Absorbency Characteristics of Kenaf Core Particles. Ph.D. Thesis, North Carolina State University, Raleigh, NC, USA, 2004.

127. Mohanty, A.K.; Misra, M.; Drzal, L.T. *Natural Fibers, Biopolymers, and Biocomposites*; CRC Press: Boca Raton, FL, USA, 2005; ISBN 0203508203.

128. Akil, H.M.; Omar, M.F.; Mazuki, A.A.M.; Safiee, S.; Ishak, Z.A.M.; Abu Bakar, A. Kenaf fiber reinforced composites: A review. *Mater. Des.* **2011**, *32*, 4107–4121. [CrossRef]

129. Mahjoub, R.; Yatim, J.M.; Mohd Sam, A.R.; Raftari, M. Characteristics of continuous unidirectional kenaf fiber reinforced epoxy composites. *Mater. Des.* **2014**, *64*, 640–649. [CrossRef]

130. Kumar, K.K.; Karunakar, C.; ChandraMouli, B. Development and Characterization of Hybrid Fibres Reinforced Composites Based on Glass and Kenaf Fibers. *Mater. Today Proc.* **2018**, *5*, 14539–14544. [CrossRef]
131. Ochi, S. Mechanical properties of kenaf fibers and kenaf/PLA composites. *Mech. Mater.* **2008**, *40*, 446–452. [CrossRef]
132. Mahjoub, R.; Yatim, J.M.; Sam, A.M.; Zulkarnain, N.A.; Raftari, M. The Use of Kenaf Fiber Reinforced Polymer to Confine the Concrete Cylinder. *Mater. Today Proc.* **2016**, *3*, 459–463. [CrossRef]
133. Jawaid, M.; Abdul Khalil, H.P.S.; Abu Bakar, A. Mechanical performance of oil palm empty fruit bunches/jute fibres reinforced epoxy hybrid composites. *Mater. Sci. Eng. A* **2010**, *527*, 7944–7949. [CrossRef]
134. Jarukumjorn, K.; Suppakarn, N. Effect of glass fiber hybridization on properties of sisal fiber-polypropylene composites. *Compos. Part B Eng.* **2009**, *40*, 623–627. [CrossRef]
135. Joseph, K.; Varghese, S.; Kalaprasad, G.; Thomas, S.; Prasannakumari, L.; Koshy, P.; Pavithran, C. Influence of interfacial adhesion on the mechanical properties and fracture behaviour of short sisal fibre reinforced polymer composites. *Eur. Polym. J.* **1996**, *32*, 1243–1250. [CrossRef]
136. Holbery, J.; Houston, D. Natural-fiber-reinforced polymer composites in automotive applications. *JOM* **2006**, *58*, 80–87. [CrossRef]
137. Ku, H.; Wang, H.; Pattarachaiyakoop, N.; Trada, M. A review on the tensile properties of natural fiber reinforced polymer composites. *Compos. Part B Eng.* **2011**, *42*, 856–873. [CrossRef]
138. Ranakoti, L.; Pokhriyal, M.; Kumar, A. Natural fibers and biopolymers characterization: A future potential composite material. *Bratisl. J. Mech. Eng. Stroj. Časopis* **2018**, *68*, 33–50.
139. Mishra, S.; Mohanty, A.K.; Drzal, L.T.; Misra, M.; Parija, S.; Nayak, S.K.; Tripathy, S.S. Studies on mechanical performance of biofibre/glass reinforced polyester hybrid composites. *Compos. Sci. Technol.* **2003**, *63*, 1377–1385. [CrossRef]
140. Nabi, S.D.; Jog, J.P. Natural Fiber Polymer Composites: A Review. *Adv. Polym. Technol.* **1999**, *18*, 351–363.
141. Jawaid, M.; Abdul Khalil, H.P.S. Cellulosic/synthetic fibre reinforced polymer hybrid composites: A review. *Carbohydr. Polym.* **2011**, *86*, 1–18. [CrossRef]
142. Sreekala, M.S.; Kumaran, M.G.; Thomas, S. Oil palm fibers: Morphology, chemical composition, surface modification, and mechanical properties. *J. Appl. Polym. Sci.* **1997**, *66*, 821–835. [CrossRef]
143. Mahjoub, R.; Bin Mohamad Yatim, J.; Mohd Sam, A.R. A review of structural performance of oil palm empty fruit bunch fiber in polymer composites. *Adv. Mater. Sci. Eng.* **2013**, *2013*. [CrossRef]
144. Davoodi, M.M.; Sapuan, S.M.; Ahmad, D.; Ali, A.; Khalina, A.; Jonoobi, M. Mechanical properties of hybrid kenaf/glass reinforced epoxy composite for passenger car bumper beam. *Mater. Des.* **2010**, *31*, 4927–4932. [CrossRef]
145. Park, J.K.; Cho, D.; Kang, T.J. A comparison of the interfacial, thermal, and ablative properties between spun and filament yarn type carbon fabric/phenolic composites. *Carbon* **2004**, *42*, 795–804. [CrossRef]
146. Huang, C.; Chen, T.; Feng, S. Finite element analysis of fatigue crack growth in CFRP-repaired four-point bend specimens. *Eng. Struct.* **2019**, *183*, 398–407. [CrossRef]
147. Babatunde Ogunbode, E.; Mohamad Yatim, J.; Yunus Ishak, M.; Abdul Hamid, H. An Evaluation of the Interfacial Bond Strength of Kenaf Fibrous Concrete and Plain Concrete Composite Cleaner and Sustainable Housing View Project Strength, Durability and Microstructural properties of concrete composites View project. *Int. J. Built Environ. Sustain.* **2019**, *6*, 1–6. [CrossRef]

 *applied sciences*

Article

Effect of Incorporating Waste Limestone Powder into Solid Waste Cemented Paste Backfill Material

Jianhua Hu [1,2], Xiaotian Ding [1,2], Qifan Ren [1,2,*], Zhouquan Luo [1,2] and Quan Jiang [1,2]

1   School of resources and safety engineering, Central South University, Changsha 410083, China; hujh21@csu.edu.cn (J.H.); 175511036@csu.edu.cn (X.D.); Lzq505@csu.edu.cn (Z.L.); jqmose@csu.edu.cn (Q.J.)
2   Hunan Key Laboratory of Mineral Resources Exploitation and Hazard Control for Deep Metal Mines, Changsha 410083, China
*   Correspondence: qifanren@csu.edu.cn; Tel.: +86-181-7513-4802

Received: 19 April 2019; Accepted: 17 May 2019; Published: 20 May 2019

Featured Application: The specific application of the material is to produce cost-saving backfill and to reduce the solid waste in mines.

Abstract: To effectively reuse waste limestone powder, which is a major solid waste around mines, we replaced limestone powder back into a part of cement in solid waste cemented paste backfill (SWCPB) and studied the parameters of pore structures. To optimize the pore microstructure characteristics of SWCPB in mines, two different components and grade tailings were selected. The samples were characterized by scanning electron microscopy (SEM) and nuclear magnetic resonance (NMR) to examine the pore properties and microstructure of SWCPB. The results showed that (1) at the later curing stage, with the optimization of pore characteristics and microstructure through the limestone powder admixture, the strength of SWCFB was guaranteed at a 20% replacement degree of cement. (2) Porosity, macropore proportion, and the average pore radius all negatively correlated with limestone powder content, which were reduced by 7.15%, 46.35%, and 16.37%, respectively. (3) Limestone powder as a crystal nucleus participated in the hydration reaction and was embedded into the product to enhance the strength.

Keywords: limestone powder; cemented paste backfill; image analysis; pore characteristics

---

1. Introduction

In order to reduce the cost of cement in concrete and to manufacture special cement for various special purposes, many worldwide scholars [1–5] have conducted extensive work on cement substitute materials and concrete additives. Recently, limestone cement has become a recent subject of research, owing to the promising properties of limestone for cost saving and cementing [6,7]. Siliceous limestone powder waste has cementitious activity as it can be reused to replace a part of the cement, leading to an improvement in the properties of concrete and thereby reducing contamination in soil as well as the air pollution [8,9]. Cemented paste backfill (CPB) is an emerging material, which is similar to concrete, that is increasingly being used in mining industries. Hu et al. [10] assessed stone powder as a replacement of cement in CPB, and they studied strength characteristics and reaction mechanism. The results showed that strength of the backfill was greatly reduced at an early stage, and was slightly reduced in the final stages. The effects of adding waste limestone powder into solid waste CPB material still need to be studied.

As a relatively new mine waste management technology, CPB has been extensively applied in underground mine operations around the world because of its significant environmental, technical, and economic benefits [11–14]. Besides, as the mining industry moves deeper into the ground, the cost of CPB has increased because of unfavorable conditions such as increased pressure in the stope and high environmental standards [15]. Currently, CPB accounts for about 40% or more cement content, although it has only 3–9 wt % of CPB [16]. Therefore, there is urgent need to find another cheap cementation material to replace part of the cement that can reduce the cost and minimize the environmental problems caused by solid waste.

Effects of pore structure on the strength and various properties of CPB are important to know [17]. In fact, researchers need to focus attention on how the pore structure changes in limestone powder. In general, total porosity (n) and pore size distribution (PSD) are the main parameters that are related to the pore structure and can be quantified by using various approaches [18–20]. Among them, scanning electron microscopy (SEM) combined with image analysis (IA) technology is one of the most common methods, which can be directly obtained from the characteristic parameters of pores. Simultaneously, analysis of nuclear magnetic resonance (NMR) spectra can be proposed for identifying material porosity, pore geometry, connectivity, and to characterize PSD material.

The present study focuses on the performance of solid waste cemented paste backfill (SWCPB) containing limestone powder as an additive (0–20%) to cement and tailings as aggregates in SWCPB. SEM-IA and NMR are used to determine relevant pore structure parameters to assess mechanical and hydraulic behaviors. Furthermore, the feasibility of reusing limestone powder in CPB will be evaluated.

2. Materials and Methods

2.1. Raw Materials

2.1.1. Composition Analysis

Tailing samples used in the present study were obtained from an underground lead–zinc mine located in the south of China. The tailings (tailing A and tailing B) from two different mineral processing areas were selected as the combined SWCPB. The binder used in this study was a complex binder made of limestone powder (0–20%) and ordinary Portland cement (OPC) (80–100%). The OPC produced by Changsha Xinxing Cement Plant with C30 grade strength was selected as the major cementing material. Abandoned limestone powder from the mine quarry was selected to replace part of the cement. The proportion of main elements in raw materials was analyzed by using a X-ray fluorescence (XRF, PANalytical B.V., Almelo, The Netherlands) spectrometer, as shown in Table 1.

Table 1. Main mineral composition of tailings A and B.

Element	Tailing A	Tailing B	Limestone Powder	
	Content/%		Compound	Concentration/%
O	34.70	44.38	CaO	57.60
Fe	23.10	12.66	SiO_2	36.87
S	15.86	8.88	Al_2O_3	1.58
Ca	14.03	9.09	MgO	1.17
Si	4.48	16.24	Fe_2O_3	1.17
Mg	2.12	0.40	SO_3	0.74
Zn	1.23	1.14	K_2O	0.27
Al	1.11	3.06	BaO	0.12
As	0.99	1.48	P_2O_5	0.085
K	0.50	0.67	Cr_2O_3	0.085
Sn	0.37	0.24	TiO_2	0.082
Pb	0.34	0.52	MnO	0.081
Sb	0.32	0.35	Na_2O	0.039
Mn	0.14	0.41	SrO	0.039

Siliceous limestone was used as raw material for limestone powder. Table 1 shows that SiO_2 accounted for about 36.87% of content, and SiO_2 reacted with $Ca(OH)_2$ to form C–H–S bonds, promoting the gelation of filling. Formation of C–H–S bonds served as the crucial part of the hydration reaction, which resulted in the limestone powder used in replacing the cement.

2.1.2. Particle Size Analysis

The main factors affecting the strength of CPB were the composition and particle size of raw materials [21]. The effect of limestone powder for SWCPB could be evaluated according to the change of particle size of SWCPB. Particle size distribution data were obtained by sieve analysis. Experimental analyses of laser particle sizing (Malvern Mastersizer 2000, Malvern Instruments Limited, Malvern, UK) were carried out. Figure 1 presents the particle size distribution data of raw materials.

The majority of particle sizes of tailing A were found to be 110 μm, and that of tailing B was 106 μm, as shown in Figure 1. Both tailings A and B lacked a particle size of around 75 μm, and the proportion of particles under 10 μm accounted for less than 30%, which resulted in the filling of insufficient fine particles in the pores between coarse particles. The distributions of particle size after crushing and grinding are shown in the Figure 1. Here, the majority of particle sizes of limestone powder came to 75 μm, which was up to 3.2%, and the particles under 10 μm were assumed for 43%. According to gradation theory, the addition of limestone powder could improve the particle size distribution of SWCPB.

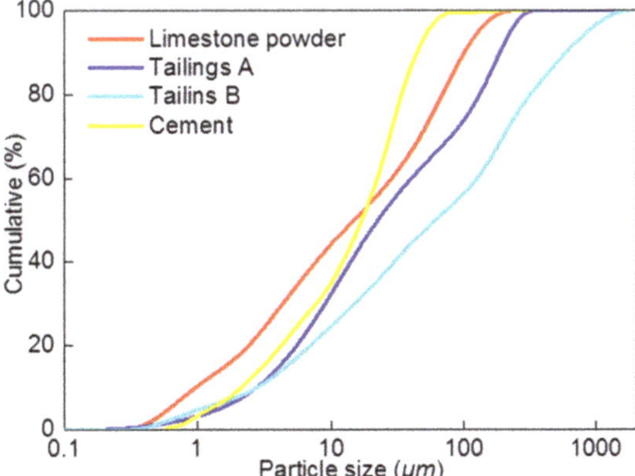

Figure 1. Particle size distribution of raw materials including limestone powder, tailing A, tailings B, and cement.

2.2. Sample Preparation

Considering actual conditions, the ratio of cementitious material to the tailings was 1:4, and the mass concentration was about 70%. In addition, pulping and filling were carried out at room temperature.

According to gradation theory, the ratio of limiting particle size (d60) to the effective particle size (d10) is referred to as the nonuniformity coefficient Cu. By avoiding discontinuous gradation, the grading of backfill aggregate should happen only when $Cu > 10$. Therefore, by introducing the coefficient of curvature, $Cc = (d30 \times d30) / (d60 \times d10)$, and by considering $Cu > 10$ and $1 < Cc < 3$, a different ratio of limestone powder replacement in the cement for the observation group was set. Meanwhile, whole cement (A1 and B1 as control groups) was considered, and the ratio of nonuniform coefficient *Cu* and curvature coefficient *Cc* were calculated, as shown in Table 2. However, the *Cc* values of mixture were below 1.00, which meant its gradation was not ideal in maximizing strength values.

Table 2. Slurry ratio and size parameter of each group. LSP: limestone stone powder; OPC: ordinary Portland cement.

Group	Sample	Mass Content (%)				Cu	Cc
		Water	Tailings	LSP	OPC		
Group A mixture	A1	30	56	0	14	14.56	0.72
	A2	30	56	1.4	12.6	14.68	0.72
	A3	30	56	2.1	11.9	14.80	0.73
	A4	30	56	2.8	11.2	14.93	0.73
Group B mixture	B1	30	56	0	14	40.02	0.71
	B2	30	56	1.4	12.6	44.36	0.64
	B3	30	56	2.1	11.9	44.10	0.64
	B4	30	56	2.8	11.2	43.83	0.65

Three samples of each group (a total of 72 samples) were used as the control to reduce error. The test sample dimensions were 70.7 mm × 70.7 mm × 70.7 mm. Test pieces with dimensions of 15 mm × 15 mm × 15 mm and with the same ratio were designed for NMR analysis. The experiment process is shown in Figure 2.

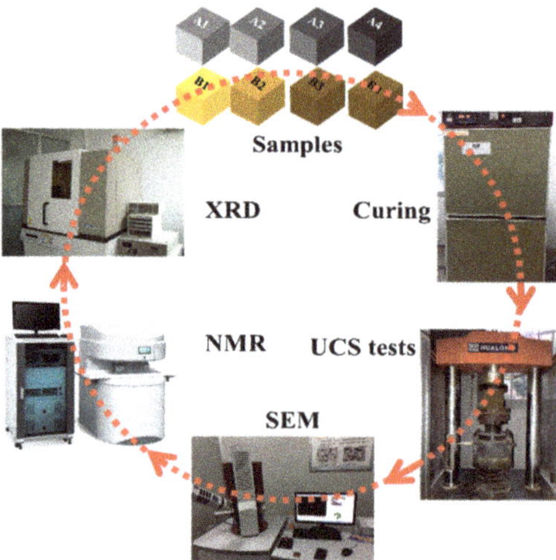

Figure 2. Experimental process with sample preparation, nuclear magnetic resonance (NMR) experiment, uniaxial compressive strength (UCS) tests, and scanning electron microscope (SEM).

2.3. Experimental Methods

2.3.1. Uniaxial Compressive Strength (UCS) Test

Uniaxial compressive strength (UCS) tests were performed on specimens by using WDW-2000 rigid hydraulic pressure servo machine (Ruite, Guilin, China). The displacement rate of loading was 0.5 mm·min^{-1}, and the test standard was GB/T50266-99.

2.3.2. Nuclear Magnetic Resonance (NMR) Analysis

An Ani-MR150 rock magnetic resonance imaging analysis system (Niumag, Shanghai, China) was used to conduct NMR tests on the samples using various mixture ratios and curing times. The average pore distribution of each test piece was calculated to obtain the microstructure distribution parameters of each sample.

2.3.3. Scanning Electron Microscopy-Image Analysis (SEM-IA)

After testing the UCS, a cube sample with a side length of 1 mm was taken from the center of each damaged sample. Then, a metal conductive film on the vacuum coater was applied for SEM tests. SEM images were used to analyze the microscopic pore structure images of each group by using a Czech TESCAN MIRA3 field-emission SEM (TESCAN, a.s., Brno, Czech Republic). Further analysis of SEM images was carried out by using MATLAB (14.0, MathWorks, Natick, MA, USA). With the FRACLAB toolbox in MATLAB, SEM images were analyzed to obtain the grayscale images, binary image, and the fractal dimensions of each sample.

3. Results

3.1. Uniaxial Compressive Strength

The strength of SWCPB directly affected ore body safety and continuity of mining. Figure 3 shows the curve strengths of SWCPB in groups A and B based on UCS test results.

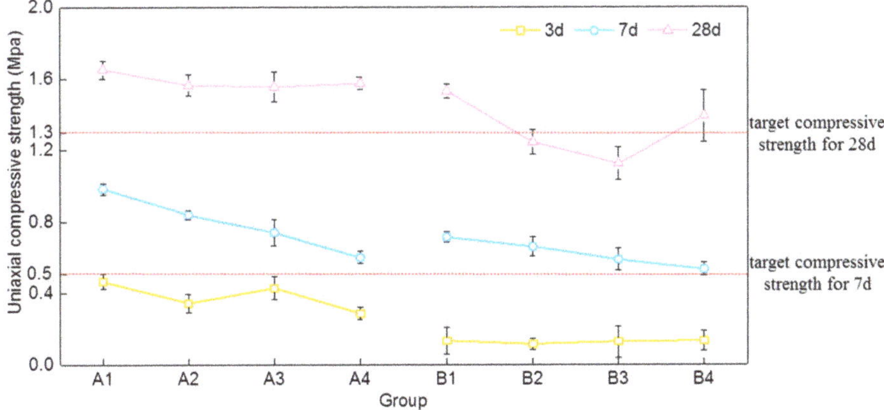

Figure 3. UCS comparison of different mixture ratios of groups A and B.

Results shown in Figure 3 indicated that the strength for three days in group A observed no obvious trend, while with seven days of curing, the strength decreased with the increase of limestone powder replaced. The strengths of A2, A3, and A4 were similar after 28 d. Similarly, for group B, the strength of B4 with 28 d was higher than B2 and B3. Comparing the strengths of group A with B, all strengths of group A were higher than those of group B. It was interesting that there was an apparent retardation in strength development as a result of cement substitution. According to [10], the hydration processes of stone powder cement tailing backfill can be divided into the following four stages: dissolution period, condensation period, infiltration period, and hardening period. At the early curing stage, the groups with cement substitution had less cement to increase the strength, and limestone powders just played a role as a crystal nucleus. However, at the later curing stage, hydration products reacted with SiO_2, which improved hydration reactions and strength.

In addition, according to mining experience, the target strength of seven days is 0.5 MPa, while for 28 days the required strength is 1.3 MPa. As a consequence, the seven-day strength of all groups met the requirements, while the 28 d strengths of B2 and B3 did not meet mining strength requirements of backfill.

3.2. NMR Analysis of Pore Properties

NMR analyses were done using an MiniMR-60 magnetic resonance imaging (MRI) analysis system manufactured by Shanghai Newmai Co. Ltd., China. As the samples were very small, three samples were tested for each group, and their mean values were used for analysis to minimize the error. NMR experiments were carried out for 28 d of curing samples of group A and B, and the pore structure parameters of SWCPB were obtained.

NMR total relaxation (T2) time is related to surface relaxation, loose relaxation of fluid procession, and diffusion relaxation caused by gradient fields [22]. For the water-saturated samples, T2 relaxation time was directly proportional to the pore size and the magnitude of the T2 curve, which directly reflected the porosity of SWCPB samples. Pore size distribution maps were plotted by considering the average amplitude of NMR, as shown in Figures 4 and 5, for groups A and B, respectively.

Three peaks of T2 relaxation time for group A after 28 d of curing can be observed in Figure 4, in which the first peak emerged at about 0.28 ms, the second peak observed at about 11.50 ms, and the third peak appeared at 204.91 ms.

The pore ratio of A1 was shown in Figure 4a, in which the total porosity was about 1.99%. The main pores in A1 were micropores (1st peak accounted for 93.3%), the mesopores accounted for a relatively small proportion (2nd peak accounted for only 1.91%), and there were also many macropores (3rd peak accounted for 4.79%). The pore ratio of A2, as depicted in Figure 4b, showed an increase in total porosity as the limestone powder consisted of fewer active ingredients than the cement. The addition of limestone powder increased the micropores but decreased the macropores, and the change of mesopores was not significant. By further increase of limestone powder to about 15%, as shown in Figure 4c, the total porosity, proportion of micropores, and mesopores were increased, while the amount of macropores was found to be reduced. When the addition of stone powder increased and reached about 20% (Figure 4d), the mesopores remained the same while the micropores increased and the macropores decreased, but the total porosity decreased overall.

The NMR distribution of T2 relaxation time for 28 d of the group B sample showed two peaks in Figure 5, the first peak appeared in the vicinity of 0.30 ms, while the second peak emerged at around 204.90 ms.

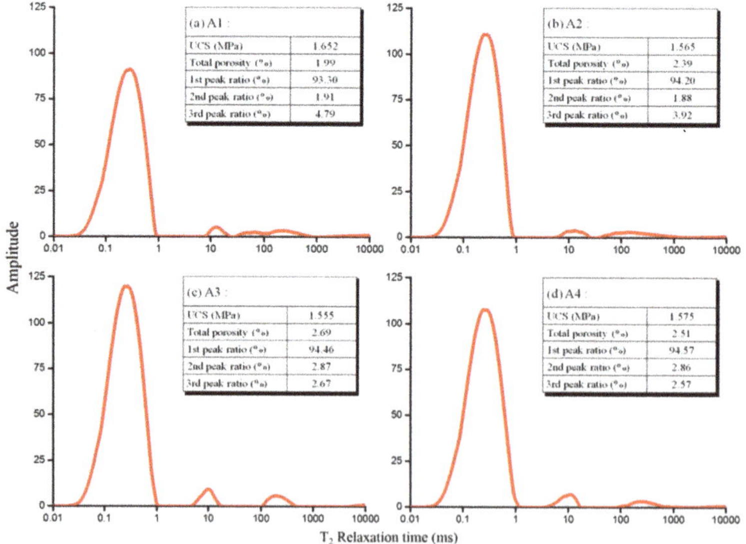

Figure 4. Pore size distribution of group A.

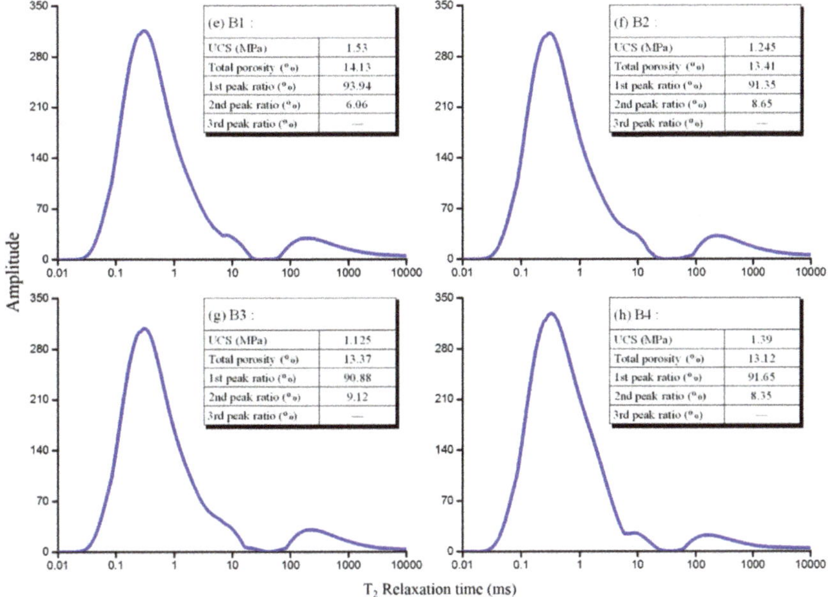

Figure 5. Pore size distribution of group B.

Due to poor grading of group B ($Cu > 40$), there was no obvious difference between the micropores and mesopores. Therefore, the curve had two peaks, and the peak areas of the first peak and the second peak in each group were assumed for 95% or higher of the total area. By increasing the limestone powder, the distribution of micropores decreased, but the macropores increased slightly while the total porosity remained constant. However, when the addition of limestone powder reached 20%, the proportion of micropores increased significantly and the macropores decreased.

3.3. SEM Analysis of the Microstructure

After the initial gelatinization process, the pore structure was formed by the cemented hydration reaction, and it was involved in transferring water, unhydrated material, and storing water. Among the three, the water storage function of the pores maintained the hydration reaction and further enhanced the strength of CPB [22–24]. In this process, the proportion of harmless pores could be increased by reducing the proportion of large pores in SWCPB and simultaneously increasing the proportion of small pores while the total porosity remains intact.

The 28 d curing samples of groups A and B were selected for SEM analysis, at 5000× magnification and a scale size of 10 μm, to study pore distribution, pore morphology, and other characteristics. Figure 6 illustrates SEM images of group A samples for 28 d curing while Figure 7 presents the samples of group B for the same.

The structures of A1, as shown in Figure 6a, were denser, in which the distribution of cementitious materials was uniform, and some of the large particle size tailings were exposed without any connection. Most of the pores were less than 3 μm in diameter. The pore structure of A2 was poorer than A1, shown in Figure 6b. Here, the porosity and pore diameter increased because the limestone powder consisted of fewer active ingredients than cement. When the particles of inert limestone powder were crystallized by the hydration process, the volume was increased, and the body of the crystal was difficult to dissolve as the pores between the tailings were filled. However, the hydration product decreased because of the decrease in cement incorporation; therefore, the microporous structure of A2 was poorer than A1. The pore structures of A3 and A4 were similar to A1, as shown in Figure 6c,d, respectively. Among them, it was found that the macropores were significantly reduced while mesopores and micropores increased, as compared to Figure 6c,d.

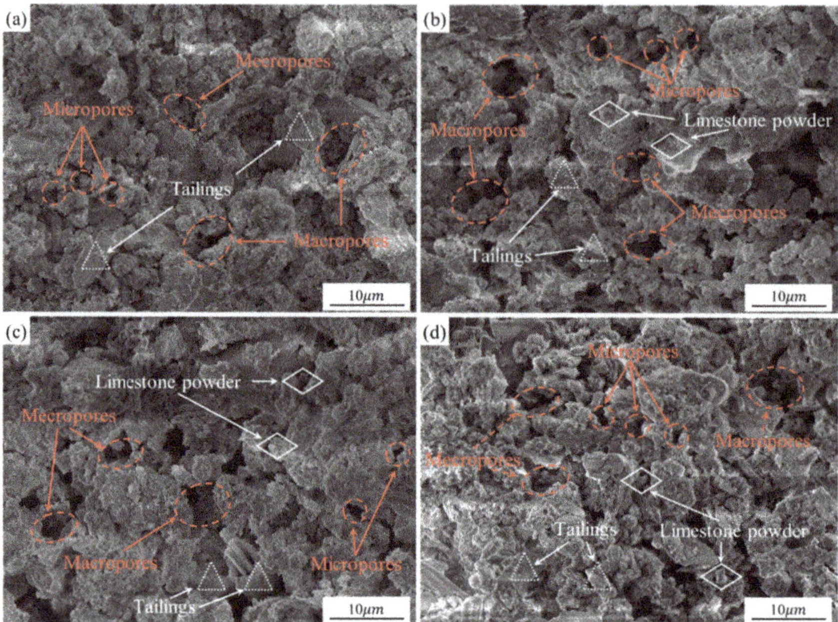

Figure 6. SEM images (5000×) of (**a**) A1, (**b**) A2, (**c**) A3, and (**d**) A4.

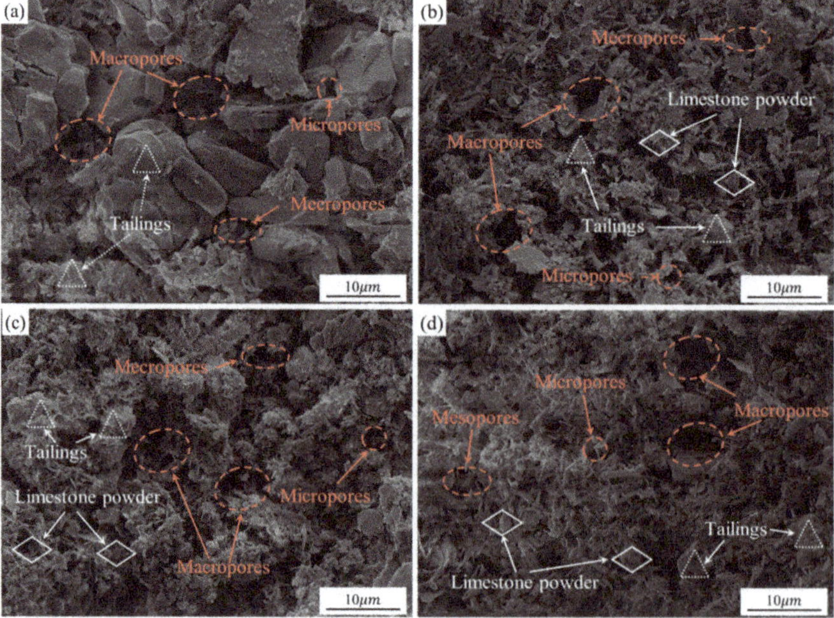

Figure 7. SEM images (5000×) of (**a**) B1, (**b**) B2, (**c**) B3, and (**d**) B4.

The SEM images of 28 d curing age samples of group B showed a small number of large-sized tailings. In the samples of group B, the exposed crystals (mainly Aft) accounted for a large proportion, and the cementitious material (mainly C–H–S gel structures) was not obvious. The structure of group B was looser than that of group A. Without limestone powder particles in backfill, a lot of coarse particles were observed, and there was no particle that filled the gaps between tailings (Figure 7a). The structure of group B4 was denser than that of B2 and B3. Groups B2 and B3 had more macropores, and a lot of macropores combined together to form a complex pore structure, as shown in Figure 7, which meant that the pore structure of B4 was simpler than those of B2 and B3. The calculated NMR porosities of B3 and B4 were almost the same (Table 3); however the average pore radii of B3 were higher than that of B4, which meant that pores of B4 were small, but the distribution was relatively uniform. In addition, group B3 had a large number of pores above 3 µm in diameter, as the number of specific particles that could fill the pores was insufficient. In B4, a large number of fine particles entered into the large pores, and the macropores were divided because there was an increase in the number of particles below 10 µm. Hence, most of the pores of B4 were found to be less than 1 µm in diameter, and the largest pore size reached less than 2 µm in diameter.

Table 3. Binary parameter statistics.

Group	Porosity (%)	Calculated Porosity (%)	Pixel Area	Counting Unit	Calculated Average Pore Radius (µm)
A1	1.99	2.17	16,900	662	1.54
A2	2.39	2.53	19,679	708	1.61
A3	2.69	2.63	20,491	738	1.61
A4	2.51	2.50	19,518	670	1.65
B1	14.13	14.12	110,065	1903	2.32
B2	13.41	12.97	101,102	2209	2.06
B3	13.37	13.97	108,841	1992	2.25
B4	13.12	14.09	109,802	2705	1.94

3.4. SEM Image Quantitative Analysis

SEM images of each group were binarized for quantitative analysis of pore topography data. SEM images (500×) were binarized by using the FRACLAB toolbox in MATLAB. By setting the nearest contrast threshold, NMR porosity was used to generate a binary image together with parallel color inversion. Figure 8 shows two samples that contrasted original SEM images and its binarized images. By implementing count instruction, the pixel count and the unit count of the bright area of binary images were calculated, followed by the calculation of porosity, and then the comparison with NMR porosity, as shown in Table 3. The size of the SEM image was 779,264 pixel units with a total size of 1024 × 761, in which the scale bar was 100 µm length and 185 pixel units. Finally, a single 500× SEM image area was calculated to be 227,710 µm^2.

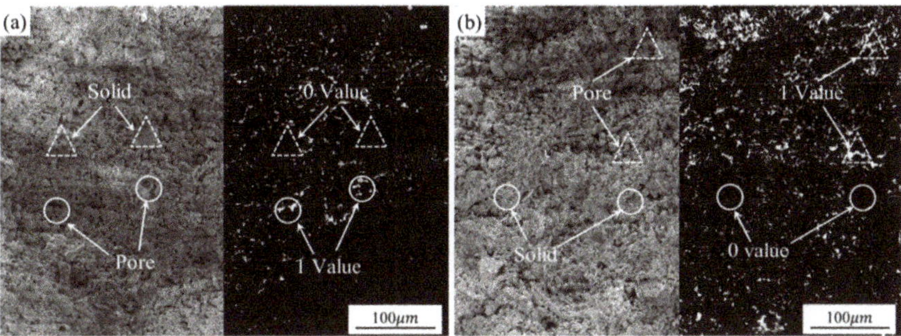

Figure 8. SEM binary image (500×) contrasts of (**a**) A4 and (**b**) B4.

The binarized images had a clearer pore structure than the original images and were easy to measure, as shown in Figure 8. Among them, A4 (Figure 8a) had a uniform pore distribution with most of the pore diameters less than 5 μm and maintaining a compact structure. In a similar way, it was analyzed that B4 had more pores than A4, while for B4 (Figure 8b) a large number of pores had sizes over 10 or even 20 μm.

Table 3 shows that porosity was calculated by using the image processing method of Figure 8. With the help of stereology principles, the calculated porosities of groups A1, A2, B3, and B4 were slightly larger than NMR porosity. The other four cases were reversed. Although there were differences between calculated porosity and NMR porosity, the values were almost the same. The calculated porosity, number of holes (counting units), and the hole radius of group B were significantly greater than in group A. The data shown in Table 3 followed that the average pore radius changed as a result of adding limestone powder. The average pore radius in group A was basically unchanged, but in group B the average pore size first decreased, then increased, but then finally decreased significantly.

4. Discussion

4.1. Pore Characteristics Analysis

For NMR results of group A (Figure 4), porosity deteriorated with the admixture of limestone powder, resulting in an increased value (AVG = 27.14%, max = 35.18%). However, optimization of pore distribution increased the proportion of small holes and decreased the proportion of large holes (AVG = 36.26%, max = 46.35%). Binarization parameters, as listed in Table 3, showed an increase in the number of pores (AVG = 6.55%, max= 11.48%) and thereby slightly increased the average pore radius (AVG = 5.41%, max = 7.14%), which was adverse. The analysis of grading parameters, as listed in Table 2, suggested a nonuniform coefficient A1 value of $Cu = 14.56$ and a curvature coefficient value of $Cc = 0.72$ (strive $Cu \geq 10$ and not too large, $1 < Cc < 3$), which increased slightly in the presence of limestone powder admixture in which grading was optimized.

For NMR results of group B (Figure 5), porosity was optimized in the presence of limestone powder with a slight decrease in value (AVG = 5.87%, max = 7.15%), but pore distribution deteriorated as macroporosity increased (AVG = 43.67 %, Max = 50.50%). Binarization parameters, as shown in Table 3, exhibited pore properties that were optimized by increasing the number of pores (AVG = 20.97%, max = 42.14%), and the average pore radius was decreased. The parameters of grading (Table 2) deteriorated after mixing with the limestone powder, in which $Cu = 40.02$ and $Cc = 0.71$ for 100% cement group, suggesting a serious loss of intermediate particle size.

The incorporation of limestone powder had a certain influence on the pore properties of SWCFB A and B, and the optimization effect on B group was apparent. The results of NMR depicted in Figure 5 showed that the porosity of 100% cement B1 was 7.1 times of that of A1. When 15% stone powder was mixed with the B group, the porosity could be reduced to 4.97 times of that of group A. In analyzing the binarization parameters from Table 3, the average pore radius ratio of B and A could be reduced to 1.18-fold from the initial 1.51-fold by incorporating 20% stone powder. In group B, the grading (Table 2) did not require any optimization but deteriorated. Hence, the proportion of particles over 100 μm in group B were assumed to be over 40% (Figure 1), which was incompatible with the particle size distribution of limestone powder.

4.2. Microstructure Analysis

From the data presented in Table 1, the existence of SiO_2 was assumed to be about 36.87%, which was involved in the hydration reaction with $Ca(OH)_2$ in the cement. The particles as crystal nuclei were mostly encapsulated in C–H–S gel and embedded in the hydration reaction product as crystalline nuclei. The C–H–S gel with stone powder had higher strength than pure C–H–S gel, but the swelling volume of the hydration reaction decreased compared to the 100% cement group by the same amount of cementitious material.

In the SEM images of group A (Figure 6), the proportion of C–H–S gel was positively correlated with the amount of cement, and the proportion of C–H–S in 100% cement group A1 (Figure 6a) was the largest. In addition, crystals of AFt, a small amount of unreacted $Ca(OH)_2$ crystals, and a certain number of exposed tailing particles were found. The structures of A2, A3, and A4 (Figure 6b–d) were similar. SEM images showed almost invisible AFt, completely reacted $Ca(OH)_2$ particles, and stone powder existed in the form of crystal nuclei. Among the three, A4 had the highest replacement amount of stone powder; its structure was denser and had reduced the porosity, indicating that the microstructure of group A can be optimized with a certain amount of limestone powder.

In group B, the SEM images (Figure 7) revealed a significant increase of AFt than those of group A, approaching towards the formation of C–H–S. $Ca(OH)_2$ almost completely reacted, and a part of the tailing particles was exposed. Limestone powder particles mostly existed among the hydration reaction structures, but some of them were exposed outside. The strength of B4 (Figure 7b) was increased by replacing a certain amount of limestone powder. As a result, the structures of B4 were improved than that of B3 (Figure 7a) with a decrease in large pores, indicating that a certain amount of limestone powder could optimize the microstructure of group B.

4.3. Macro-Strength Characterization

Before and after the incorporation of limestone powder, the UCS of each samples of group B (Figure 3) at each age and mixture ratio were lower than that of the corresponding group A. By analyzing the pore characteristics of A and B, the average porosity (Table 3) of group B was found as 5.64 times, the average proportion of macropores was 2.31 times, and the average pore radius (Table 3) was 1.34 times more than that of group A. The mean nonuniformity (Table 2) was 2.92 times of group A. Parameters obtained from the pore characteristics did not show any beneficial effect. The particle size of C–H–S gel (Figure 6 and 7) increased by the incorporation of limestone powder, but the content of C–H–S gel of group B was significantly lower than that of group A as the microstructural function of limestone powder was not fully demonstrated. In summary, optimization of pore characteristics and microstructure through a limestone powder admixture optimized the strength of SWCFB in group B, but it was to a limited extent. It cannot change disadvantageous aspects of the tailings in group B.

After replacing a part of cement with limestone powder, the strength after 28 d in group A (Figure 3) still maintained a certain level; the strengths of A2, A3, and A4 were found to be in close ranges of 1.565 ± 0.01 MPa; and the proportion of stone powder in the cementitious material could reach 20%. By analyzing the pore characteristics of group A (Figure 4), the porosity and average pore radii of A2, A3, and A4 (Table 3) increased with the replacement of limestone powder, but the proportion of macropores decreased significantly, in which grading (Table 2) was optimized. In contrast, for the microstructure (Figure 6), with an increase in stone powder content, the proportion of C–H–S did not decrease. A2, A3, and A4 were similar in structure. The number of pores in A4 was fewer, and the structure was dense. According to another constant shear test result, the larger the amount of stone powder, the longer it takes for the slurry to reach equilibrium state, whereas the values of equilibrium shear stress and equilibrium viscosity are smaller. That means limestone powder can reduce the dynamic viscosity of slurry and can enhance the slurry-conveying performance. As a consequence, fewer macropores will be formed in backfill after slurries solidify [25]. The strength of samples with cement substitution remained high even after replacing the limestone powder because the proportion of macropores in group A was reduced and the microstructure in the presence of limestone powder admixtures was optimized.

5. Conclusions

To effectively reuse waste limestone powder, portion of cement in SWCPB was replaced by limestone powder. This paper studied the parameters of its pore structures and strength characteristics. The main conclusions are as follows:

(1) The strengths of SWCPB had negative correlations with limestone powder content after three and seven curing days; however, after 28 curing days, limestone powder content did not have a significant impact on strength. Except for groups B2 and B3, the strengths of the other groups can meet mining requirements.

(2) Porosity, macropore proportion, and the average pore radius all negatively correlated with limestone powder content, which were reduced by 7.15%, 46.35%, and 16.37%, respectively. Limestone powder in backfill can reduce the number of pores and the values of average pore radius.

(3) Limestone powder as a crystal nucleus participated in the hydration reaction and was embedded into the product to enhance the strength of SWCPB. Thereby, the pore distribution of backfill was optimized. With optimization of pore characteristics and the microstructure through the limestone powder admixture, the strength of SWCFB can be optimized to a certain extent.

Author Contributions: Conceptualization, J.H. and X.D.; Methodology, Q.R.; Validation, Q.R. and Q.J.; Analysis, X.D. and Q.R.; data curation, Q.R. and Q.J.; Writing—original draft preparation, X.D.; Writing—review and editing, J.H.; Supervision, J.H.; Project administration, J.H. and Z.L.

Funding: The research was supported by (1) The National Key Research and Development Program of China, grant number 2017YFC0602901; (2) The National Natural Science Foundation of China, grant number 41672298.

Acknowledgments: We thank the Gaofeng Mine's management and staff for their valuable support. We thank instructional support specialist Modern Analysis and Testing Central of Central South University. Finally, we thank the two anonymous reviewers for their helpful comments.

Conflicts of Interest: The authors declare no conflict of interest.

References

1. Al-Kheetan, M.J.; Rahman, M.M.; Chamberlain, D.A. Chamberlain. Influence of early water exposure on modified cementitious coating. *Constr. Build. Mater.* **2017**, *141*, 64–71. [CrossRef]

2. Al-Kheetan, M.J.; Rahman, M.M.; Chamberlain, D.A. A novel approach of introducing crystalline protection material and curing agent in fresh concrete for enhancing hydrophobicity. *Constr. Build. Mater.* **2018**, *160*, 644–652. [CrossRef]

3. Kupwade-Patil, K.; Palkovic, S.D.; Bumajdad, A.; Soriano, C.; Buyukozturk, O. Use of silica fume and natural volcanic ash as a replacement to Portland cement: Micro and pore structural investigation using NMR, XRD, FTIR and X-ray microtomography. *Constr. Build. Mater.* **2018**, *58*, 574–590.

4. Celik, K.; Meral, C.; Gursel, A.P.; Mehta, P.K.; Horvath, A.; Monteiro, P.J.M. Mechanical properties, durability, and life-cycle assessment of self-consolidating concrete mixtures made with blended portland cements containing fly ash and limestone powder. *Cem. Concr. Compos.* **2015**, *56*, 59–72. [CrossRef]

5. De Weerdt, K.; Ben Haha, M.; le Saout, G.; Kjellsen, K.O.; Justnes, H.; Lothenbach, B. Hydration mechanisms of ternary Portland cements containing limestone powder and fly ash. *Cem. Concr. Res.* **2011**, *41*, 279–291. [CrossRef]

6. Wang, D.; Shi, C.; Farzadnia, N.; Shi, Z.G.; Jia, H.F.; Ou, Z.H. A review on use of limestone powder in cement-based materials: Mechanism, hydration and microstructures. *Constr. Build. Mater.* **2018**, *181*, 659–672. [CrossRef]

7. Sua-Iam, G.; Makul, N. Utilization of limestone powder to improve the properties of self-compacting concrete incorporating high volumes of untreated rice husk ash as fine aggregate. *Constr. Build. Mater.* **2013**, *38*, 455–464. [CrossRef]

8. Pokharel, M.; Fall, M. Combined influence of sulphate and temperature on the saturated hydraulic conductivity of hardened cemented paste backfill. *Cem. Concr. Compos.* **2013**, *38*, 21–28. [CrossRef]

9. Benzaazoua, M.; Ouellet, J.; Servant, S.; Newman, P.; Verburg, R. Cementitious backfill with high sulfur content: Physical, chemical and mineralogical characterization. *Cem. Concr. Res.* **1999**, *29*, 719–725. [CrossRef]

10. Hu, J.H.; Ren, Q.F.; Jiang, Q.; Gao, R.G.; Zhang, L.; Luo, Z.Q. Strength characteristics and the reaction mechanism of stone powder cement tailings backfill. *Adv. Mater. Sci. Eng.* **2018**, *2018*, 8651239. [CrossRef]

11. Fall, M.; Pokharel, M. Coupled effects of sulphate and temperature on the strength development of cemented tailings backfills: Portland cemen-paste backfill. *Cem. Concr. Compos.* **2010**, *32*, 819–828. [CrossRef]

12. Kesimal, A.; Yilmaz, E.; Ercikdi, B.; Alp, I.; Deveci, H. Effect of properties of tailings and binder on the short- and long-term strength and stability of cemented paste backfill. *Mater. Lett.* **2005**, *59*, 3703–3709. [CrossRef]

13. Zhou, K.P.; Gao, R.G.; Gao, F. Particle Flow Characteristics and Transportation Optimization of Superfine Unclassified Backfilling. *Miner* **2017**, *7*, 6. [CrossRef]

14. Ercikdi, B.; Kuekci, G.; Yilmaz, T. Utilization of granulated marble and waste bricks as mineral admixture in cemented paste backfill of sulphide-rich tailings. *Constr. Build. Mater.* **2015**, *93*, 573–583. [CrossRef]

15. Ye, G.; Liu, X.; de Schutter, G.; Poppe, A.M.; Taerwe, L. Influence of limestone powder used as filler in SCC on hydration and microstructure of cement pastes. *Cem. Concr. Res.* **2007**, *29*, 94–102. [CrossRef]

16. Hassani, F.P.; Ouellet, J.; Hossein, M. Strength development in underground high sulphate paste backfill operation. *CIM Bull.* **2001**, *1050*, 57–62.

17. Ouellet, S.; Bussiere, B.; Aubertin, M.; Benzaazoua, M. Characterization of cemented paste backfill pore structure using SE and IA analysis. *Bull. Eng. Geol. Environ.* **2008**, *67*, 139–152. [CrossRef]

18. Davy, C.A.; Adler, P.M. Three-scale analysis of the permeability of a natural shale. *Phys. Rev.* **2017**, *96*, 063116. [CrossRef] [PubMed]

19. Oh, B.H.; Jang, S.Y.; Shin, Y.S. Experimental investigation of the threshold chloride concentration for corrosion initiation in reinforced concrete structures. *Mag. Concr. Res.* **2001**, *55*, 441–504. [CrossRef]

20. Shang, J.L.; Hu, J.H.; Zhou, K.P.; Luo, X.W.; Aliyu, M. Porosity increment and strength degradation of low-porosity sedimentary rocks under different loading conditions. *Int. J. Rock Mech. Min. Sci.* **2015**, *75*, 216–223. [CrossRef]

21. Zhou, K.P.; Li, J.L.; Xu, Y.J.; Zhang, Y.M. Determination of pore structure characteristics of rock based on nuclear magnetic resonance. *J. Central South Univ. (Sci. Technol.)* **2012**, *43*, 4796–4800.

22. Wu, Z.W. High performance concrete-green concrete. *Chin. Concr. Cem. Prod.* **2000**, *2*, 1–4.

23. Zhou, S. Research on pore structure fractal characteristics of cement concrete. *Concrete* **2016**, *315*, 56–58.

24. Liu, G.Q.; Liu, S.X.; Huang, Q.J.; Zhong, Y.L.; Zhong, S.; Qin, X. Metallography and material microstructure quantitative analysis technology. *Chin. J. Stereol. Image Anal.* **2002**, *7*, 248–251.

25. Al-Kheetan, M.J.; Rahman, M.M.; Chamberlain, D.A. Development of hydrophobic concrete by adding dualcrystalline admixture at mixing stage. *Struct. Concr.* **2018**, *19*, 1504–1511. [CrossRef]

Article

The Influence of Thermo-Mechanical Activation of Bentonite on the Mechanical and Durability Performance of Concrete

Safi Ur Rehman [1,*], Muhammad Yaqub [1], Muhammad Noman [2], Babar Ali [1], Muhammad Nasir Ayaz Khan [3], Muhammad Fahad [4], Malik Muneeb Abid [5] and Akhtar Gul [6]

[1] Department of Civil Engineering, University of Engineering and Technology, Taxila 47050, Pakistan; yaqub_structure@yahoo.com (M.Y.); babar.ali@students.uettaxila.edu.pk (B.A.)
[2] Department of Civil Engineering, International Islamic University, Islamabad 44000, Pakistan; muhammad.noman@iiu.edu.pk
[3] Department of Civil Engineering, HITEC University, Taxila 47050, Pakistan; ayaz.khan@scetwah.edu.pk or nasir.ayaz@hitecuni.edu.pk
[4] Department of Civil Engineering, University of Engineering and Technology, Peshawar 25120, Pakistan; fahadkhan@uetpeshawar.edu.pk
[5] Department of Civil Engineering, College of Engineering and Technology, University of Sargodha, Sargodha 40100, Pakistan; muneeb.abid@uos.edu.pk
[6] Department of Civil Engineering, University of Engineering and Technology, Peshawar (Bannu Campus) 28100, Pakistan; akhtarwazir@uetpeshawar.edu.pk
* Correspondence: cesafiurrehman@gmail.com or safi.rehman@students.uettaxila.edu.pk

Received: 27 September 2019; Accepted: 19 November 2019; Published: 17 December 2019

Abstract: Despite presenting a very high global warming toll, Portland cement concrete is the most widely used construction material in the world. The eco-efficiency, economy, and the overall mechanical and durability performances of concrete can be improved by incorporating supplementary cementitious materials (SCMs) as partial substitutions to ordinary Portland cement (OPC). Naturally found bentonite possesses pozzolanic properties and has very low carbon footprint compared to OPC. By applying activation techniques, the reactivity of bentonite can be improved, and its incorporation levels can be maximized. In this study, the influence of mechanical and thermo-mechanical activation of bentonite is investigated on properties of concrete. Bentonite was used for 0%, 10%, 15%, 20%, 25%, 30%, and 35% mass replacements of OPC. Mechanical (compressive strength and split tensile strength) and durability (water absorption, sorptivity coefficient, and acid attack resistance) properties were studied. Results of experimental testing revealed that, concrete containing bentonite showed good mechanical performance, while durability was significantly improved relative to control mix. Application of thermo-mechanical activation can enhance the incorporation levels of bentonite in concrete. At 15% and 25%, bentonite produced optimum results for mechanical and thermo-mechanical activation, respectively. Bentonite inclusion is more beneficial to the durability than the mechanical strength of concrete.

Keywords: bentonite; compressive strength; sorptivity; acid attack resistance; durability of concrete; eco-efficient binder; low-cost binder

---

1. Introduction

In developing countries like Pakistan, the demands for housing and infrastructure development are increasing each day due to rapid urbanization and population growth. Being a major construction material, the demand for cement has increased abruptly in the recent past to fulfill the infrastructure needs. The production of cement exceeded 4.1 billion tons in 2018 alone, which makes it the most used material after water [1]. Large scale, global, ordinary Portland cement (OPC) production has always been a big threat to the environment, resources, and economy. Nowadays, cement manufacturing accounts for more than 5% of worldwide CO_2 emissions [2]. Moreover, the cost of cement has increased almost 150% in the short span of 10 years [3].

The use of supplementary cementitious materials (SCMs) as replacements of OPC is of great interest for limiting the environmental impact of construction industry [4,5]. It is reported that CO_2 emissions can be minimized by replacing OPC with SCMs [6–9]. SCMs, which do not require any additional clinker processing, can significantly reduce the CO_2 emissions per ton of cementitious materials [4]. The use of natural pozzolans in binder or concrete can result in various beneficial properties, such as low hydration heat, improved later-strength, low permeability, high sulfate-attack resistance, and low alkali-silica reactivity [10,11]. It is estimated that concrete works in a building cost about 31% of the total, and OPC is the main cost-controlling constituent of concrete [12]. SCMs can be utilized for low-cost concrete construction as well [3,13].

Different types of SCMs such as fly ash, bentonite, metakaolin, silica fume, calcined shale, etc., have been used to improve the performance of cement-based composites [14–16]. Till now, SCMs have been extensively used (60%) in ready-mixed concrete owing to their pozzolanic nature [17]. SCMs give comparable results in terms of mechanical performance and show outstanding performances in terms of durability [10,18]. By blending SCMs with OPC, the penetration of harmful chemicals into concrete can be controlled. After the hydration of OPC, a primary by-product, calcium hydroxide (CH), is produced, which is the main target of corrosive media to deteriorate the concrete [19,20]. Supplementary pozzolanic material consumes the CH during the pozzolanic reaction and forms additional calcium silicate hydrate (C-S-H) gel, leading to a durable concrete [21–25]. Pozzolanic cement blends show resistance to thermal cracking due to its low heats of hydration. Moreover, these blended cements have improved ultimate compressive strengths and low permeability because of pore refinement [20,26,27].

In certain locations, natural SCMs or pozzolans are abundant and widely used in countries such as Italy, Germany, Greece, and China as supplements to OPC [28]. Natural SCMs have a wider range of composition and larger variability in physical characteristics compared to traditional or industrial SCMs. The applications of natural SCMs are limited and controlled by their local availability and competition with the industrially produced SCMs (by-products). In future, the use of natural SCMs will expand more because of their extensive deposits all over the world and proven technical advantages [29]. Clay minerals are hydrous aluminum phyllosilicates, composed of repeated two dimensional, tetrahedral (T) and octahedral (O), sheets in different ratios [30]. The two sheets are hold together by an exchangeable interlayer cation coordinated to H_2O molecules in the interlayer region.

When T-sheets and O-sheets combine 2:1, the smectite group of clay minerals is formed. Bentonite clay (montmorillonite mineral) belongs to this group [31]. Bentonite is a natural pozzolana found in different regions of the world. The word bentonite was first proposed by Knight in 1898 [32]. Bentonite is also found in different areas of Pakistan, including the province Khyber Pakhtunkhwa [13]. Bentonite is basically an impure clay mainly consisting of montmorillonite mineral. Bentonite clay fulfills all the requirements of ASTMC-618C [33] to be used as a cement replacement material, as shown in Tables 1 and 2. Moreover, by heating this clay, its pozzolanic reactivity can be enhanced further compared to unheated condition [31,34]. By using bentonite clay as a supplement to OPC, an eco-friendly, economical, and durable concrete can be achieved [3,13].

Pozzolana can be activated by different methods; i.e., mechanical, thermal, and chemical activation. Mechanical activation means to ground the SCMs into powder form to increase their fineness and surface areas. Previous studies found that the pozzolanic reactivity of SCMs can be increased by mechanical activation (grinding) [35,36]. As in a pozzolanic reaction, the silica (SiO_2) part of pozzolan reacts with CH in the presence of moisture and forms CSH gel. It was reported that the pozzolanic reaction is determined by rate of dissolution of SiO_2 in water, and the dissolution rate of SiO_2 is proportional to its surface area [37,38]. Despite increasing surface area prolonging the grinding of a material, it also increases the number of active centers at the edges of the material structure which are more energetic than the normal structure; the pozzolanic reactivity of the material depends on the number of active centers [39,40]. Thermal activation means heat treatment (calcination). It can be heating of the pozzolan or the high temperature curing of the specimens made with pozzolans. It was reported that thermal treat of clays produce a metastable state due to structural disorder and an amorphous reactive structure is formed [41,42]. Different clays have different calcination effects on their pozzolanic reactivities. Mielenz et al. [43] studied thermal treatment effect on the pozzolanic reactivity of 70 different types of pozzolana and reported the optimum activation temperature range of 650–870 °C. In previous studies, the optimum activation temperature for Bentonite was reported as the range of 800–830 °C [28,31]. In our previous research [34], the optimum activation temperature of the bentonite clay was found to be 800 °C.

Table 1. Physical properties of ordinary Portland cement (OPC) and bentonite clay.

Physical Properties	OPC	Bentonite	ASTM C618 Class N Requirements (%)
% Retained #325 mesh	-	11.4	34 maximum
Blaine fineness ($cm^2\ g^{-1}$)	3152	2571	-
Specific gravity ($g\ cm^{-3}$)	3.05	2.81	-
Average Particle size	20 μm	4 to 5 μm	-
Strength Activity Index (%)	-	-	-
7 days	-	84.4 [34]	75 minimum
28 days	-	85.3 [34]	75 minimum

Table 2. Chemical composition of OPC and bentonite clay.

Chemical Composition (wt.%)	OPC	Bentonite	ASTM C618 Class N Requirements (%)
Na_2O	-	1.39	5 maximum
MgO	2.31	2.70	-
Al_2O_3	9.78	18.32	-
SiO_2	18.8	56.6	-
K_2O	-	0.67	-
CaO	59	3.1	-
TiO_2	-	0.98	-
Fe_2O_3	3.44	6.1	-
SO_3	2.85	-	-
$(SiO_2) + (Al_2O_3) + (Fe_2O_3)$	-	80.93	70 minimum
Loss on ignition (LOI)	-	7.1	10 maximum

Many studies have been conducted in Pakistan for utilizing industrial wastes, such as bagasse ash [44], rice husk ash [45], fly ash [46,47], etc., as SCMs. Some natural pozzolana have also been studied [3,13]. In the existing literature, very few studies have been conducted to explore the potential of bentonite clay as partial replacement of cement. A study of bentonite on the mechanical and durability properties of concrete needs to be done in more detail. Memon et al., [13] studied sawabi bentonite clay (3%, 6%, 9%, 12%, 15%, 18%, and 21% by mass) as an OPC replacement and reported positive results of bentonite on the fresh and mechanical properties of concrete. Also, further study was recommended [13] to explore the mechanical and thermal activation of bentonite clay and to study its effects on the properties of concrete. Moreover, high replacement levels of bentonite clay need to be explored. This study explored the mechanical and thermo-mechanical activation of bentonite clay to enhance its incorporation level in concrete. The performance of concrete was studied, both for mechanically activated bentonite (MAB) and thermo-mechanically activated bentonite (TMAB) (at the levels of 10%, 15%, 20%, 25%, 30%, and 35% by mass of OPC), in terms of compressive strength, split tensile strength, water absorption, sorptivity, and acid attack resistance up to 180 days. The results concluded that 15% mechanically-activated and 25% thermo-mechanically activated bentonite can be used (significantly) as substitutes of OPC, with good mechanical and even better durability properties for concrete.

2. Materials and Methods

2.1. Materials

2.1.1. Binder

Ordinary Portland cement (OPC) Type-I was used in the research as a binder. Its physical properties and chemical composition are given in the Tables 1 and 2, respectively.

Bentonite clay taken from district Sawabi, KPK, Pakistan, was used in this research. The clay collected from the natural source was in crude form and could not be used directly in concrete, so after collection, the clay was transferred to ball mill machine to obtain the passing size of number 200 sieve with a nominal sieve opening of 0.074 mm and packed in polythene bags protected from light and moisture. Physical properties and the chemical composition of bentonite clay are illustrated in Tables 1 and 2, respectively. The Bentonite clay fulfils the requirements of ASTM C618-12 [33] to be used as natural pozzolana.

2.1.2. Aggregates

Locally available Lawrencepur sand was used as a fine aggregate. Its properties are given in Table 3. Its particle size distribution as per ASTM C136-06 [48] is shown in Figure 1a. Margalla crush was used as coarse aggregate. Its properties are also given in Table 3. Its particle size distribution as per ASTM C136-06 is shown in Figure 1b.

Table 3. General properties of aggregates.

Property	Fine Aggregate	Coarse Aggregate
Maximum nominal size (mm)	4.75	22.5
Minimum nominal size (mm)	0.075	4.75
Saturated surface dry water absorption (%)	1.40	1.12
Fineness modulus	2.45	-
Abrasion value (%)	-	24.56

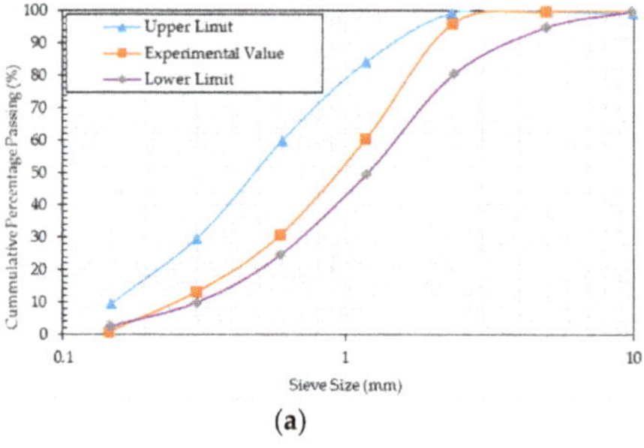
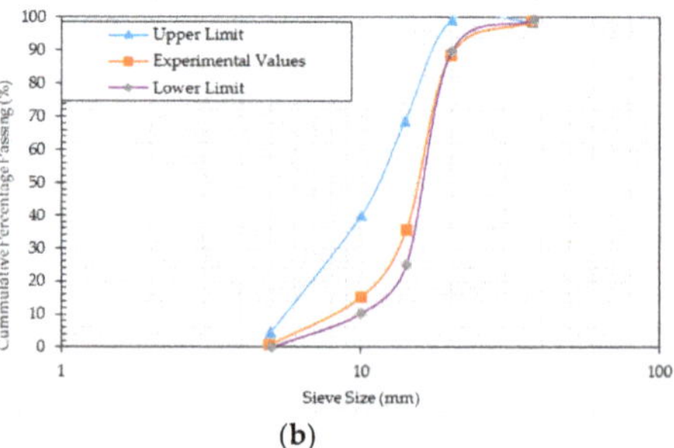

Figure 1. Particle size distribution of the (a) fine aggregate and (b) coarse aggregate.

2.2. Activation of Bentonite

2.2.1. Mechanical Activation (Grinding)

The clay collected from the natural source was first oven-dried for 24 h at 100 °C. After drying, the clay was ground in Loss Angeles abrasion machine. A total of 4500 revolutions were performed on each 5 kg batch of the clay to maintain uniform thickness; the powdered clay was then passed through sieve number 200 with a nominal sieve opening of 0.074 mm and packed in polythene bags to protect it from moisture.

2.2.2. Thermal Activation (Heating)

After mechanical grinding, bentonite clay was heated at 800 °C for 3 h. For control and uniform heating, the temperature was allowed to reach the target limit before placing the 5 kg clay batch in the furnace. Then the clay was put in the furnace for the required duration. The furnace took 24 h to reach room temperature. In the meantime, the sample remained in the furnace. The clay, after reaching room temperature, was packed into PVC bags in order to protect it from moisture.

2.3. Composition of Concrete Mixes

A total of 13 concrete mixes were produced in this study. Mechanically activated bentonite (MAB) and thermo-mechanically activated bentonite (TMAB) were used as 0, 10, 15, 20, 25, 30, and 35% by mass replacements of OPC. A constant water-cementitious material ratio of 0.5 was chosen for all mixes. The nomenclature and the composition of each mixture are presented in Table 4.

Table 4. Nomenclature and the compositions of mixtures.

Mix ID	Type of Activation	Bentonite by Weight of Cement (%)	Cement (kg/m^3)	Bentonite (kg/m^3)	Water (kg/m^3)	Fine Aggregate (kg/m^3)	Coarse Aggregate (kg/m^3)
B0 (CON)	Control	0	425	0	213	635	1270
B10/G	MAB	10	382.5	42.5	213	635	1270
B15/G	MAB	15	361.25	63.75	213	635	1270
B20/G	MAB	20	340	85	213	635	1270
B25/G	MAB	25	318.75	106.25	213	635	1270
B30/G	MAB	30	297.5	127.5	213	635	1270
B35/G	MAB	35	276.25	148.75	213	635	1270
B10/G/T	TMAB	10	382.5	42.5	213	635	1270
B15/G/T	TMAB	15	361.25	63.75	213	635	1270
B20/G/T	TMAB	20	340	85	213	635	1270
B25/G/T	TMAB	25	318.75	106.25	213	635	1270
B30/G/T	TMAB	30	297.5	127.5	213	635	1270
B35/G/T	TMAB	35	276.25	148.75	213	635	1270

Where "B" denotes "bentonite"; "G" denotes "grinding"; "T" denotes "thermal"; "MAB" denotes "mechanically activated bentonite"; and "TMAB" denotes "thermo-mechanically activated bentonite".

Mixing of all concrete mixes was done in a mechanical mixer of 0.15 m^3 capacity at the speed of 35 revolutions per minute. First, all solid ingredients were mixed for about 4 min. Subsequently, water was added, and mixing of concrete continued for the next 4 min. The total duration of mixing was kept to 8 min for all mixes.

2.4. Specimen Preparation and Testing

To assess the performance of concrete modified with bentonite clay, the following methodology was adopted. To evaluate mechanical performance compression and split tensile tests were performed. For compression testing, cylinders of 150 mm diameter and 300 mm height were cast as per ASTM C39 [49] for three different curing durations; i.e., 7, 28, and 180 days. For split tensile testing, cylinders of 150 mm diameter and 300 mm height were cast as per ASTM C496 [50] for curing durations of 7, 28, and 180 days. For each test 3 specimens were cast and tested. The cylinders were demolded after 24 h of casting. After demolding, all the cylinders were cured in water for the required durations.

For durability assessment, water absorption, sorptivity coefficient, and H_2SO_4-acid attack resistance were determined. For sorptivity 50 mm thick and 100 mm diameter discs were prepared and tested as per ASTM C1585 [51]. Similarly, for water absorption, 50 mm thick and 100 mm diameter discs were tested as per ASTM C642 [52]. To evaluate acid attack resistance, for each mix a cubic specimen of 100 mm was cast and cured in normal water for 14 days to gain some strength. After 14 days each specimen was kept in 4% sulfuric acid (H_2SO_4) solution for 7, 28, 56, and 90 days. To maintain 4% concentration, the acid solution was changed every week. The acid attack resistance was measured in terms of mass loss (%).

3. Results and Discussion

3.1. Mechanical Properties

3.1.1. Compressive Strength

Compressive strength is an important parameter of cement-based materials. Many durability and strength parameters of concrete can be predicted accurately using its compressive strength [53,54]. The compressive strength of each mix at 7, 28, and 180-days is presented in Figure 2. The general trend in results indicates that an increasing level of mechanically activated bentonite (MAB) reduces the compressive strength. The use of 10% MAB gives optimum results after 28 and 180-days compared to the control mix (CON). Compressive strength at all ages is badly affected by increasing levels of MAB. After 180 days, mixes with 10%–20% MAB gave better compressive strength than the strength of the control mix at day 28. The improvements in the compressive strength of concrete with the inclusion of 10% MAB can be ascribed to (1) its smaller particle size than OPC, which improves the particle size distribution in the binder matrix and (2) the reaction of bentonite particles with free CH (a primary hydration product) to produce secondary CSH-gels [21]. The reduction in compressive strength with an increasing incorporation level of MAB can be ascribed to reduction in overall CH content in the binder owing to lower calcium oxide percentage of bentonite which reduces the chance of MAB consumption in pozzolanic reaction. Naturally, bentonite is found in consolidated form; therefore, without a suitable treatment/activation method, it cannot be used directly in "as is" condition as a supplement for OPC. Memon et al., [13] reported that an MAB replacement of up to 21% does not affect the compressive strength of concrete compared to the control mix.

Concrete mixes with thermo-mechanically activated bentonite (TMAB) showed higher compressive strength than MAB mixes for a given age and replacement level. After 28-days, 10% incorporation of MAB and TMAB showed 1% and 4.3% higher compressive strengths than the control (CON), respectively, whereas, after 180-days, incorporation of 10% MAB and TMAB resulted in 3.5% and 9.8% higher compressive strengths than CON, respectively. Using the thermo-mechanical activation process, the replacement level of bentonite can be enhanced from 10% to 20% to produce a higher compressive strength than CON at 28-days. Clays may consist of some fixed crystalline structures and many of them possess pozzolanic properties. Calcination of a clay mineral alters its crystalline structure into an amorphous one and increases its pozzolanic reactivity considerably [55]. The heating of clay, shale or other mineral changes their structures to alumina or quasi-amorphous silica, which have good pozzolanic reactivities with CH [56].

Strength activity indices (SAIs) of all mixes at days 7 and 28 are displayed in Figure 3. ASTM C618 [33] states that for any material to be considered pozzolanic, the SAI should be at least 75% of the control blend at both day 7 and day 28. From the SAI values, it can be seen that MAB can be used up to 15% with a SAI higher than 75%. Whereas for TMAB up to a 25% level, SAI is well above 75% for both days 7 and 28. It can be concluded that thermo-mechanical activation produces mixes with more compressive strength than mechanical activation, which can help to maximize the level of bentonite in concrete production.

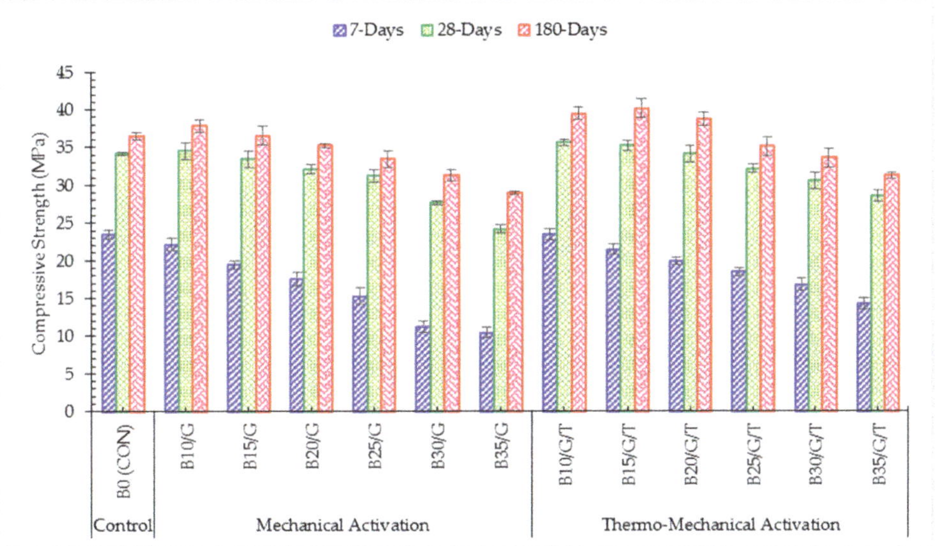

Figure 2. Compressive strength of each mix at 7, 28 and 180-days e clay.

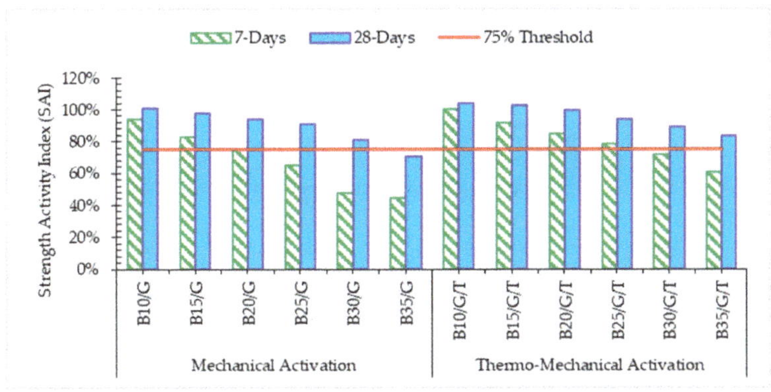

Figure 3. Strength activity indices of mixes after 7 and 28 days.

A relative analysis of compressive strength is also illustrated in the Figure 4. Compressive strength of the control mix (B0 (CON)) at day 28 was taken as the reference. It can be observed from Figure 4, that the day 28 compressive strength of concrete with 10%–20% MAB almost overlaps with the compressive strength of the reference mix and a notable decrease is noted for 20%–35% MAB. Similarly, for day 28's compressive strength of concrete modified with 10%–25% TMAB, it overlaps or nearly coincides with the compressive strength of the reference mix and a decline was observed for 25%–35% MAB. At day 180, mixes with 10%–20% MAB and 10%–25% MAB show significantly higher compressive strength than the reference.

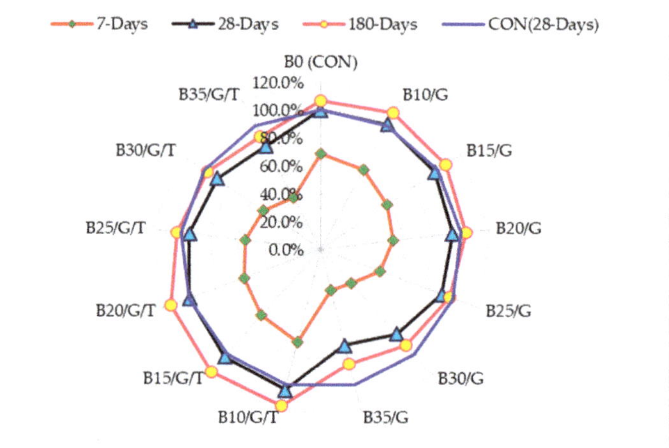

Figure 4. Compressive strength of each mix relative to the day 28 strength of reference mix (CON).

3.1.2. Split Tensile Strength

Results of split tensile testing are shown in Figure 5. At day 7, the tensile strengths of all MAB modified mixes were lower than that of the CON. Similarly to the compressive strength results, MAB incorporation of up to 10% produced maximal compressive strength after 28 and 180 days compared to CON. It is also worth mentioning here, that split tensile strength is more negatively influenced than compressive strength with increasing the incorporation of bentonite. This might be due to different natures of failure for the specimens under compression and split tensile tests. At higher levels of MAB, the filling effect of unreacted bentonite particles can play a good role in resisting compressive stresses, whereas, under splitting tension unreacted particles of bentonite may offer low resistance to failure (two failure surfaces can easily separate due to low cohesion in the binder); therefore, high levels of bentonite are more detrimental to tensile strength than compressive strength.

TMAB showed significant improvement over MAB, due to increased reactivity of clay minerals in pozzolanic reaction. Split tensile strength was reduced by 12% at a 35% level of TMAB after 28 days, whereas after 180 days, all TMAB incorporated mixes exhibited comparable or higher tensile strengths than the CON mix. MAB showed 3% and TMAB 5% higher split tensile strength than CON mix at day 180. But at higher levels, TMAB mixes performed 10% better than MAB. TMAB may have more reactive alumina and silica to react with CH, so have large pozzolanic reactivity [57,58]. Therefore, higher levels of TMAB show more strength development than those of the MAB.

Relative analysis of the split tensile strength results is presented in Figure 6. 28 days' strength of the control mix (B0 (CON)) was taken as the benchmark. At day 28, there is no notable surpassing form the loop of reference mix/benchmark. For mixes modified with MAB (10%–15%), the replacement overlapped the reference loop. For mixes modified with TMAB (at 10%–25% level), they coincided with the benchmark. At day 180, mixes with MAB (10%–15%) outperformed the reference mix, while all the mixes with TMAB, except B35/G/T, beat the reference mix and crossed the loop. Overall no significant difference was evident for the split tensile strength of TMAB incorporated concrete.

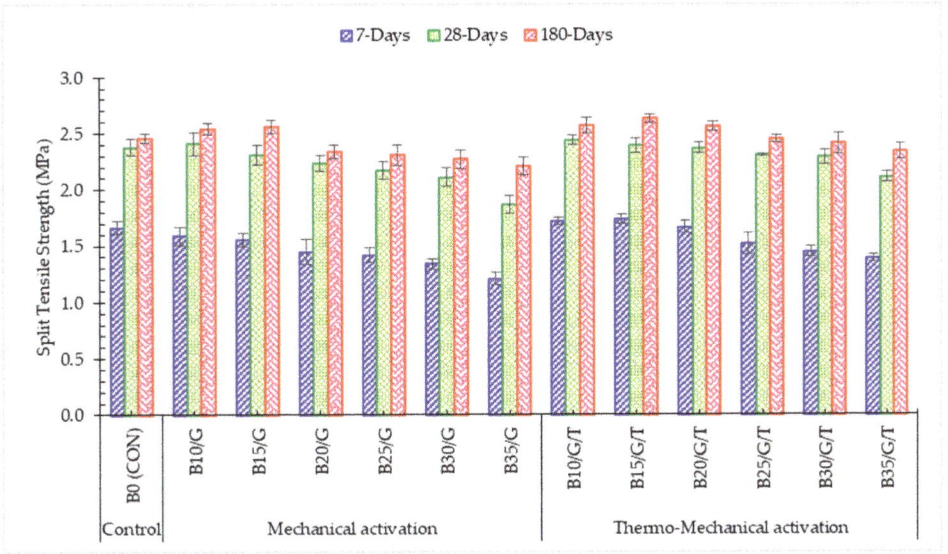

Figure 5. Split tensile strength of each mix after 7, 28, and 180 days.

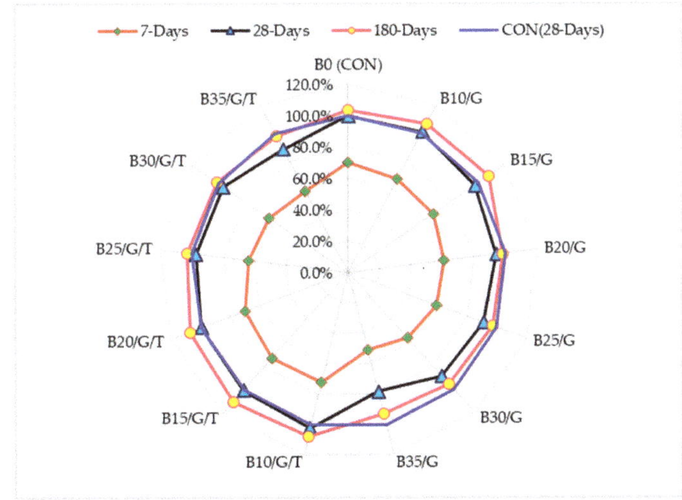

Figure 6. Split tensile strength of each mix relative to day 28's strength of reference mix (CON).

As, already mentioned, variation in compressive strength and split tensile strength with increasing incorporation level of bentonite follows a similar trend. Therefore, a strong relationship exists between compressive strength and split tensile strength. In Figure 7, the compressive strength of each mix at a given age and replacement level of bentonite is plotted against the corresponding value of split tensile strength. Equation (1) can be used to predict the cylindrical split tensile strength of concrete mix with a coefficient of determination (R^2) higher than 0.9.

$$f_{ts} = 0.40 \times \sqrt{f_c} \left(\text{for 7, 28, and 180 days, } R^2 = 0.92 \right), \tag{1}$$

where f_{ts} (MPa) is split tensile strength and f_c (MPa) is compressive strength of cylindrical specimen.

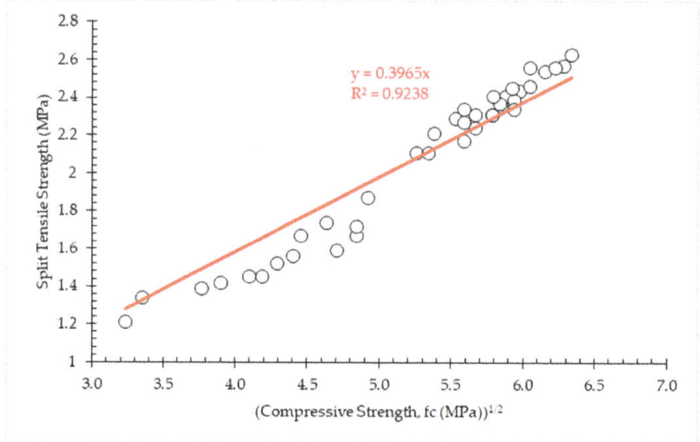

Figure 7. Correlation between experimental values of compressive strength and split tensile strength.

3.2. Durability Properties

3.2.1. Water Absorption (WA)

WA is the measure of water accessible porosity of a material. WA is the indirect measurement durability of concrete, since water can carry harmful chemicals into concrete which may react with its constituents and result in changes in the properties of the material. WA tests were conducted on each mix after 28 and 180 days, and the results are given in Figure 8.

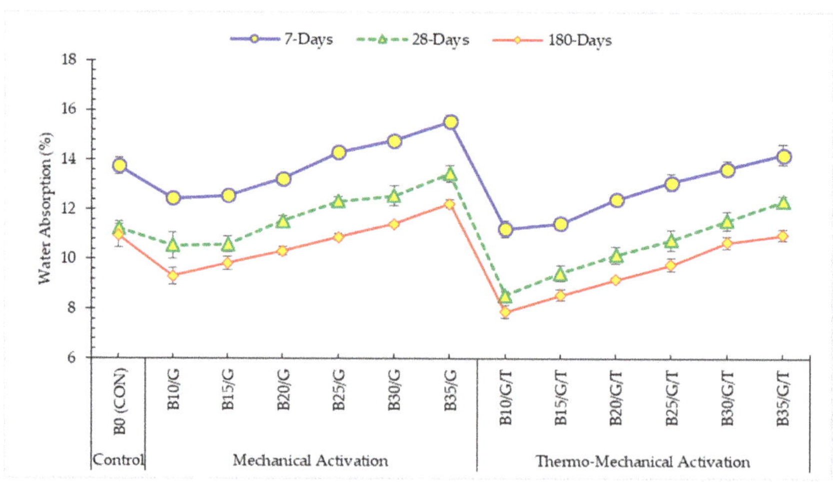

Figure 8. Water absorption test results after 7, 28, and 180 days.

A general trend indicates that the addition of 10%–15% MAB reduces the WA of concrete after 28 days, compared to CON. After 180 days, 10%–20% MAB reduces the WA of concrete compared to CON. This behavior can be ascribed to an improvement in particle size distribution which results in a reduction in effective pore volume of concrete. Reduction in pore volume results in reduction in WA at all ages, whereas chemical reaction between bentonite particles and free CH can decrease the length of interconnected pores. Studies [59,60] have reported that filler effect of pozzolans (fly ash, GGBS, metakaolin, etc.) decreases the permeability of concrete, despite reducing the early age mechanical strength. After 180 days, the mix with 25% MAB shows WA comparable to that of the CON.

After 28 days, mixes with TMAB up to 25%, show higher resistance to WA compared to CON. After 180 days, a mix with any level of TMAB incorporated shows higher WA resistance than CON. This can be ascribed to improved reactivity of bentonite particles and better distribution of particles in the binder matrix. It was noticed that for a given age and replacement level of bentonite, TMAB-added mixes outperformed MAB by a notable margin. For example, at a 10% replacement of bentonite after 28 days, the MAB-added mix showed a 7% lower WA value compared to Con, whereas TMAB-added mix's value was 24% lower.

The experimental relationship between WA and compressive strength values is presented in Figure 9. Generally, compressive strength value increases with decreasing WA. An inverse-linear relationship is formed is between two parameters, as given by Equation (2). The proposed relationship has good predictability, showing an R^2 of 0.80.

$$\text{WA } (\%) = 18.45 \times e^{-0.017 \times f_c} \left(\text{for 7, 28, and 180 days, } R^2 = 0.80\right), \tag{2}$$

where WA is water absorption in percentage, and f_c is cylindrical compressive strength of concrete in MPa.

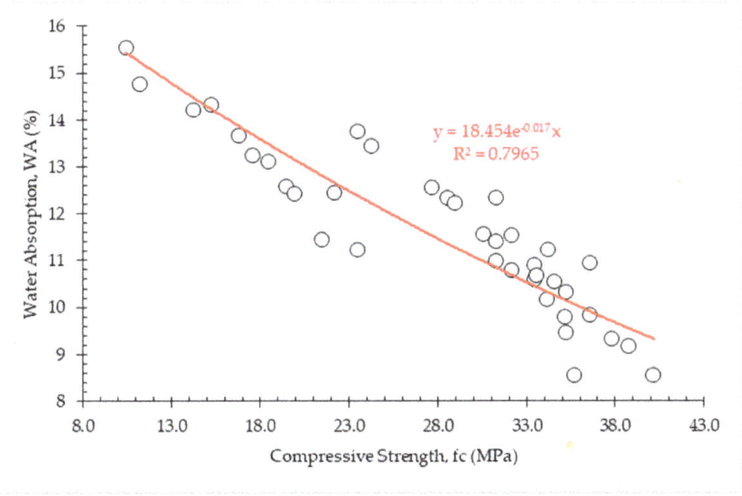

Figure 9. Relationship between experimental values of compressive strength (f_c) and water absorption.

3.2.2. Sorptivity Coefficient (SC)

SC can be defined as the absorption time rate of water by hydraulic cement concrete. It is one of the main properties of concrete for assessing its durability [61]. SC values are given in Figure 10. A similar trend like WA was observed for the results of SC with the varying incorporation levels of MAB and TMAB. For the same age and incorporation level of bentonite, TMAB mixes showed more resistance to sorptivity than MAB-added mixes. Since both WA and SC are functions of porosity and microstructural development of concrete, no significant difference in the results of SC and WA was observed with varying dosages of bentonite. After 180 days, almost all bentonite-added mixes showed higher resistance to sorptivity than CON, whereas after 28 days, up to 25% replacement of bentonite (either MAB or TMAB) with OPC lowered the sorptivity compared to the CON mix.

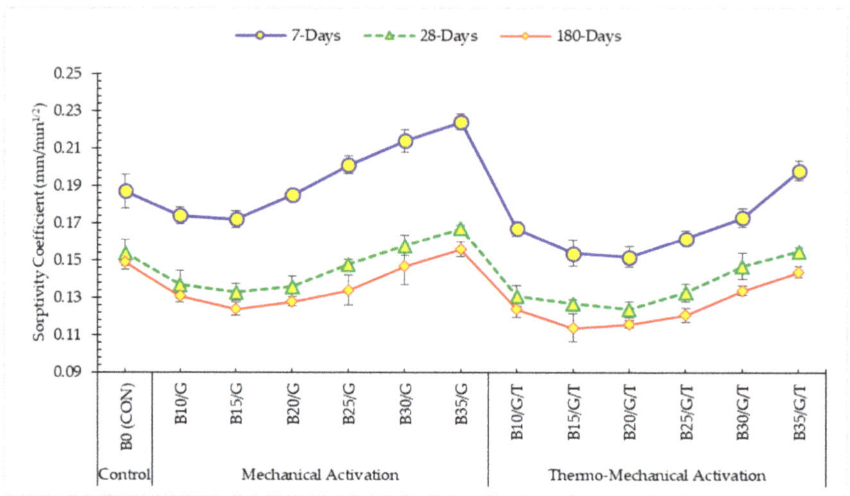

Figure 10. Sorptivity coefficient (SC) test results at days 7, 28, and 180.

Similarly to WA, SC has a good relationship with compressive strength; see Figure 11. Both SC and compressive strength are inversely related to each other in an exponential equation; see Equation (3). Since microstructural development and density due to bentonite inclusion leads to reduction in volume and connectivity of pores, these developments can also improve the compressive strength of concrete. Therefore, permeability-related durability and strength parameters are inversely related to each other.

$$SC\left(mm/min^{0.5}\right) = 0.255 \times e^{-0.019 \times f_c}\ \left(\text{for 7, 28, and 180 days, } R^2 = 0.85\right), \tag{3}$$

where SC is sorptivity coefficient and f_c (MPa) is cylindrical compressive strength of concrete.

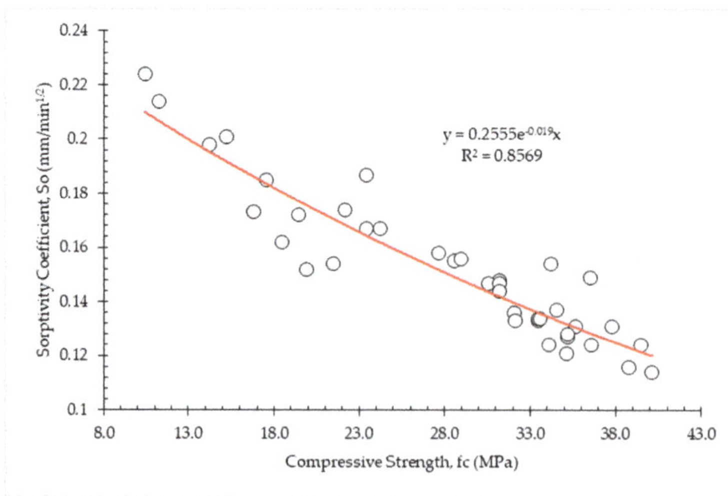

Figure 11. Relationship between experimental values of compressive strength (f_c) sorptivity coefficient (SC).

3.2.3. Acid Attack Resistance (H_2SO_4 Attack)

Acids have a strong tendency to react with the CHs of concrete in order to produce a calcium salt and water. There are many aggressive acids, such as acetic acids, nitric acids, hydrochloric acids, and sulfuric acids (H_2SO_4). Concrete structures handling the acidic wastewaters of different industries are extremely vulnerable to deterioration. Therefore, it is very necessary to decrease the overall CH content of concrete. In this research, H_2SO_4 was used for acid attacks on concrete mixes with various levels of MAB and TMAB. H_2SO_4 reacts with CH and forms calcium sulfate which causes deterioration of the microstructure (due to increased internal pressure) and leads to loss in mass and mechanical strength of concrete. The results of acid attack are presented in terms of loss in mass of the specimens after 7, 28, 56, and 90 days for each mix, see Figure 12.

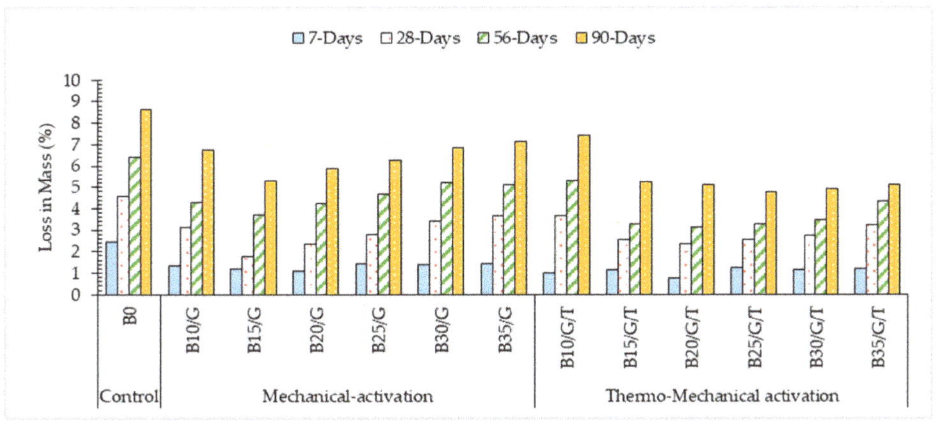

Figure 12. Loss in mass of bentonite-added mixes due to H_2SO_4 attack.

The general trend in results indicates that the incorporation of either MAB or TMAB results in significant improvement in the H_2SO_4 attack resistance of concrete. Increasing the incorporation level of bentonite not only consumes the CH in a pozzolanic reaction, but results in an overall reduction of the net CH content of binder [21,22]. OPC contains a very high content of CaO, whereas bentonite is rich in SiO_2 and Al_2O_3, therefore, increasing MAB/TMAB level results in the reduction of CH (a primary hydration product of cement). It was also noticed that for a given age and incorporation level, TMAB-added mixes were more resistant to acid attack than MAB-added mixes. Not only does bentonite reduce the CH content of concrete, but it reduces the pore volume. Reduction in pore volume can also control the penetration of SO_4^{-2} ions from the solution into the microstructure that may cause the deterioration of concrete.

4. Conclusions

In this research, the influences on mechanical strength and durability of concrete, of various levels (0%, 10%, 15%, 20%, 25%, 30%, and 35% mass replacements of OPC) of mechanically activated (MAB) and thermo-mechanically activated (TMAB) bentonite clays, were investigated. The following conclusions can be drawn from the experimental results:

Compressive strength is badly affected when bentonite incorporation is increased beyond 25%. The maximum incorporation level of MAB is 10% to produce optimum performance after 28 and 180 days compared to the control mix and 20% MAB produces a comparable performance at 180 days. Contrastingly, maximum incorporation level of TMAB is 20%, in order to produce optimum performance after 28 and 180 days, and 25% to produce comparable results after 180 days. Split tensile strength varies similarly to compressive strength with varying incorporation levels of bentonite. MAB/TMAB added mixes showed more strength development between 28 and 180 days than the control mix.

In general, more positive influence of bentonite incorporation was noticed on durability properties; i.e., WA, SC, and H_2SO_4 attack resistance. Due to improved reactivity, TMAB mixes show notably improved durability properties over MAB mixes for a given level of bentonite. With an up to 20% level of bentonite, both MAB and TMAB mixes indicate high durability in terms of SC and WA after 28 and 180 days. All mixes incorporating MAB/TMAB, showed very high resistance to acid attack due to reductions in the overall CH contents of the concretes by both pozzolanic reactions and net reductions in CaO contents of binders.

Author Contributions: Conceptualization, S.U.R. and M.Y.; methodology, S.U.R. and B.A.; formal analysis, M.N. and M.M.A.; investigation, M.N.A.K.; resources, A.G., M.N.A.K. and M.F.; writing—original draft preparation, S.U.R.; writing—review and editing, M.Y., B.A. and M.N.; supervision, M.Y.

Funding: The authors received no specific funding for this work.

Conflicts of Interest: The authors have no conflict of interest to declare.

References

1. Koushkbaghi, M.; Kazemi, M.J.; Mosavi, H.; Mohsenid, E. Acid resistance and durability properties of steel fiber-reinforced concrete incorporating rice husk ash and recycled aggregate. *Constr. Build. Mater.* **2019**, *202*, 266–275. [CrossRef]

2. Oh, D.-Y.; Noguchi, T.; Kitagaki, R.; Park, W.J. CO_2 emission reduction by reuse of building material waste in the Japanese cement industry. *Renew. Sustain. Energy Rev.* **2014**, *38*, 796–810. [CrossRef]

3. Mirza, J.; Riaz, M.; Naseer, A.; Rehman, F.; Khan, A.N.; Ali, Q. Pakistani bentonite in mortars and concrete as low cost construction material. *Appl. Clay Sci.* **2009**, *45*, 220–226. [CrossRef]

4. Lothenbach, B.; Scrivener, K.; Hooton, R.D. Supplementary cementitious materials. *Cem. Concr. Res.* **2011**, *41*, 1244–1256. [CrossRef]

5. Juenger, M.; Provis, J.L.; Elsen, J.; Matthes, W.; Hooton, R.D.; Duchesne, J.; Courard, L.; He, H.; Michel, F.; Snellings, R.; et al. Supplementary cementitious materials for concrete: Characterization needs. *MRS Online Proc. Libr. Arch.* **2012**, *1488*. [CrossRef]

6. Mehta, P.K. Greening of the concrete industry for sustainable development. *Concr. Int.* **2002**, *24*, 23–28.

7. Khotbehsara, M.M.; Mohseni, E.; Yazdi, M.A.; Sarker, P.; Ranjbar, M.M. Effect of nano-CuO and fly ash on the properties of self-compacting mortar. *Constr. Build. Mater.* **2015**, *94*, 758–766. [CrossRef]

8. Mohseni, E.; Ranjbar, M.M.; Yazdi, M.A.; Hosseiny, S.S.; Roshandel, E. The effects of silicon dioxide, iron (III) oxide and copper oxide nanomaterials on the properties of self-compacting mortar containing fly ash. *Mag. Concr. Res.* **2015**, *67*, 1112–1124. [CrossRef]

9. Kurad, R.; Silvestre, J.D.; de Brito, J.; Ahmed, H. Effect of incorporation of high volume of recycled concrete aggregates and fly ash on the strength and global warming potential of concrete. *J. Clean. Prod.* **2017**, *166*, 485–502. [CrossRef]

10. Malhotra, V.M.; Mehta, P.K. *Pozzolanic and Cementitious Materials*; CRC Press: Boca Raton, FL, USA, 2014. [CrossRef]

11. Sabir, B.B.; Wild, S.; Bai, J. Metakaolin and calcined clays as pozzolans for concrete: A review. *Cem. Concr. Compos.* **2001**, *23*, 441–454. [CrossRef]

12. Adu-Boateng, A.O.; Bediako, M. The use of clay as pozzolana for building purposes in Ghana. *Build. Integr. Solut.* **2006**, 1–10. [CrossRef]

13. Memon, S.A.; Arsalan, R.; Khan, S.; Lo, T.Y. Utilization of Pakistani bentonite as partial replacement of cement in concrete. *Constr. Build. Mater.* **2012**, *30*, 237–242. [CrossRef]

14. Bakharev, T. Thermal behaviour of geopolymers prepared using class F fly ash and elevated temperature curing. *Cem. Concr. Res.* **2006**, *36*, 1134–1147. [CrossRef]

15. Mwiti, M.J.; Thiong'o, J.K.; Muthengia, W.J. *Thermal Resistivity of Chemically Activated Calcined Clays-Based Cements, in Calcined Clays for Sustainable Concrete*; Springer: Dordrecht, The Netherlands, 2018; pp. 327–333. [CrossRef]

16. Martirena, F.; Favier, A.; Scrivener, K. Calcined clays for sustainable concrete. In Proceedings of the 2nd International Conference on Calcined Clays for Sustainable Concrete, La Havana, Cuba, 5–7 December 2017; Springer: Heidelberg ,Germany, 2017; Volume 16. [CrossRef]

17. PCA (Portland Cement Association). *Survey of Mineral Admixtures and Blended Cements in Ready Mixed Concrete*; Portland Cement Association: Skokie, IL, USA, 2000.

18. Siddique, R. *Waste Materials and by-Products in Concrete*; Springer Science & Business Media: Berlin/Heidelberg, Germany, 2007.

19. Bai, J.; Wild, S.; Sabir, B.B. Chloride ingress and strength loss in concrete with different PC–PFA–MK binder compositions exposed to synthetic seawater. *Cem. Concr. Res.* **2003**, *33*, 353–362. [CrossRef]

20. Khan, M.U.; Ahmad, S.; Al-Gahtani, H.J. Chloride-induced corrosion of steel in concrete: an overview on chloride diffusion and prediction of corrosion initiation time. *Int. J. Corros.* **2017**, *2017*. [CrossRef]

21. Sarfo-Ansah, J.; Atiemo, E.; Boakye, K.A.; Adjei, D.; Adjaottor, A.A. Calcined clay Pozzolan as an admixture to mitigate the alkali-silica reaction in concrete. *J. Mater. Sci. Chem. Eng.* **2014**, *2*, 20. [CrossRef]

22. Pierkes, R.; Schulze, S.E.; Rickert, J. Durability of concretes made with calcined clay composite cements. In *Calcined Clays for Sustainable Concrete*; Springer: Dordrecht, The Netherlands, 2018; Volume 16, pp. 366–371. [CrossRef]

23. Barış, K.E.; Tanaçan, L. Durability of steam cured pozzolanic mortars at atmospheric pressure. In *Calcined Clays for Sustainable Concrete*; Springer: Dordrecht, The Netherlands, 2018; Volume 16, pp. 46–53. [CrossRef]

24. Díaz, E.; González, R.; Rocha, D.; Alujas, A.; Martirena, F. Carbonation of concrete with low carbon cement LC3 exposed to different environmental conditions. In *Calcined Clays for Sustainable Concrete*; Springer: Dordrecht, The Netherlands, 2018; Volume 16, pp. 141–146. [CrossRef]

25. Maraghechi, H.; Avet, F.; Scrivener, K. Chloride transport behavior of LC 3 binders. In *Calcined Clays for Sustainable Concrete*; Springer: Dordrecht, The Netherlands, 2018; Volume 16, pp. 306–309. [CrossRef]

26. Berrocal, C.G.; Lundgren, K.; Löfgren, I. Corrosion of steel bars embedded in fibre reinforced concrete under chloride attack: state of the art. *Cem. Concr. Res.* **2016**, *80*, 69–85. [CrossRef]

27. Amin, N.-U.; Alam, S.; Gul, S. Effect of thermally activated clay on corrosion and chloride resistivity of cement mortar. *J. Clean. Prod.* **2016**, *111*, 155–160. [CrossRef]

28. Snellings, R.; Mertens, G.; Elsen, J. Supplementary cementitious materials. *Rev. Mineral. Geochem.* **2012**, *74*, 211–278. [CrossRef]

29. Thomas, M. *Supplementary Cementing Materials in Concrete*; CRC Press: Boca Raton, FL, USA, 2013. [CrossRef]

30. Brigatti, M.F.; Galan, E.; Theng, B.K.G. Structure and mineralogy of clay minerals. In *Developments in Clay Science*; Elsevier: Amsterdam, The Netherlands, 2013; Volume 5A, pp. 21–81. [CrossRef]

31. Garg, N.; Skibsted, J. Thermal activation of a pure montmorillonite clay and its reactivity in cementitious systems. *J. Phys. Chem. C* **2014**, *118*, 11464–11477. [CrossRef]

32. Ahmad, Z.; Siddiqi, R.A. *Minerals and Rocks for Industry*; Geological Survey of Pakistan Quetta: Quetta, Pakistan, 1995; pp. 202–245.

33. ASTM. *ASTM, C618 Standard Specification for Coal Fly Ash and Raw or Calcined Natural Pozzolan for Use in Concrete*; ASTM: West Conshohocken, PA, USA, 2012. [CrossRef]

34. Rehman, S.U.; Yaqub, M.; Ali, T.; Shahzada, K.; Khan, S.W.; Noman, M. Durability of Mortars Modified with Calcined Montmorillonite Clay. *Civ. Eng. J.* **2019**, *5*. [CrossRef]

35. Vizcayno, C.; De Gutierrez, R.M.; Castello, R.; Rodriguez, E.; Guerrero, C.E. Pozzolan obtained by mechanochemical and thermal treatments of kaolin. *Appl. Clay Sci.* **2010**, *49*, 405–413. [CrossRef]

36. Alexander, K.M. Reactivity of ultrafine powders produced from siliceous rocks. *ACI J. Proc.* **1960**, *57*, 557–570. [CrossRef]

37. Greenberg, S.A. Reaction between silica and calcium hydroxide solutions. I. Kinetics in the temperature range 30 to 85°. *J. Phys. Chem.* **1961**, *65*, 12–16. [CrossRef]

38. O'Connor, T.L.; Greenberg, S.A. The kinetics for the solution of silica in aqueous solutions. *J. Phys. Chem.* **1958**, *62*, 1195–1198. [CrossRef]

39. Dave, N.G. Pozzolanic wastes and their activation to produce improved lime pozzolana mixtures. In Proceedings of the 2nd Australian Conference on Engineering Materials, Sydney, Australia, 6–8 July 1981; pp. 623–638.

40. Gregg, S.J. *The Surface Chemistry of Solids*; Chapman and Hall Ltd.: London, UK, 1961.

41. Fabbri, B.; Gualtieri, S.; Leonardi, C. Modifications induced by the thermal treatment of kaolin and determination of reactivity of metakaolin. *Appl. Clay Sci.* **2013**, *73*, 2–10. [CrossRef]
42. Castillo, R.; Fernández, R.; Antoni, M.; Scrivener, K.; Alujas, A.; Martirena, J.F. Activación de arcillas de bajo grado a altas temperaturas. *Revista Ingeniería de Construcción* **2010**, *25*, 329–352. [CrossRef]
43. Mielenz, R.C.; Witte, L.P.; Glantz, O.J. Effect of calcination on natural pozzolans. *ASTM Int.* **1950**. [CrossRef]
44. Akram, T.; Memon, S.A.; Obaid, H. Production of low cost self compacting concrete using bagasse ash. *Constr. Build. Mater.* **2009**, *23*, 703–712. [CrossRef]
45. Memon, S.A.; Shaikh, M.A.; Akbar, H. Utilization of rice husk ash as viscosity modifying agent in self compacting concrete. *Constr. Build. Mater.* **2011**, *25*, 1044–1048. [CrossRef]
46. Akram, T.; Memon, S.A.; Akram, R. Utilization of fly ash and rice husk ash as high volume replacement of cement. In *International Conference on Advances in Cement Based Materials and Applications in Civil Infrastructure ACBM-ACI*; Lahore, Pakistan, 2017; pp. 247–256.
47. Ali, B.; Qureshi, L.A. Combined effect of fly ash and glass fibers on mechanical performance of concrete. *NED Univ. J. Res.* **2018**, *15*, 91–100.
48. ASTM. *Standard Test Method for Sieve Analysis of Fine and Coarse Aggregates*; ASTM: West Conshohocken, PA, USA, 2006. [CrossRef]
49. C39-17, ASTMC39/standard test method for compressive strength of cylindrical concrete specimens. In *Annual Book of Standards*; ASTM: West Conshohocken, PA, USA, 2017. [CrossRef]
50. C496M-17, ASTMC496/standard test method for split tensile strength of cylindrical concrete specimens. In *Annual Book of Standards*; ASTM: West Conshohocken, PA, USA, 2017. [CrossRef]
51. C1585-13, ASTMC1585/standard test method for measurement of rate of absorption of water by hydraulic-cement concretes. In *Annual Book of Standards*; ASTM: West Conshohocken, PA, USA, 2013. [CrossRef]
52. C642-13, ASTMC642/standard test method for density, absorption, and voids in hardened concrete. In *Annual Book of Standards*; ASTM: West Conshohocken, PA, USA, 2013. [CrossRef]
53. Velandia, D.F.; Lynsdale, C.J.; Provis, J.L.; Ramirez, F. Effect of mix design inputs, curing and compressive strength on the durability of Na_2SO_4-activated high volume fly ash concretes. *Cem. Concr. Compos.* **2018**, *91*, 11–20. [CrossRef]
54. Ali, B.; Qureshi, L.A. Durability of recycled aggregate concrete modified with sugarcane molasses. *Constr. Build. Mater.* **2019**, *229*, 116913. [CrossRef]
55. Shi, C. An overview on the activation of reactivity of natural pozzolans. *Can. J. Civ. Eng.* **2001**, *28*, 778–786. [CrossRef]
56. Day, R.L. Pozzolans for use in low cost housing: A state of the art report. In *Department of Civil Engineering*; Research Report No. CE92-1; The University of Calgary: Calgary, AB, Canada, 1990.
57. Tironi, A.; Trezza, M.A.; Scian, A.N.; Irassar, E.F. Assessment of pozzolanic activity of different calcined clays. *Cem. Concr. Compos.* **2013**, *37*, 319–327. [CrossRef]
58. He, C.; Makovicky, E.; Osbaeck, B. Thermal treatment and pozzolanic activity of Na-and Ca-montmorillonite. *Appl. Clay Sci.* **1996**, *10*, 351–368. [CrossRef]
59. Kurda, R.; de Brito, J.; Silvestre, J.D. Water absorption and electrical resistivity of concrete with recycled concrete aggregates and fly ash. *Cem. Concr. Compos.* **2019**, *95*, 169–182. [CrossRef]
60. Kou, S.-C.; Poon, C.-S.; Agrela, F. Comparisons of natural and recycled aggregate concretes prepared with the addition of different mineral admixtures. *Cem. Concr. Compos.* **2011**, *33*, 788–795. [CrossRef]
61. Bentz, D.P.; Trezza, M.A.; Scian, A.N.; Irassar, E.F. *Transport Properties and Durability of Concrete: Literature Review and Research Plan*; NIST: Gaithersburg, MD, USA, 1999. [CrossRef]

Review

A Systematic Review of the Discrepancies in Life Cycle Assessments of Green Concrete

Hisham Hafez [1,*], Rawaz Kurda [2,3,*], Wai Ming Cheung [1] and Brabha Nagaratnam [1]

[1]   Department of Mechanical and Construction Engineering, University of Northumbria, Newcastle upon Tyne NE1 8ST, UK; wai.m.cheung@northumbria.ac.uk (W.M.C.); brabha.nagaratnam@northumbria.ac.uk (B.N.)
[2]   CERIS, Civil Engineering, Architecture and Georesources Department, Instituto Superior Técnico, Universidade de Lisboa, Av. Rovisco Pais, 1049-001 Lisbon, Portugal
[3]   Department of Civil Engineering, Technical Engineering College, Erbil Polytechnic University, Erbil 44001, Kurdistan-Region, Iraq
*   Correspondence: hisham.hafez@northumbria.ac.uk (H.H.); Rawaz.kurda@tecnico.ulisboa.pt (R.K.); Tel.: +44-7493280699 (H.H.); +351-965722755 (R.K.)

Received: 30 September 2019; Accepted: 7 November 2019; Published: 10 November 2019

Abstract: It is challenging to measure the environmental impact of concrete with the absence of a consensus on a standardized methodology for life cycle assessment (LCA). Consequently, the values communicated in the literature for "green" concrete alternatives vary widely between 84 and 612 kg eq CO_2/m^3. This does not provide enough evidence regarding the acclaimed environmental benefits compared to ordinary Portland cement concrete knowing that the average for the latter was concluded in this study to be around 370 kg eq CO_2/m^3. Thus, the purpose of this study was to survey the literature on concrete LCAs in an attempt to identify the potential sources of discrepancies and propose a potential solution. This was done through examining 146 papers systematically and attributing the sources of error to the four stages of an LCA: scope definition, inventory data, impact assessment and results interpretations. The main findings showed that there are 13 main sources of discrepancies in a concrete LCA that contribute to the incompatibility between the results. These sources varied between (i) user-based choices such as depending on a cradle-to-gate scope, selecting a basic volume-based functional unit and ignoring the impact allocation and (ii) intrinsic uncertainty in some of the elements, such as the means of transportation, the expected service life and fluctuations in market prices. The former affects the reliability of a study, and hence, a concrete LCA methodology should not allow for any of the uncertainties. On the other hand, the latter affects the degree of uncertainty of the final outcome, and hence, we recommended conducting scenario analyses and communicating the aggregated uncertainty through the selected indicators.

Keywords: green concrete; life cycle assessment; environmental impact assessment; inventory data; allocation; functional unit; service life

---

1. Introduction

For every living individual, around 4 tonnes of conventional concrete, comprised primarily of ordinary Portland cement (OPC) and naturally sourced aggregates (NA), were produced in 2015 [1]. Due to its inherent strength and durability properties, concrete is the second most used substance on Earth after water [2]. Unfortunately, the use of concrete is associated with immense negative environmental impacts. The current production rate of more than 4 billion tonnes of OPC annually is responsible for 7% of the global CO_2 emissions [3]. It also risks depleting natural resources, since more than 50 billion tonnes of aggregates are being extracted annually [4]. Concrete has an environmental impact of 320 kg eq CO_2/m^3 on average as will be concluded in the next section of this paper, of which 90% is attributable to OPC [5]. Although this is less than that of steel and most polymers per unit mass [6], the intensive use of OPC concrete results in alarming environmental hazards. In China for example, the over-reliance on concrete alone resulted in approximately 1.5 billion tonnes of greenhouse gases (GHG) emissions in 2014 [7], which represents around 20% of the total produced in the same year [8]. Nevertheless, projections indicate that the growing global urbanization will double the demand on concrete by 2050 [9].

Hence, recent research has been directed to meet the "2015 Paris climate conference" guidelines of enhancing the sustainability of concrete [10]. As seen in Figure 1, there are five main families of concrete types found in the literature that are considered more eco-friendly [11–49]. Those are the types that were chosen to be studied in this paper as a sample for green concrete. First, there is the use of natural materials, such as bacteria, and agricultural waste, such as hemp [11], in order to create natural biological concrete (BioC) with reduced environmental impact. Second, there is recycling aggregate concrete (RAC), where construction and demolition wastes (CDWs) are used as aggregates in concrete. This reduces the landfill potential of concrete by 50–75% and its embodied carbon by 10–30% [12,13]. Blended cement concrete (BCC), where OPC in the binder, is partially replaced with various pozzolanic materials called supplementary cementitious materials (SCM), and is considered the third family. Examples of these are secondary materials such as fly ash (FA), which is a by-product of coal combustion; ground granulated blastfurnace slag (GGBS) which is a by-product of steel manufacturing; and silica fume (SF), which is generated from glass manufacturing. Additionally, some primary materials that are manufactured with lower energy demands than OPC can be used to partially replace it, such as calcined clay (CC) and lime. The mechanical and durability properties of the resulting concretes vary significantly between the different types of materials and the percentages by which OPC is being replaced, and similarly, the environmental impact varies [50]. For example, the embodied emissions of concrete could decrease by up to 30% and 60% with the incorporation of 35% and 70% of FA and GGBS, respectively [51]. In order to totally replace OPC, alkali activated concretes (AAC), are made with precursors of 100% FA, (CC) or GGBS that are activated using an alkaline solution from usually sodium hydroxide or sodium silicate. This is the forth concrete type selected in this study. AAC causes 70–75% less GHG emissions compared to OPC concrete [52]. The fifth family is high performance concrete (HPA), which is recognized as a concrete type with enhanced mechanical and durability properties compared to OPC concrete (OPCC). This allows for a reduction in the required volume of concrete in certain applications. Additionally, the concrete mixes are prepared with fillers such as lime powder to increase the particle packing of the mixing components, which minimizes the required amount of binder [15]. This potentially yields a binder with less environmental impact than OPCC.

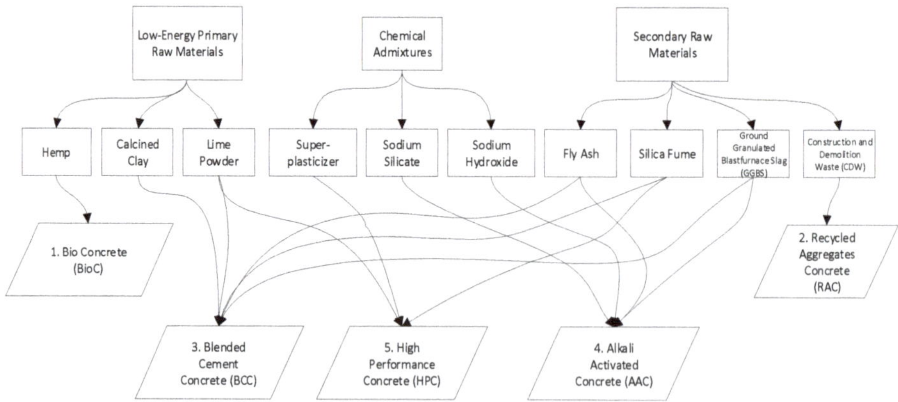

Figure 1. Green concrete strategies and types.

Therefore, the starting point of studying a green concrete type is to create alternative concrete types that reduce the environmental impact of OPCC [52]. Life cycle assessment (LCA) is the most widely-accepted tool to assess and compare these acclaimed environmental benefits [53]. According to ISO 14040:2006, LCA is defined as "the compilation and evaluation of the inputs, outputs and potential environmental impacts of a product system throughout its life cycle." An LCA study is divided into four main stages: (i) Scope and goal definition. (ii) Defining the inventory for the life cycle processes. (iii) Characterizing and measuring the life cycle impact. (iv) Interpretation of the results [54]. First, goal and scope definition involves outlining the system boundary, the functional unit (FU) selection and any assumptions and/or limitations that need to be considered. A system boundary of a concrete product could be cradle-to-gate, which means including all processes and emissions until the production of its different constituents, or cradle-to-grave which includes the "Use" and "End-of-Life" phases as per Figure 2 [55]. A Cradle-to-Cradle LCA scope is that which assumes that all waste generated will be recycled in the future [23].

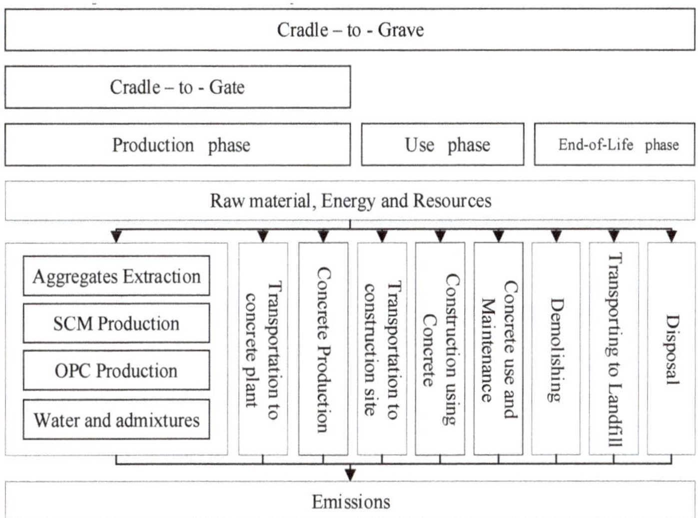

Figure 2. Different system boundaries of green concrete life cycle assessment (LCA) [21].

A FU is the basis for quantifying the inputs and outputs between alternatives. Hence, its selection needs to be reflective of the nature of the LCA subjects [56]. The second LCA stage includes collecting the data of energy and emissions associated with the aforementioned scope. The data needed for standard processes can mostly be obtained from primary sources or found in databases such as Ecoinvent and European reference Life Cycle Database (ELCD) [57]. A further source of inventory data is the environmental product declarations (EPD) of the concrete's raw materials, which are produced by the local manufacturers according to a local binding legal framework [16]. At this stage, it is also important to decide on allocation, which is basically portioning the environmental burden of the original process to the product under study [17]. The third and final stage of an LCA is to calculate the environmental impact of the product being studied. This is performed by adding up the individual impacts of all the associated processes as per ISO 14040:2006 to calculate an environmental impact indicator, a number that makes the output of the impact assessment study more understandable to the user [58]. According to Menoufi [59], there are two main types of indicators: mid-point indicators, which correlate with the estimated impact of a specific change in the environment, such as global warming potential, and end-point indicators, which correlate that same impact to damages via cause–effect changes, such as human health.

As explained before, the LCA of green concrete cannot be easily assessed. In fact, there are a lot of uncertainties in the assessment process due to lack of a standardized methodology. To highlight this issue, an attempt was made by this study to compare the absolute values for the environmental impacts of the aforementioned green concrete types to that of OPCC, by considering around 300 different mixes from 39 journal papers [11–49]. Using the most predominant environmental impact indicator, global warming potential (GWP), the impact per unit volume of the concrete mixes varied between 110 and 600 kg eq CO_2/m^3, as shown in Figure 3. Assuming the results from these values have a normal distribution, the mean value would be around 320 kg eq CO_2/m^3 while the standard deviation would be around 90 kg eq CO_2/m^3. This is indicative of large discrepancies that could challenge the original argument that these concrete mixes cause less environmental impact compared to OPCC. Nonetheless, the absolute values communicated for the OPCC using the same indicator (GWP) were found to also vary widely. Upon reviewing 80 mixes from 20 papers, the mean value was found to be around 370 kg eq CO_2/m^3 with a standard deviation of around 110 kg eq CO_2/m^3, as shown in Figure 3. Therefore, this systematic review paper was presented in order to critically examine each stage of the LCA studies carried out by researchers in the domain of green concrete in order to identify the sources of the discrepancies. Huijbregts [60] attributes these large discrepancies to the uncertainties involved in the current use of LCA methodology. Hafliger [61] claim that the source of these uncertainties are modelling choices by the user of the system boundary, FU and source of data. On the other hand, Menoufi [27] differentiates between the uncertainties due to the nature of the inventory data used and those from choices such as the impact allocation and FU. The first affects the precision of the results due to the fact that the elements included in the study include a percentage of uncertainty, while the latter affects the reliability of the study. Hence, in this paper, which is the first of its kind to systematically tackle the sources of discrepancies in concrete LCA studies, the same categorization will be followed. Hence, the objective of this review was to provide the concrete LCA user with a coherent, state-of-the-art guide to avoid the sources of reliability error and to solve the issues caused by the sources of uncertainty.

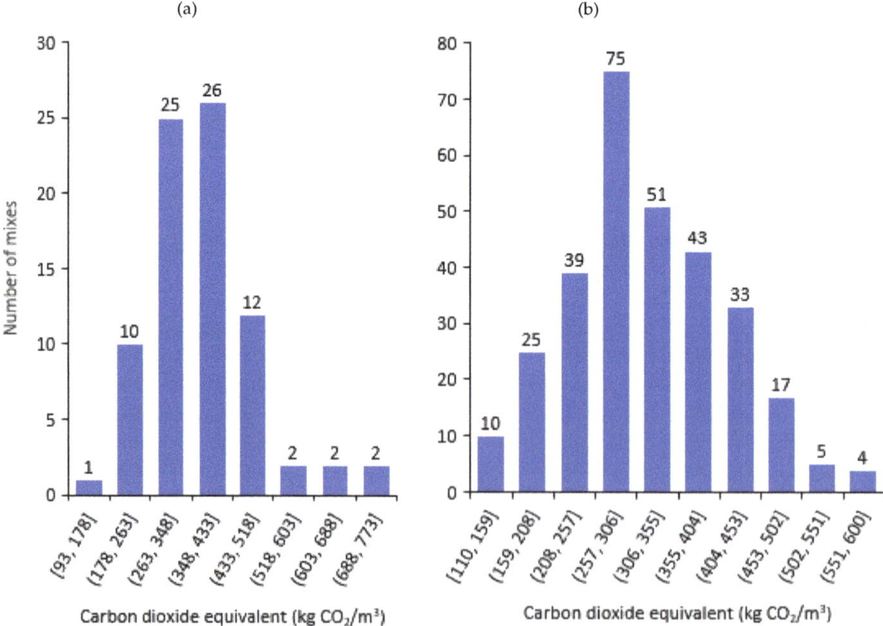

Figure 3. A review of the values for the equivalent kg CO_2/m^3 of: (**a**) 80 Ordinary Portland Cement Concrete (OPCC) mixes and (**b**) 300 mixes of all 5 concrete types studied from the literature.

2. Methodology of the Systematic Review

The scope of the systematic review was to examine the methodology of LCA studies performed on green concrete, along with OPCC, in order to track down the sources of discrepancies. A total of 11,000 references were found after searching online databases such as Science Direct, Taylor and Francis and Scopus using a combination of the following keywords:

- LCA;
- Concrete;
- Cement;
- SCM;
- Sustainable;
- Methodology.

First, the references were filtered based on their titles, then on their abstracts, and finally, on full paper analysis. The point of preference was that the study included either an LCA study on an OPCC or any of the green concrete types in the scope (exploratory) or a review of the environmental impact of any of these types; or it could be a paper that studies the methodology of a concrete LCA. The 146 references [14–159] selected were divided into categories: 107 "exploratory" articles, 23 "methodology" articles, and 16 reviews. The distribution of the exploratory LCA references between the six concrete types studied was found to be almost 50% BCC, as shown in Figure 4a. In addition, most of the references were published during the last 10 years (2009–2019) as seen in Figure 4b. The countries of origin of the publications are shown in Figure 4c. The method followed in order to come up with the sources of discrepancies was examining the methodology of each of the 107 "exploratory" studies and pointing out anomalies across the four LCA stages. Hence, the review is divided into four subsections, for each of the sources found in the corresponding LCA stages as follows.

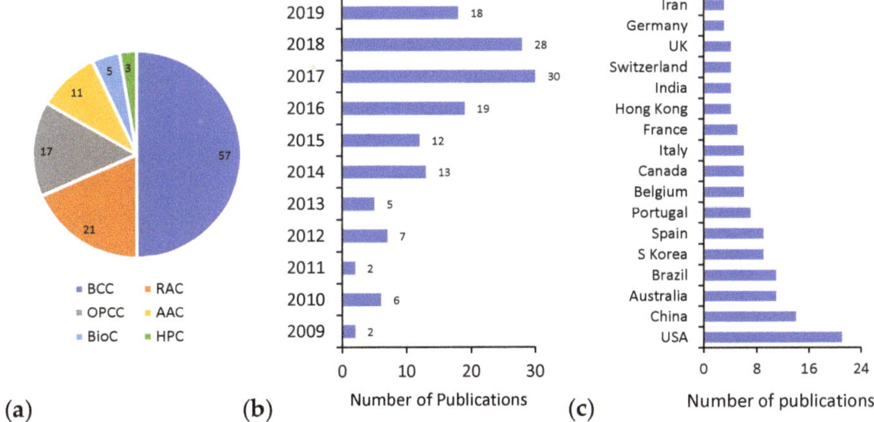

Figure 4. Number of LCA studies per (**a**) type of concrete, (**b**) published year and (**c**) country.

3. Sources of Discrepancies in LCA Stages

3.1. Stage 1: LCA Scope

3.1.1. System Boundary

More than 75% of the studies (107) we reviewed, in which an LCA study was actually conducted, used a cradle-to-gate system boundary, as shown in Figure 5. A cradle-to-gate system boundary would limit the scope of the processes and resulting emissions and energy studied in the LCA up until the production stage, excluding the use and end-of-life stages. As stated by Wu et al. [24] the ISO 24067, released in 2013 to provide a benchmark for the LCA methodology, specifies that for a user to exclude the use and end-of-life stages while conducting an LCA, there needs to be enough evidence that the results will not be affected by this. Hence, it would not be acceptable to cut-off the use and end-of-life phases from the scope due to the following reasons (i–iii):

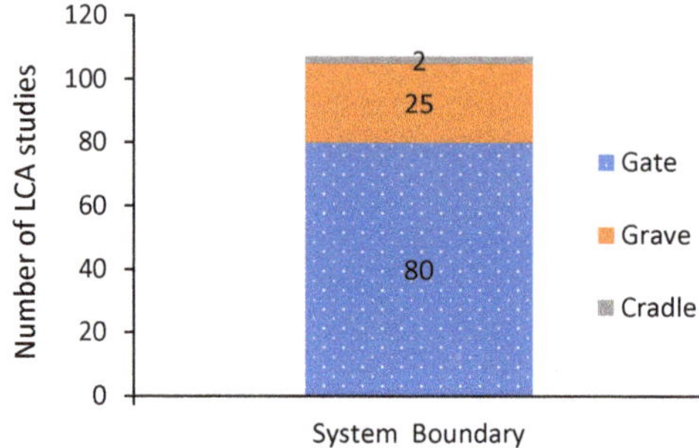

Figure 5. Meta-analysis of the selected system boundary for each of the LCA studies reviewed in this paper.

(i) Throughout its "use" phase, concrete exhibits carbonation, a process by which carbon dioxide is absorbed by the concrete from the exposed environment reacting with the calcium compounds in its matrix, forming carbonates [62]. A justifiable cut-off percentage is when the processes affects less than 1% of the total environmental impact according to Wu et al. [16] and 5% according to Gursel et al. [63]. However, through the carbonation process, concrete can absorb, throughout its whole service life, 13–48% of the carbon dioxide it emitted during the production phase [62]. This value of the captured carbon dioxide, denoted as carbon sequestered, varies depending on the concrete type, exposure conditions and geometry [44]. Out of the 107 references, only seven included the sequestered carbon in the LCA study, with varying values, as seen in Table 1. OPCC can capture up to 47% of its embodied carbon during use and end-of-life phases, while BCC can capture only up to 22% [39]. In all cases, it is apparent that the sequestered carbon ought to be included in an LCA study to allow for its value to be deducted from the carbon emissions in the remaining processes.

Table 1. The different values for carbon sequestration from the papers reviewed.

References	First Author's Last Name	Year	Kg CO_2/m^3
[62]	Collins	2010	5
[29]	Garcia-Segura	2013	61
[64]	Kim and Chae	2016	172
[65]	Lee	2013	10
[46]	Panesar and Churchill	2010	30 *
[66]	Zhang	2019	39 *
[44]	Souto-Matrinez	2017	60

* Calculated based on a percentage of the 320 eq CO_2/m^3 average value.

(ii) By omitting the use phase, the user is also assuming that all concrete mixes being compared will sustain the required service life. However, the findings in Table 2 show proof from the literature that this assumption is not true. Depending on the exposure conditions, concrete cover and the concrete mix, there is a high probability that a reinforced green concrete mix is unable to fulfil its required service life, especially one above 60 years [61]. Hence, according to Panesar et al. [56], there should be at least a 20% increase in the environmental impact of concrete to account for the potential maintenance that allows it to fulfil its service life requirements. Nevertheless, Tae el al. [15] set the expected CO_2 emissions for concrete at 0.60 eq kg CO_2/m^3/year in service. Additionally, it was also proposed that a replacement factor N, which is a ratio between the reference service life of the concrete member studied and its predicted service life, needs to be included as a multiplying factor to the environmental impact of the concrete under study to make the calculation more relative, and therefore, reliable [61]. That, then, shows that there is a major underestimation of the environmental impact—the relative and absolute environmental impacts—of concrete if the maintenance/replacement impact is not considered as a part of the "use" phase of the LCA.

Table 2. Service life predictions for different concrete types from the literature.

Reference		[36]			[46]		
Concrete Type		**OPCC**	**25%GGBS**	**50%GGBS**	**15% FA**	**35%FA**	**50%FA**
	Cover (mm)	**Service Life (Years)**					
	65	62	124	200	-	-	-
Carbonation	50	124	200	200	-	-	-
	35	200	200	200	-	100	100
	65	33	68	114	-	-	-
Chloride Penetration	50	67	138	200	51	-	60
	35	91	186	200	-	-	-

(iii) Having an end-of-life phase (which could be achieved by considering a cradle-to-grave boundary or a cradle-to-cradle one) included in the LCA system boundary is a prerequisite to studying RAC. De Schepper et al. [23] assumed that the aggregates used in a concrete mix were fully recyclable. By selecting a cradle-to-cradle system boundary, it was calculated that avoiding the landfilling of concrete reduces the environmental impact compared to OPCC by 4%–15%. Ding et al. [34], they included the avoided-landfilling potential for recycled CDW in the LCA and the result was that the environmental indicator CMR, consumption of natural resources, decreased by 46%. Apart from that, when the service life of a concrete product ends, the demolition process requires energy. Whether the waste will be re-used or not is an unknown at the LCA study stage. However, as seen in Table 3, the energy and impact required for the demolition of concrete constitutes 2%–10% of an average 320 kg eq CO_2/m^3 of concrete, which means that it should not be ignored.

Table 3. Some values for the environmental impact of demolishing concrete from the literature reviewed.

References	First Author's Last Name	Year	kg CO_2/m^3
[28]	Garcez	2017	25
[29]	Garcia-Segura	2013	5
[67]	Lopez-Gayyare	2015	13
[62]	Collins	2010	5

3.1.2. Functional Unit Selection

The second part of the LCA scope that the user selects is the FU. According Panesar et al. [56], a functional unit is the element that dictates how the inputs and outputs of the LCA are quantified. There are four main levels for functional units of concrete: a whole structure or a building, which was selected by around 6% of the 107 studies reviewed, as seen in Figure 6; or a component of a structure such as a beam, column or a bridge girder, and only 16% of the studies opted for that. The most famous level of detail (LOD) studied in concrete LCA is the material unit for concrete, which was selected by around 70% of the references. A unit-based FU for concrete is divided into: volume-based, volume-based while considering the strength of concrete and volume-based while considering both the strength and service life of concrete. Panesar et al. [56] claims that it is not accurate to call a unit volume a FU since it is not indicative of enough comparable functional properties. It should be called a declared unit instead. However, as seen in Figure 6, 65% of the references reviewed in this paper where a unit-based FU was selected did not consider strength nor durability; 25% considered strength and only 10% considered both. According to Zhang et al. [68], depending on the type of unit-based FU selected, the results for the LCA study might vary up to 30%. In order to further investigate this through our systematic review, the following examples were prepared.

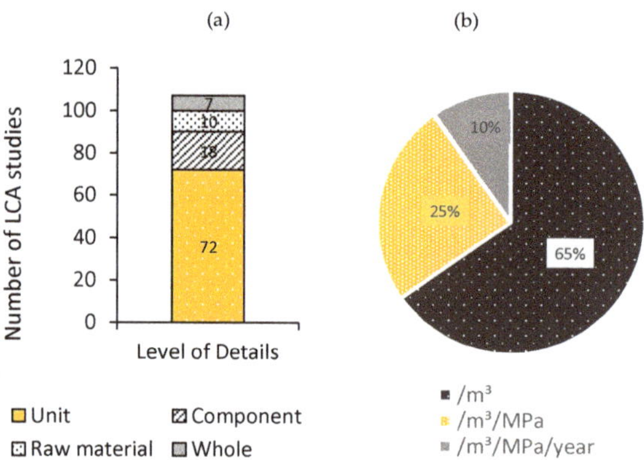

Figure 6. The division of the FU in the LCA studies reviewed based on (**a**) Level of Detail and (**b**) type of unit volume.

In terms of the difference in strength between the concrete alternatives in question, for RAC, maintaining the same binder, replacing fresh aggregates with coarse and/or fine recycled aggregates from CDW, will decrease the strength of the resulting mix (coarse aggregate (RC) and fine aggregate (RF) respectively) [58]. This decrease in strength of the concrete incorporating the recycled aggregates can sometimes be larger than the associated decrease in environmental impact. Hence, as shown in Figure 7, it is clear that when a FU of kg eq CO_2/MPa was used instead of just kg eq CO_2, the environmental impact of the mixes with the recycled aggregates turned out to be larger than that of OPCC, not less [69].

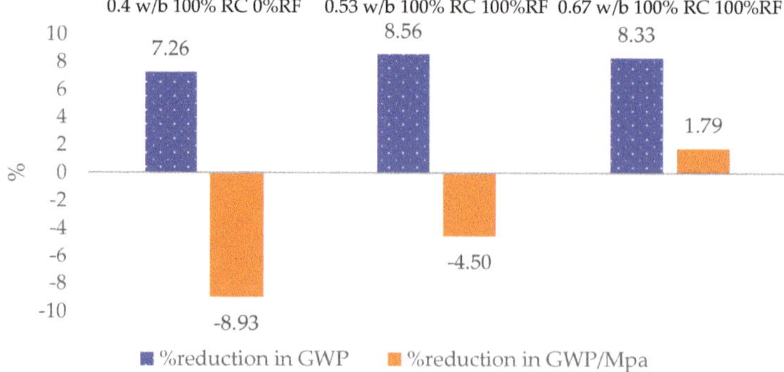

Figure 7. A comparison between the impact of a recycling aggregate concrete (RAC) using volume-based FU and a FU normalized to strength [69].

For BCC types, the results from Celik et al. [21] suggested that the optimum replacement of OPC with FA in terms of minimizing environmental impact in a BCC mix is 70%. However, adding FA beyond a certain threshold would significantly decrease the compressive strength of concrete [35]. As seen in Figure 8, when the same GWP results were modelled using a FU of kg eq CO_2/m^3/MPa instead of a volume-based kg eq CO_2/m^3, the optimum replacement percentage dropped to only 40%. Smaller gaps between both FU results were found when examining the results for GGBS from Bilim et al. [19], as seen in Figure 8. This could be due to the fact that the results for the compressive strength were tested after 90 days of curing instead of the common 28, at which the pozzolanic reaction would mature and most BCC mixes would achieve comparable strength to that of OPCC [70].

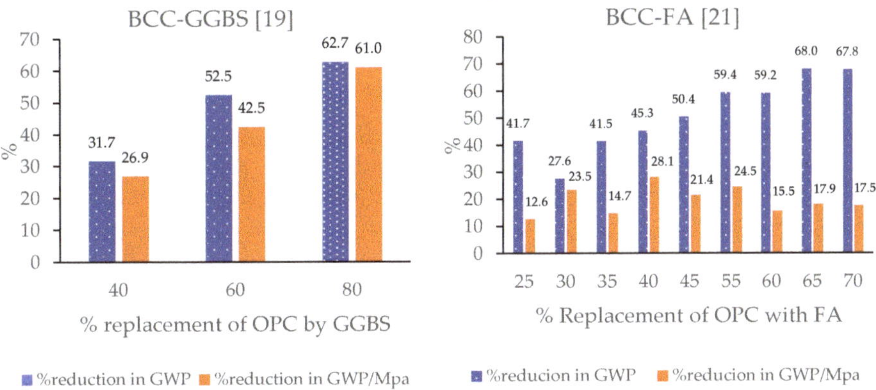

Figure 8. A comparison between the impact of blended cement concrete (BCC) using a volume-based FU and a FU normalized to strength.

According to Mahima et al. [71], premature concrete deterioration, due to carbonation and chloride penetration, is responsible for a loss of 2.2 trillion USD, which is equivalent to 3% of the world's gross domestic product (GDP). This means that durability is a more detrimental factor to the performance of concrete than that of the compressive strength. Thus, the durability of concrete is an essential factor that needs to be included when the LCA of reinforced BC concrete is compared to reinforced OPC concrete. Panesar et al. [56] defined a FU where the volume of the BC concrete is multiplied by its compressive strength and the chloride ion penetration resistance, and is compared to the FU of an equivalent OPC concrete. Similarly, Celik et al. [21] and Kurda et al. [69] used experimental data of the different BCC mixes in terms of compressive strength and chloride penetration to compare the performance of BCC with OPCC. However, for the absolute environmental impact values to be credible, these durability properties need to be translated into the service life to describe a performance parameter of concrete [59]. Heede and De Belie [47] accounted for a 100 year timeframe as the service life of concrete, but only carbonation was used to determine the service life. Sagastume-Gutierrez et al. [57] devised a FU that divides the volume of cement by the number of years of durability from both chloride penetration and carbonation. Furthermore, an accurate methodology was proposed by Gettu et al. [30], where the test results of thirty different BC concrete mixes were incorporated into a FU (A-indices) that converts carbonation and chloride penetration parameters into expected service-life predictions. However, in all of the aforementioned, concretes with more than 100 years of durability will have a better environmental impact using both indices, while the specified service life for the mix is only 100 years. The same applies to compressive strength. Not capping the performance nor the durability of the concrete being studied, though it maximizes the sustainability potential according to Muller et al. [72], impinges upon the performance base specifications of the concrete. Instead, performance-based specifications similar to those in the framework proposed by Hafez et al. [55] should be adopted. Finally, according to Sagastume-Gutiérrez et al. [57], the comparison between the environmental impacts of two construction materials can be reliable, only after considering the combined effects of mechanical and durability characteristics.

The user's choice for a unit-based LOD could be based on a personal preference or absence of the necessary details about the project, such as the quantity of concrete per member or per a whole structure. However, the benefit of studying the member or a whole structure in which the concrete mix would be a part adds is that it another important factor to the equation, which is the optimization of the total volume. For example, increasing the prescribed strength requirement for concrete from 25 MPa to 50 MPa in a solid slab building would decrease the volume of concrete needed by around 15%, especially with columns [28]. It is recommended, therefore, to run a parametric analysis on the concrete under study based on compressive strength, strength, service life and the resulting volume, and hence, the combined environmental impact.

3.2. Stage 2: Inventory Data

The second source of uncertainties and unreliability in LCA results after the scope definition is LCA inventory (LCI). This is the data collection stage, in which the input and output factors, including energy, raw materials, products and waste, are analysed for the LCA of concrete. The LCI for a concrete mix mainly include: (a) upstream processes: those involved in the production of each of the constituents and its transportation to the concrete production plant; (b) core processes which involve the energy and emissions required for mixing concrete and transportation to site; and (c) downstream processes needed for the demolition or any other end-of-life scenario [24]. Out of all the inputs/outputs data from these processes, it is primarily important to quantify the emissions and energy-use rather than the oil use, waste generated and the rest [63]. Upon reviewing the necessary literature, it shows that LCI data is a major contributor to the uncertainty in a concrete LCA study due to the following reasons (i–iv):

(i) The LCI source has no standards as to where and how to get LCI data for a LCA of concrete. Anand and Amor [73] stated that concrete inventory data should come from primary sources for reliability purposes or secondary sources if the former is not available. Primary data could be lab results, governmental reports or EPDs from the building industry to which the user has access. EPDs are standardized documents to communicate the environmental performance of a product that are accredited by local authorities [74]. On the other hand, secondary data could be from accredited environmental databases, such as EcoInvent, GaBi and ELCD database or just using previously published data from the literature. By examining the 107 papers that actually included an LCA study, it was found that more than half, as seen in Figure 9, opted for the use of secondary sources for inventory data, which could not be reliable enough to describe the special scenarios being modelled in the concrete LCA. Although EcoInvent (which was developed by the Ecoinvent Centre, a competence centre of the Swiss Federal Institutes) and GaBi (which was created by Thinkstep Inc.) are updated annually to reflect any changes in the inventory data included, the environmental impact of concrete when modelled using the Ecoinvent database and EPDs has a variability of up to 20% [61]. Hence, it is suggested that the priority in the source of upstream processes of a concrete mix is for EPDs and in the case of several EPDs, an average should be taken. The reason is that EPDs are done in accordance with the same process, an LCA, under the guidance and supervision of local authorities such as the Green Building Council of Australia's concrete [75]. This would contribute to standardized processes and more efficient error tracking of concrete LCAs.

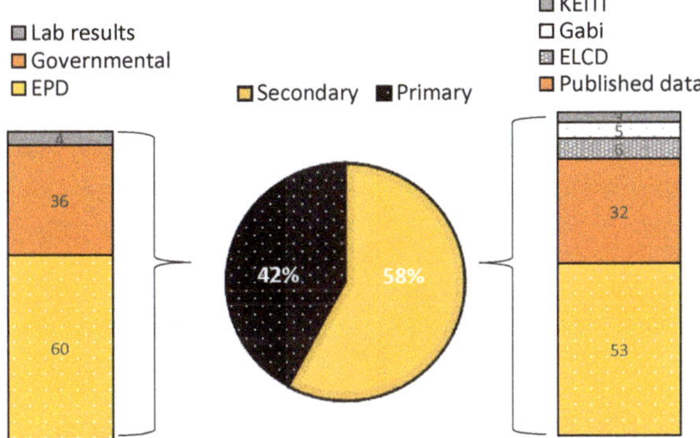

Figure 9. The meta data for the primary versus secondary sources for LCA studies reviewed and the databases used.

(ii) Apart from the reliability issue of the choice of the suitable LCI for the data, the existing data in each of the LCI sources contain large uncertainties. Looking into 25 papers, the inventory impact for OPC was found to vary between 691 kg eq CO_2/tonne and 1452 eq kg CO_2/tonne as shown in Figure 10. The reason could be that the OPC production process is different in efficiency between one producer and the other [76]. Additionally, the upstream process for OPC production depends on the electricity mix of the country of origin. For example, in the U.S. about 8% of the OPC used is imported and the upstream inventories of the imported clinker specific to the country of origin, as well as the energy consumed in transporting the OPC to the US, would increase the resulting impact of the OPC over the local alternatives [63]. The electricity mix in China almost has twice the environmental impact as that of Malaysia, Indonesia and Thailand due to, for example, the higher dependency on fossil fuel in electricity generation [31]. In all cases, the discrepancy in the impact of cement attributed has a great impact on the final environmental impact calculated through an LCA for a concrete alternative.

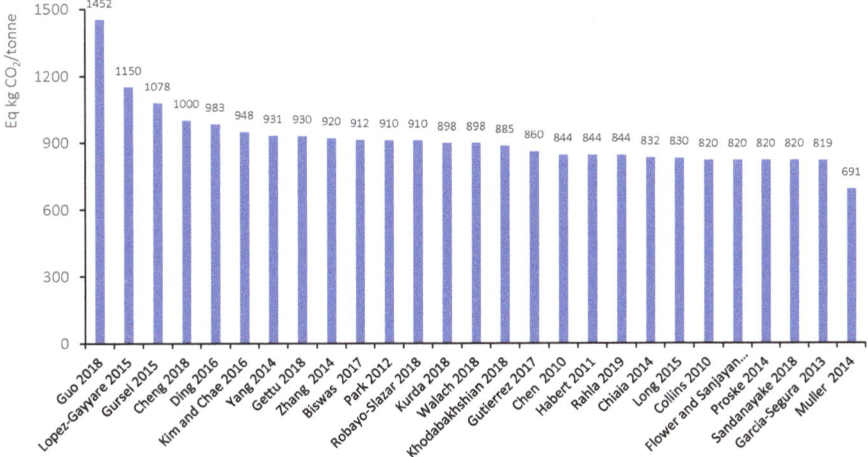

Figure 10. A review of the global warming potential (GWP) of a tonne of ordinary Portland cement (OPC) reported in 25 papers.

(iii) The variability in inventory data is not only in the OPC, but as seen in Table 4, it is the same with the rest of the concrete mix components. The most variable components found were the supplementary cementitious materials, such as FA, GGBS and SF whether in BCC or AAC mixes. The reason behind this could be attributed to the case of environmental impact allocation. Impact allocation is the process of portioning the environmental burden of the original process to the waste material being recycled in the product under study [17]. According to the EU directive 2008, waste can be considered to be a by-product when its further use is certain, it is produced as an integral part of a production process, it can be used without any further processing other than normal industrial practice and its further use is lawful [77]. All four points apply to FA, GGBS and SF; hence, they ought to be considered by-products, not waste. This means that they ought to be allocated a percentage of the environmental burden of their original production processes, which are coal combustion, steel production and glass manufacturing, respectively [19]. The first impact allocation scenario is "mass allocation" where the percentage allocation is based on the relative mass between the waste material as a by-product and the total mass (the effective mass of electricity + the mass of FA) as shown in Equation (1). The second scenario is "economic allocation" in which the percentage allocated is based on the relative market value between the final product, which is FA, and electricity, as per Equation (2) [77].

Table 4. Inventory data for different concrete components from the literature we reviewed.

Ref.	1st Author's Name	Year	eq kg CO_2/Tonne									
			Steel	Coarse NA	Fine NA	FA	SF	GGBS	Lime	SH	SS	SP
[78]	Yang	2014		3.2	2.3			27				
[30]	Gettu	2018				265		144	516			
[79]	Zhang	2014		14	41	27						720
[20]	Biswas	2017	1470	5	2.4		72	156				1130
[37]	Park	2012		4	1	20		27				250
[39]	Robayo-Slazar	2018		1.1	0.5	9		277		1359	793	
[69]	Kurda	2018		29	2	4						
[80]	Walach	2018		3	13		4					1840
[77]	Chen	2010				350		19				
[5]	Habert	2011		4.3	2.4	5		17	35		1140	749
[81]	Rahla	2019				210	1580	134				
[82]	Chiaia	2014	150	2.5	2.5				19			720
[83]	Long	2015		6	1	9		19	17			720
[62]	Collins	2010		46	14	27		143				
[84]	Flower and Sanjayan	2007		36	14	27		143				
[85]	Proske	2014	874	7	2.3	11			29			772
[52]	Sandanayake	2018		40	14					1425	780	
[29]	Garcia-Segura	2013	920	4	4	4		52				220
[18]	Al-Ayish	2018	370	2.4	1.7			88				
	Mean		694.8	11.547368	6.44737	74.46	552	95.85	105.5	1392	904.3	833.2
	St dev		486.2	14.489436	9.75525	118.2	890.9	78.61	201.2	46.7	204.2	441.7
	St. dev (% of mean)		69.98	125.47825	151.306	158.7	161.4	82.01	190.7	3.35	22.58	53.01

$$\text{Mass Allocation} = \frac{(m)_{by-product}}{(m)_{main\ product} + (m)_{by-product}} \tag{1}$$

$$\text{Economic Allocation} = \frac{(\text{\euro}.\, m)_{by-product}}{(\text{\euro}.\, m)_{main\ product} + (\text{\euro}.\, m)_{by-product}} \tag{2}$$

Hence, it is seen as a reliability requirement for LCA studies, including SCMs, to include an impact allocation scenario. However, upon reviewing the literature, it was found that out of 59 exploratory LCA studies of green concrete involving SCMs, only 14 (25%) of the LCA studies included an allocation scenario. Eight papers included economic allocation scenarios while six included both of them. According to Marinkovic et al., in case the difference between the price of main and secondary process generating the SCM product is more than 25%, economic allocation should be applied. However, the fact that the fluctuation in market prices of the raw materials should now be a part of the environmental impact assessment of concrete through the economic allocation, creates room for further discrepancies in the results, as seen in Table 5. It is recommended to keep the LCA results of concrete limited to the time frame for which the prices of the raw materials still stand unchanged.

Table 5. Market prices of different concrete components from the literature.

Ref.	1st Author's Name	Year	Country	FA (£/Tonne)	GGBS (£/Tonne)	SF (£/Tonne)	Electricity (£/kWh)
[53]	Anastasiou	2015	Greece		3.50		
[86]	Chen	2019	France	35	23		0.12
[77]	Chen	2010	France	20	40		0.1
[75]	Crossin	2012	Australia		100		
[87]	Gursel	2015	USA			890	
[5]	Habert	2011	Switzerland	25	45		0.12
[51]	Jiang	2014	USA		74		
[88]	Khodabakhshian	2018	Iran			500	
[89]	Li	2015	China		20		
[17]	Marinkovic	2017	Serbia	3.5			0.05
[34]	Igancio	2018	Spain	38		1140	
[56]	Panesar	2019	Canada	135			
[37]	Park	2012	South Korea	33	41		
[81]	Rahla	2019	Portugal	28	37	430	
[90]	Seto	2017	Canada	107			0.07
[45]	Teixeira	2015	Portugal	21			0.22
[91]	Tucker	2017	USA				0.09
[92]	Wang	2017	China	10			0.11
[93]	Yuan	2017	China				0.11
[59]	Zhang	2014	China	9			
	Mean			38.7	42.6	740.0	0.11
	St dev			40.3	29.0	334.8	0.05
	St. dev (% of mean)			104.2	68.2	45.2	43.1

(iv) Another source of discrepancy in LCAs of concrete with regard to inventory data is the impact attributed to the transportation of raw materials to the concrete manufacturing plant. It could vary between 5% and 20% depending on the location of the raw materials relative to the concrete batch plant [94]. While examining a sample of the papers being reviewed, it was apparent, as seen in Table 6, that the transportation distances vary widely between the different studies. A study by Panesar et al. [95] concluded that the critical distance for importing FA that would still yield a BCC mix with a positive overall environmental impact profile compared to OPCC is around 900 km as opposed to the 3000 km proposed by Hafez et al. [160] and 5700 km by O'brien et al. [96]. Additionally, Turk et al. [13] argue that if the recycled aggregates are sourced from a landfill that is more than 230 km from the concrete batch plant, the RAC produced would have a higher environmental impact compared to OPCC, which is a larger figure than the 145 km concluded by Anastasiou et al. [53]. In all cases, it is recommended to perform a scenario analysis in each LCA study for concrete in case the sources for the raw materials are not exact, to determine the sensitivity of the output relative to the change in distances.

Table 6. A sample of the inventory data for the transportation distances of concrete raw materials in km.

Ref.	Author	Year	Kilometres						
			OPC	Aggregates	FA	GGBS	SP	SH	SS
[53]	Anastasiou	2015	50	50	50				
[20]	Biwas	2017	25	1		0.01			
[93]	Chrysostomou	2017	45	36			39		
[75]	Crossin	2012	90	30	945	9317			
[98]	Maria	2018	50	50		10			
[24]	Ding	2016		100					
[29]	Garcia-Segura	2013	32	12	180	1640	724		
[64]	Kim and Chae	2016	106	32			77		
[69]	Kurda	2018	60	65	160		15		
[89]	Li	2015				177			
[17]	Marinkovic	2017	100	100	50			50	20
[39]	Robayo-Slazar	2018	6	10	10	493		53	192
[41]	Salas	2018	5	5				12	64
[13]	Turk	2015	50	1					
[46]	Van den Heede and De Belie	2010	113	193	38		118		
[78]	Yang	2014	277	43	322	339	70		
	mean		72.1	48.5	195.9	1498.0	173.8	38.3	92.0
	St dev		68.5	51.0	299.2	3205.7	271.8	22.9	89.4
	St. dev (% of mean)		95.0	105.2	152.7	214.0	156.4	59.6	97.1

3.3. Stage 3: Impact Assessment

The third stage of an LCA is the assessment of the impact of the concrete mix by simply multiplying the functional unit by the aggregates impact of the concrete from the three life phases. As seen in Equations (3) and (4), the emissions and energy use are calculated by adding up all the emissions and energy uses of the products and processes involved in the production, use and end-of life stages.

$$\text{Total emissions for mix } (x) = FU_x \times \sum_{n=1}^{3} \text{ emissions of all products and processes of stage n} \qquad (3)$$

$$\text{Total energy use for mix } (x) = FU_x \times \sum_{n=1}^{3} \text{ energy use of all products and processes of stage n} \qquad (4)$$

In order to contextualize the information about the concrete mix being studied, an environmental impact indicator is needed. An impact assessment method is vital to producing judgements on the severity of the impact of concrete on the three main areas of protection: (i) ecosystem quality, (ii) human health and (iii) natural resources [59]. This is done through three steps: characterization of the impact, which is a must-do; then, normalization and weighing, which are both optional [49]. According to Sayagh et al. [99], there are two main types of indicators: mid-point indicators, which correlate the calculated impact to a specific change in the environment, such as global warming potential, and end-point indicators which correlate the same increase to damage occurring later-on in the cause–effect chain, such as human health. The significance of this differentiation is that the same comparison between products or processes could result in different scores if looked upon by a mid-point or an end-point indicator, due to the exaggeration of damage that happens to reach the latter [100]. By examining the references reviewed, it was noticed that out of 107 papers, as seen in Figure 11 only six chose to rely on end-point indicators to present the concrete LCA findings.

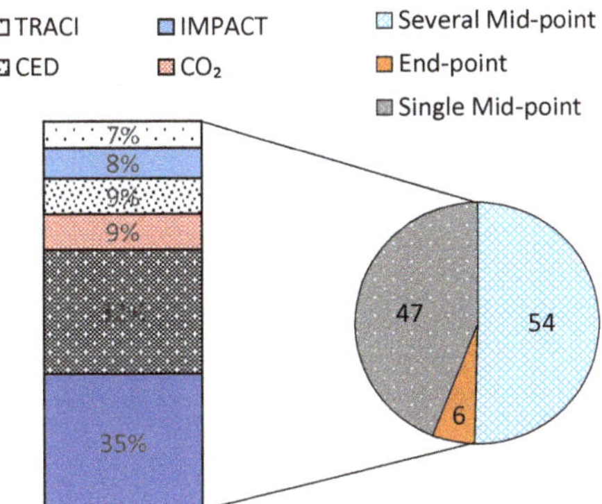

Figure 11. Meta-data for the indicators used in the concrete LCA studies reviewed.

In terms of relying on mid-point indicators to represent the environmental profile of concrete, the majority of the papers only selected global warming potential. The first reliability issue is that 10% of the papers opted to use the carbon dioxide emissions value as equivalent to the global warming potential [37,64,78,79,101,102]. In fact, according to the two most established midpoint environmental methodologies: CML and the Tool for the Reduction and Assessment of Chemical and other environmental Impacts (TRACI), GWP is based on the aggregation of carbon dioxide emissions, methane and nitrogen dioxide. CML was developed in 1992 by the Institute of Environmental Sciences of the University of Leiden, and TRACI was prepared by the US Environmental Protection Agency's (US EPA's) National Risk Management Research Laboratory in 2003.

The second reliability issue with this choice is that GWP is not the only significant environmental impact indicator. As seen in Figure 11, more than 10% of the authors opted to use, along with GWP, the cumulative energy demand (CED). More than half of the authors opted to present several midpoints, according to the following methodologies: CML (32%), Integrated Material Profile And Costing Tool (IMPACT) (10%) and TRACI (7%). These midpoint indicators include, besides GWP, ozone depletion potential (ODP), acidification potential (AP), eutrophication potential (EP), and abiotic depletion potential (ADPE), among others. Considering these indicators would include emissions such as carbon monoxide, sulphates and ammonia [67]. This concludes that in order to provide a reliable assessment of the environmental impact of a concrete alternative, it is advisable to calculate it using several midpoint indicators. However, it remains up to the user to remove any that are deemed irrelevant. For example, Passuelo et al. [103] argue that since the ODP potential of 1 tonne of geopolymer concrete is almost equal to 1.34×10^{-5}, equivalent to operating a household lamp for 2.5 years, this is not a significant indicator to consider.

3.4. Stage 4: Interpretation of Results

The fourth and final stage of an LCA is that when the user analyses the assessed outcome from stage 3 to judge the environmental impact of the concrete alternative studied. In this stage, two main problems were observed. It was noticed that almost 90% of the papers reviewed opted to produce deterministic results for the environmental impact of concrete. Regardless of the indicator chosen, even if the user avoids all the systematic errors explained earlier to make the study reliable, it has been established that some of the most significant elements of a concrete LCA are intrinsically uncertain, such as (i) predicting the expected service life of a mix based on the exposure conditions and mixing proportions; (ii) uncertainty in upstream data regarding some raw materials, depending on the source or database used; (iii) the forecasted energy use in futuristic activities, such as demolition and/or maintenance of the concrete alternative; (iv) predicting the amount of carbon the concrete mix is able to sequester depending on the surface area and atmospheric conditions; and (v) the characterization factors for the mid-point indicators selected. Hence, the results of a concrete LCA should be communicated in a probabilistic fashion. In order to quantify these uncertainties, scenario analysis should be performed on each of the individual elements. Upon quantifying the aforementioned uncertainties, the user could then model the impact assessment indicators using Monte Carlo simulation, and through running the model repeatedly, that would generate a probabilistic aggregated indicator. The model would also allow the user to choose the uncertainty distribution and define the confidence level over the model output accordingly.

4. Conclusions

This review paper aimed at analysing the process of preparing a LCA for concrete and identifying potential sources of discrepancies. The purpose was to standardize the concrete LCA methodology in order to be able to judge the environmental impact in absolute terms across industries, which is a prerequisite for the rising demands to cut down the environmental impact of concrete to combat the rising global warming issues. As seen in Table 7 upon reviewing 146 references, it is apparent that there are 13 most-common sources of discrepancies that spread across the four stages of an LCA study. More than half of these sources are user-based choices, such as depending on a cradle-to-gate scope, selecting a basic volume-based functional unit, ignoring the impact allocation and relying on a single indicator. Throughout the paper, it was shown that based on these choices, the results of the concrete LCA study are not deemed reliable and are not comparable either to other concrete studies, nor in absolute terms to other products and services. Hence, it was concluded that the errors need to be addressed by following an inclusive methodology. In order to cater for the intrinsic uncertainty in some of the elements of the methodology proposed, such as the means of transportation, the expected service life and fluctuations in market prices, it is advised to perform scenario analyses on each of these elements and run a Monte Carlo simulation to aggregate the uncertainty in the final presented LCA outcome.

Table 7. Summary of the sources of error from the review.

#	LCA Stage	Source of Discrepancy	Description	Outcome	Frequency	Category	Impact on LCA Outcome	Potential Solution
1	Stage 1: LCA Scope	System Boundaries	Disregarding the "use" phase	Ignoring potential carbon sequestration by concrete	90%	Reliability	2–20% overestimation	Deduct the sequestered carbon by predicting the carbonation potential of concrete throughout its expected service life
2			Disregarding the "use" phase	Assuming same service life for concrete alternatives	75%	Reliability	Underestimation (variable)	Include maintenance and/or replacement impact depending on the difference between the required and expected service life of concrete
3				Ignoring the operational energy consumed by concrete	75%	Precision	1–10% underestimation	Add an estimate of the operational energy consumed by concrete throughout its service life
4			Disregarding the "end-of-life" phase	Ignoring the impact of demolishing concrete	75%	Precision	1–10% underestimation	Add an estimate of the emissions and energy required to demolish the concrete alternative at the end of its service life
5				Ignores the avoided landfill impact for CDW waste	95%	Precision	1–10% overestimation	Deduct the impact from avoiding landfilling the waste that is being recycled as aggregates in the concrete mix
6		Functional Unit	Selecting a volume based FU	Ignores the functional properties of concrete	65%	Reliability	Variable	Select a functional unit that reflects performance based specifications such as strength and predicted service life
7			Selecting a unit level of details	Missing out on the potential of optimizing the total volume required	70%	Precision	Variable	If possible, select a "whole structure" LoD to run different scenarios optimizing the impact based on strength and volume of concrete

Table 7. *Cont.*

#	LCA Stage	Source of Discrepancy	Description	Outcome	Frequency	Category	Impact on LCA Outcome	Potential Solution
8		Data source	Relying on secondary sources	Not reliable and sometimes irrelevant data	45%	Reliability	Variable	Whenever possible, rely on primary sources such as EPDs and certified lab results
9	Stage 2: Inventory Data	Data variability	Variability in upstream data	Higher uncertainty in the values for impact and cost of transportation and raw materials	NA	Precision	Variable	Whenever needed, perform scenario analyses to measure the sensitivity of the outcome to the potential variability in upstream data
10		Impact allocation	Impact allocation for SCM	Ignoring the impact allocation for SCMs recycled in concrete	75%	Reliability	Variable	If the difference between the price of main and secondary process generating the SCM product is > 25%; economic allocation should be applied
11	Stage 3: Impact Assessment	Indicators	Using end-point indicators	Carries large uncertainties in correlating cause-effect environmental impact relationships	5%	Reliability	Variable	Use a combination of mid-point indicators to calculate the impact for concrete such as GWP, EP, ODP, AP, CED
12			Using CO_2 as GWP	It ignores the impact associated with methane and nitrogen dioxide as GHG emissions	10%	Reliability	10–30% underestimation	Use a standard methodology such as CML or TRACI that characterizes the different GHG emissions contributing to the GWP indicator
13	Stage4: Interpretation of results	Absolute judgements	Deterministic LCA outcome	Ignores the uncertainties aforementioned in the nature of the upstream data	90%	Reliability	NA	Perform scenario analyses, quantify the uncertainties in each of the elements then do a Monti Carlo simulation to aggregate the uncertainties of the indicators

Author Contributions: Conceptualization, H.H.; methodology, H.H.; software, H.H.; validation, H.H.; formal analysis, H.H.; investigation, H.H.; resources, H.H.; data curation, H.H.; writing—original draft preparation, H.H.; writing—review and editing, H.H., W.M.C., B.N. and R.K.; visualization, H.H.; supervision, W.M.C., B.N. and R.K.; project administration, W.M.C., B.N. and R.K.; funding acquisition, W.M.C., B.N. and R.K.

Funding: This work is funded by the faculty of Engineering and Environment at Northumbria University, UK.

Acknowledgments: None.

Conflicts of Interest: The authors declare no conflict of interests.

References

1. Miller, S.A. Supplementary cementitious materials to mitigate greenhouse gas emissions from concrete: Can there be too much of a good thing? *J. Clean. Prod.* **2018**, *178*, 587–598. [CrossRef]
2. Serres, N.; Braymand, S.; Feugeas, F. Environmental evaluation of concrete made from recycled concrete aggregate implementing life cycle assessment. *J. Build. Eng.* **2016**, *5*, 24–33. [CrossRef]
3. Colangelo, F.; Forcina, A.; Farina, I.; Petrillo, A. Life Cycle Assessment (LCA) of different kinds of concrete containing waste for sustainable construction. *Buildings* **2018**, *8*, 70. [CrossRef]
4. Kurda, R.; Silvestre, J.D.; de Brito, J. Lifecycle assessment of concrete made with high volume of recycled concrete aggregates and fly ash. *Resour. Conserv. Recycl.* **2018**, *139*, 407–417. [CrossRef]
5. Habert, G.; de Lacaillerie, J.B.D.; Roussel, N. An environmental evaluation of geopolymer based concrete production: Reviewing current research trends. *J. Clean. Prod.* **2011**, *19*, 1229–1238. [CrossRef]
6. Ashby, M.F. *Materials and the Environment: Eco-informed Material Choice*; Elsevier Science: Amsterdam, The Netherlands, 2012; Chapter 10.
7. Miller, S.A.; Horvath, A.; Monteiro, P.J.M. Readily implemenTable techniques can cut annual CO_2 emissions from the production of concrete by over 20%. *Envion. Res. Lett.* **2016**, *11*, 074029. [CrossRef]
8. Yuli, S.; Dabo, G.; Heran, Z.; Jiamin, O.; Yuan, L.; Jing, M.; Zhifu, M.; Zhu, L.; Qiang, Z. China CO_2 emission accounts 1997–2015. *Sci. Data* **2018**, *5*. [CrossRef]
9. Miller, S.A.; John, V.M.; Pacca, S.A.; Horvath, A. Carbon dioxide reduction potential in the global cement industry by 2050. *Cem. Concr. Res.* **2017**, *114*. [CrossRef]
10. Viñuales, J.E.; Depledge, J.; Reiner, D.M.; Lees, E. Climate policy after the Paris 2015 climate conference. *Clim. Policy* **2017**, *17*, 1–8. [CrossRef]
11. Pretot, S.; Collet, F.; Garnier, C. Life cycle assessment of a hemp concrete wall: Impact of thickness and coating. *Build. Environ.* **2014**, *72*, 223–231. [CrossRef]
12. Shan, X.; Zhou, J.; Chang, V.W.C.; Yang, E.-H. Life cycle assessment of adoption of local recycled aggregates and green concrete in Singapore perspective. *J. Clean. Prod.* **2017**, *164*, 918–926. [CrossRef]
13. Turk, J.; Cotič, Z.; Mladenovič, A.; Šajna, A. Environmental evaluation of green concretes versus conventional concrete by means of LCA. *Waste Manag.* **2015**, *45*, 194–205. [CrossRef] [PubMed]
14. Tait, M.W.; Cheung, W.M. A comparative cradle-to-gate life cycle assessment of three concrete mix designs. *Int. J. Life Cycle Assess.* **2016**, *21*, 847–860. [CrossRef]
15. Tae, S.; Baek, C.; Shin, S. Life cycle CO_2 evaluation on reinforced concrete structures with high-strength concrete. *Environ. Impact Assess. Rev.* **2011**, *31*. [CrossRef]
16. Wu, P.; Xia, B.; Zhao, X. The importance of use and end-of-life phases to the life cycle greenhouse gas (GHG) emissions of concrete—A review. *Renew. Suatain. Energy Rev.* **2014**, *37*, 360–369. [CrossRef]
17. Marinković, S.; Dragaš, J.; Ignjatović, I.; Tošić, N. Environmental assessment of green concretes for structural use. *J. Clean. Prod.* **2017**, *154*, 633–649. [CrossRef]
18. Al-Ayish, N.; During, O.; Malaga, K.; Silva, N.; Gudmundsson, K. The influence of supplementary cementitious materials on climate impact of concrete bridges exposed to chlorides. *Constr. Build. Mater.* **2018**, *188*, 391–398. [CrossRef]
19. Bilim, C.; Atiş, C.D.; Tanyildizi, H.; Karahan, O. Predicting the compressive strength of ground granulated blast furnace slag concrete using artificial neural network. *Adv. Eng. Softw.* **2009**, *40*, 334–340. [CrossRef]
20. Biswas, W.K.; Alhorr, Y.; Lawania, K.K.; Sarker, P.K.; Elsarrag, E. Life cycle assessment for environmental product declaration of concrete in the Gulf States. *Sustain. Cities Soc.* **2017**, *35*, 36–46. [CrossRef]

21. Celik, K.; Meral, C.; Gursel, A.P.; Mehta, P.K.; Horvath, A.; Monteiro, P.J.M. Mechanical properties, durability, and life-cycle assessment of self-consolidating concrete mixtures made with blended portland cements containing fly ash and limestone powder. *Cem. Concr. Compos.* **2015**, *56*, 59–72. [CrossRef]
22. Cheng, S.; Shui, Z.; Yu, R.; Zhang, X.; Zhu, S. Durability and environment evaluation of an eco-friendly cement-based material incorporating recycled chromium containing slag. *J. Clean. Prod.* **2018**, *185*, 23–31. [CrossRef]
23. De Schepper, M.; van den Heede, P.; van Driessche, I.; de Belie, N. Life cycle assessment of completely recyclable concrete. *Materials* **2014**, *7*, 6010–6027. [CrossRef] [PubMed]
24. Ding, T.; Xiao, J.; Tam, V.W.Y. A closed-loop life cycle assessment of recycled aggregate concrete utilization in China. *Waste Manag.* **2016**, *56*, 367–375. [CrossRef] [PubMed]
25. Einsfeld, R.A.; Velasco, M.S.L. Fracture parameters for high-performance concrete. *Cem. Concr. Res.* **2006**, *36*, 576–583. [CrossRef]
26. Felekoğlu, B.; Türkel, S.; Baradan, B. Effect of water/cement ratio on the fresh and hardened properties of self-compacting concrete. *Build. Environ.* **2007**, *42*, 1795–1802. [CrossRef]
27. Fan, C.; Miller, S.A. Reducing greenhouse gas emissions for prescribed concrete compressive strength. *Constr. Build. Mater.* **2018**, *167*, 918–928. [CrossRef]
28. Garcez, M.R.; Rohden, A.B.; de Godoy, L.G.G. The role of concrete compressive strength on the service life and life cycle of a RC structure: Case study. *J. Clean. Prod.* **2018**, *172*, 27–38. [CrossRef]
29. García-Segura, T.; Yepes, V.; Alcalá, J. Life cycle greenhouse gas emissions of blended cement concrete including carbonation and durability. *Int. J. Life Cycle Assess.* **2014**, *19*, 3–12. [CrossRef]
30. Gettu, R.; Pillai, R.; Santhanam, M.; Basavaraj, A.; Rathnarajan, S.; Dhanya, B. Sustainability-based decision support framework for choosing concrete mixture proportions. *Mater. Struct.* **2018**, *51*, 1–16. [CrossRef]
31. Gursel, A.P.; Maryman, H.; Ostertag, C. A life-cycle approach to environmental, mechanical, and durability properties of "green" concrete mixes with rice husk ash. *J. Clean. Prod.* **2016**, *112*, 823–836. [CrossRef]
32. Kleijer, A.L.; Lasvaux, S.; Citherlet, S.; Viviani, M. Product-specific Life Cycle Assessment of ready mix concrete: Comparison between a recycled and an ordinary concrete. *Resour. Conserv. Recycl.* **2017**, *122*, 210–218. [CrossRef]
33. Miller, S.A.; Monteiro, P.J.M.; Ostertag, C.P.; Horvath, A. Comparison indices for design and proportioning of concrete mixtures taking environmental impacts into account. *Cem. Concr. Compos.* **2016**, *68*, 131–143. [CrossRef]
34. Ignacio, J.N.; Víctor, Y.; José, V.M. Life Cycle Cost Assessment of Preventive Strategies Applied to Prestressed Concrete Bridges Exposed to Chlorides. *Sustainability* **2018**, *10*, 845. [CrossRef]
35. Oner, A.; Akyuz, S.; Yildiz, R. An experimental study on strength development of concrete containing fly ash and optimum usage of fly ash in concrete. *Cem. Concr. Res.* **2005**, *35*, 1165–1171. [CrossRef]
36. Panesar, D.K.; Churchill, C.J. The influence of design variables and environmental factors on life-cycle cost assessment of concrete culverts. *Struct. Infrastruct. Eng.* **2010**, *9*, 1–13. [CrossRef]
37. Park, J.; Tae, S.; Kim, T. Life cycle CO_2 assessment of concrete by compressive strength on construction site in Korea. *Renew. Sustain. Energy Rev.* **2012**, *16*, 2940–2946. [CrossRef]
38. Poon, C.S.; Lam, L.; Wong, Y.L. A study on high strength concrete prepared with large volumes of low calcium fly ash. *Cem. Concr. Res.* **2000**, *30*. [CrossRef]
39. Robayo-Salazar, R.; Mejía-Arcila, J.; de Gutiérrez, R.M.; Martínez, E. Life cycle assessment (LCA) of an alkali-activated binary concrete based on natural volcanic pozzolan: A comparative analysis to OPC concrete. *Constr. Build. Mater.* **2018**, *176*, 103–111. [CrossRef]
40. Rohden, A.B.; Garcez, M.R. Increasing the sustainability potential of a reinforced concrete building through design strategies: Case study. *Case Stud. Constr. Mater.* **2018**, *9*, e00174. [CrossRef]
41. Salas, D.A.; Ramirez, A.D.; Ulloa, N.; Baykara, H.; Boero, A.J. Life cycle assessment of geopolymer concrete. *Constr. Build. Mater.* **2018**, *190*, 170–177. [CrossRef]
42. Sandanayake, M.; Gunasekara, C.; Law, D.; Zhang, G.; Setunge, S. Greenhouse gas emissions of different fly ash based geopolymer concretes in building construction. *J. Clean. Prod.* **2018**, *204*, 399–408. [CrossRef]
43. Siddique, R. Performance characteristics of high-volume Class F fly ash concrete. *Cem. Concr. Res.* **2004**, *34*, 487–493. [CrossRef]

44. Souto-Martinez, A.; Delesky, E.A.; Foster, K.E.O.; Srubar, W.V., III. A mathematical model for predicting the carbon sequestration potential of ordinary portland cement (OPC) concrete. *Constr. Build. Mater.* **2017**, *147*, 417. [CrossRef]

45. Teixeira, E.R.; Mateus, R.; Camões, A.F.; Bragança, L.; Branco, F.G. Comparative environmental life-cycle analysis of concretes using biomass and coal fly ashes as partial cement replacement material. *J. Clean. Prod.* **2016**, *112*, 2221–2230. [CrossRef]

46. Van Den Heede, P.; de Belie, N. *Durability Related Functional Units for Life Cycle Assessment of High-Volume Fly Ash Concrete*; UWM Center for By-Products Utilization: Milwaukee, WI, USA, 2010; pp. 583–594.

47. Van Den Heede, P.; de Belie, N. Accelerated and natural carbonation of concrete with high volumes of fly ash: Chemical, mineralogical and microstructural effects. *R. Soc. Open Sci.* **2018**, *6*, 181665. [CrossRef] [PubMed]

48. Yazdanbakhsh, A.; Bank, L.; Baez, T.; Wernick, I. Comparative LCA of concrete with natural and recycled coarse aggregate in the New York City area. *Int. J. Life Cycle Assess.* **2018**, *23*, 1163–1173. [CrossRef]

49. Zhang, Y.-R.; Wu, W.-J.; Wang, Y.-F. Bridge life cycle assessment with data uncertainty. *Int. J. Life Cycle Assess.* **2016**, *21*, 569–576. [CrossRef]

50. Dhanya, B.S.; Santhanam, M.; Gettu, R.; Pillai, R.G. Performance evaluation of concretes having different supplementary cementitious material dosages belonging to different strength ranges. *Constr. Build. Mater.* **2018**, *187*, 984–995. [CrossRef]

51. Jiang, M.; Chen, X.; Rajabipour, F.; Hendrickson, C.T. Comparative Life Cycle Assessment of Conventional, Glass Powder, and Alkali-Activated Slag Concrete and Mortar. *J. Infrastruct. Syst.* **2014**, *20*. [CrossRef]

52. Guo, Z.; Tu, A.; Chen, C.; Lehman, D.E. Mechanical properties, durability, and life-cycle assessment of concrete building blocks incorporating recycled concrete aggregates. *J. Clean. Prod.* **2018**, *199*, 136–149. [CrossRef]

53. Anastasiou, E.K.; Liapis, A.; Papayianni, I. Comparative life cycle assessment of concrete road pavements using industrial by-products as alternative materials. *Resour. Conserv. Recycl.* **2015**, *101*, 1–8. [CrossRef]

54. Teh, S.H.; Wiedmann, T.; Castel, A.; de Burgh, J. Hybrid life cycle assessment of greenhouse gas emissions from cement, concrete and geopolymer concrete in Australia. *J. Clean. Prod.* **2017**, *152*, 312–320. [CrossRef]

55. Hafez, H.; Cheung, W.M.; Nagaratnam, B.; Kurda, R. A Proposed Performance Based Approach for Life Cycle Assessment of Reinforced Blended Cement Concrete. In Proceedings of the 5th SCMT Conference, Kingston University, Kingston, UK, 14–17 July 2019; pp. 50–61.

56. Panesar, D.; Seto, K.; Churchill, C. Impact of the selection of functional unit on the life cycle assessment of green concrete. *Int. J. Life Cycle Assess.* **2017**, *22*, 1969–1986. [CrossRef]

57. Sagastume Gutiérrez, A.; Eras, J.J.C.; Gaviria, C.A.; van Caneghem, J.; Vandecasteele, C. Improved selection of the functional unit in environmental impact assessment of cement. *J. Clean. Prod.* **2017**, *168*, 463–473. [CrossRef]

58. Bjørn, A.; Hauschild, M. Introducing carrying capacity-based normalisation in LCA: Framework and development of references at midpoint level. *Int. J. Life Cycle Assess.* **2015**, *20*, 1005–1018. [CrossRef]

59. Menoufi, K.A.I. *Life Cycle Analysis and Life Cyle Impact Assessment Methodologies: A State of the Art*; Castell, A., Cabeza, L.F., Eds.; Universitat De Lleida, Escola Politècnica: Liberia, Spain, 2011.

60. Huijbregts, M. Application of uncertainty and variability in LCA. *Int. J. Life Cycle Assess.* **1998**, *3*, 273–280. [CrossRef]

61. Häfliger, I.-F.; John, V.; Passer, A.; Lasvaux, S.; Hoxha, E.; Saade, M.R.M.; Habert, G. Buildings environmental impacts' sensitivity related to LCA modelling choices of construction materials. *J. Clean. Prod.* **2017**, *156*, 805–816. [CrossRef]

62. Collins, F. Inclusion of carbonation during the life cycle of built and recycled concrete: Influence on their carbon footprint. *Int. J. Life Cycle Assess.* **2010**, *15*, 549–556. [CrossRef]

63. Gursel, A.P. Life-Cycle Assessment of Concrete: Decision-Support Tool and Case Study Application. *Int. J. Life Cycle Assess.* **2014**, *19*. [CrossRef]

64. Kim, T.; Chae, C.U. Evaluation analysis of the CO_2 emission and absorption life cycle for precast concrete in Korea. *Sustainability (Switzerland)* **2016**, *8*, 663. [CrossRef]

65. Lee, S.; Park, W.; Lee, H. Life cycle CO_2 assessment method for concrete using CO_2 balance and suggestion to decrease $LCCO_2$ of concrete in South-Korean apartment. *Energy Build.* **2012**, *58*. [CrossRef]

66. Zhang, Y.; Zhang, J.; Lu, M.; Wang, J.D.; Gao, Y. Considering uncertainty in life-cycle carbon dioxide emissions of fly ash concrete. *Proc. Inst. Civ. Eng. Eng. Sustain.* **2019**, *172*, 198–206. [CrossRef]

67. Gayarre, F.L.; Pérez, J.G.; Pérez, C.L.; López, M.S.; Martínez, A.L. Life cycle assessment for concrete kerbs manufactured with recycled aggregates. *J. Clean. Prod.* **2016**, *113*, 41–53. [CrossRef]

68. Zhang, Y.; Luo, W.; Wang, J.; Wang, Y.; Xu, Y.; Xiao, J. A review of life cycle assessment of recycled aggregate concrete. *Constr. Build. Mater.* **2019**, *209*, 115–125. [CrossRef]

69. Kurda, R.; de Brito, J.; Silvestre, J. CONCRETop—A multi-criteria decision method for concrete optimization. *Environ. Impact Assess. Rev.* **2019**, *74*, 73. [CrossRef]

70. Hedayatinia, F.; Delnavaz, M.; Emamzadeh, S.S. Rheological properties, compressive strength and life cycle assessment of self-compacting concrete containing natural pumice pozzolan (Book review). *Constr. Build. Mater.* **2019**, *206*, 122–129. [CrossRef]

71. Mahima, S.; Moorthi, P.; Bahurudeen, A.; Gopinath, A. Influence of chloride threshold value in service life prediction of reinforced concrete structures. *Sādhanā* **2018**, *43*, 1–19. [CrossRef]

72. Müller, H.S.; Haist, M.; Vogel, M. Assessment of the sustainability potential of concrete and concrete structures considering their environmental impact, performance and lifetime. *Constr. Build. Mater.* **2014**, *67*, 321–337. [CrossRef]

73. Anand, C.K.; Amor, B. Recent developments, future challenges and new research directions in LCA of buildings: A critical review. *Renew. Suatain. Energy Rev.* **2017**, *67*, 408–416. [CrossRef]

74. Del Borghi, A. LCA and communication: Environmental Product Declaration (Editorial). *Int. J. Life Cycle Assess.* **2013**, *18*, 293. [CrossRef]

75. Crossin, E. *Comparative Life Cycle Assessment of Concrete Blends*; RMIT University: Melbourne, Australia, 2012.

76. Huntzinger, D.N.; Eatmon, T.D. A life-cycle assessment of Portland cement manufacturing: Comparing the traditional process with alternative technologies. *J. Clean. Prod.* **2009**, *17*, 668–675. [CrossRef]

77. Chen, C.; Habert, G.; Bouzidi, Y.; Jullien, A. Ventura, A. LCA allocation procedure used as an incitative method for waste recycling: An application to mineral additions in concrete. *Resour. Conserv. Recycl.* **2010**, *54*, 1231–1240. [CrossRef]

78. Yang, K.-H.; Seo, E.-A.; Jung, Y.-B.; Tae, S.-H. Effect of Ground Granulated Blast-Furnace Slag on Life-Cycle Environmental Impact of Concrete. *J. Korea Concr. Inst.* **2014**, *26*, 13–21. [CrossRef]

79. Zhang, Y.R.; Liu, M.H.; Xie, H.B.; Wang, Y.F. Assessment of CO_2 emissions and cost in fly ash concrete. In *Environment, Energy and Applied Technology, Proceedings of the 2014 International Conference on Frontier of Energy and Environment Engineering (ICFEEE 2014), Beijing, China, 6–7 December 2014*; CRC Press: Boca Raton, FL, USA, 2014; p. 327.

80. Wałach, D.; Dybeł, P.; Sagan, J.; Gicala, M. Environmental performance of ordinary and new generation concrete structures—A comparative analysis. *Environ. Sci. Pollut. Res.* **2019**, *26*, 3980–3990. [CrossRef] [PubMed]

81. Rahla, K.M.; Mateus, R.; Bragança, L. Comparative sustainability assessment of binary blended concretes using Supplementary Cementitious Materials (SCMs) and Ordinary Portland Cement (OPC). *J. Clean. Prod.* **2019**, *220*, 445–459. [CrossRef]

82. Chiaia, B.; Fantilli, A.P.; Guerini, A.; Volpatti, G.; Zampini, D. Eco-mechanical index for structural concrete. *Constr. Build. Mater.* **2014**, *67*, 386–392. [CrossRef]

83. Long, G.; Gao, Y.; Xie, Y. Designing more sustainable and greener self-compacting concrete. *Constr. Build. Mater.* **2015**, *84*, 301–306. [CrossRef]

84. Flower, D.; Sanjayan, J. Green house gas emissions due to concrete manufacture. *Int. J. Life Cycle Assess.* **2007**, *12*, 282–288. [CrossRef]

85. Proske, T.; Hainer, S.; Rezvani, M.; Graubner, C.-A. Eco-friendly concretes with reduced water and cement content—Mix design principles and application in practice. *Constr. Build. Mater.* **2014**, *67*, 413–421. [CrossRef]

86. Chen, X.; Wang, H.; Najm, H.; Venkiteela, G.; Hencken, J. Evaluating engineering properties and environmental impact of pervious concrete with fly ash and slag. *J. Clean. Prod.* **2019**, *237*, 117714. [CrossRef]

87. Gursel, A.P.; Ostertag, C.P. Impact of Singapore's importers on life-cycle assessment of concrete. *J. Clean. Prod.* **2016**, *118*, 140–150. [CrossRef]

88. Khodabakhshian, A.; de Brito, J.; Ghalehnovi, M.; Shamsabadi, E.A. Mechanical environmental and economic performance of structural concrete containing silica fume and marble industry waste powder. *Constr. Build. Mater.* **2018**, *169*, 237–251. [CrossRef]

89. Li, C.; Nie, Z.; Cui, S.; Gong, X.; Wang, Z.; Meng, X. The life cycle inventory study of cement manufacture in China. *J. Clean. Prod.* **2014**, *72*, 204–211. [CrossRef]

90. Seto, K.; Panesar, D.; Churchill, C. Criteria for the evaluation of life cycle assessment software packages and life cycle inventory data with application to concrete. *Int. J. Life Cycle Assess.* **2017**, *22*, 694–706. [CrossRef]

91. Tucker, E.L.; Ferraro, C.C.; Laux, S.J.; Townsend, T.G. Economic and life cycle assessment of recycling municipal glass as a pozzolan in portland cement concrete production. *Resour. Conserv. Recycl.* **2018**, *129*, 240–247. [CrossRef]

92. Wang, J.; Wang, Y.; Sun, Y.; Tingley, D.D.; Zhang, Y. Life cycle sustainability assessment of fly ash concrete structures. *Renew. Sustain. Energy Rev.* **2017**, *80*, 1162–1174. [CrossRef]

93. Yuan, X.; Tang, Y.; Li, Y.; Wang, Q.; Zuo, J.; Song, Z. Environmental and economic impacts assessment of concrete pavement brick and permeable brick production process—A case study in China. *J. Clean. Prod.* **2018**, *171*, 198–208. [CrossRef]

94. Kim, T.; Lee, S.; Chae, C.U.; Jang, H.; Lee, K. Development of the CO_2 emission evaluation tool for the life cycle assessment of concrete. *Sustainability (Switzerland)* **2017**, *9*, 2116. [CrossRef]

95. Panesar, D.K.; Kanraj, D.; Abualrous, Y. Effect of transportation of fly ash: Life cycle assessment and life cycle cost analysis of concrete. *Cem. Concr. Compos.* **2019**, *99*, 214–224. [CrossRef]

96. O'Brien, K.; Ménaché, J.; O'Moore, L. Impact of fly ash content and fly ash transportation distance on embodied greenhouse gas emissions and water consumption in concrete. *Int. J. Life Cycle Assess.* **2009**, *14*, 621–629. [CrossRef]

97. Chrysostomou, C.; Kylili, A.; Nicolaides, D.; Fokaides, P.A. Life Cycle Assessment of concrete manufacturing in small isolated states: The case of Cyprus. *Int. J. Sustain. Energy* **2017**, *36*, 825–839. [CrossRef]

98. Maria, A.; Salman, M.; Dubois, M.; Acker, K. Life cycle assessment to evaluate the environmental performance of new construction material from stainless steel slag. *Int. J. Life Cycle Assess.* **2018**, *23*, 2091–2109. [CrossRef]

99. Sayagh, S.; Ventura, A.; Hoang, T.; François, D.; Jullien, A. Sensitivity of the LCA allocation procedure for BFS recycled into pavement structures. *Resour. Conserv. Recycl.* **2010**, *54*, 348–358. [CrossRef]

100. Maia de Souza, D.; Lafontaine, M.; Charron-Doucet, F.; Chappert, B.; Kicak, K.; Duarte, F.; Lima, L. Comparative life cycle assessment of ceramic brick, concrete brick and cast-in-place reinforced concrete exterior walls. *J. Clean. Prod.* **2016**, *137*, 70–82. [CrossRef]

101. De Matos, P.R.; Sakata, R.D.; Prudêncio, L.R. Eco-efficient low binder high-performance self-compacting concretes. *Constr. Build. Mater.* **2019**, *225*, 941–955. [CrossRef]

102. Pillai, R.G.; Gettu, R.; Santhanam, M.; Rengaraju, S.; Dhandapani, Y.; Rathnarajan, S.; Basavaraj, A.S. Service life and life cycle assessment of reinforced concrete systems with limestone calcined clay cement (LC3). *Cem. Concr. Res.* **2019**, *118*, 111. [CrossRef]

103. Passuello, A.; Rodríguez, E.D.; Hirt, E.; Longhi, M.; Bernal, S.A.; Provis, J.L.; Kirchheim, A.P. Evaluation of the potential improvement in the environmental footprint of geopolymers using waste-derived activators. *J. Clean. Prod.* **2017**, *166*, 680–689. [CrossRef]

104. Boesch, M.E.; Hellweg, S. Identifying improvement potentials in cement production with life cycle assessment. *Environ. Sci. Technol.* **2010**, *44*, 9143–9149. [CrossRef] [PubMed]

105. Broun, R.; Menzies, G.F. Life Cycle Energy and Environmental Analysis of Partition Wall Systems in the UK. *Proced. Eng.* **2011**, *21*, 864–873. [CrossRef]

106. Budelmann, H.; Holst, A.; Wachsmann, A. *Durability Related Life-Cycle Assessment of Concrete Structures: Mechanisms, Models, Implementation*; IALCCE: Vienna, Austria, 2012; pp. 75–86.

107. Chau, C.K.; Leung, T.M.; Ng, W.Y. A review on Life Cycle Assessment, Life Cycle Energy Assessment and Life Cycle Carbon Emissions Assessment on buildings. *Appl. Energy* **2015**, *143*, 395–413. [CrossRef]

108. Cheung, J.; Roberts, L.; Liu, J. Admixtures and sustainability. *Cem. Concr. Res.* **2018**, *114*, 79–89. [CrossRef]

109. Colangelo, F.; Petrillo, A.; Cioffi, R.; Borrelli, C.; Forcina, A. Life cycle assessment of recycled concretes: A case study in southern Italy. *Sci. Total Environ.* **2018**, *615*, 1506–1517. [CrossRef] [PubMed]

110. D'Alessandro, A.; Fabiani, C.; Pisello, A.L.; Ubertini, F.; Materazzi, A.L.; Cotana, F. Innovative concretes for low-carbon constructions: A review. *Int. J. Low Carbon Technol.* **2017**, *12*, 289–309. [CrossRef]

111. Damineli, B.L.; Kemeid, F.M.; Aguiar, P.S.; John, V.M. Measuring the eco-efficiency of cement use. *Cem. Concr. Compos.* **2010**, *32*, 555–562. [CrossRef]

112. Densley Tingley, D.; Davison, B. Developing an LCA methodology to account for the environmental benefits of design for deconstruction. *Build. Environ.* **2012**, *57*, 387–395. [CrossRef]

113. Dobbelaere, G.; de Brito, J.; Evangelista, L. Definition of an equivalent functional unit for structural concrete incorporating recycled aggregates. *Eng. Struct.* **2016**, *122*, 196–208. [CrossRef]

114. Dong, Y.H.; Ng, S.T.; Kwan, A.H.K.; Wu, S.K. Substituting local data for overseas life cycle inventories e a case study of concrete products in Hong Kong. *J. Clean. Prod.* **2015**, *108*, 414–422. [CrossRef]

115. Estanqueiro, B.; Silvestre, J.D.; de Brito, J.; Pinheiro, M.D. Environmental life cycle assessment of coarse natural and recycled aggregates for concrete. *Eur. J. Environ. Civ. Eng.* **2018**, *22*, 429–449. [CrossRef]

116. Evangelista, B.L.; Rosado, L.P.; Penteado, C.S.G. Life cycle assessment of concrete paving blocks using electric arc furnace slag as natural coarse aggregate substitute. *J. Clean. Prod.* **2018**, *178*, 176–185. [CrossRef]

117. Evangelista, P.P.A.; Kiperstok, A.; Torres, E.A.; Gonçalves, J.P. Environmental performance analysis of residential buildings in Brazil using life cycle assessment (LCA). *Constr. Build. Mater.* **2018**, *169*, 748–761. [CrossRef]

118. Fantilli, A.P.; Tondolo, F.; Chiaia, B.; Habert, G. Designing reinforced concrete beams containing supplementary cementitious materials. *Materials* **2019**, *12*, 1248. [CrossRef] [PubMed]

119. Ferreiro-Cabello, J.; Fraile-Garcia, E.; Martinez-Camara, E.; Perez-de-La-Parte, M. Sensitivity analysis of Life Cycle Assessment to select reinforced concrete structures with one-way slabs. *Eng. Struct.* **2017**, *132*, 586–596. [CrossRef]

120. Font, A.; Borrachero, M.V.; Soriano, L.; Monz, J.; Mellado, A.; Pay, J. New eco-cellular concretes: Sustainable and energy-efficient materials. *Green Chem.* **2018**, *20*, 4684–4694. [CrossRef]

121. Fraile-Garcia, E.; Ferreiro-Cabello, J.; Martinez-Camara, E.; Jimenez-Macias, E. Repercussion the use phase in the life cycle assessment of structures in residential buildings using one-way slabs. *J. Clean. Prod.* **2017**, *143*, 191–199. [CrossRef]

122. Franco de Carvalho, J.M.; Melo, T.V.d.; Fontes, W.C.; Batista, J.O.D.S.; Brigolini, G.J.; Peixoto, R.A.F. More eco-efficient concrete: An approach on optimization in the production and use of waste-based supplementary cementing materials (Book review). *Constr. Build. Mater.* **2019**, *206*, 397–409. [CrossRef]

123. Gómez de Cózar, J.C.; Martínez, A.G.; López, Í.A.; Alfonsea, M.R. Life cycle assessment as a decision-making tool for selecting building systems in heritage intervention: Case study of Roman Theatre in Itálica, Spain. *J. Clean. Prod.* **2019**, *206*, 27–39. [CrossRef]

124. Göswein, V.; Rodrigues, C.; Silvestre, J.D.; Freire, F.; Habert, G.; König, J. Using anticipatory life cycle assessment to enable future sustainable construction. *J. Ind. Ecol.* **2019**. [CrossRef]

125. Gursel, A.; Masanet, E.; Horvath, A.; Stadel, A. Life-cycle inventory analysis of concrete production: A critical review. *Cem. Concr. Compos.* **2014**, *51*. [CrossRef]

126. Hong, T.; Ji, C.; Park, H. Integrated model for assessing the cost and CO_2 emission (IMACC) for sustainable structural design in ready-mix concrete. *J. Environ. Manag.* **2012**, *103*, 1–8. [CrossRef] [PubMed]

127. Hong, T.-H.; Ji, C.-Y.; Jang, M.-H.; Park, H.-S. Predicting the CO_2 Emission of Concrete Using Statistical Analysis. *J. Constr. Eng. Proj. Manag.* **2012**, *2*, 53–60. [CrossRef]

128. Hossain, M.; Poon, C.; Lo, I.; Cheng, J. Evaluation of environmental friendliness of concrete paving eco-blocks using LCA approach. *Int. J. Life Cycle Assess.* **2016**, *21*, 70–84. [CrossRef]

129. Hossain, M.U.; Poon, C.S.; Lo, I.M.C.; Cheng, J.C.P. Comparative environmental evaluation of aggregate production from recycled waste materials and virgin sources by LCA. *Resour. Conserv. Recycl.* **2016**, *109*, 67–77. [CrossRef]

130. Iezzi, B.; Brady, R.; Sardag, S.; Eu, B.; Skerlos, S. Growing bricks: Assessing biocement for lower embodied carbon structures. *Proced. CIRP* **2019**, *80*, 470–475. [CrossRef]

131. Josa, A. Comparative analysis of the life cycle impact assessment of available cement inventories in the EU. *Cem. Concr. Res.* **2007**, *37*. [CrossRef]

132. Juhart, J.; David, G.-A.; Saade, M.R.M.; Baldermann, C.; Passer, A.; Mittermayr, F. Functional and environmental performance optimization of Portland cement-based materials by combined mineral fillers. *Cem. Concr. Res.* **2019**, *122*, 157–178. [CrossRef]

133. Kurda, R.; Silvestre, J.D.; de Brito, J.; Ahmed, H. Optimizing recycled concrete containing high volume of fly ash in terms of the embodied energy and chloride ion resistance. *J. Clean. Prod.* **2018**, *194*, 735–750. [CrossRef]

134. Liu, Y.; Shi, C.; Zhang, Z.; Li, N. An overview on the reuse of waste glasses in alkali-activated materials. *Resour. Conserv. Recycl.* **2019**, *144*, 297–309. [CrossRef]

135. Luo, W.; Sandanayake, M.; Zhang, G. Direct and indirect carbon emissions in foundation construction—Two case studies of driven precast and cast-in-situ piles (Report). *J. Clean. Prod.* **2019**, *211*, 1517. [CrossRef]

136. Miller, S.A.; Monteiro, P.J.M.; Ostertag, C.P.; Horvath, A. Concrete mixture proportioning for desired strength and reduced global warming potential. *Constr. Build. Mater.* **2016**, *128*, 410–421. [CrossRef]

137. Mohammadi, J.; South, W. Life cycle assessment (LCA) of benchmark concrete products in Australia. *Int. J. Life Cycle Assess.* **2017**, *22*, 1588–1608. [CrossRef]

138. Nikbin, I.M.; Aliaghazadeh, M.; Charkhtab, A.S.; Fathollahpour, A. Environmental impacts and mechanical properties of lightweight concrete containing bauxite residue (red mud). *J. Clean. Prod.* **2018**, *172*, 2683–2694. [CrossRef]

139. Omar, W.M.S.W.; Doh, J.-H.; Panuwatwanich, K.; Miller, D. Assessment of the embodied carbon in precast concrete wall panels using a hybrid life cycle assessment approach in Malaysia. *Sustain. Cities Soc.* **2014**, *10*, 101–111. [CrossRef]

140. Pradhan, S.; Tiwari, B.R.; Kumar, S.; Barai, S.V. Comparative LCA of recycled and natural aggregate concrete using Particle Packing Method and conventional method of design mix. *J. Clean. Prod.* **2019**, *228*, 679–691. [CrossRef]

141. Provis, J.L. Alkali-activated materials. *Cem. Concr. Res.* **2017**. [CrossRef]

142. Röyne, F. *Life Cycle Assessment of BioZEment—Concrete Production Based on Bacteria*; RISE—Research Institutes of Sweden, Built Environment, Energy and Circular Economy: Stockholm, Sweden, 2017.

143. Saade, M.R.M.; Passer, A.; Mittermayr, F. A Preliminary Systematic Investigation onto Sprayed Concrete's Environmental Performance. *Proced. CIRP* **2018**, *69*, 212–217. [CrossRef]

144. Säynäjoki, A.; Heinonen, J.; Junnila, S.; Horvath, A. Can life-cycle assessment produce reliable policy guidelines in the building sector? *Environ. Res. Lett.* **2017**, *12*, 013001. [CrossRef]

145. Säynäjoki, A.; Heinonen, J.; Junnonen, J.-M.; Junnila, S. Input–output and process LCAs in the building sector: Are the results compatible with each other? *Carbon Manag.* **2017**, *8*, 155–166. [CrossRef]

146. Seto, K.E.; Churchill, C.J.; Panesar, D.K. Influence of fly ash allocation approaches on the life cycle assessment of cement-based materials. *J. Clean. Prod.* **2017**, *157*, 65–75. [CrossRef]

147. Shrivastava, S.; Shrivastava, R.L. A systematic literature review on green manufacturing concepts in cement industries. *Int. J. Qual. Reliab. Manag.* **2017**, *34*, 68–90. [CrossRef]

148. Silva, R.V.; Neves, R.; de Brito, J.; Dhir, R.K. Carbonation behaviour of recycled aggregate concrete. *Cem. Concr. Compos.* **2015**, *62*, 22–32. [CrossRef]

149. Sinka, M.; van den Heede, P.; de Belie, N.; Bajare, D.; Sahmenko, G.; Korjakins, A. Comparative life cycle assessment of magnesium binders as an alternative for hemp concrete. *Resour. Conserv. Recycl.* **2018**, *133*, 288–299. [CrossRef]

150. Soleimani, M.; Shahandashti, M. Comparative process-based life-cycle assessment of bioconcrete and conventional concrete. *J. Eng. Des. Technol.* **2017**, *15*, 667–688. [CrossRef]

151. Suárez Silgado, S.; Valdiviezo, L.C.; Domingo, S.G.; Roca, X. Multi-criteria decision analysis to assess the environmental and economic performance of using recycled gypsum cement and recycled aggregate to produce concrete: The case of Catalonia (Spain). *Resour. Conserv. Recycl.* **2018**, *133*, 120–131. [CrossRef]

152. Tam, V.W.Y.; Soomro, M.; Evangelista, A.C.J. A review of recycled aggregate in concrete applications (2000–2017). *Constr. Build. Mater.* **2018**, *172*, 272–292. [CrossRef]

153. Tempest, B.; Sanusi, O.; Gergely, J.; Ogunro, V.; Weggel, D. Compressive Strength and Embodied Energy Optimization of Fly Ash Based Geopolymer Concrete. In Proceedings of the 2009 World of Coal Ash (WOCA) Conference, Lexington, KY, USA, 4–7 May 2009.

154. Tošić, N.; Marinković, S.; Dašić, T.; Stanić, M. Multicriteria optimization of natural and recycled aggregate concrete for structural use. *J. Clean. Prod.* **2015**, *87*, 766–776. [CrossRef]

155. Turner, L.K.; Collins, F.G. Carbon dioxide equivalent (CO_2-e) emissions: A comparison between geopolymer and OPC cement concrete. *Constr. Build. Mater.* **2013**, *43*, 125–130. [CrossRef]

156. Vieira, D.R.; Calmon, J.L.; Coelho, F.Z. Life cycle assessment (LCA) applied to the manufacturing of common and ecological concrete: A review. *Constr. Build. Mater.* **2016**, *124*, 656–666. [CrossRef]

157. Wijayasundara, M.; Mendis, P.; Crawford, R.H. Methodology for the integrated assessment on the use of recycled concrete aggregate replacing natural aggregate in structural concrete. *J. Clean. Prod.* **2017**, *166*, 321–334. [CrossRef]

158. Zhong, Y.; Wu, P. Economic sustainability, environmental sustainability and constructability indicators related to concrete- and steel-projects. *J. Clean. Prod.* **2015**, *108*, 748–756. [CrossRef]
159. Zingg, S.; Habert, G.; Lämmlein, T.; Lura, P.; Denarié, E.; Hajiesmaeili, A. Environmental Assessment of Radical Innovation in Concrete Structures. In Proceedings of the 2016 Sustainable Built Environment (SBE) Regional Conference, Zürich, Switzerland, 15–17 June 2016.
160. Hafez, H.; Kurda, R.; Cheung, W.M.; Nagaratnam, B. Comparative life cycle assessment between imported and recovered fly ash for blended cement concrete in the UK. *J. Clean. Prod.* **2019**, *244*, 118722. [CrossRef]

Review

A Critical Review on the Influence of Fine Recycled Aggregates on Technical Performance, Environmental Impact and Cost of Concrete

Hisham Hafez [1],*, Reben Kurda [2], Rawaz Kurda [3],*, Botan Al-Hadad [4], Rasheed Mustafa [5] and Barham Ali [6]

[1]   Department of Mechanical and Construction Engineering, University of Northumbria, Newcastle upon Tyne NE18ST, UK
[2]   Department of Information System Engineering, Erbil Technical Engineering College, Erbil Polytechnic University, Erbil, Kurdistan Region 44001, Iraq; Reben.kurda@epu.edu.iq
[3]   Department of Civil Engineering, Technical Engineering College, Erbil Polytechnic University, Erbil, Kurdistan Region 44001, Iraq
[4]   Erbil Technology Institute, Erbil Polytechnic University, Erbil, Kurdistan Region 44001, Iraq; botan@epu.edu.iq
[5]   Department of Environmental Engineering, College of Engineering, Knowledge University, Erbil, Kurdistan 44001, Iraq; rasheed1954@yahoo.com
[6]   Civil Engineering Department, Engineering Faculty, Tishk International University, Erbil, Kurdistan Region 44001, Iraq; barham.haydar@tiu.edu.iq
*   Correspondence: hisham.hafez@northumbria.ac.uk (H.H.); rawaz.kurda@epu.edu.iq (R.K.); Tel.: +44-7493280699 (H.H.); +351-965722755 (R.K.)

Received: 30 December 2019; Accepted: 27 January 2020; Published: 4 February 2020

Abstract: The aim of this critical review is to show the applicability of recycled fine aggregates (RFA) in concrete regarding technical performance, environmental impact, energy consumption and cost. It is not possible to judge the performance of concrete by considering one dimension. Thus, this study focussed on the fresh and hardened (e.g., mechanical and durability) properties and environmental and economic life cycle assessment of concrete. Most literature investigated showed that any addition of recycled fine aggregates from construction and demolition waste as a replacement for natural fine aggregates proves detrimental to the functional properties (quality) of the resulting concrete. However, the incorporation of recycled fine aggregates in concrete was proven to enhance the environmental and economic performance. In this study, an extensive literature review based multi criteria decision making analysis framework was made to evaluate the effect of RFA on functional, environmental, and economic parameters of concrete. The results show that sustainability of RFA based concrete is very sensitive to transportation distances. Several scenarios for the transportation distances of natural and recycled fine aggregates and their results show that only if the transportation distance of the natural aggregates is more than double that of RFA, e the RFA based concrete alternatives would be considered as more sustainable.

Keywords: fine recycled aggregates; construction and demolition waste; recycled aggregate concrete; life cycle assessment; sustainability; optimization

---

1. Introduction

Due to the rising need for urbanization, concrete usage is expected to be doubled by 2050 [1]. Given the fact that aggregates comprise around 70% by volume of concrete, alarming depletion rates of non-renewable natural resources such as river sand and limestone are expected [2]. Already, there are more than 30 billion tons of natural aggregates consumed annually for producing concrete globally.

The average increase in aggregates use is estimated at 5% per year globally, which means that the total could be around 66.3 billion tonnes by 2022 [3]. According to study [4], recycled aggregate (RA) still represents only three percent of the total aggregates demand. In Europe, construction and demolition waste (CDW) occupies one third of the waste produced, but significant differences can be found in the level of recycling among different countries [2]. Similarly, the construction industry is the second largest in generating waste, either during construction or demolition [5]. Four billion tonnes of CDW are generated in China yearly [6] and around 900 million tonnes in Europe [7]. Surprisingly, only 5% of the currently generated CDW is being recycled in concrete in any form [4]. Accordingly, special attention is devoted to the potential of recycling CDW as aggregates in concrete to simultaneously reduce the threat of depleting natural resources and avoid the landfilling of CDW [2,8,9].

In line with the European Union Directive [10] that encourages the reuse and recycling of waste materials, researchers examined the feasibility of creating sustainable concrete by replacing naturally sourced aggregates (NA) in concrete with recycled aggregates from CDW in terms of mechanical properties, durability, environmental impact (EI) and cost [11–16]. Generally, it was agreed among researchers that in addition to the potential of being a cheaper alternative, producing recycled aggregates concrete (RAC) could reduce the EI of concrete specifically in terms of "landfill use". However, it was also concluded that the technical performance would be adversely affected. Using RA, instead of NA, leads to a reduction of the mechanical properties and durability of a RAC compared to conventional concrete [17–21].

Regardless of the numerous studies found, there was no consensus found on the replacement ratio of fine NA by fine RA (RFA) from CDW that would achieve the optimum sustainability-potential of the resulting concrete mix. The sustainable development potential of a material depends on three main parameters: its functionality, its EI and economic viability [22]. Therefore, this paper will attempt to assess the optimum sustainability potential, based on a multi criteria decision analysis framework, which combines these three sources. However, the scope of the paper will only focus on RFA from CDW (RFA) being incorporated, with different percentages, in concrete replacing naturally sourced fine aggregates. Recently, there was a shift in the direction of the research concerning recycled aggregates focusing more on RFA made with glass, lightweight materials, etc. The reason is that these studies [23–39] recommended not to include CDW as RFA anymore just on the basis of its functional properties. For that reason, in this study, the applicability of RFA in concrete in terms of all parameters (quality, environmental impact, energy consumption, toxicity and cost) was considered. Regression models were developed based on data obtained from the literature, and then, a multi criteria decision making analysis framework was used to combine the results for the mentioned parameters and to find the optimum incorporation level of RFA in concrete. However, it is important to highlight that the statistical assumption that the raw data used in the regression models follow a normal distribution was not verified. The reason is that the sample sizes were too small to eliminate any of the data as will be shown later.

2. State-of-Art Review

2.1. General Parameters Affecting the Performance of RFA in Concrete

2.1.1. Recycling Process

The process by which CDW is turned into aggregates, whether coarse or fine, suitable for reintegration in concrete can be divided in several stages, namely separation by metal detectors, initial screening, crushing, secondary screening, secondary crushing and then sieving and separating [40]. This number of stages of crushing has a significant effect on the quality of the RFA produced and subsequently on the RAC which will include it. Source concretes subjected to a secondary crushing procedure in an impact crusher normally result in RFA with less attached mortar than those following only a primary crushing procedure. Thus, due to the hardened cement paste having higher porosity than that of NA, as the content of adhered-mortar increases, so does the RA's water absorption [41].

Kim and Yun [42] obtained RFA from demolished concrete of current compressive strength (21 MPa). By repeatedly crushing and sorting, the authors used water absorption ratios of 6% and 8% to evaluate the effect of water absorption ratio of the RFA on the bond strength between reinforcing bar and RAC. Lee [43] investigated two different crushing processes using a jaw crusher and an impact crusher. The work resulted in oven-dried density of 2390 kg/m^3 and 2280 kg/m^3 and water absorption of 6.59% and 10.35%, respectively. The process of washing the RA also has a strong effect on their WA. A study by Wegen and Haverkort [44] showed that, after RA were washed, the water absorption of the aggregates dropped by 35% to 55%. This can be explained by the fact that very small particles were removed by the water after washing the aggregates, which significantly affected the water absorption of the fine RA. Song and Ryou [45] washed fine RFA with different washing stages and used a combination of chemical and physical processes. The process of washing the RFA had several effects; the water absorption fell from 6% to 2%, the impurity content fell from 0.5% to 0.2% and the ratio of absolute volume increased from 62% to 65%. Hence, it could be concluded that it is advisable to use jaw crushers rather than impact ones and to do several cycled of washing and crushing the CDW. However, it should be noted that the more cycles done, the higher the energy use is, which will then reflect on the environmental performance of the resulting RFA as discussed in Section 2.3.

2.1.2. Particle Size

The recycling process though as explained in Section 2.1.1 could be very thorough, it does not detach the old mortar from the aggregates. Hence, as seen in Figure 1, NA have only one transition zone between the aggregate and the cement matrix, and it is a porous narrow band which forms at the cement past/aggregate interface called the interfacial transition zone (ITZ). For RFA there are two ITZ phases: a new ITZ between the new cement paste and the old mortar from the RFA and an old ITZ between the old cement paste and the original aggregate NA [46]. Moreover, the crushing process itself results in micro cracks in the adhered mortar/original aggregate. These pores and cracks in the RFA absorb water and lead to high water content, which then leads to weaker concrete.

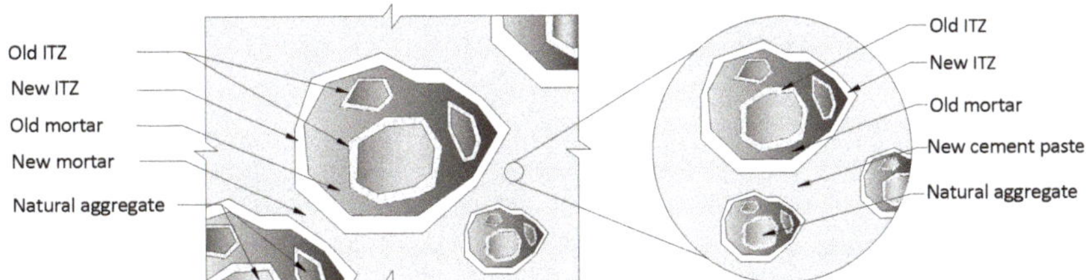

Figure 1. Structure of full replacement of fine and coarse naturally sourced aggregates (NA) with fine and coarse recycled aggregate (RA) and their transition zones.

According to study of Lima et al. [47], the higher the minimum particle size of the RFA is, the higher it would contain old adhered cement binder. Fineness modulus is a standard index that defines how fine or coarse a sample of sand is. A fineness modulus value smaller than 2.8 indicates fine sand, while a larger number indicates coarse sand. Evidence shows that the fineness modulus of the RFA incorporated in concrete has a significant effect on several functional parameters. First, for RFA with similar oven-dried density and lower fineness modulus (<2.8), the following studies [15,47–52] obtained higher water absorption compared to the results of the studies, which used RFA with higher fineness modulus (>2.8) [42,53–66]. Similarly, Geng and Sun [67] studied the carbonation behaviour of RFA concrete with different minimum particle sizes and incorporation level of RFA with two design compositions. In the first design, the water content was kept constant and in the second design the workability was kept constant. The authors reported that as the minimum particle sizes of RFA decreases, the carbonation depth of RFA based concrete increase. Contrary to the previous statement,

Geng and Sun [67] studied the effect of RFA's particle size and reported that, as seen in Figure 2, the compressive strength of concrete for the same incorporation levels of RFA decreased with increasing content of smaller fine RFA particles. The researchers cited the fact that larger RFA particles have higher adhesion properties due to the presence of a higher % of old adhered binder.

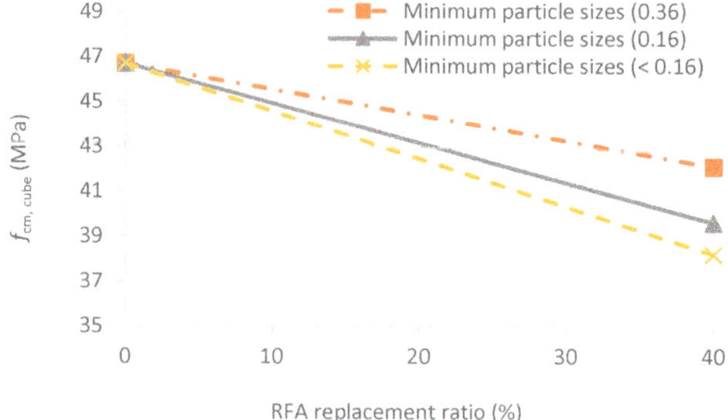

Figure 2. Influence of minimum fine recycled fine aggregates (RFA) particles size on 28 days strength of concrete [67].

2.1.3. Quality of the Source Material

Solyman [66] analysed the influence incorporation of different types and levels of fine RA on concrete's modulus of elasticity and workability of the resulting RAC. As seen in Figure 3, among 9 different types of CDW, it was concluded that using 80% old concrete +20% old bricks as RFA yields the least drop in workability and in elastic modulus with the higher incorporation rates. The reason could also be related to optimizing the water absorption potential.

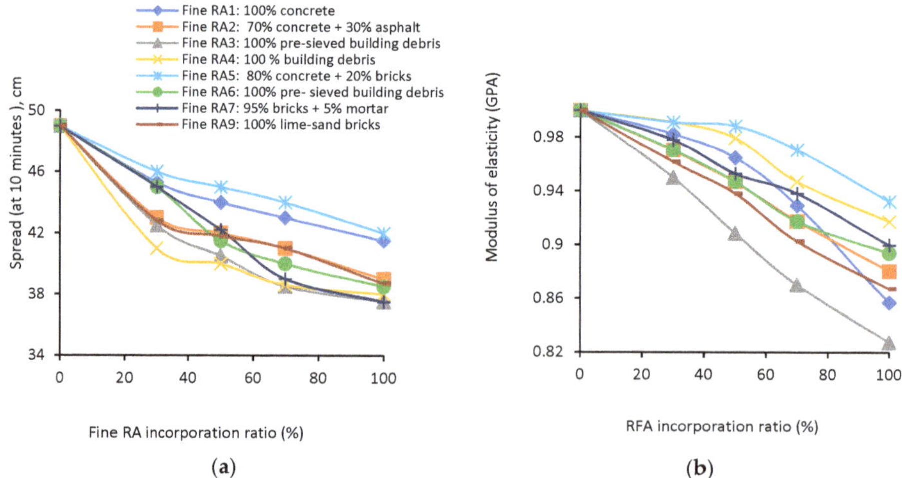

Figure 3. Effect of replacing different incorporation ratios and types of fine RA content on (**a**) workability and (**b**) relative modulus of elasticity of concrete [66].

2.2. The Effect of Varying RFA Incorporation Ratio on Concrete Performance

2.2.1. Workability

Figure 4 presents the relationship between the concrete slump and the incorporation levels of RFA, sourced from these studies [64,65,67]. A reasonable explanation for this may be that RFA absorbs more water from the mix than NA [47,68,69]. Moreover, the minimum particle size of RFA affects concrete's workability because smaller RFA have higher water absorption than the bigger ones from the same source concrete. Greater surface-area and higher water absorption of the fine particle sizes of RA have more potential for higher water-demand and decrease workability as a result [67]. The RFA fineness level and its effect on workability are other aspects that should be considered.

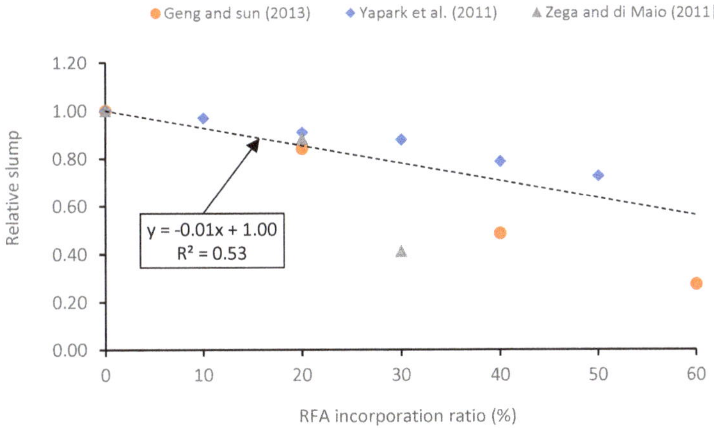

Figure 4. Relationship between RFA concrete and slump for similar w/c ratio.

Yaprak et al. [64] obtained slump values between 85 and 165 mm for fresh concrete that were made with different incorporation levels of RFA. The data of this study confirm that the slump of RFA concrete decreases as the incorporation level of RFA increases. This discrepancy might be due to the texture, shape, water absorption and dust content of the crushed RFA when compared to the natural sand. Geng and Sun [67] kept the water/binder (w/b) ratio constant at 0.40 and reported that RFA based concrete has a higher water adsorption leading to poor slump and workability at higher RFA incorporation ratios.

For normal RFA concrete without using superplasticizers (SP), these studies [54,70,71] showed that the w/c ratio needs to be increased as the incorporation ratio of RFA increased in the concrete mixes to get the same target slump. This could be achieved by adding 15% extra water to the RFA concrete [72]. The following studies by [73–76] confirmed this trend. Nevertheless, Leite [50] reported that the workability of concrete with high fineness of RFA may not decrease. According to Figure 4, it could be concluded from the literature that given a concrete mix has a starting value for slump equal to S_i, the incorporation of a X% of RFA as a replacement to fine NA would result in a slump value of Y_{slump} in Equation (1).

$$Y_{slump} = Si \times (1 - 0.05X) \tag{1}$$

2.2.2. Bulk Fresh Density

Generally, the density of RFA concrete is lower than that of the conventional concrete and their density decrease with increasing incorporation of RFA due to the presence of lower density residual-cement mortars attached to aggregate particles [73]. Figure 5 shows the influence of increasing incorporation of RFA on the bulk density of concrete mixes in the fresh state, sourced from these studies [54,64,70,71,77]. In order to avoid sacrificial pseudo-replication error [78], for each study, the data of high- and low-strength concrete mixes are separated then the global trends for each concrete are drawn. The fresh-bulk density of low-strength concrete mixes were lower than that of high-strength concrete mixes, and the fresh density of both concrete types decreased when replacing fine NA with RFA. The results show that the rate of reduction in the fresh density by increasing incorporation of RFA was similar for high- and normal-strength concrete.

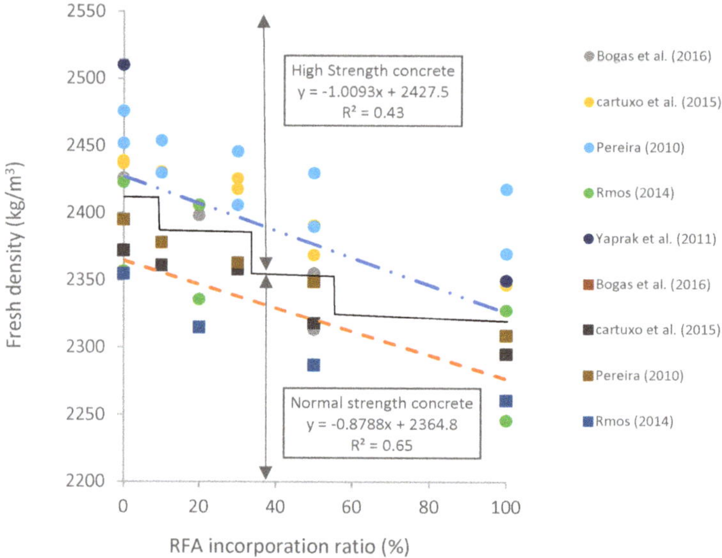

Figure 5. The effect of incorporating RFA on the fresh bulk density of high- and normal-strength concrete.

Bogas et al. [77] and Ramos [71] obtained similar results and both investigated three concrete mix families. The first family was normal-strength concrete without admixtures, the second family was high-strength concrete without air entrainment, and the third family was high-strength concrete with air entrainment. The fresh density of the three families decreased between 1% and 5% as the replacement ratio of fine NA with RA increased. The reduction range due to incorporating RFA did not change for high- and low-strength concrete. Moreover, the use of SP increased the fresh density of concrete by about 3% as a consequence of lowering the w/c ratio and increased the compactness of concrete as a result, while the use of air enter admixture offset the gain in density obtained by using the SP. Yaprak et al. [64] obtained unit weight values of fresh concrete mixes. The values changed between 2350 kg/m³ and 2510 kg/m³. The unit-weight value decreased as the RFA content in the concrete increased. The reason behind this was that the specific gravity of RFA was lower than that of fine NA.

2.2.3. Compressive Strength

The compressive strength of concrete is one of the most significant parameters that determine the capacity of a structural member to withstand certain loads in service. The compressive strength of RAC normally depends on the age of concrete, RFA incorporation level, additives, admixtures, w/c, quality of the source material and moisture content, type, and size of the RFA [79]. There are two contrary results drawn from different researches work on this subject. Figure 6 summarizes the results of the following studies and shows the influence of increasing incorporation of RFA on the compressive strength of concrete mixes over time. The majority of the researchers arrived to the same conclusion, i.e., the incorporation of RFA in concrete mixes is harmful in terms of compressive strength [49,64,67,71,77,80]. This is related to several factors, the most effective being the water content needed to increase the w/c in RAC mixes to archive the same workability to that of NA concrete. This can be related to the high-water absorption of the RFA and its texture and angular shape. Nevertheless, the results of these studies [15,42,70,71] show that RFA does not significantly influence the compressive strength of the low-strength concrete mixes [15,42,70,71] because the ultimate strength of low-strength concrete mainly depends on the quality of paste rather than aggregates.

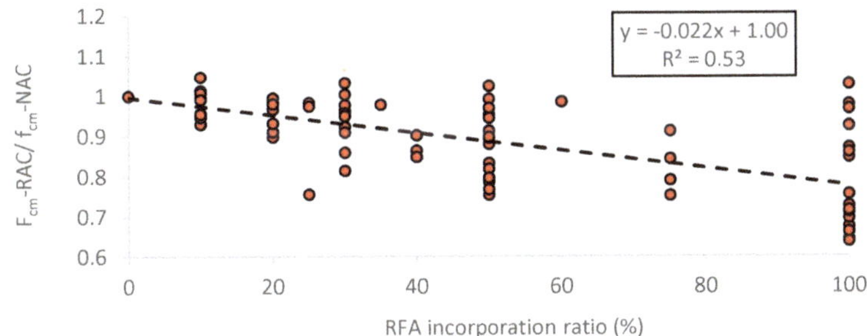

Figure 6. Effect of incorporation of RFA in concrete on compressive strength over time.

Evangelista and de Brito [15] reported that the compressive strength of concrete with up to 30% of RFA may not be jeopardized. In addition, the loss for 100% of RFA was only 7%. This behaviour can be explained by the fact that RFA increased the total cement content (during the crushing process of source concrete, some hydration products that trapped the original cement particles may break and release the non-hydrated cement particles), which can reach as much as 25% of its weight [81]. Kou and Poon [49] have demonstrated that replacing 25% to 50% of the fine NA with RFA does not affect the compressive strength of self-compacted concrete. Ahmed [82] confirmed that the compressive strength of concrete with replacement of up to 50% of fine NA with RFA was similar to that of the reference concrete, or even slightly higher. For higher replacement ratios, the maximum loss of compressive strength was 13% and 22% at 28 and 56 days, respectively. The results of Leite's [50] study show that the compressive strength of concrete mixes increased as the incorporation levels of RFA increased. The researcher believes that concrete gained strength by incorporating RFA because the roughness of RFA is bigger than that of fine NA. This causes a better bond and increases the stiffness of cement paste because RFA has more porosity than fine NA, and it allows the acceleration of cement hydration crystals in its pores. To conclude, although with a low statistical significance, based on the aforementioned literature findings, it could be generalized that if a mix has an initial compressive strength value equivalent to F_{ci}, the replacement of fine NA with a percentage X of RFA results in a strength Y in Equation (2).

$$Y = F_{ci} - 0.022X \tag{2}$$

2.2.4. Modulus of Elasticity

Several investigations found a relationship between the modulus of elasticity of RFA concrete and its compressive strength. Based on these results, the authors of this study have derived another equation (Figure 7). There is a gradual increase of the modulus of elasticity of RFA concrete when its compressive strength increases for normal-strength concrete, but for high-strength concrete, the graph shows that there is only a slight increase when compressive strength increases. Hence, the regression model in Table 1 was established.

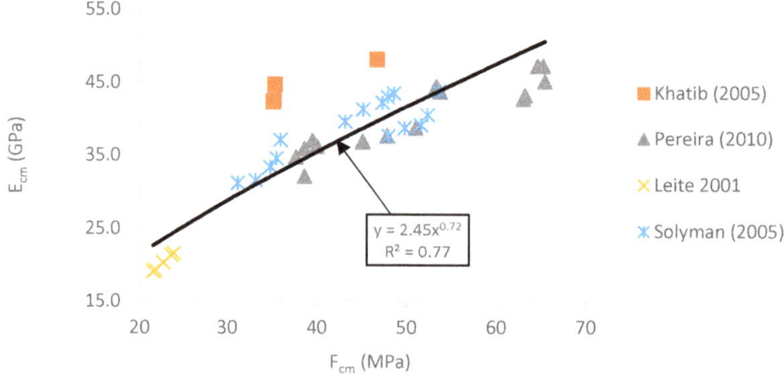

Figure 7. Relationship between compressive strength and modulus of elasticity for increasing incorporation levels of RFA concrete.

Table 1. Relationship between modulus of elasticity and compressive strength of RFA concrete according to various equations.

Equation	Notes	Reference
$E_{cm} = a \cdot f_c^{\frac{1}{3}} \cdot ((1-r) \cdot \rho_{FNA} + r \cdot \rho_{FRA}) \cdot \left[\frac{\left(\frac{w}{c}\right)_{RC0}}{\left(\frac{w}{c}\right)} \right]^b$	Taking into account the w/c ratio; replacement ratio; quality RFA. Effective w/c ratio considered 0.55 and correlation factors a and b equal to 4.228 and 0.22, respectively, with a coefficient of determination $R^2 = 0.916$. The equation was developed based on Model code [83]	[84]
$E_c = 2.58\ f_{ck}^{0.63}$	Relationship between compressive strength and modulus of elasticity. The equation can be used to determine RFA and CRA concrete, taking into account the modulus of elasticity of RFA concrete 10% lower than that of CRA concrete	[85]
$Y_E = 2.58 - 0.0159F_c^2 + 1.9107F_c - 13.276$	The relationship between compressive strength and modules of elasticity. With a coefficient of determination $R^2 = 0.80$	Literature review

2.2.5. Carbonation Resistance

The content and type of aggregates affect the pores system of hardened concrete. Generally, it is agreed that RA's incorporation decreases the density of concrete because of the adhered mortar. Thus, it can be said that RAC has more pores than NAC, which allow the atmospheric CO_2 to diffuse into the hardened concrete more easily and, when the carbonation reaction takes place, the alkalinity of concrete reduces (Figure 8). Moreover, the influence of RAC on carbonation of concrete may depend on the incorporation level, crushing-procedure and quality of the RA, and concrete exposure to different curing conditions, admixture use, degree of hydration over time, and mineral additions use [79].

There are grounds to suppose that carbonation increases as the replacement level of fine NA with RFA increases when the binder content is constant in the concrete mixes. The following studies [67,86] all agreed with the previous supposition. The carbonation depth increases with the porosity's increment [87]. As the incorporation level of RFA increases, the porosity of RFA concrete increases because the dry particle density of RFA is much lower than that of fine NA due to adhered mortar. Broadly speaking, it can be said that the pores of RFA are much greater than those of fine NA. However, Levy and Helene [88] obtained lower carbonation depth for 20% and 50% replacement of fine NA with RFA. The authors reported that this can be explained by increased cement content to achieve the same compressive strength of NAC. But for 100% replacement of fine NA with RFA, carbonation increased even when the cement content increased in the mixes (Figure 9).

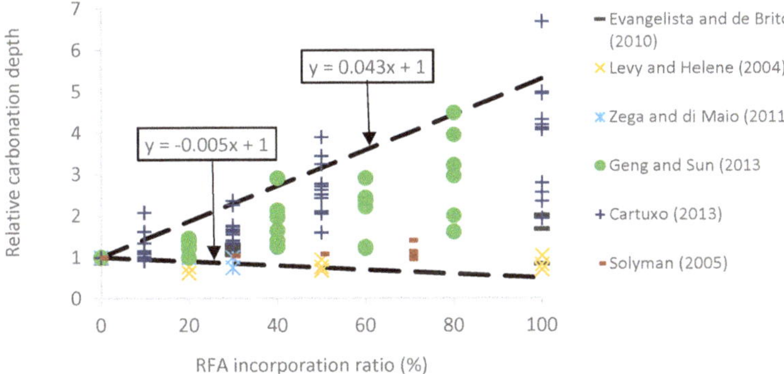

Figure 8. Detail of different particles densities in the same RAC mix.

Figure 9. Relative carbonation depth for various RFA incorporation levels regardless of the age of mixes.

Figure 10 shows that carbonation increased as the replacement ratio of fine NA with RFA increased while for the same mixes compressive strength decreased. Equation (3) can be deducted based on the relationship between RFA incorporation and concrete compressive strength and carbonation with a coefficient of determination R^2 equal to 0.83.

$$Carbonation\ depth_{fine\ RCA} = Carbonation\ depth_{Ref.} \cdot \left(18.273 \cdot \left(\frac{Fcm_{fine\ RCA}}{Fcm_{Ref.}}\right)^2 - 43.301 \cdot \frac{Fcm_{fine\ RCA}}{Fcm_{Ref.}} + 26.092\right) \qquad (3)$$

The data seem to suggest that carbonation depth for fixed w/c ratio steadily increases up to 40% of replacement of fine NA with RFA. For higher values, the trend shows steep increases in the carbonation depth. For constant workability, the carbonation depth of concrete mixes increased at least 16.5% with increasing incorporation of RFA at 28 days. The water content is one of the main factors that affects RFA concrete's carbonation. Therefore, the w/c ratio should be considered at first, especially when the incorporation level of RFA exceeds 40%.

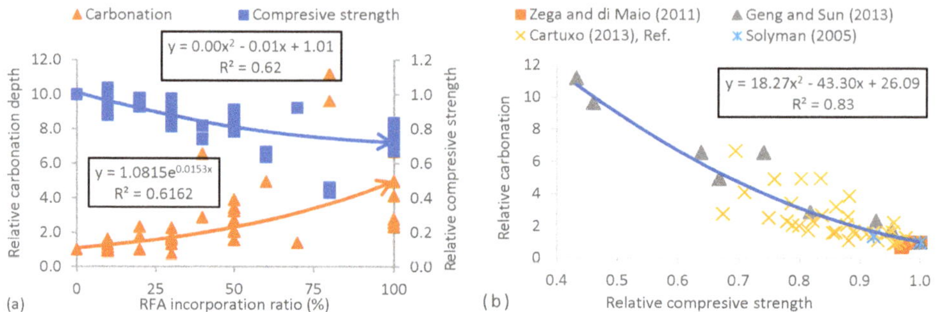

Figure 10. Relative carbonation of RA concrete vs. (**a**) RFA incorporation level and concrete relative compressive strength (**b**) relative concrete compressive strength.

Evangelista and de Brito [89] measured the carbonation depth at different ages for concrete samples made with 30% and 100% of RFA. Their study confirms that carbonation depth increases as the incorporation of RFA increases. The carbonation depth at 14, 21 and 91 days increased by 100%, 100% and 70%, respectively, when fine NA were fully replaced. From the above observations, it follows that the carbonation rate of concrete with RFA relative to conventional concrete decreases for later ages (Figure 11). Hence, given a conventional concrete mix has an expected 90 days carbonation depth of D_{90D}, the mixes with incorporation % X would have a carbonation depth Y_{cd} in Equation (4).

$$Y_{CD} = D_{90D} + 0.041X \tag{4}$$

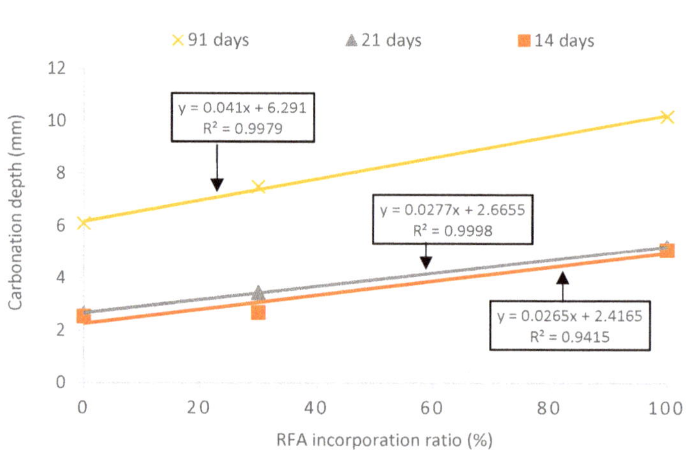

Figure 11. Carbonation depth over time of concrete mixes versus fine NA with RFA incorporation ratio.

2.2.6. Chloride Penetration Resistance

The chloride ion penetration resistance of concrete is highly affected by the pore content in concrete, similarly to carbonation resistance. Generally, it can be said that as the incorporation of RFA increases the chloride penetration resistance decreases because RFA increases concrete's porosity. The source of the deleterious action of chloride ions is generally classified into two types: internal agents, such as the mixing water, aggregate or admixtures, e.g., in very low temperature calcium chloride is used as an accelerator, and agents that may come from external sources, e.g., marine environment, pools and de-icing salts [90].

In his study, Silva [79] listed several factors that affect the chloride ion penetration of RAC, such as the RA replacement level, quality of the RA, exposure to different curing conditions, mixing procedure, admixture use, and use of mineral additions. The author reported that RA leads to higher chloride ion penetration depths than NA. The aggregates subjected to primary plus secondary crushing procedure are better than those from primary only crushing procedure because aggregates subjected to secondary cursing procedure have lower adhered mortar content, which leads to lower aggregate porosity. Thus, the RA allow lower permeability and chloride ion penetration. The resistance of RAC to chloride ion penetration approaches the resistance of NAC with the passage of time. The addition of materials and steam curing have a positive effect on RAC performance in terms of chloride ion penetration. The following studies [86,89] concluded that incorporating 10% of RFA may have a positive effect on the resistance to chloride ion penetration. For higher values, similarly to carbonation, the chloride ion penetration resistance decreases as the incorporation of RFA increases (Figure 12). The regression model displayed in the graph shows that given a conventional concrete mix exhibiting a certain chloride penetration potential P_{Cl1}, the predicted value is relative to that with the addition of a % X of RFA P_{Cl2} (Equation (5)).

$$Y_{cl} \ (P_{Cl2}/P_{Cl1}) = 1 + 0.0033X \tag{5}$$

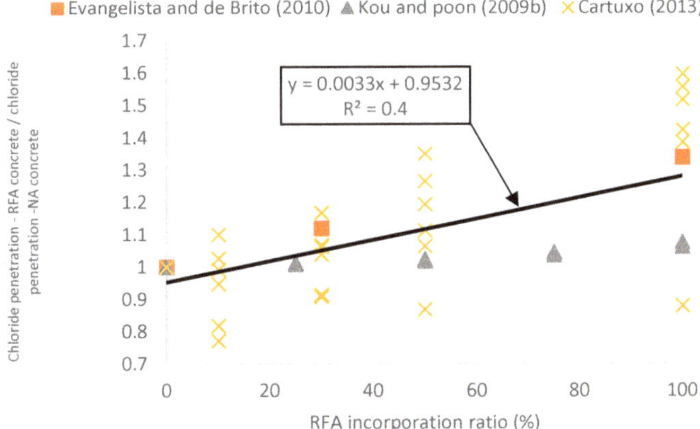

Figure 12. Influence of increasing the RFA content on chloride penetration resistance.

2.3. The Effect of Varying RFA Incorporation Ratio on Environmental Impact of Concrete

2.3.1. Toxicity of Raw Materials for Concrete

An established method to assess the risk of toxicity of construction materials is the quantification, using standard tests, of the possibility of leaching potentially harmful substances to the environment. An indication would be to test the traces of heavy metals, TOC (total organic carbon), the phenol index (carboxyl, halogen, hydroxyl, methoxyl or sulfonic acid), NOx, SOx, and the pH. Regarding the recycled aggregates from CDW, Barbudo et al. [91] concluded that aggregates from CDW are only susceptible to minimal heavy metal leaching. However, high concentrations of SO_3 compounds, which can cause the pollution of superficial and/or ground water, were found in mixed RA containing either ceramic particles or gypsum. Another study [92] on the leaching characteristic of unbound RFA showed that "leached heavy metals did not exceed the Norwegian drinking water criteria".

2.3.2. Life Cycle Assessment of RFA in Concrete

A life cycle assessment (LCA) is an agreed methodology to identify and quantify the EI and energy consumption paving the way for making improvements to the environmental sustainability of the process or product understudy [93,94]. An LCA can be classified into 4 stages: (1) the system boundary. A system boundary of a concrete product could be Cradle-to-Gate, which means including all processes and emissions until the production of its different constituents or Cradle-to-Grave which includes the "Use" and "End-of-Life" phases as per Table 2 [95]. In most of the references reviewed in this paper, a cradle-to-gate boundary was selected on the strict condition that the CDW are allocated the avoidance of landfill from their previous life cycle.

Table 2. Life cycle assessment (LCA) system boundary.

LCA Boundaries	Life Cycle Stages		Life Cycle Stage
Cradle to gate	Product stage (A1–A3)	A1	Raw material extraction and processing, processing of secondary material input
		A2	Transport to the manufacturer
		A3	Manufacturing
Gate to grave	Construction process stage (A4–A5)	A4	Transport to the building site
		A5	Installation into the building
	Use stage—information modules related to the building fabric (B1–B5)	B1	Use or application of the installed product
		B2	Maintenance
		B3	Repair
	Use stage—information modules related to the operation of the building (B4–B5)	B4	Operational energy use
		B5	Operational water use
	End-of-life stage (C1–C4)	C1	De-construction, demolition
		C2	Transport to waste processing
		C3	Waste processing for reuse, recovery and/or recycling (3R)
		C4	Disposal

(2) System Inventory. In this stage, the data required for the definition of the EI of the products and processes involved in the predefined scope are acquired either first hand from the site or through EI declarations or databases such as Ecoinvent and ELCD [96]. (3) The third stage it to multiply the inventory data by the functional unit specified and determine the impact using environmental indicators [1]. Mostly, the following eight midpoint environmental categories are considered: Abiotic Depletion Potential ADP, Acidification Potential AP, Eutrophication Potential EP, Global Warming Potential GWP, Photochemical Ozone CPOCP, and consumption of primary energy, renewable (PE-RE) and non-renewable (PE-NRe). The former six categories are quantified using the CML baseline method, while the Cumulative Energy Demand method for the last two [97]. (4) The final stage is to analyse the concluded data and communicate it with the stakeholders.

Evangelista and de Brito [98] found that the EI (ADP, GWP, ODP, AP, EP and POCP) decrease between 6% and 8% and between 19% and 23%, when 30% and 100% of fine NA are replaced with RFA, respectively. According to these studies [99,100], impacts from NAC and RAC can be similar when the additional cement content of RAC is below 10%. However, these may not be a sustainable strategy because the EI of cement is significantly higher than that of the aggregates [101]. Estanqueiro et al. [2] carried out a calculation of the EI of NA (Scenario i) and RA in the manufacture of concrete using, for the latter, a recycling fixed (Scenario ii) and mobile plant (Scenario iii) and concluded that the use of RA in the production of concrete is more favourable than the use of NA only in terms of land use and respiratory inorganic-impact categories, resulting mainly from the exploitation of the quarry. This study also concluded, however, that coarse RA can present a better EI than coarse NA if fine RFA were also used in concrete production instead of being sent to a landfill. Additionally, they also reported that advantage of RFA in terms of EI are mainly dependent on transportation distances. The summary of the literature findings of the EI of RFA concrete compared to NA concrete could be seen in Table 3.

Table 3. Impact assessment values for producing 1 kg of different aggregates types.

Source	Country	ADP kg Sb eq	GWP $\text{kg CO}_2\text{ eq}$	ODP $\text{kg CFC}^{-11}\text{ eq}$	POCP $\text{kg C}_2\text{H}_4\text{ eq}$	AP $\text{kg SO}_2\text{ eq}$	EP $\text{kg PO}_4^{-3}\text{ eq}$	Pe-NRe MJ
Natural Fine Aggregates								
Braga [102]	Portugal	3.37×10^{-10}	9.87×10^{-3}	1.71×10^{-11}	2.80×10^{-6}	4.58×10^{-5}	1.08×10^{-5}	1.35×10^{-1}
		1.24×10^{-9}	2.79×10^{-2}	2.26×10^{-10}	9.06×10^{-6}	1.59×10^{-4}	3.54×10^{-5}	3.92×10^{-1}
		1.09×10^{-9}	2.44×10^{-2}	2.43×10^{-10}	7.83×10^{-6}	1.44×10^{-4}	3.18×10^{-5}	3.44×10^{-1}
		1.39×10^{-9}	3.14×10^{-2}	2.09×10^{-10}	1.03×10^{-5}	1.75×10^{-4}	3.90×10^{-5}	4.41×10^{-1}
Tošić et al. [103]	Serbia		1.43×10^{-3}		2.78×10^{-7}	1.64×10^{-5}	2.02×10^{-6}	1.48×10^{-05}
			2.12×10^{-3}		4.15×10^{-7}	2.42×10^{-5}	3.01×10^{-6}	2.19×10^{-5}
Korre and Durucan [16]	UK		9.30×10^{-4}	1.06×10^{-10}	4.58×10^{-7}	5.85×10^{-6}		4.35×10^{-7}
			3.29×10^{-3}	4.50×10^{-10}	1.20×10^{-6}	1.89×10^{-5}		1.07×10^{-6}
			2.16×10^{-3}	3.19×10^{-10}	7.35×10^{-1}	1.20×10^{-5}		6.87×10^{-7}
			1.85×10^{-3}	2.14×10^{-10}	9.85×10^{-7}	1.03×10^{-5}		5.90×10^{-7}
			3.79×10^{-2}	8.50×10^{-6}	5.40×10^{-5}	6.77×10^{-4}		1.04×10^{-4}
			3.80×10^{-2}	1.78×10^{-10}	5.40×10^{-5}	6.77×10^{-4}		1.04×10^{-4}
Marinkovic' et al. [18]	Serbia		1.43×10^{-3}		2.82×10^{-7}	1.64×10^{-5}	2.02×10^{-6}	
Sjunnesson [104]	Sweden		1.60×10^{-3}		1.70×10^{-6}	7.80×10^{-7}		3.00×10^{-2}
			7.00×10^{-4}		3.80×10^{-10}	5.00×10^{-5}		1.24×10^{-3}
Average		1.01×10^{-9}	1.23×10^{-2}	8.50×10^{-7}	4.90×10^{-2}	1.36×10^{-4}	2.58×10^{-5}	1.68×10^{-1}
Recycled Fine Aggregates								
Braga [102]	Portugal	2.12×10^{-10}	7.44×10^{-3}	1.60×10^{-10}	2.14×10^{-6}	4.05×10^{-5}	9.28×10^{-6}	1.08×10^{-1}
Tošić et al. [103]	Serbia		2.28×10^{-3}		7.03×10^{-7}	2.49×10^{-5}	3.01×10^{-6}	2.59×10^{-5}
			3.38×10^{-3}		1.18×10^{-6}	3.61×10^{-5}	4.34×10^{-6}	3.95×10^{-5}
Korre and Durucan [16]	UK		2.42×10^{-3}	2.83×10^{-10}	8.00×10^{-7}	1.21×10^{-5}	7.06×10^{-7}	
Marinkovic' et al. [18]	Serbia		1.74×10^{-3}		3.40×10^{-7}	2.00×10^{-5}	2.47×10^{-6}	
Average		2.12×10^{-10}	3.45×10^{-3}	2.22×10^{-10}	1.03×10^{-6}	2.67×10^{-5}	3.96×10^{-6}	3.60×10^{-2}

The EI of producing aggregates could be attributed to two main processes; the energy of extracting natural aggregates versus recycling CDW in addition to the transportation to the concrete batch plant. Although the EI assessment data vary widely, there is a unanimous agreement in the references that the EI of recycled aggregates in concrete is proportional to the transportation distances whether a mobile or a normal stationary plant was used (since the distance of transporting CDW would still be the same). Marinkovic' et al. [18] compared the EI of two types of concrete mixes, one using NA and the other using RA, in the Serbian context. The study was based on two transportation scenarios: the first had the concrete batch plant only 15 km away from the recycling plant for RA, while the second considered it as 100 km. The study concluded that the EI of NA and RAC are mostly dependent on travel-distances and transport-type of aggregates between construction sites and recycling plants. Moreover, when transport distances of RA are smaller than that of NA, the EI of the resulting RAC is less than that of NAC even if the former had 3% extra cement in the mix [18].

It is known that the transportation distance changes between the supplier of raw material and the concrete plant based on the region considered. Therefore, a sensitive analysis is recommended before considering the transportation scenario. For example, the method developed by study of Göswein et al. [105] can be used to obtain a sensitive analysis. According to ELCD core database V3.0, two main types of lorries (medium sized lorry transport, low impact, maximum capacity of 17.3 tonnes; articulated lorry transport: high impact, maximum capacity of 27 tonnes) are used to transport raw materials to concrete plant (Table 4).

Table 4. Impact–assessment results to transport one kg·km (ELCD-core database V3.0).

Lorry/Maximum Capacity (Tonnes)	Baseline CML Method						Cumulative Energy Demand
	ADP	GWP	ODP	POCP	AP	EP	P X10-NRe
	kg Sb eq	kg CO_2 eq	kg CFC^{-11} eq	kg C_2H_4 eq	kg SO_2 eq	kg PO_4^{-3} eq	MJ
Articulated-lorry transport/27 t	1.98×10^{-12}	4.98×10^{-5}	1.01×10^{-13}	1.59×10^{-8}	2.24×10^{-7}	5.14×10^{-8}	6.73×10^{-4}
Lorry–transport/17.3 t	2.62×10^{-12}	6.57×10^{-5}	1.33×10^{-13}	2.24×10^{-8}	3.11×10^{-7}	7.20×10^{-8}	9.27×10^{-4}

Accordingly, it could be concluded that, as shown in Table 5, the effect of the incorporation % of RFA on the resulting EI of concrete is linear to the difference in transportation distances between natural aggregates and RFA. The equations for the different indicators Y could now be calculated as the impact for 1 kg of fine aggregates concrete based on % incorporation (X) of RFA where D1 and D2 are the distance of transportation in km for natural aggregates and RFA, respectively.

Table 5. Predictive models of different mid-point environmental indicators for the % of incorporation X of RFA in concrete.

Environmental Indicators		NA (Average)	NA/km	RFA (Average)	RFA/km	Equation for Y (Impact for 1 kg of Fine Aggregates Concrete Based on % Incorporation (X) of RFA
ADP	kg Sb eq	1.01×10^{-9}	9.47×10^{-12}	2.12×10^{-10}	1.70×10^{-12}	$Y = (D2 \times 1.7 \times 10^{-12} - D1 \times 9.47 \times 10^{-12}) X + D1 \times 9.47 \times 10^{-12}$
GWP	kg CO_2 eq	1.23×10^{-2}	1.15×10^{-4}	3.45×10^{-3}	2.76×10^{-5}	$Y = (D2 \times 2.76 \times 10^{-5} - D1 \times 1.15 \times 10^{-4}) X + D1 \times 1.15 \times 10^{-4}$
ODP	kg CFC^{-11} eq	8.50×10^{-7}	7.97×10^{-9}	2.22×10^{-10}	1.78×10^{-12}	$Y = (D2 \times 1.78 \times 10^{-12} - D1 \times 7.97 \times 10^{-9}) X + D1 \times 7.97 \times 10^{-9}$
POCP	kg C_2H_4 eq	4.90×10^{-2}	4.59×10^{-4}	1.03×10^{-6}	8.24×10^{-9}	$Y = (D2 \times 8.24 \times 10^{-9} - D1 \times 4.59 \times 10^{-4}) X + D1 \times 4.59 \times 10^{-4}$
AP	kg SO_2 eq	1.36×10^{-4}	1.28×10^{-6}	2.67×10^{-5}	2.14×10^{-7}	$Y = (D2 \times 2.14 \times 10^{-7} - D1 \times 1.28 \times 10^{-6}) X + D1 \times 1.28 \times 10^{-6}$
EP	kg PO_4^{-3} eq	2.58×10^{-5}	2.42×10^{-7}	3.96×10^{-6}	3.17×10^{-8}	$Y = (D2 \times 3.17 \times 10^{-8} - D1 \times 2.42 \times 10^{-7}) X + D1 \times 2.47 \times 10^{-7}$
PE-NRe	MJ	1.68×10^{-1}	1.58×10^{-3}	3.60×10^{-2}	2.88×10^{-4}	$Y = (D2 \times 2.88 \times 10^{-4} - D1 \times 1.58 \times 10^{-3}) X + D1 \times 2.88 \times 10^{-3}$

2.4. The Effect of Varying RFA Incorporation Ratio on Economic Impact of Concrete

It is significant to consider the economic impact of a material if the objective is to assess its sustainability. The cost of construction materials is the second most influential factor (after functionality) in the decision-making process regarding selection of alternatives [25]. According to Hafez et al. [95], the eco-costs/value ratio of conventional concrete could be compared to that of other Green concrete mixes on the bases of the net present value considering that the study takes into consideration a replacement ration (N) that compares the service life of the alternatives being considered as well as the interest rate and the expected inflation rate during the time period of the forecasted cashflow [97]. A simpler way of quantifying the economic burden of concrete is to calculate its market price and compare it to the other alternatives using Equation (6) [106].

$$\text{Cost of concrete alternative per unit volume } (x) = \sum_{Y=1}^{n} \text{mass of constituent Y per unit volumen} * \text{market price of Y per unit mass} \tag{6}$$

Braga [102] concluded that using limestone instead of granite aggregates results in 50% cost reduction while using recycled aggregates results in 80% savings. However, similar to the findings from 2.3, the economics of replacing sand with RFA was found to be linearly proportional with the transportation distances. Since the scope of this paper is limited to fine aggregate replacement, the following equation could be used to estimate the cost instead:

$$\text{Cost of concrete} \quad \text{constituent (Y)} = Cost\ of\ raw\ materials\ production + Cost\ of\ raw\ materials\ transportation \tag{7}$$

According to Tošić et al. [103], the cost of RFA and sand excluding transportation is 11.5 and 3 €/tonne, respectively, while the cost of transportation is 0.02 €/tonne/km. Yang et al. [107] cited the transportation cost of aggregates as 0.08 €/tonne/km so the average taken will be 0.05 €/tonne/km. Moreover, Braga et al. [102] noted the cost of RFA and sand excluding transportation is 4.5 and 4 €/tonne, respectively, so the average for that is taken in this study as 8 €/tonne for RFA and 3.5 €/tonne for sand. It is important to state that in all the calculations of environmental and economic impact, the transportation means is assumed to be a large truck. For the environmental impact, the distances are assumed to be double the actual geographic distances to allow for an empty return trip but not for the economic impact calculations. Hence, the following equation could be descriptive of the cost (Y) of a tonne of the fine aggregate portion of the concrete alternative studied with the incorporation % X of RFA where D1 and D2 are the distance of transportation in km for natural aggregates and RFA, respectively, ignoring the economy of scale and the variability in market prices (Equation (8))

$$Y_{economic} = 3.5 \times D1 + X \times (0.045 \times D2 - 3.5 \times D1) \tag{8}$$

3. Sustainability Assessment of RFA Incorporation Ratio Based on Multi Criteria Decision Analysis

The objective of this section is to generate a framework for assessing the optimum replacement ratio of sand or natural fine aggregates with RFA from CDW. A multi criteria decision analysis framework was developed by Mateus et al. [108] as shown in Figure 13, which combines the functional, economic and environmental parameters of concrete to assess its sustainability. This is achieved in four main steps: First, the alternatives that are to be compared need to be defined. Second, based on this data, the different alternatives are evaluated for their functional, environmental and economic parameters. Third, these impact factors would be normalized according to the equation below for a factor between 0 and 1. Finally, based on user-defined weights, these three factors are to be aggregated together to form one sustainability index upon which the alternatives are to be ranked, and the optimal one is selected.

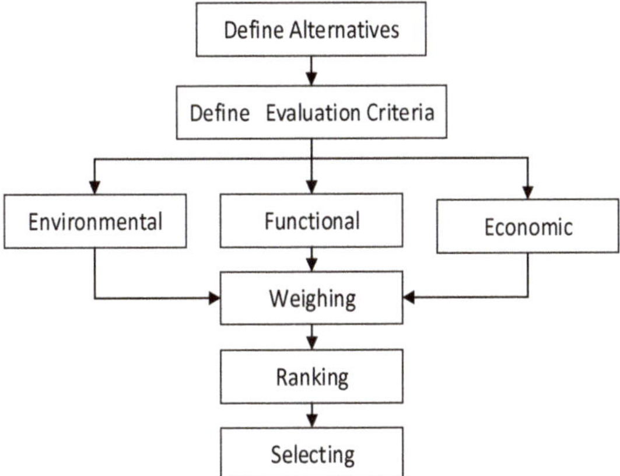

Figure 13. The support method for the assessment of the relative sustainability of building technologies proposed multi criteria decision analysis framework for sustainability assessment.

In light of the presented findings from the literature, the selected parameters and corresponding predictive models were summarized in Table 6. The authors of this paper chose to include only a selected few functional and environmental properties of concrete but rather followed the trend set by similar studies such as Silgado et al. [9] and Rahla et al. [109].

Table 6. A summary of the selected parameters, their weights and the regression models generated form the literature.

Parameter	Empirical Function of Incorporation Ratio X (%) for the RFA in the Concrete Mix	Weights (%)	
Y (Functional)	$Y_{fc} = F_{ci} - 0.022X$	25	33.3
	$Y_E = 2.58 - 0.0159F_c^2 + 1.9107F_c - 13.276$	25	
	$Y_{CD} = D_{90D} + 0.041X$	25	
	$Y_{cl} (P_{Cl2}/P_{Cl1}) = 1 + 0.0033X$	25	
Y (Environmental)	$Y_{ADP} = (D2 \times 1.7 \times 10^{-12} - D1 \times 9.47 \times 10^{-12}) X + D1 \times 9.47 \times 10^{-12}$	14.3	33.3
	$Y_{GWP} = (D2 \times 2.65 \times 10^{-5} - D1 \times 1.16 \times 10^{-4}) X + D1 \times 1.16 \times 10^{-4}$	14.3	
	$Y_{ODP} = (D2 \times 1.77 \times 10^{-12} - D1 \times 2 \times 10^{-12}) X + D1 \times 2 \times 10^{-12}$	14.3	
	$Y_{POCP} = D2 \times 8.26 \times 10^{-9} - D1 \times 9.42 \times 10^{-8}) X + D1 \times 9.42 \times 10^{-8}$	14.3	
	$Y_{AP} = (D2 \times 2.14 \times 10^{-7} - D1 \times 1.27 \times 10^{-6}) X + D1 \times 1.27 \times 10^{-6}$	14.3	
	$Y_{EP} = (D2 \times 3.17 \times 10^{-8} - D1 \times 2.41 \times 10^{-7}) X + D1 \times 2.41 \times 10^{-7}$	14.3	
	$Y_{Pe\text{-}NRe} = (D2 \times 2.88 \times 10^{-4} - D1 \times 1.57 \times 10^{-3}) X + D1 \times 1.57 \times 10^{-3}$	14.3	
Y (Economic)	Yeconomic $= 3.5 \times D1 + X \times (0.045 \times D2 - 3.5 \times D1)$	100	33.3

Three scenarios were assumed for the transportation distances since it proved as a vital variable in the comparison, especially in the environmental and economic impact. The assumed values for D1, the transportation distance between the natural aggregates source and the concrete batch plan, were 150, 100 and 50 km. Similarly, the values for D2, the distance between RFA and the concrete batch plan, were assumed as 50, 100 and 150 km, in that order.

The conventional concrete mix was assumed to have the following functional characteristics with zero RFA added:

- Compressive strength = 30 MPa
- Expected maximum carbonation Depth = 20 mm
- Permeability to chlorides = 10×10^{-12} m^2/s

The 11 alternatives representing possible values for X were modelled using the framework described and the results were summarized in Table 7 as follows.

Table 7. A summary of the sustainability assessment using the multicriteria decision analysis framework.

Alternatives	X (%)	Functional	Scenario 1			Scenario 2			Scenario 3		
			Environmental	Economic	Single Score	Environmental	Economic	Single Score	Environmental	Economic	Single Score
NA	0%	1.00	0.13	1.00	0.71	0.27	1.00	0.76	0.43	1.00	0.81
1	10	0.98	0.14	0.98	0.70	0.28	0.89	0.72	0.44	0.78	0.74
2	20	0.97	0.16	0.96	0.69	0.30	0.81	0.69	0.47	0.64	0.69
3	30	0.95	0.17	0.93	0.69	0.32	0.74	0.67	0.49	0.54	0.66
4	40	0.93	0.19	0.91	0.68	0.34	0.68	0.65	0.52	0.47	0.64
5	50	0.92	0.22	0.89	0.68	0.37	0.63	0.64	0.56	0.42	0.63
6	60	0.90	0.25	0.88	0.68	0.41	0.58	0.63	0.61	0.37	0.63
7	70	0.89	0.30	0.86	0.68	0.47	0.54	0.63	0.66	0.34	0.63
8	80	0.87	0.38	0.84	0.70	0.55	0.51	0.65	0.74	0.31	0.64
9	90	0.86	0.52	0.83	0.73	0.70	0.48	0.68	0.84	0.28	0.66
10	100	0.84	1.00	0.81	0.88	1.00	0.45	0.76	1.00	0.26	0.70

The results present enough evidence to support a claim that the incorporation of any increment of RFA to replace natural fine aggregates has an adverse effect on the overall functional performance of concrete. Nevertheless, it could be also concluded as per the comparison shown in Figure 14, regardless of the transportation distance, and due to the much higher purchase cost assumed for RFA, any incorporation % would yield a more expensive concrete alternative.

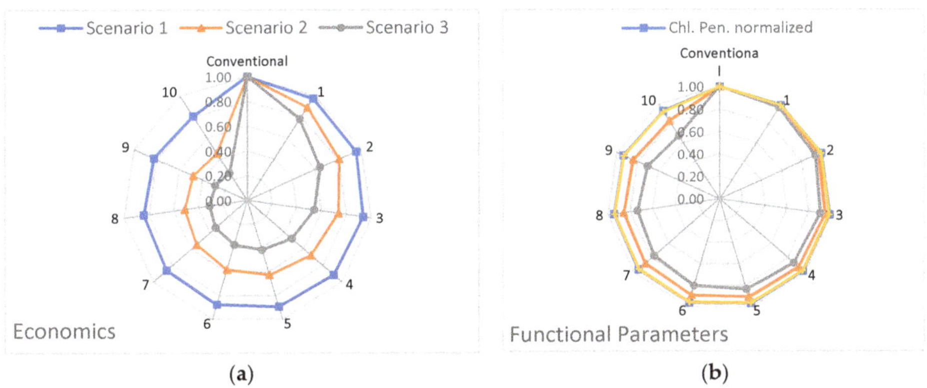

Figure 14. A comparison between the normalized values for the (**a**) economic and (**b**) functional parameters of the RFA based concrete alternatives.

On the other hand, the EI of the studied alternatives incorporating a range of RCA% (10% to 100%) showed sensitivity to the relative transportation distance of natural aggregates and RFA. Agreeing with the aforementioned data from the literature, when the distances travelled by NA double that assumed for RFA, which is described in this study as scenario 1, the concrete alternatives with RFA showed enhance environmental performance. As shown in Figure 15, mid-point indicators such as abiotic depletion potential (ADP) and the consumption of non-renewable primary energy (Pe-NRe) testify that alternative 10, which suggests the full incorporation of RFA instead of natural sand, is 80%, 60% and 40% better after normalization compared to the NA conventional concrete alternative considering scenarios 1, 2 and 3, respectively.

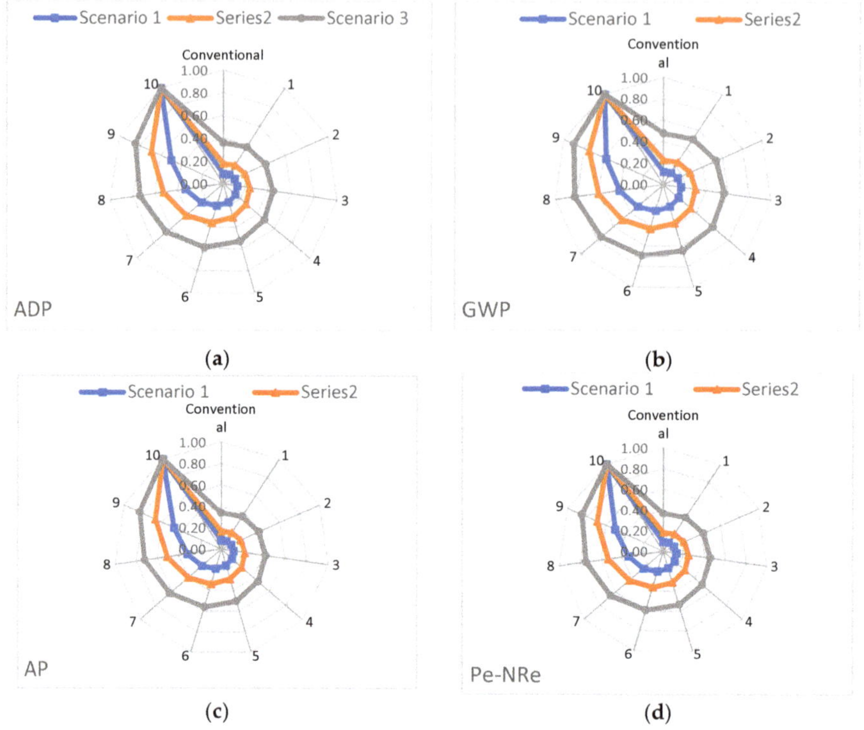

Figure 15. EI assessment of the proposed scenarios and RFA concrete alternatives using the following Mid-point indicators: (**a**) ADP, (**b**) GWP, (**c**) AP and (**d**) Pe-NRe.

4. Conclusions

This study aimed at investigating, through an extensive literature review, the performance of concrete in which RFA from CDW are used to replace, partially or totally, natural fine aggregates. The aim was to model the combined functional, environmental and economic performance parameters of different incorporation levels of RFA to judge the optimum value that achieves sustainability of concrete. A multi criteria decision analysis framework was applied, and several boundaries were assumed for the most significant variables: compressive strength, carbonation depth, chloride penetration and the transportation distances for the aggregates. Results showed that due to the significant decrease in the functional and economic parameters of concrete with any level of RFA incorporation, it is only possible to achieve a more overall sustainable concrete mix when the transportation distance of natural aggregates is at least double that of the RFA.

Author Contributions: Conceptualization, R.K. (Rawaz Kurda); methodology, R.K. (Rawaz Kurda); investigation, H.H. and R.K. (Rawaz Kurda); resources, R.K. (Rawaz Kurda); data curation, R.K. (Rawaz Kurda), H.H. and R.K. (Reben Kurda); writing—original draft preparation, R.K. (Rawaz Kurda) and H.H.; writing—review and editing, H.H., R.K. (Rawaz Kurda), R.K. (Reben Kurda) and B.A.-H.; visualization, H.H, R.K. (Reben Kurda), R.K. (Rawaz Kurda), B.A.-H., R.M. and B.A.; supervision, H.H., R.K. (Reben Kurda), R.K. (Rawaz Kurda), B.A.-H., R.M. and B.A.; project administration, H.H., R.K. (Reben Kurda), R.K. (Rawaz Kurda), B.A.-H., R.M. and B.A. In addition, data analysis, computing and optimization procedure were conducted by R.K. from the department of Information Systems Engineering, Erbil Technical Engineering College, Erbil Polytechnic University. All authors have read and agreed to the published version of the manuscript.

Funding: This work received no funding.

Conflicts of Interest: The authors declare no conflict of interests.

References

1. Wu, P.; Xia, B.; Zhao, X. The importance of use and end-of-life phases to the life cycle greenhouse gas (GHG) emissions of concrete—A review. *Renew. Sustain. Energy Rev.* **2014**, *37*, 360–369. [CrossRef]
2. Estanqueiro, B.; Silvestre, J.; de Brito, J.; Pinheiro, M. Environmental life cycle assessment of coarse natural and recycled aggregates for concrete. *Eur. J. Environ. Civ. Eng.* **2018**, *22*, 429–449. [CrossRef]
3. Freedonia. *World Construction Aggregates, in No 2838*; The Freedonia Group: Cleveland, OH, USA, 2012. Available online: https://www.freedoniagroup.com/industry-study/world-construction-aggregates-2838.htm (accessed on 5 January 2020).
4. Dhir, R.; Paine, K. Value Added Sustainable Use of Recycled and Secondary Aggregates in Concrete. *Indian Concr. J.* **2010**, *84*, 7–26.
5. Turk, J.; Cotič, Z.; Mladenovič, A.; Šajna, A. Environmental evaluation of green concretes versus conventional concrete by means of LCA. *Waste Manag.* **2015**, *45*, 194–205. [CrossRef] [PubMed]
6. Guo, Z.; Tu, A.; Chen, C.; Lehman, D.E. Mechanical properties, durability, and life-cycle assessment of concrete building blocks incorporating recycled concrete aggregates. *J. Clean. Prod.* **2018**, *199*, 136–149. [CrossRef]
7. Ossa, A.; García, J.L.; Botero, E. Use of recycled construction and demolition waste (CDW) aggregates: A sustainable alternative for the pavement construction industry. *J. Clean. Prod.* **2016**, *135*, 379–386. [CrossRef]
8. Rodríguez-Robles, D.; van den Heede, P.; de Belie, N. 9—Life Cycle Assessment Applied to Recycled Aggregate Concrete. In *New Trends in Eco-Efficient and Recycled Concrete*; De Brito, J., Agrela, F., Eds.; Woodhead Publishing: Cambridge, UK, 2019; pp. 207–256.
9. Suárez Silgado, S.; Valdiviezo, L.C.; Domingo, S.G.; Roca, X. Multi-criteria decision analysis to assess the environmental and economic performance of using recycled gypsum cement and recycled aggregate to produce concrete: The case of Catalonia (Spain). *Resour. Conserv. Recycl.* **2018**, *133*, 120–131. [CrossRef]
10. EU. Directive 2008/98/CE N, The European Parliament and the Council on waste and repealing certain Directives. *Off. J. Eur. Union* **2008**, *28*. Available online: https://eur-lex.europa.eu/legal-content/EN/TXT/?uri=celex%3A32008L0098 (accessed on 5 January 2020).
11. Arora, S.; Singh, P. Analysis of flexural fatigue failure of concrete made with 100% coarse recycled concrete aggregates. *Constr. Build. Mater.* **2016**, *102*, 782–791. [CrossRef]
12. Silva, R.; De Brito, J.; Dhir, R. Establishing a relationship between modulus of elasticity and compressive strength of recycled aggregate concrete. *J. Clean. Prod.* **2016**, *112*, 2171–2186. [CrossRef]
13. Tošić, N.; Marinković, S.; Pecić, N.; Ignjatović, I.; Dragaš, J. Long-term behaviour of reinforced beams made with natural or recycled aggregate concrete and high-volume fly ash concrete. *Constr. Build. Mater.* **2018**, *176*, 344–358. [CrossRef]
14. Tošić, N.D.; Marinković, S.B.; Ignjatović, I.S.; Bajat, B.J.; Pejović, M.P. *Experimental Setup for Measuring Long-Term Behavior of Green Reinforced Concrete Beams*; Springer International Publishing: Cham, Switzerland, 2018.
15. Evangelista, L.; de Brito, J. Mechanical behaviour of concrete made with fine recycled concrete aggregates. *Cem. Concr. Compos.* **2007**, *29*, 397–401. [CrossRef]
16. Korre, A.; Durucan, S. *Life Cycle Assessment of Aggregates. EVA025-Final Report: Aggregates Industry Life Cycle Assessment Model: Modelling Tools and Case Studies*; Waste & Resources Action Programme: Oxon, UK, 2009.
17. Kurda, R. Sustainable Development of Cement-Based Materials: Application to Recycled Aggregates Concrete. Ph.D. Thesis, Instituto Superior Técnico—University of Lisbon, Lisbon, Portugal, 2017. Available online: http://bibliotecas.utl.pt/cgi-bin/koha/opac-detail.pl?biblionumber=520154 (accessed on 5 December 2019).
18. Marinković, S.; Radonjanin, V.; Malesev, M.; Ignjatovic, I. Comparative environmental assessment of natural and recycled aggregate concrete. *Waste Manag.* **2010**, *30*, 2255–2264. [CrossRef] [PubMed]
19. Gonçalves, T.; Silva, R.V.; de Brito, J.; Fernández, J.M.; Esquinas, A.R. Mechanical and durability performance of mortars with fine recycled concrete aggregates and reactive magnesium oxide as partial cement replacement. *Cem. Concr. Compos.* **2020**, *105*, 103420. [CrossRef]
20. Silva, R.V.; de Brito, J. Reinforced recycled aggregate concrete slabs: Structural design based on Eurocode 2. *Eng. Struct.* **2020**, *204*, 110047. [CrossRef]
21. Masood, B.; Elahi, A.; Barbhuiya, S.; Ali, B. Mechanical and durability performance of recycled aggregate concrete incorporating low calcium bentonite. *Constr. Build. Mater.* **2020**, *237*, 117760. [CrossRef]

22. Gettu, R.; Pillai, R.G.; Santhanam, M.; Basavaraj, A.S.; Rathnarajan, S.; Dhanya, B. Sustainability-based decision support framework for choosing concrete mixture proportions. *Mater. Struct.* **2018**, *51*, 165. [CrossRef]

23. Kurda, R.; de Brito, J.; Silvestre, J.D. A comparative study of the mechanical and life cycle assessment of high-content fly ash and recycled aggregates concrete. *J. Build. Eng.* **2020**, 101173. [CrossRef]

24. Pedro, D.; de Brito, J.; Evangelista, L. Structural concrete with simultaneous incorporation of fine and coarse recycled concrete aggregates: Mechanical, durability and long-term properties. *Constr. Build. Mater.* **2017**, *154*, 294–309. [CrossRef]

25. Kurda, R.; de Brito, J.; Silvestre, J.D. CONCRETop method: Optimization of concrete with various incorporation ratios of fly ash and recycled aggregates in terms of quality performance and life-cycle cost and environmental impacts. *J. Clean. Prod.* **2019**, *226*, 642–657. [CrossRef]

26. Boudali, S.; Kerdal, D.E.; Ayed, K.; Abdulsalam, B.; Soliman, A.M. Performance of self-compacting concrete incorporating recycled concrete fines and aggregate exposed to sulphate attack. *Constr. Build. Mater.* **2016**, *124*, 705–713. [CrossRef]

27. Wang, Y.; Zhang, H.; Geng, Y.; Wang, Q.; Zhang, S. Prediction of the elastic modulus and the splitting tensile strength of concrete incorporating both fine and coarse recycled aggregate. *Constr. Build. Mater.* **2019**, *215*, 332–346. [CrossRef]

28. Behera, M.; Minocha, A.K.; Bhattacharyya, S.K. Flow behavior, microstructure, strength and shrinkage properties of self-compacting concrete incorporating recycled fine aggregate. *Constr. Build. Mater.* **2019**, *228*, 116819. [CrossRef]

29. Vinay Kumar, B.M.; Ananthan, H.; Balaji, K.V.A. Experimental studies on utilization of recycled coarse and fine aggregates in high performance concrete mixes. *Alex. Eng. J.* **2018**, *57*, 1749–1759. [CrossRef]

30. Kurda, R.; de Brito, J.; Silvestre, J. Water absorption and electrical resistivity of concrete with recycled concrete aggre-gates and fly ash. *Cem. Concr. Compos.* **2019**, *95*, 169–182. [CrossRef]

31. Fan, C.-C.; Huang, R.; Hwang, H.; Chao, S.-J. Properties of concrete incorporating fine recycled aggregates from crushed concrete wastes. *Constr. Build. Mater.* **2016**, *112*, 708–715. [CrossRef]

32. Velay-Lizancos, M.; Martinez-Lage, I.; Azenha, M.; Granja, J.; Vazquez-Burgo, P. Concrete with fine and coarse recycled aggregates: E-modulus evolution, compressive strength and non-destructive testing at early ages. *Constr. Build. Mater.* **2018**, *193*, 323–331. [CrossRef]

33. Singh, N.; Singh, S.P. Evaluating the performance of self compacting concretes made with recycled coarse and fine aggregates using non destructive testing techniques. *Constr. Build. Mater.* **2018**, *181*, 73–84. [CrossRef]

34. Geng, Y.; Zhao, M.; Yang, H.; Wang, Y. Creep model of concrete with recycled coarse and fine aggregates that accounts for creep development trend difference between recycled and natural aggregate concrete. *Cem. Concr. Compos.* **2019**, *103*, 303–317. [CrossRef]

35. Leite, M.B.; Santana, V.M. Evaluation of an experimental mix proportion study and production of concrete using fine recycled aggregate. *J. Build. Eng.* **2019**, *21*, 243–253. [CrossRef]

36. Zhang, H.; Wang, Y.; Lehman, D.E.; Geng, Y.; Kuder, K. Time-dependent drying shrinkage model for concrete with coarse and fine recycled aggregate. *Cem. Concr. Compos.* **2020**, *105*, 103426. [CrossRef]

37. Kurda, R.; de Brito, J.; Silvestre, J. Carbonation of concrete made with high amount of fly ash and recycled concrete aggregates for utilization of CO_2. *J. CO_2 Util.* **2019**, *29*, 12–19. [CrossRef]

38. Wang, Y.; Liu, F.; Xu, L.; Zhao, H. Effect of elevated temperatures and cooling methods on strength of concrete made with coarse and fine recycled concrete aggregates. *Constr. Build. Mater.* **2019**, *210*, 540–547. [CrossRef]

39. Sasanipour, H.; Aslani, F. Durability properties evaluation of self-compacting concrete prepared with waste fine and coarse recycled concrete aggregates. *Constr. Build. Mater.* **2020**, *236*, 117540. [CrossRef]

40. Gayarre, F.L.; Pérez, J.G.; Pérez, C.L.-C.; López, M.S.; Martínez, A.L. Life cycle assessment for concrete kerbs manufactured with recycled aggregates. *J. Clean. Prod.* **2016**, *113*, 41–53. [CrossRef]

41. Etxeberria, M.; Vázquez, E.; Marí, A.; Barra, M. Influence of amount of recycled coarse aggregates and production process on properties of recycled aggregate concrete. *Cem. Concr. Res.* **2007**, *37*, 735–742. [CrossRef]

42. Kim, S.; Yun, H. Evaluation of the bond behavior of steel reinforcing bars in recycled fine aggregate concrete. *Cem. Concr. Compos.* **2014**, *46*, 8–18. [CrossRef]

43. Lee, S. Influence of recycled fine aggregates on the resistance of mortars to magnesium sulfate attack. *Waste Manag.* **2009**, *29*, 2385–2391. [CrossRef]

44. Wegen, G.; Haverkort, R. Recycled construction and demolition waste as a fine aggregate for concrete. In Proceedings of the International Symposium on Sustainable Construction: Use of Recycled Concrete Aggregate, London, UK, 11–12 November 1998; pp. 333–346.

45. Song, I.H.; Ryou, J.S. Hybrid techniques for quality improvement of recycled fine aggregate. *Constr. Build. Mater.* **2014**, *72*, 56–64. [CrossRef]

46. Chisholm, D. *Best Practice Guide for the Use of Recycled Aggregates in New Concrete*; Cement & Concrete Association of New Zealand: Wellington, New Zealand, 2011; pp. 31–34.

47. Lima, C.; Caggiano, A.; Faella, C.; Martinelli, E.; Pepe, M.; Realfonzo, R. Physical properties and mechanical behaviour of concrete made with recycled aggregates and fly ash. *Constr. Build. Mater.* **2013**, *47*, 547–559. [CrossRef]

48. Chan, C. Use of Recycled Aggregate in Shotcrete and Concrete 193. Ph.D. Thesis, University of British Columbia, Vancouver, BC, Canada, 1998.

49. Kou, C.; Poon, S. Properties of self-compacting concrete prepared with coarse and fine recycled concrete aggregates. *Cem. Concr. Compos.* **2009**, *31*, 622–627. [CrossRef]

50. Leite, M. Evaluation of the Mechanical Properties of Concrete Produced with Recycled Aggregates from Construction and Demolition Wastes. Ph.D. Thesis, Federal University of Rio Grande do Sul, Rio Grande do Sul, Porto Alegre, Brasil, 2001.

51. Müeller, A.; Winkler, A. Characteristics of processed concrete rubble. In *Sustainable Construction: Use of Recycled Concrete Aggregate*; Dhir, R.K., Henderson, N.A., Limbachiya, M.C., Eds.; Thomas Telford: London, UK, 1998; pp. 109–119.

52. Sérifou, M.; Sbartaï, M.; Yotte, S.; Boffoué, O.; Emeruwa, E.; Bos, F. A Study of Concrete Made with Fine and Coarse Aggregates Recycled from Fresh Concrete Waste. *J. Constr. Eng.* **2013**, *2013*, 317182. [CrossRef]

53. Carro-López, D.; González-Fonteboa, B.; de Brito, J.; Martínez-Abella, F.; González-Taboada, I.; Silva, P. Study of the rheology of self-compacting concrete with fine recycled concrete aggregates. *Constr. Build. Mater.* **2015**, *96*, 491–501. [CrossRef]

54. Cartuxo, F.; De Brito, J.; Evangelista, L.; Jiménez, R.; Ledesma, F. Rheological behaviour of concrete made with fine recycled concrete aggregates—Influence of the superplasticizer. *Constr. Build. Mater.* **2015**, *89*, 36–47. [CrossRef]

55. Evangelista, L.; Guedes, M.; De Brito, J.; Ferro, C.; Pereira, F. Physical, chemical and mineralogical properties of fine recycled aggregates made from concrete waste. *Constr. Build. Mater.* **2015**, *86*, 178–188. [CrossRef]

56. Hasaba, S.; Kawamura, M.; Toriik, K.; Takemoto, K. Drying shrinkage and durability of concrete made of recycled concrete aggregates. *Transl. Jpn. Concr. Inst.* **1981**, *3*, 55–60.

57. Kikushi, M.; Dosho, Y.; Narikawa, M.; Miura, T. Application of recycled aggregate concrete for structural concrete. Part 1—Experimental study on the quality of recycled aggregate and recycled aggregate concrete. In *Sustainable Construction: Use of Recycled Concrete Aggregate, Proceedings of the International Symposium on Sustainable Construction: Use of Recycled Concrete Aggregate, London, UK, 11–12 November 1998*; University of Dundee: Dundee, Scotland, 1998; pp. 55–68.

58. Levy, S.; Helene, P. Durability of concrete mixed with fine recycled aggregates. *Exacta* **2007**, *5*, 25–34. [CrossRef]

59. Lin, Y.; Tyan, Y.; Chang, T.; Chang, C. An assessment of optimal mixture for concrete made with recycled concrete aggregates. *Cem. Concr. Res.* **2004**, *34*, 1373–1380. [CrossRef]

60. Schoon, J.; De-Buysser, K.; Driessche, I.; Belie, N.D. Fines extracted from recycled concrete as alternative raw material for Portland cement clinker production. *Cem. Concr. Compos.* **2015**, *58*, 70–80. [CrossRef]

61. Sim, J.; Park, C. Compressive strength and resistance to chloride ion penetration and carbonation of recycled aggregate concrete with varying amount of fly ash and fine recycled aggregate. *Waste Manag.* **2011**, *31*, 2352–2360. [CrossRef]

62. Uygunoğlu, T.; Topcu, I.; Gencel, O.; Brostow, W. The effect of fly ash content and types of aggregates on the properties of pre-fabricated concrete interlocking blocks (PCIBs). *Constr. Build. Mater.* **2012**, *30*, 180–187. [CrossRef]

63. Wang, Z. The Effects of Aggregate Moisture Conditions on Rheological Behaviors of High-Workability Mortar Prepared with Fine Recycled-Concrete Aggregate. Master's Thesis, Graduate Council of Texas State University, San Marcos, TX, USA, 2012.

64. Yaprak, H.; Aruntas, H.; Demir, I.; Simsek, O.; Durmus, G. Effects of the fine recycled concrete aggregates on the concrete properties. *Int. J. Phys. Sci.* **2011**, *6*, 2455–2461.

65. Zega, J.; Maio, A.D. Use of recycled fine aggregate in concretes with durable requirements. *Waste Manag.* **2011**, *31*, 2336–2340. [CrossRef]

66. Solyman, M. Classification of Recycled Sands and Their Applications as Fine Aggregates for Concrete and Bituminous Mixtures. Klassifizierung von Recycling-Brechsanden und Ihre Anwendungen Für Beton und Für Straßenbaustoffe 196. Ph.D. Thesis, Kassel University Press, Kassel, Germany, 2005. Available online: https://books.google.pt/books?id=CCyAbpG3OTwC&printsec=frontcover#v=onepage&q&f=false (accessed on 5 January 2020).

67. Geng, J.; Sun, J. Characteristics of the carbonation resistance of recycled fine aggregate concrete. *Constr. Build. Mater.* **2013**, *49*, 814–820. [CrossRef]

68. Kim, K.; Shin, M.; Cha, S. Combined effects of recycled aggregate and fly ash towards concrete sustainability. *Constr. Build. Mater.* **2013**, *48*, 499–507. [CrossRef]

69. Somna, R.; Jaturapitakkul, C.; Amde, A.M. Effect of ground fly ash and ground bagasse ash on the durability of recycled aggregate concrete. *Cem. Concr. Compos.* **2012**, *34*, 848–854. [CrossRef]

70. Pereira, P. Structural Concrete with Incorporated Recycled Concrete Fine Aggregates: Influence of Superplasticizers on the Mechanical Behaviour (Translated from Portuguese) 144. Master's Thesis, Instituto Superior Técnico, Lisbon, Portugal, 2010.

71. Ramos, D. Freeze Thaw Resistance of Concrete Produced with Fine Recycled Concrete Aggregates 135. Master's Thesis, Universidade de Lisboa/Instituto Superior Técnico, Lisbon, Portugal, 2014. (In Portuguess).

72. Mukai, T.; Kikuchi, M.; Koizumi, H. *Fundamental Study on Bond Properties between Recycled Aggregate Concrete and Steel Bars*; Cement Association of Japan: Tokyo, Japan, 1978; Volume 32.

73. Hansen, C.; Narud, H. Strength of recycled concrete made from crushed concrete coarse aggregate. *Concr. Int.* **1983**, *5*, 79–83.

74. Buck, A. Recycled concrete as a source of aggregate. *ACI Mater. J.* **1977**, *74*, 212–219.

75. Malhotra, M. Use of recycled concrete as a new aggregate. In Proceedings of the Symposium on Energy and Resource Conservation in the Cement and Concrete Industry, Ottawa, ON, Canada, 21–23 October 1978. CANMET Report no. 76-8.

76. Ravindarajah, S.; Tam, T. Properties of concrete made with crushed concrete as coarse aggregate. *Mag. Concr. Res. March* **1985**, 29–38. [CrossRef]

77. Bogas, A.; de Brito, J.; Ramos, D. Freeze-thaw resistance of concrete produced with fine recycled concrete aggregates. *J. Clean. Prod.* **2016**. [CrossRef]

78. Hurlbert, S.H. The ancient black art and transdisciplinary extent of pseudoreplication. *J. Comp. Psychol.* **2009**, *123*, 434. [CrossRef] [PubMed]

79. Silva, R. Use of Recycled Aggregates from Construction and Demolition Wastes in the Production of Structural Concrete 274. Ph.D. Thesis, Department of Civil Engineering, Universidade de Lisboa, Instituto Superior Técnico, Lisboa, Portugal, 2015.

80. Khatib, M. Properties of concrete incorporating fine recycled aggregate. *Cem. Concr. Res.* **2005**, *35*, 763–769. [CrossRef]

81. Katz, A. Properties of concrete made with recycled aggregate from partially hydrated old concrete. *Cem. Concr. Res.* **2003**, *33*, 703–711. [CrossRef]

82. Ahmed, S. Properties of concrete containing recycled fine aggregate and fly ash. In Proceedings of the Concrete 2011 Conference, Perth, Australia, 12 October 2011.

83. EuroLightCon. *Tensile Strength as Design Parameter—Economic Design and Construction with Light Weight Aggregate Concrete.* Document BE96-3942/R32; European Union—Brite EuRam III: Luxembourg, 2000.

84. Pereira, P.; Evangelista, L.; de Brito, J. The effect of superplasticizers on the mechanical performance of concrete made with fine recycled concrete aggregates. *Cem. Concr. Compos.* **2012**, *34*, 1044–1052. [CrossRef]

85. Cabral, A.; Schalch, V.; Molin, D.; Ribeiro, J. Mechanical properties modelling of recycled aggregate concrete. *Constr. Build. Mater.* **2010**, *24*, 421–430. [CrossRef]

86. Cartuxo, F. Concrete with Fine Recycled Concrete Aggregates: Influence of Superplasticizers on the Durability Related Performance (Translated from Portuguese) 232. Master's Thesis, Universidade de Lisboa/Instituto Superior Técnico, Lisboa, Portugal, 2013.

87. Basheer, L.; Cleland, D.; Kropp, J. Assessment of the durability of concrete from its permea-tion properties—A review. *Constr. Build. Mater.* **2001**, *15*, 93–103. [CrossRef]

88. Levy, M.; Helene, P. Durability of recycled aggregates concrete: A safe way to sustainable development. *Cem. Concr. Res.* **2004**, *34*, 1975–1980. [CrossRef]

89. Evangelista, L.; de Brito, J. Durability performance of concrete made with fine recycled concrete aggregates. *Cem. Concr. Compos.* **2010**, *32*, 9–14. [CrossRef]

90. Neville, A.M.; Brooks, J.J. *Concrete Technology*, 2nd ed.; Pearson, Prentice Hall: London, UK, 2010; ISBN 9780273732198.

91. Barbudo, A.; Galvín, A.P.; Agrela, F.; Ayuso, J.; Jiménez, J.R. Correlation analysis between sulphate content and leaching of sulphates in recycled aggregates from construction and demolition wastes. *Waste Manag.* **2012**, *32*, 1229–1235. [CrossRef]

92. Engelsen, C.; Lund, O.; Breedveld, G.; Mehus, J.; Petkovic, G.; Håøya, A. Leaching characteristic of unbound recycled aggregates: Preliminary study and ongoing research. In Proceedings of the 5th International Conference on the Environmental and Technical Implications of Construction with Alternative Materials, San Sebastian, Spain, 4–6 June 2003. Available online: https://www.vegvesen.no/_attachment/110361/binary/192280 (accessed on 10 January 2020).

93. Durão, V.; Silvestre, J.D.; Ricardo, M.; de Brito, J. Assessment and communication of the environmental performance of construction products in Europe: Comparison between PEF and EN 15804 compliant EPD schemes. *Resour. Conserv. Recycl.* **2020**, *156*, 104703. [CrossRef]

94. Pinheiro, M. *Environment and Sustainable Construction*; Rev.; Correia, F., Branco, F., Guedes, M., Eds.; Instituto do Ambiente: Amadora, Portugal, 2006. (In Portuguess)

95. Hafez, H.; Kurda, R.; Cheung, W.M.; Nagaratnam, B. A Systematic Review of the Discrepancies in Life Cycle Assessments of Green Concrete. *Appl. Sci.* **2019**, *9*, 4803. [CrossRef]

96. Sagastume Gutiérrez, A.; Eras, J.J.C.; Gaviria, C.A.; van Caneghem, J.; Vandecasteele, C. Improved selection of the functional unit in environmental impact assessment of cement. *J. Clean. Prod.* **2017**, *168*, 463–473. [CrossRef]

97. Kurda, R.; Silvestre, J.D.; de Brito, J.; Ahmed, H. Optimizing recycled concrete containing high volume of fly ash in terms of the embodied energy and chloride ion resistance. *J. Clean. Prod.* **2018**, *194*, 735–750. [CrossRef]

98. Evangelista, L.; de Brito, J. Environmental life cycle assessment of concrete made with fine recycled concrete aggregates. In *SB07 Lisbon—Sustainable Construction, Materials and Practices: Challenge of the Industry for the New Millennium*; In-House Publishing: Rotterdam, The Netherlands, 2008; 6p.

99. Weil, M.; Jeske, U.; Schebek, L. Closed-loop recycling of construction and demolition waste in Germany in view of stricter environmental threshold values. *Waste Manag. Res.* **2006**, *24*, 197–206. [CrossRef] [PubMed]

100. Knoeri, C.; Sanyé-Mengual, E.; Althaus, H. Comparative LCA of recycled and conventional concrete for structural applications. *Int. J. Life Cycle Assess.* **2013**, *18*, 909–918. [CrossRef]

101. Kurda, R.; Silvestre, J.D.; de Brito, J. Life cycle assessment of concrete made with high volume of recycled concrete aggregates and fly ash. *Resour. Conserv. Recycl.* **2018**, *139*, 407–417. [CrossRef]

102. Braga, A. Comparative Analysis of the Life Cycle Assessment of Conventional and Recycled Aggregate Concrete 112. Master's Thesis, Instituto Superior Técnico—University of Lisbon, Lisbon, Portugal, 2015. Available online: https://fenix.tecnico.ulisboa.pt/cursos/mec/dissertacao/846778572210846 (accessed on 10 December 2019). (In Portuguese).

103. Tošić, N.; Marinković, S.; Dašić, T.; Stanić, M. Multicriteria optimization of natural and recycled aggregate concrete for structural use. *J. Clean. Prod.* **2015**, *87*, 766–776. [CrossRef]

104. Sjunnesson, J. Life Cycle Assessment of Concrete. Master's Thesis, Department of Technology and Society, Lund University, Lund, Sweden, 2005.

105. Göswein, V.; Gonçalves, A.; Silvestre, J.D.; Freire, F.; Habert, G.; Kurda, R. Transportation matters—Does it? GIS-based comparative environmental assessment of concrete mixes with cement, fly ash, natural and recycled aggregates. *Resour. Conserv. Recycl.* **2018**, *137*, 1–10. [CrossRef]

106. Zhang, Y.; Liu, M.; Xie, H.; Wang, Y. Assessment of CO_2 emissions and cost in fly ash concrete. In *Environment, Energy and Applied Technology, Proceedings of the 2014 International Conference on Frontier of Energy and Environment Engineering (ICFEEE 2014), Taiwan, 6–7 December 2014*; CRC Press: Boca Raton, FL, USA, 2015.

107. Yang, K.-H.; Seo, E.-A.; Jung, Y.-B.; Tae, S.-H. Effect of ground granulated blast-furnace slag on life-cycle environmental impact of concrete. *J. Korea Concr. Inst.* **2014**, *26*, 13–21. [CrossRef]
108. Mateus, R.; Neiva, S.; Bragança, L.; Mendonça, P.; Macieira, M. Sustainability assessment of an innovative lightweight building technology for partition walls–comparison with conventional technologies. *Build. Environ.* **2013**, *67*, 147–159. [CrossRef]
109. Rahla, K.M.; Mateus, R.; Bragança, L. Comparative sustainability assessment of binary blended concretes using Supplementary Cementitious Materials (SCMs) and Ordinary Portland Cement (OPC). *J. Clean. Prod.* **2019**, *220*, 445–459. [CrossRef]

Article

Predicting Compressive Strength of Cement-Stabilized Rammed Earth Based on SEM Images Using Computer Vision and Deep Learning

Piotr Narloch [1,*], Ahmad Hassanat [2,3], Ahmad S. Tarawneh [4], Hubert Anysz [1,*], Jakub Kotowski [5] and Khalid Almohammadi [3]

1   Faculty of Civil Engineering, Warsaw University of Technology, Al. Armii Ludowej 16, 00-637 Warsaw, Poland
2   Computer Science Department, Mutah University, 61710 Karak, Jordan; hasanat@mutah.edu.jo
3   Computer Science Department, Community College, University of Tabuk, 71941 Tabuk, Saudi Arabia; kalmohammadi@ut.edu.sa
4   Department of Algorithms and Their Applications, Eötvös Loránd University, 1053 Budapest, Egvetem ter 1-3 Hungary; Ahmad.trwh@gmail.com
5   Faculty of Geology, University of Warsaw, Żwirki i Wigury 93, 02-089 Warsaw, Poland; j.kotowski@uw.edu.pl
*   Correspondence: p.narloch@il.pw.edu.pl (P.N.); h.anysz@il.pw.edu.pl (H.A.); Tel.: +48-691-660-184 (P.N.); +48-606-668-288 (H.A.)

Received: 1 October 2019; Accepted: 22 November 2019; Published: 27 November 2019

Abstract: Predicting the compressive strength of cement-stabilized rammed earth (CSRE) using current testing machines is time-consuming and costly and may harm the environment due to the samples' waste. This paper presents an automatic method using computer vision and deep learning to solve the problem. For this purpose, a deep convolutional neural network (DCNN) model is proposed, which was evaluated on a new in-house scanning electron microscope (SEM) image database containing 4284 images of materials with different compressive strengths. The experimental results show reasonable prediction results compared to other traditional methods, achieving 84% prediction accuracy and a small (1.5) oot Mean Square Error (RMSE). This indicates that the proposed method (with some enhancements) can be used in practice for predicting the compressive strength of CSRE samples.

Keywords: deep learning; convolutional neural network; SEM images; rammed earth; cement-stabilized rammed earth; cement stabilization

---

1. Introduction

1.1. Aim and Scope of the Research

This article aims to present that machine learning can predict the compressive strength of building materials with a high degree of accuracy based on scanning electron microscope (SEM) images. The methodology of predicting the compressive strength of cement-stabilized rammed earth (CSRE) is described and the results achieved

CSRE is a material used to build construction walls from locally available, inorganic soil found under the layer of humus. Most importantly, CSRE is a highly sustainable construction material. In the study [1], resulting from an analysis using the Building Research Establishment Environmental Assessment Method (BREEAM), external building partitions containing a load-bearing rammed earth layer obtained the highest A+ rating. According to Hall and Swaney [2], such a rating would also be obtained by a partition with a load-bearing layer of rammed earth stabilized with cement. However, designing CSRE of sufficiently high compressive strength is difficult because this feature depends on several mixture properties and technological issues, such as a compaction method [3], density of elements [4–6], porosity of elements [6], particle size distribution of the soil [4,5,7], mineral composition of the soil [8], content and type of cement used as a stabilizer [6,9,10], moisture content of the soil–cement mixture during construction [4,9,11–13], moisture content of element during service [11,12], exposure conditions of the construction elements [5,14], and age of the construction elements [6,15,16].

A multitude of factors hampers the practical application of rammed earth technology. Even if a method of predicting the compressive strength of CSRE based on these factors is developed, assembling a structure using locally sourced soil would require detailed laboratory tests on a case-by-case basis, to determine the basic parameters of that soil—e.g the mineral composition and particle size distribution of the soil. Other properties of the soil–cement mixture, such as the moisture content, may change unexpectedly due to changes in atmospheric conditions prevailing at the construction site. It is proposed to model the compressive strength of CSRE partitions based on SEM images using machine learning and deep learning techniques instead of the whole process involving laboratory testing and the creation of models of relationships between mixture properties and parameters. The proposed method is presented in the article.

1.2. Method of Erecting Monolithic CSRE Walls

The method of erecting monolithic CSRE walls (see Figure 1) includes the following stages [17,18]:

- Laying concrete foundations and assembling on them a formwork for the designed CSRE walls.
- Preparing the soil-cement mixture. I If necessary, the particle size composition of the locally available soil is adjusted. Next, Portland cement is added. The components are mixed in an air-dry state until they reach a uniform consistency. Just before the planned ramming, water is added in a quantity that gives the mixture its predetermined optimal moisture content.
- Ramming the moist soil mixture in the formwork in layers. Obtaining the required compaction depends on the method of ramming, energy used for the compaction process, and the thickness of the CSRE layer. The effectiveness of the compaction process should be verified experimentally under construction conditions.
- Curing of the finished wall in the formwork for a minimum of one day, followed by demolding.

Further curing of the wall after demolding should be continued until obtaining full serviceable properties by CSRE structure., The minimum curing time is 28 days in the air temperatures above 0 °C (Celsius scale). During this period, it is recommended to cover the partition against precipitation and excessive sun exposure.

Figure 1. Wall made from compacted earth. The compacted layers of mixed soil are shown.

2. Related Work

It was found that various machine learning techniques were used to predict the state of various materials' features based on SEM images of these materials. Lahoti et al. used machine learning-based classifiers and statistical analyses to predict the compressive strength of metakaolin-based geopolymers. They applied random forests; Naive Bayes, and the k-nearest neighbors classifiers, while the statistical analysis was based on an analysis of variance (ANOVA) [19].

The prediction of the compressive strength of concrete modified with glass cullet was conducted by Mirzahosseini and co-workers [20], who used genetic programming to build the prediction models of the compressive strength. Their sensitivity and parametric analyses confirmed that the compressive strength is very sensitive to curing time and temperature, as well as to the surface area of the particles.

Qin et. al. employed a deep learning technique to establish an end-to-end mapping of the nonlinear relationship between SEM images of the cemented paste backfill and its mechanical strengths. However, the average accuracy of the prediction of the mechanical strength was very low [21]. There more examples of this approach e.g. [22,23].

Perhaps the most related study to this paper is the work of Gallagher and co-workers [24], which explores the application of computer vision and machine learning methods to predict the compressive strength of consolidated molecular solid (triaminotrinitrobenzene (TATB)) samples based on their SEM images. However, the TATB samples are not as complex as the CSRE ones, and therefore the reported results may not be able to be generalized to the CSRE, the core of this paper.

The deep learning techniques allow for a wide range of their applications and better solutions for many problems in the computer vision domain, as shown by [25–35], and others. All of these works recommend the use of deep learning techniques to find better solutions in the area of computer optics, in the computer vision domain, which is the main motivation for using the deep learning technique in this paper.

To the best of the authors' knowledge, this work is the first attempt to the problem of predicting the compressive strength of CSRE based on SEM images using computer vision and deep learning. The main contribution of this paper is twofold. It proposes a new deep learning approach to solve the research problem; secondly, it describes a new SEM image database of the CSRE samples, which was created for this study.

3. Materials and Methods

3.1. Materials

Through adding the appropriate amount of dry ingredients (seven types of loams with different mineral compositions, as well as sand of pure quartz and gravel of 75% quartz and carbonate crumbs), seven artificial soils with two different particle size distributions labeled LC and MC (LC with 4% and MC with 16% of clay fraction, see Figure 2) were obtained. These artificial soil mixtures simulated the inorganic soils with different mineral compositions that could be obtained locally in Europe and were used as the main component of CSRE. Then, cement CEM I 42.R was added in an amount of 6% or 9% by weight of the artificial soil. Finally, water was added to dry mixtures to obtain optimum moisture content (OMC). The compositions of all 14 mixtures are shown in Table 1.

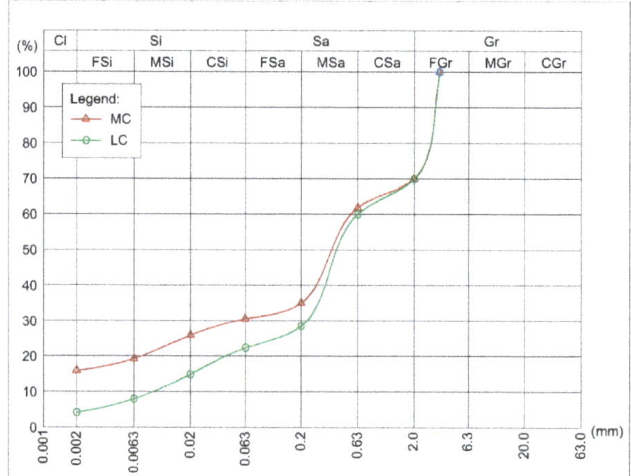

Figure 2. The particle size distribution of soil mixtures used in the cement-stabilized rammed earth (CSRE) compressive strength tests.

Table 1. Mineral compositions of soil mixtures given in percentages. The table also gives the percentage of water and CEM I 42.5 R cement additions.

Mixture symbol	Montmorillonite (%)	Beidellite (%)	Kaolinite (%)	Illite (%)	Goethite (%)	Siderite (%)	Calcite (%)	Organic substance (%)	Quartz and others (%)	Cement addition (%)	OMC (%)
LC II 6% LC II 9%	0.0	2.6	0.4	0.8	0.3	0.0	6.8	0.2	88.9	6 9	7
LC VII 6% LC VII 9%	0.0	2.3	1.2	0.0	0.8	0.7	0.0	0.3	94.6	6 9	7
LC XI 6% LC XI 9%	0.0	1.8	0.4	2.7	0.0	0.5	0.0	0.0	94.6	6 9	7
MC III 6% MC III 9%	0.0	6.6	1.9	0.0	0.9	0.0	13.1	0.1	77.3	6 9	8
MC IV 6% MC IV 9%	0.0	0.0	21.8	0.0	0.3	0.0	0.0	0.1	77.8	6 9	8
MC V 6% MC V 9%	0.0	0.0	21.1	0.0	0.0	0.0	0.0	0.1	78.7	6 9	8
MC X 6% MC X 9%	3.0	4.1	6.9	2.9	0.0	1.1	0.0	0.4	81.7	6 9	8

3.2. Preparation of Samples

All CSRE samples were prepared at the construction materials laboratory (Warsaw University of Technology, Warsaw, Poland). In addition to the soil mixture properties listed in Section 3.1, the shape of the samples and the compaction technique also affect the compressive strength of the CSRE. Therefore, cubic $100 \times 100 \times 100$ mm samples for compressive strength tests were formed as described in the work of Hall and Djerbib [36]. The shaping of the samples was carried out by ramming the moist soil-cement mixture in three equal layers. Each layer was compacted by freely lowering a 6.5 kg rammer from a height of 30 cm to the surface of the mixture. Samples were removed from steel molds after 24 h. From the moment of compaction until the compressive strength test, the samples were cured in conditions of a temperature of 20 °C (±1 °C) and relative humidity of 95% (± 2%). Ten samples were prepared for each of the 14 soil-cement mixtures.

Initially, samples were formed with a height of 105 mm. The day before the planned compressive strength test, they were cut to a height of 100 mm (Figure 3). Thin sections to be examined using an SEM were made from the cut fragments of the samples.

Figure 3. Part of CSRE samples on which the compressive strength tests were made.

3.3. Methods

3.3.1. CSRE Compressive Strength Test

CSRE retains the layer structure both in the monolithic wall and in molded samples. For this reason, the method of testing the compressive strength by loading the sample in the direction of its formation was considered representative. Tests on the compressive strength of CSRE samples were carried out in a testing machine with a measuring range of 0–3000 kN, characterized by a measurement error of less than 1%.

The results of the median of compressive strength of each of 14 tested series are shown in Figure 4. Comparing the results of the LC (4% clay fraction) and MC (16% clay fraction) series, it can be seen that the particle size of the soil has a great impact on the CSRE compressive strength. As expected, differences were observed between the compressive strengths of samples with the same particle size and cement content and with different mineral composition.

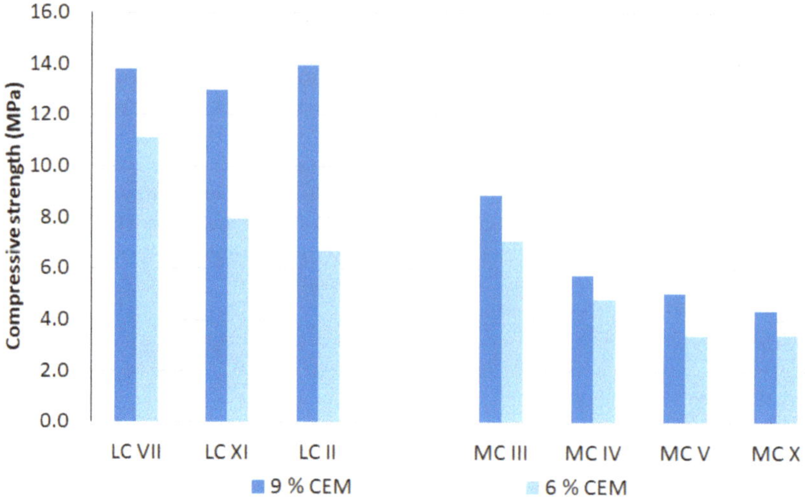

Figure 4. Compressive strength of the CSRE sample series (median).

3.3.2. SEM Methodology

To present the exact texture of the samples at the micro area, images were taken using an SEM. The magnifications used allowed the observation of small structures within the material matrix as well as small clusters of clay minerals.

Thin sections were sputtered with a thin layer of carbon coating ($\approx$20 nm) to remove charging effects from the surface of the samples (Figure 5). The observations were carried out in a high vacuum on a Carl Zeiss SIGMA VP FE-SEM (Carl Zeiss Microscopy Ltd., Cambridge, the United Kingdom) with an accelerating voltage of 20 kV. The high voltage used reduced the impact of carbon coating and surface contamination on the quality of obtained backscattered electron (BSE) imaging. Thin sections were mapped under 150× magnification, and acquired images had a relatively low resolution (300 × 225 px) as required for deep learning purposes. The area of the sample presented in each photo was approx. 2.112 mm^2 (a rectangle of 1.258 × 1.678 mm). Some samples of the SEM images are presented in Figure 6.

Figure 5. Carbon-sputtered thin sections of four different CSRE samples.

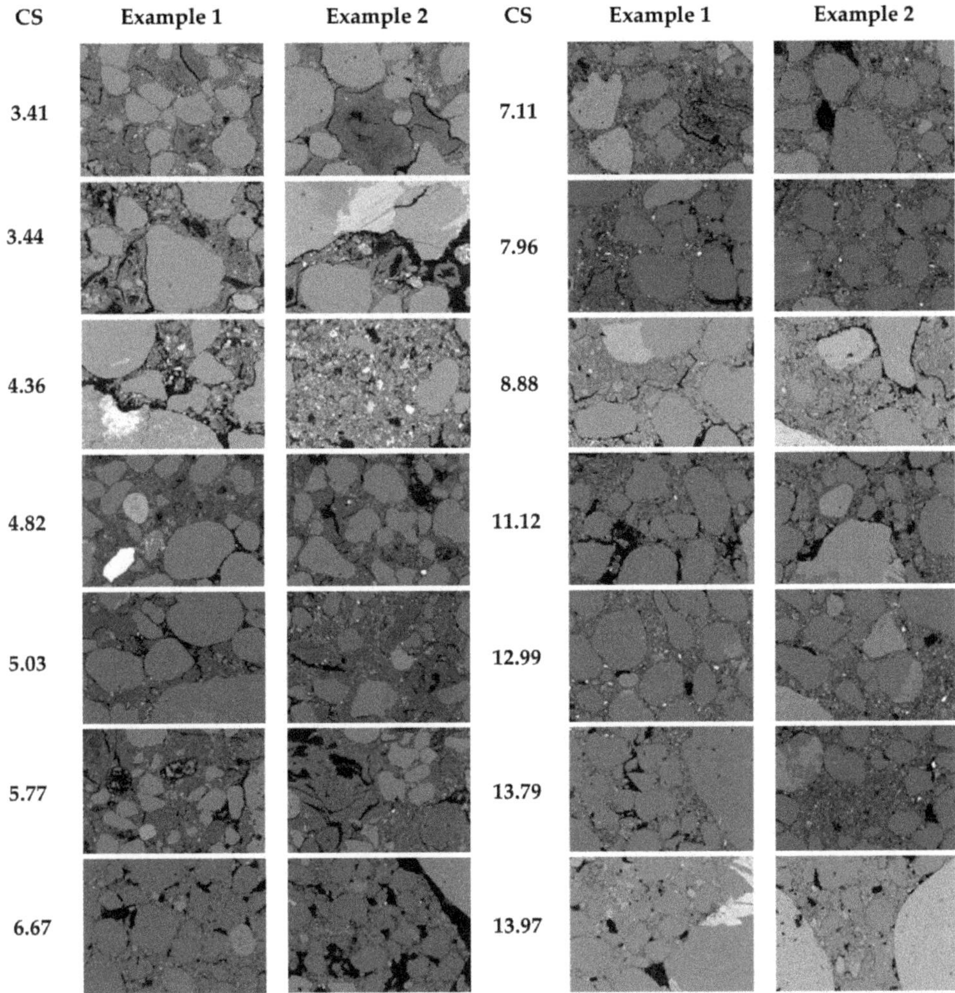

Figure 6. Random samples from the in-house SEM images database. CS – compressive strength (MPa).

3.3.3. Deep Learning Methodology

To predict the compressive strength of the CSRE samples from their SEM images, a deep convolutional neural network (DCNN) regression model was built. The architecture of the proposed model consisted of 24 layers, and the input images were resized and mapped to be 250 × 250. The other DCNN architecture details are described in Figure 7. The proposed neural network was trained using stochastic gradient descent with momentum (SGDM) optimizer built-in Matlab 2019a software (Mathworks, Natick, MA, USA), the mini-batch size assigned to be 128.

1	**Image Input**	**250 x 250 x 3 images with "zerocenter normalization"**
2	**Convolution**	**16 5 x 5 convolutions with stride [1 1] and padding 'same'**
3	**Batch Normalization**	**Batch normalization**
4	**ReLU**	**ReLU**
5	**Max Pooling**	**5 x 5 max pooling with stride [3 3] and padding [0 0 0 0]**
6	**Convolution**	**32 5 x 5 convolutions with stride [1 1] and padding "same"**
7	**Batch Normalization**	**Batch normalization**
8	**ReLU**	**ReLU**
9	**Max Pooling**	**5 x 5 max pooling with stride [3 3] and padding [0 0 0 0]**
10	**Convolution**	**64 3 x 3 convolutions with stride [1 1] and padding "same"**
11	**Batch Normalization**	**Batch normalization**
12	**ReLU**	**ReLU**
13	**Max Pooling**	**5 x 5 max pooling with stride [3 3] and padding [0 0 0 0]**
14	**Convolution**	**128 3 x 3 convolutions with stride [1 1] and padding "same"**
15	**Batch Normalization**	**Batch normalization**
16	**ReLU**	**ReLU**
17	**Max Pooling**	**3x3 max pooling with stride [2 2] and padding [0 0 0 0]**
18	**Convolution**	**512 3 x 3 convolutions with stride [1 1] and padding "same"**
19	**Batch Normalization**	**Batch normalization**
20	**ReLU**	**ReLU**
21	**Max Pooling**	**3 x 3 max pooling with stride [2 2] and padding [0 0 0 0]**
22	**Dropout**	**50% dropout**
23	**Fully Connected**	**1 fully connected layer**
24	**Regression Output**	**Mean-squared-error**

Figure 7. The architecture of the proposed deep convolutional neural network (DCNN) model.

4. Results and Discussion

To validate and test the proposed DCNN model, a 10-fold cross-validation approach was used. The training and validation process of the proposed model is presented in Figure 8.

As shown in Figure 8, the model—which is designed to be validated every 428 iterations to ensure that the network was not overfitting—starts to fit the data early in the process. The final root mean square error (RMSE) in the training process is approximately 1.39.

To measure the prediction accuracy of the proposed DCNN, variable α was set to represent the gap between the predicted and the actual values. α is assumed to be an acceptable margin of error; for example, if α is set at 0.5 and the actual value of one sample is 1.7, this prediction would be accepted as correct if the predicted value was in the range 1.7 ± 0.5. In this work, the prediction accuracy is measured using various α values. The prediction accuracy and its corresponding α value are presented in Table 2. The final prediction accuracy was calculated by averaging the accuracy of all folds.

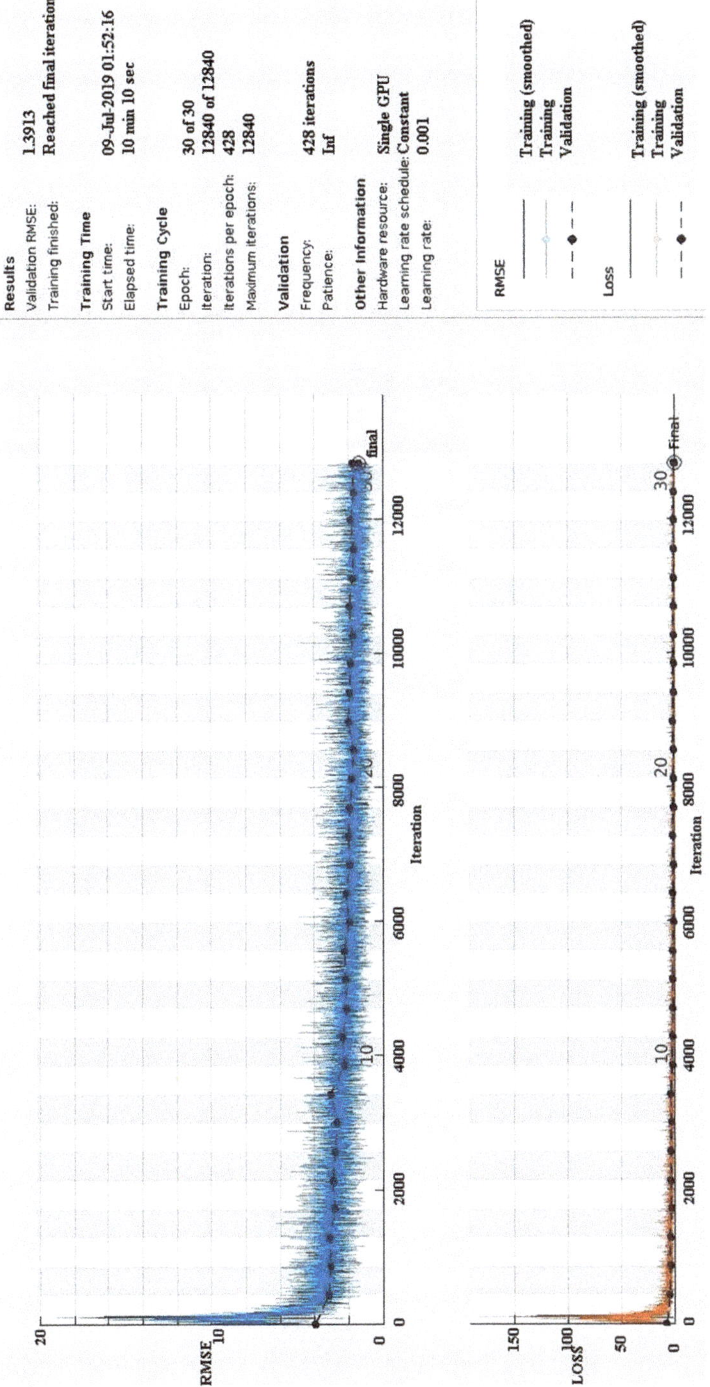

Figure 8. The training process of the proposed DCNN model.

Table 2. Prediction accuracy of the proposed DCNN with their α values.

α	Prediction Accuracy
0.4	0.31
0.6	0.46
0.8	0.57
1	0.64
1.2	0.71
1.4	0.75
1.6	0.79
1.8	0.81
2	0.84

As can be seen in Figure 9 and Table 2, the prediction accuracy starts to be acceptable (accuracy >50%) when α is assigned to be greater than 0.8 of the tested α values. Moreover, if it is recalled that the minimum compressive strength of the CSRE samples was 3.41 MPa and the maximum was 13.97 MPa, then $\alpha = 2$ might be considered as an acceptable error threshold. Then a reasonable prediction accuracy of 84% is achieved. The accuracy of the prediction is proportional to the margin of error; however, the margin of error cannot be significantly increased to claim a higher prediction accuracy. Therefore, other measures such as the RMSE can be used to acquire a better understanding of the prediction system's performance [37].

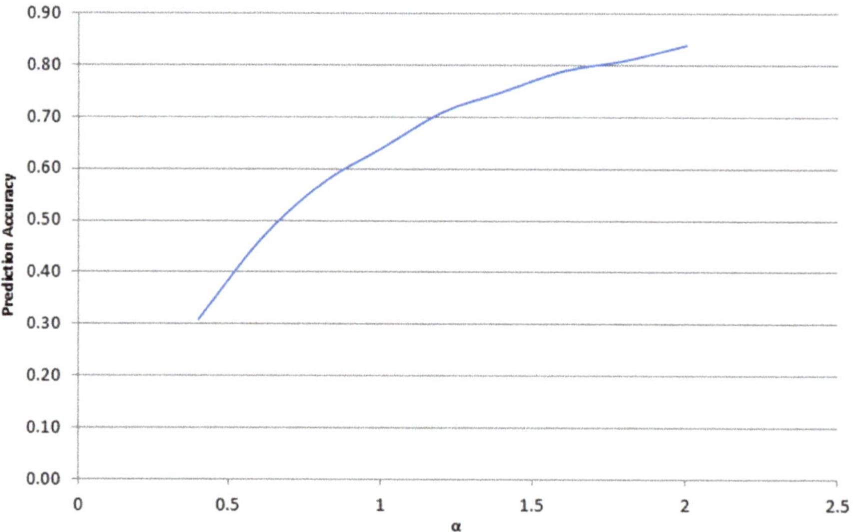

Figure 9. Prediction accuracy of the proposed DCNN as a function of α value.

Furthermore, to provide more information about the predictions of the proposed deep learning model, the covariance and correlation coefficient are calculated as results of the experiments. Table 3 demonstrates the correlation coefficient and the covariance for each fold separately.

As it can be observed in Table 3, the correlation is close to 0.9, which statistically means that the predictions and the actual values have a very strong relationship between them. Moreover, the covariance shows that these sets of data have a positive relationship.

Table 3. Correlation coefficient and covariance for each fold separately.

Fold No.	Covariance	Correlation Coefficient
Fold 1	8.742	0.857
Fold 2	10.08	0.897
Fold 3	9.588	0.914
Fold 4	9.827	0.915
Fold 5	9.385	0.853
Fold 6	10.21	0.934
Fold 7	9.411	0.910
Fold 8	8.654	0.896
Fold 9	8.622	0.890
Fold 10	9.353	0.914
Average	9.39	0.898

In addition to the prediction, which is based on regression, the proposed DCNN can offer deep features that can be used by another regression method to obtain the prediction results. Two more regression methods are used: the random forest and linear regression features. In the case of calculating the prediction accuracy of the DCNN, the RMSE is one of the important measures that is usually used to measure the performance of a prediction system [37]. The number of deep features obtained by the proposed DCNN was quite large (4096 features); therefore, a principal component analysis (PCA) was used to reduce the number of features [25,38,39], keeping 95% of the data variance to attain only 572 features. The average RMSE values (over 10 folds) of predicting the CSRE samples' compressive strength achieved by each method are presented in Table 4.

Table 4. Average root mean square error (RMSE) of predicting the compressive strength of CSRE samples using the proposed DCNN and the deep features obtained.

Method	No. of Features	RMSE (MPa)
DCNN	4096	1.5
Random forest with PCA on the deep features	572	3.1185
Linear regression with PCA on the deep features	572	2.5057

It can be noticed (based on the results presented in Table 4) that the proposed DCNN with a regression model based on an artificial neural network, applied on all the obtained features (4096), outperforms the other regression models when compared. This is due to the use of all the deep features obtained from the DCNN, unlike the PCA, which reduces the quantity of the features. However, using a large number of features it is a time-consuming process.

To compare the proposed DCNN prediction model with other state-of-the-art traditional computer vision methods, the selection of the features from the SEM images was made with the use of the following algorithms: histogram of oriented gradients (HOG) [40], local binary pattern (LBP) [41], and scale-invariant feature transform (SIFT) [42]. The values of RMSE calculated for the compressive strength (of CSRE samples) predictions made with the use of these traditional features extraction methods are presented in Table 5.

Table 5. RMSE of compressive strength (of CSRE samples) predictions achieved in a random forest and linear regression, fed with the features extracted traditionally from the SEM images.

Method	No. of Features	RMSE (MPa)	
		Random Forest	Linear Regression
HOG	72	3.2192	3.2217
LBP	10	2.4807	2.8518
SIFT	100	3.0682	4.6951

It can be noticed (see Table 5) that none of the traditional methods could handle the prediction better than the proposed DCNN. This may be attributed to the representative power of the deep features obtained by the proposed model compared to that of the handcrafted low-level features (HOG, LBP, and SIFT). A similar effect is achieved by many researchers, e.g. in [43–45].

5. Conclusions

In addition to creating a new in-house SEM image database to be used for evaluating the model proposed in this paper, we provide a new deep learning model for predicting the compressive strength of CSRE based on SEM images using computer vision and deep learning. The comparative results of the experiments conducted on the SEM image database to predict the compressive strength of CSRE samples show that the proposed DCNN model outperforms other traditional computer vision methods, obtaining reasonable prediction results and comparatively low RMSE. Therefore, the method can be recommended for practical use as time, and cost-saving tool, as well as the environment protecting one, if applied instead of the current strength-testing machines.

Despite the reasonable performance of the proposed DCNN, the overall prediction accuracy process is still imperfect because it is the error threshold dependent. Such imperfect performance may be attributed to the number of SEM images used to train the deep features. This deficit can be alleviated by obtaining more SEM images of a higher number of CSRE samples to feed the DCNN. Higher prediction accuracy is expected then. Moreover, using the image enhancement techniques at a preprocessing stage may enhance the prediction results. It is considered to verify these statements in future analysis. Prospective future works will also include the use of DCNN to solve other problems described in [46–48]. It is expected to overcome the problem of prediction speed. It should increase with the use of more efficient indexing techniques, such as presented in [49–51].

Author Contributions: Conceptualization, H.A. and P.N.; methodology, A.S.T. and P.N.; software, A.S.T. and A.H.; validation, A.S.T., P.N. and A.H.; formal analysis, A.H. and K.A.; investigation, H.A., A.H., A.S.T., P.N. and J.K.; resources, P.N. and H.A.; data creation, P.N. and J.K.; writing—original draft preparation, A.S.T., P.N., K.A. and A.H.; writing—review and editing, A.H., K.A., and H.A.; visualization, P.N. and A.S.T.

Funding: This research received no external funding.

Acknowledgments: This research was supported by the Cryo-SEM laboratory, the National Multidisciplinary Laboratory of Functional Nanomaterials NanoFun project no. POIG.02.02.00-00-025/09 at the Faculty of Geology, University of Warsaw. A.S.T. would like to thank the Tempus Public Foundation for sponsoring his Ph.D. study. Also, A.S.T.'s work falls under the project EFOP-3.6.3-VEKOP-16-2017-00001 (Talent Management in Autonomous Vehicle Control Technologies) and is supported by the Hungarian Government, and co-financed by the European Social Fund. Moreover, the authors would like to thank the Institute for Computer Science and Control (SZTAKI) for providing an advanced server for training the proposed model.

Conflicts of Interest: The authors declare no conflict of interest.

References

1. Bre. Bre Group. 2008. Available online: https://www.bregroup.com/greenguide/ggelement.jsp?buildingType=Offices&category=1019&parent=6&elementType=10166 (accessed on 23 September 2019).
2. Hall, M.R.; Swaney, W. European modern earth construction. In *Modern Earth Buildings: Materials, Engineering, Construction and Applications*; Hall, M.R., Lindsay, R., Krayenhoff, M., Eds.; Woodhead Pub Ltd.: Oxford, UK, 2012; pp. 650–687.
3. Reddy, B.V.; Kumar, P.P. Cement stabilised rammed earth. Part A: Compaction characteristics and physical properties of compacted cement stabilised soils. *Mater. Struct.* **2011**, *44*, 681–693. [CrossRef]
4. Anysz, H.; Narloch, P. Designing the composition of cement stabilized rammed earth using artificial neural networks. *Materials* **2019**, *12*, 1396. [CrossRef] [PubMed]
5. Ciancio, D.; Jaquin, P.; Walker, P. Advances on the assessment of soil suitability for rammed earth. *Constr. Build. Mater.* **2013**, *42*, 40–47. [CrossRef]

6. Consoli, N.C.; Festugato, L.; da Rocha, C.G.; Cruz, R.C. Key parameters for strength control of rammed sand–cement mixtures: Influence of types of portland cement. *Constr. Build. Mater.* **2013**, *49*, 591–597. [CrossRef]

7. Lina, H.; Zhenga, S.; Lourençoa, S.; Jaquin, P. Characterization of coarse soils derived from igneous rocks for rammed earth. *Eng. Geol.* **2017**, *228*, 137–145. [CrossRef]

8. Narloch, P.L.; Woyciechowski, P.; Jęda, P. The influence of loam type and cement content on the compressive strength of rammed earth. *Arch. Civ. Eng.* **2015**, *61*, 73–88. [CrossRef]

9. Hall, M.; Allinson, D. Influence of cementitious binder content on moisture transport in stabilised earth materials analysed using 1-dimensional sharp wet front theory. *Build. Environ.* **2009**, *44*, 688–693. [CrossRef]

10. Hall, M.; Allinson, D. Assessing the moisture-content-dependent parameters of stabilised earth materials using the cyclic-response admittance method. *Energy Build.* **2008**, *40*, 2044–2051. [CrossRef]

11. Bui, Q.B. Assessing the rebound hammer test for rammed earth material. *Sustainability* **2017**, *9*, 1904. [CrossRef]

12. Bui, Q.B.; Morel, J.C.; Hans, S.; Walker, P. Effect of moisture content on the mechanical characteristics of rammed earth. *Constr. Build. Mater.* **2014**, *54*, 163–169. [CrossRef]

13. Beckett, C.; Ciancio, D. Effect of compaction water content on the strength of cement-stabilized rammed earth materials. *Can. Geotech. J.* **2014**, *51*, 583–590. [CrossRef]

14. Arrigoni, A.; Grillet, A.C.; Pelosato, R.; Dotelli, G.; Beckett, C.T.; Woloszyn, M.; Ciancio, D. Reduction of rammed earth's hygroscopic performance under stabilisation: An experimental investigation. *Build. Environ.* **2017**, *115*, 358–367. [CrossRef]

15. Bui, Q.B.; Morel, J.C. First Exploratory Study on the Ageing of Rammed Earth Material. *Materials* **2015**, *8*, 1–15. [CrossRef] [PubMed]

16. Bui, Q.B.; Morel, J.C.; Reddy, B.V.V.; Ghayad, W. Durability of rammed earth walls exposed for 20 years to natural weathering. *Build. Environ.* **2009**, *44*, 912–919. [CrossRef]

17. Narloch, P.L.; Lidner, M.; Kunicka, M.; Bielecki, M. Flexural tensile strength of construction elements made out of cement stabilized rammed earth. *Procedia Eng.* **2015**, *111*, 589–595. [CrossRef]

18. Narloch, P.L.; Woyciechowski, P.P.; Dmowska, E.; Halemba, K. Durability assessment of monolithic rammed earth walls. *Arch. Civ. Eng.* **2015**, *61*, 73–88. [CrossRef]

19. Lahoti, M.; Narang, P.; Tan, K.H.; Yang, E.H. Mix design factors and strength prediction of metakaolin-based geopolymer. *Ceram. Int.* **2017**, *43*, 11433–11441. [CrossRef]

20. Mirzahosseini, M.; Jiao, P.; Barri, K.; Riding, K.; Alavi, A.H. New machine learning prediction models for compressive strength of concrete modified with glass cullet. *Eng. Comput.* **2019**, *36*, 876–898. [CrossRef]

21. Qin, X.; Cui, S.; Liu, L.; Wang, P.; Wang, M.; Xin, J. Prediction of Mechanical Strength Based on Deep Learning Using the Scanning Electron Image of Microscopic Cemented Paste Backfill. *Adv. Civ. Eng.* **2018**, *2018*, 1–7. [CrossRef]

22. Drumetz, L.; Mura, M.D.; Meulenyzer, S.; Lombard, S.; Chanussot, J. Semiautomatic classification of cementitious materials using scanning electron microscope images. *J. Electron.* **2015**, *24*, 061109. [CrossRef]

23. Hughes, A.; Liu, Z.; Raftari, M.; Reeves, M.E. A workflow for characterizing nanoparticle monolayers for biosensors: Machine learning on real and artificial SEM images. *Peerj Prepr.* **2014**. [CrossRef]

24. Gallagher, B.; Rever, M.; Loveland, D.; Mundhenk, T.N.; Beauchamp, B.; Robertson, E.; Han, T. Predicting Compressive Strength of Consolidated Molecular Solids Using Computer Vision; Deep Learning. *arXiv* **2019**, arXiv:1906.02130.

25. Tarawneh, A.S.; Chetverikov, D.; Hassanat, A.B. Pilot Comparative Study of Different Deep Features for Palmprint Identification in Low-Quality Images. In Proceedings of the Ninth Hungarian Conference on Computer Graphics and Geometry, Budapest, Hungary, 10 March 2018.

26. Saritha, R.R.; Paul, V.; Kumar, P.G. Content based image retrieval using deep learning process. *Clust. Comput.* **2018**, *22*, 1–14. [CrossRef]

27. Tzelepi, M.; Tefas, A. Deep convolutional learning for Content Based Image Retrieval. *Neurocomputing* **2018**, *275*, 2467–2478. [CrossRef]

28. Cheng, G.; Zhou, P.; Han, J. Learning rotation-invariant convolutional neural networks for object detection in VHR optical remote sensing images. *IEEE Trans. Geosci. Remote Sens.* **2016**, *54*, 7405–7415. [CrossRef]

29. Lei, Z.; Yi, D.; Li, S.Z. Learning stacked image descriptor for face recognition. *IEEE Trans. Circuits Syst. Video Technol.* **2016**, *26*, 1685–1696. [CrossRef]

30. Kappeler, A.; Yoo, S.; Dai, Q.; Katsaggelos, A.K. Video super-resolution with convolutional neural networks. *IEEE Trans. Comput. Imaging* **2016**, *2*, 109–122. [CrossRef]
31. Gao, S.; Zhang, Y.; Jia, K.; Lu, J.; Zhang, Y. Single sample face recognition via learning deep supervised autoencoders. *IEEE Trans. Inf. Forensics Secur.* **2015**, *10*, 2108–2118. [CrossRef]
32. Zhao, L.; Hu, Q.; Wang, W. Heterogeneous feature selection with multi-modal deep neural networks and sparse group lasso. *IEEE Trans. Multimed.* **2015**, *17*, 1936–1948. [CrossRef]
33. Liu, D.; Wang, Z.; Wen, B.; Yang, J.; Han, W.; Huang, T.S. Robust single image super-resolution via deep networks with sparse prior. *IEEE Trans. Image Process.* **2016**, *25*, 3194–3207. [CrossRef]
34. Dong, C.; Loy, C.C.; He, K.; Tang, X. Image super-resolution using deep convolutional networks. *IEEE Trans. Pattern Anal. Mach. Intell.* **2016**, *38*, 295–307. [CrossRef] [PubMed]
35. Goh, H.; Thome, N.; Cord, M.; Lim, J. Learning deep hierarchical visual feature coding. *IEEE Trans. Neural Netw. Learn. Syst.* **2014**, *25*, 2212–2225. [CrossRef] [PubMed]
36. Hall, M.; Djerbib, Y. Rammed earth sample production: Context, recommendations and consistency. *Constr. Build. Mater.* **2004**, *18*, 281–286. [CrossRef]
37. Lu, Y.; Tian, Z.; Peng, P.; Niu, J.; Li, W.; Zhang, H. GMM clustering for heating load patterns in-depth identification and prediction model accuracy improvement of district heating system. *Energy Build.* **2019**, *190*, 49–60. [CrossRef]
38. Hassanat, A.B.; Tarawneh, A.S. Fusion of Color and Statistic Features for Enhancing Content-Based Image Retrieval Systems. *J. Theor. Appl. Inf. Technol.* **2016**, *88*, 644–655.
39. Hassanat, A.B.A.; Prasath, V.B.S.; Al-kasassbeh, M.; Tarawneh, A.S.; Al-shamailh, A.J. Magnetic energy-based feature extraction for low-quality fingerprint images. *Signal Image Video Process.* **2018**, *12*, 1471–1478. [CrossRef]
40. Dalal, N.; Triggs, B. *Histograms of Oriented Gradients for Human Detection*; CVPR: Washington, DC, USA, 2005; pp. 886–893.
41. He, D.-C.; Wang, L. Texture Unit, Texture Spectrum, And Texture Analysis. *IEEE Trans. Geosci. Remote Sens.* **1990**, *28*, 509–512.
42. Lowe, D.G. Object recognition from local scale-invariant features. In Proceedings of the International Conference on Computer Vision, Kerkyra, Greece, 20 September 1999; pp. 1150–1157.
43. Wang, J.; Chen, Y.; Hao, S.; Peng, X.; Hu, J. Deep learning for sensor-based activity recognition: A survey. *Pattern Recognit. Lett.* **2019**, *119*, 3–11. [CrossRef]
44. Tarawneh, A.S.; Hassanat, A.B.; Celik, C.; Chetverikov, D.; Rahman, M.S.; Verma, C. Deep Face Image Retrieval: A Comparative Study with Dictionary Learning. In Proceedings of the 10th International Conference on Information and Communication Systems (ICICS), Irbid, Jordan, 11 June 2019; pp. 185–192.
45. Xia, Y.; Wulan, N.; Wang, K.; Zhang, H. Detecting atrial fibrillation by deep convolutional neural networks. *Comput. Biol. Med.* **2018**, *93*, 84–92. [CrossRef]
46. Hassanat, A.B.A. On Identifying Terrorists Using Their Victory Signs. *Data Sci. J.* **2018**, *17*, 1–13. [CrossRef]
47. Hassanat, A.B.; Btoush, E.; Abbadi, M.A.; Al-Mahadeen, B.M.; Al-Awadi, M.; Mseidein, K.I.; Almseden, A.M.; Tarawneh, A.S.; Alhasanat, M.B.; Prasath, V.S.; et al. Victory sign biometrie for terrorists identification: Preliminary results. In Proceedings of the 2017 8th International Conference on Information and Communication Systems (ICICS), Irbid, Jordan, 4–6 April 2017; pp. 182–187.
48. Hassanat, A.B.; Prasath, V.B.S.; Al-Mahadeen, B.M.; Alhasanat, S.M.M. Classification and gender recognition from veiled-faces. *Int. J. Biom.* **2017**, *9*, 347–364. [CrossRef]
49. Hassanat, A. Furthest-Pair-Based Decision Trees: Experimental Results on Big Data Classification. *Information* **2018**, *9*, 284. [CrossRef]
50. Hassanat, A. Furthest-pair-based binary search tree for speeding big data classification using k-nearest neighbors. *Big Data* **2018**, *6*, 225–235. [CrossRef]
51. Hassanat, A. Norm-Based Binary Search Trees for Speeding Up KNN Big Data Classification. *Computers* **2018**, *7*, 54. [CrossRef]

Article

Short-Term Deformability of Three-Dimensional Printable EVA-Modified Cementitious Mortars

Jaeheum Yeon

Department of Engineering and Technology, Texas A&M University-Commerce, Commerce, TX 75429, USA;
jaeheum.yeon@tamuc.edu; Tel.: +1-903-468-8115

Received: 24 September 2019; Accepted: 30 September 2019; Published: 8 October 2019

Abstract: This study experimentally examined the deformability of cementitious mortars modified with ethylene-vinyl acetate (EVA) for use in extrusion-based additive construction. The research was based on the author's previous study of the properties of fresh EVA-modified cementitious mixtures for use in additive construction via extrusion. The particular focus was on these mortars' short-term deformation factors, including the modulus of elasticity, drying shrinkage, and thermal expansion. The experimental results indicate that as the EVA/cement ratio was increased, the compressive strength and elastic modulus tended to decrease but the maximum compressive strain increased. At 28 days, the drying shrinkage tended to increase as the EVA/cement ratio was increased. The coefficient of thermal expansion was also found to increase as the EVA/cement ratio was increased. A very high correlation was found between these three deformation factors and the EVA/cement ratio. Given these results, it was determined that the addition of EVA powder to EVA-modified cementitious mortars used in extrusion-based additive construction could adversely affect their short-term deformation factors. However, increasing the EVA/cement ratio resulted in a decrease in the modulus of elasticity, thereby reducing the level of stress caused by drying shrinkage and thermal expansion. This effect will eventually lead to improvements in the degree of extensibility, thereby offsetting the negative impacts. However, it is still desirable to minimize the EVA/cement ratio to the extent that adequate properties for the fresh material can be obtained.

Keywords: EVA-modified cementitious mortars; additive construction; short-term deformability; modulus of elasticity; drying shrinkage; coefficient of thermal expansion

---

1. Introduction

Construction processes are intrinsically labor-intensive and accompanied by a high risk of accidents, and thus would benefit from the introduction of automated solutions. The construction industry, however, lags behind other fields in implementing such automation. Encouragingly, an additive construction method intended for concrete structures has recently been developed and applied to small building projects, such as pedestrian bridges [1–3]. Cementitious concrete is one of the most widely used construction materials worldwide, and is mostly in the form of ready-mixed concrete. Cementitious concrete placement, however, typically requires a formwork. Formwork installation costs a great deal in terms of material, human labor, and equipment resources. Also, the construction process is inevitably long term, due to the labor intensity of formwork installation and removal. In addition, the waste produced by formworks may have a negative impact on the environment [4].

The additive construction method was introduced to address these chronic issues associated with concrete work. This process is a computer-controlled construction technology that is used to build structures by layering extruded cementitious concrete without a formwork [5]. Also known as 3D concrete printing (3DCP), this process was first successfully applied in the construction industry by Khoshnevis [6], who developed a contour crafting (CC) method in which a fresh mixture is first extruded in a layer, and then additional layers are added on top. This free-form construction process is possible because the method does not require formwork to construct the cement concrete structure [7].

Compared to conventional construction technologies, 3DCP is commonly viewed as a sustainable design solution that offers almost unlimited possibilities for implementing geometrically complex designs. The technology is advantageous in various ways, such as in reducing construction cost and time, minimizing environmental degradation. The technology can also be used to streamline environmentally friendly construction processes, reduce industrial waste, and decrease energy consumption resulting from producing the raw materials used in formwork [8].

The 3DCP process consists of three components: a concrete printer, 3D modeling software, and printing material. The procedure draws from three specialized areas, including the mechanical, 3D design, and concrete materials fields [9]. Printing materials comprise the primary concern of the present study, especially cementitious mixtures. In recent relevant research, the most frequently examined materials for 3DCP contained Portland cement, sand, fly ash, and silica fume as base materials, along with small amounts of additives such as superplasticizers and viscosity-modifying agents [4,10–13]. Comparatively speaking, the information available on specific additive construction methods using redispersible ethylene-vinyl acetate (EVA) powder is very limited, even though EVA provides excellent adhesive strength and dynamic cracking resistance [14].

In a previous study conducted by the author, EVA-modified cementitious mortars were produced with various EVA/cement ratios. Then, the properties of fresh EVA-modified cementitious mixtures intended for use in additive construction via extrusion were experimentally investigated [15]. It was clear from this earlier work that EVA-modified cementitious mortars are applicable to the 3DCP process. However, it is still unknown whether or not three-dimensional printable EVA-modified cementitious mortars would be sufficiently stable after the 3DCP process. Thus, in the present study, three-dimensional printable EVA-modified cementitious mortars were experimentally investigated with regard to various properties associated with dimensional stability, such as elasticity and drying and thermal shrinkage. This work provides fundamental research data that will assist in the adoption of EVA-modified cementitious mortars for additive construction applications.

2. Summary of the Author's Previous Research

The author, together with collaborators, studied the properties of fresh EVA-modified cementitious mortars as possible materials for use in additive construction [15]. EVA is easier to handle than polymer in a liquid form (i.e., latex or emulsion types), especially when producing mixtures onsite. This is because EVA comes in a powder form. Also, a premixed package that is ready to use onsite can be produced in the factory. Hence, EVA was selected as the object of this series of studies. The main results obtained are as follows.

Determining the optimal flow of EVA-modified cementitious mixtures is of the utmost importance for determining the optimum mix ratio. In the previous study, the author employed a trial-and-error procedure to investigate the ideal flow that would meet all buildability requirements. The flow of each EVA-modified cementitious mixture was tested at 5% intervals ranging from 50% to 75%, in order to determine the optimum state. Through this trial-and-error process, the optimal flow was determined to be 65%, as shown in Figures 1 and 2. This flow is considerably lower than the 110% ± 5% level, which is the standard flow range applied when producing specimens for compressive strength testing under American Society for Testing and Materials (ASTM) C109/C109M-02: Testing Method for Compressive Strength of Hydraulic Cement Mortar [16].

Figure 1. View of the 3D concrete printing (3DCP) process for buildability testing [15].

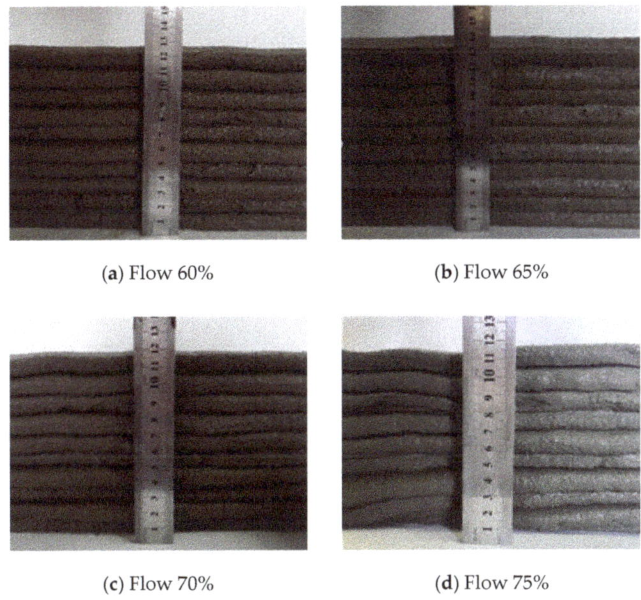

(**a**) Flow 60%

(**b**) Flow 65%

(**c**) Flow 70%

(**d**) Flow 75%

Figure 2. Test results determining the optimal flow [15].

2.1. Flowability and Open Time

Vebe time, compacting factor, slump, and flow test methods are all flowability test methods. Among these, the flow test method is preferable if the workability of the mixture to be tested is to be very high [17]. The results of the flow test conducted for this study showed that the flow increased when the EVA/cement ratio was increased, as shown in Figure 3. This means that the level of the flow loss decreased when the EVA/cement ratio was increased. This result was quite favorable because it indicated that flowability could be secured. Also, the test results showed that the flow consistency improved due to a dispersing effect of the surfactants in the polymers. This effect originated from the ball-bearing action of the polymer particles and entrained air when the EVA powder (which was re-dispersible) was dispersed in the water as the mixture was produced [18].

Open time is the minimum amount of time that a material can be used without a loss of performance. In 3DCP, open time starts at the beginning of extrusion and ends at the time at which there is no more extrusion due to decreased flowability. Hence, open time is the best way to show changes in a mixture's workability over time. Open time can be determined by plotting horizontal lines, as shown in Figure 3. In Figure 3, (1) is the reference point at a 65% flow, and (2) is the reference point at a 50% flow, indicating that the mixture can no longer be extruded through the 3DCP process. According to the test results, a longer open time was secured when the EVA/cement ratio was increased. These results demonstrate that the time identified was sufficient to complete the 3DCP process. This extended open time originated from a delay in the initial setting. In other words, the initial hydration reaction of the cement was inhibited by the formation of a polymer film [19].

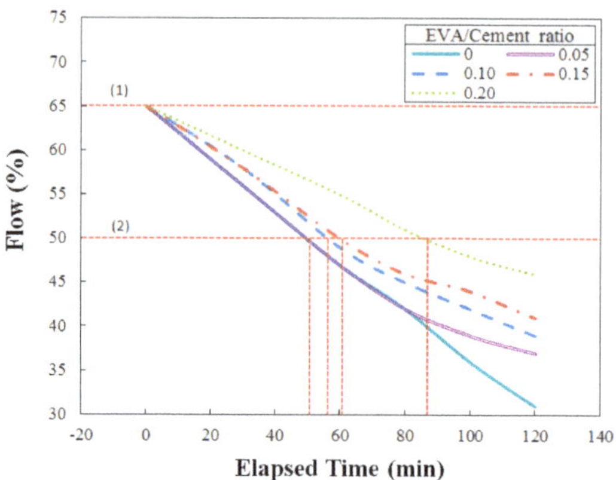

Figure 3. Elapsed time versus flow [15].

2.2. Buildability

Buildability was evaluated by measuring the stacked height and its vertical deformation over time after stacking 10 layers with a unit length of 50 cm per layer. According to the buildability test results, the EVA-modified cementitious mortars proposed in this study showed excellent buildability because the stacked heights experienced a minimal decrease, even though changes were made to the EVA/cement ratio. Among the EVA/ratios tested in this study, the buildability was the most stable when the EVA/ratio was 0.15; no vertical deformation was observed. Figure 4 shows the results of the buildability test when the flow of the EVA-modified cementitious mortar samples was 65%. Based on the observations made, no cracks occurred on the surface of the stacked layers at the point where the direction of the nozzle head was changed. However, there were cracks when the EVA/cement ratio was 0 (i.e., plain mortar). The best buildability was observed when the EVA/cement ratio was 0.15—not only were there no cracks, but also the surface at the point where the direction of the nozzle head was changed was smooth. According to the data collected, the occurrence of cohesion due to the viscosity provided excellent resistance to both bleeding and segregation, even though the polymer-modified cementitious mortars had more substantial flowability characteristics as compared to ordinary cementitious mortars [18]. Similar results were observed in a buildability test completed in the earlier study.

(**a**) E/C: 0   (**b**) E/C: 0.05   (**c**) E/C: 0.10   (**d**) E/C: 0.15   (**e**) E/C: 0.20

Figure 4. Results of the buildability test at a 65% flow and different ethylene-vinyl acetate (EVA)/cement ratios [15].

In prior research by the author, the properties of fresh EVA-modified cementitious mortars were experimentally investigated with regard to whether those mixtures might be applicable to the 3DCP process. According to the results, it was experimentally determined that EVA-modified cementitious mortars could be employed as 3DCP material. However, the dimensional stability of the EVA-modified cementitious mortars tested could not be estimated because at the time, there had been no study examining the material's deformation properties. Thus, the present research experimentally investigated their dimensional stability.

3. Materials and Methods

3.1. Materials

The materials used were the same as those employed in the previous study. Ordinary Portland cement, silica sand, fly ash, silica fume, superplasticizer, and viscosity modifying agent were used. Ordinary Portland cement was used as the main binder, fly ash was employed to improve flowability and to inhibit the heat of hydration in the early hydration stage. Also, silica fume was employed to improve the strength.

The characteristics of the materials used in this study are shown in Tables 1–6. The EVA used as a modifier was a white powder. Its product data and chemical constitution are presented in Table 7 and Figure 5, respectively.

Table 1. Properties of ordinary Portland cement (Type I).

Density (g/cm^3)	Specific Surface (cm^2/g)	Chemical Composition (%)						
		CaO	SiO_2	Al_2O_3	Fe_2O_3	SO_3	MgO	Ig. loss
3.14	3630	64.10	17.00	4.44	3.88	2.97	2.34	2.76

Table 2. Properties of silica sand.

Size(mm)	Apparent Density	Purity (%)	Water Content (%)
0.08	1.57	97.3	≤0.1

Table 3. Properties of fly ash.

Density (g/cm^3)	Specific Surface (cm^2/g)	Chemical Composition (%)						
		SiO_2	Al_2O_3	CaO	Fe_2O_3	SO_3	MgO	Ig. loss
2.22	3651	51.90	21.80	8.25	6.93	1.02	0.89	3.20

Table 4. Properties of silica fume.

Bulk Density-Densified (kg/m^3)	Specific Surface (cm^2/g)	Chemical Composition (%)						
		SiO_2	Al_2O_3	CaO	MgO	Fe_2O_3	SO_3	Ig. loss
600–700	157,700	96.70	0.29	0.25	0.15	0.10	-	2.39

Table 5. Properties of superplasticizer.

Specific Gravity (20 °C)	pH	Alkali (%)	Chloride (%)
1.05 ± 0.05	5.0 ± 2.0	≤0.01	≤0.01

Table 6. Properties of viscosity modifying agent.

Appearance	Bulk Density (kg/m^3)	Moisture Content (%)	Particle Size (0.074 mm, %)
White powder	430	≤12	≥95

Table 7. Product data for the EVA powder.

Solids Content (%)	Ash Content (%)	Bulk Density (kg/m^3)	Particle Size after Redispersion (μm)	Minimum Film Forming Temp (°C)	Protective Colloid
98–100	9–13	470–570	0.5–8.0	4	Polyvinyl alcohol (PVA)

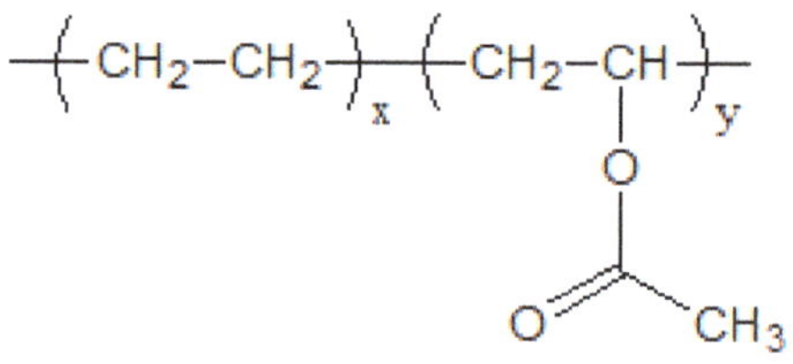

$$\left(\!CH_2\!-\!CH_2\!\right)_x \left(\!CH_2\!-\!CH\!\right)_y$$

Figure 5. Chemical constitution of EVA.

3.2. Method

3.2.1. Mixture

As mentioned above, this study was based on the author's previous research, entitled Fresh Properties of EVA-Modified cementitious Mixtures for use in Additive Construction by Extrusion [15]. Hence, the mix proportions applied in the present work were the same as the mix proportions used in the author's prior research. Details related to the mix proportions are summarized in Table 8.

Table 8. Mix proportions of EVA-modified cementitious mortars (kg/m^3).

EVA/Cement Ratio	W/C Ratio	EVA	Cement	Water	Silica Sand	Fly Ash	Silica Fume	Super-Plasticizer	Viscosity Modifying Agent
0	0.45	0	642	289	1377	184	92	6	0.3
0.05	0.46	32	638	294	1368	182	91	6	0.3
0.10	0.51	63	635	324	1360	181	91	6	0.3
0.15	0.52	95	631	328	1351	180	90	6	0.3
0.20	0.55	125	627	345	1343	179	90	6	0.3

3.2.2. Preparation of Specimens

Three test specimens were produced for each EVA/cement ratio (i.e., test variable). Among each set of three test results produced for each EVA/cement ratio, the middle value was selected because all of the results appeared continuously. Casting the specimens with mortar directly extruded into the formwork created many voids, resulting in significant data errors. Hence, ASTM C109/C109M-02: Standard Test Method for Compressive Strength of Hydraulic Cement Mortars [16] was applied to produce the specimens. These specimens were cured at a temperature of 23 ± 2 °C and a relative humidity of 65% ± 5%. Also, the cylindrical specimens used for the compressive stress–strain tests were ground for planeness before the tests were conducted.

3.2.3. Test of Modulus of Elasticity

The modulus of elasticity was measured with respect to ASTM C 469M-14: Standard Test Method for Static Modulus of Elasticity and Poisson's Ratio of Concrete in Compression [20]. Cylindrical specimens (50 mm in diameter, 100 mm long) were employed in the tests. Loads were applied via the load-controlled method and the rate of application of compressive stress was 0.25 MPa/s. The elastic modulus equation applied in the present study is shown in Equation (1), and the input data were extracted from compressive stress–strain curves. The strain was measured using an electric resistance-type strain gauge and data logger (Tokyo Sokki, TDS-602). A schematic of the system is presented in Figure 6.

$$E = \frac{S_2 - S_1}{\varepsilon_2 - 0.000050},\tag{1}$$

where E is the chord modulus of elasticity (MPa), S_1 is the stress corresponding to a longitudinal strain of 0.00005 (MPa), S_2 is the stress corresponding to 40% of the ultimate load (MPa), and ε_2 is the longitudinal strain produced by stress S_2.

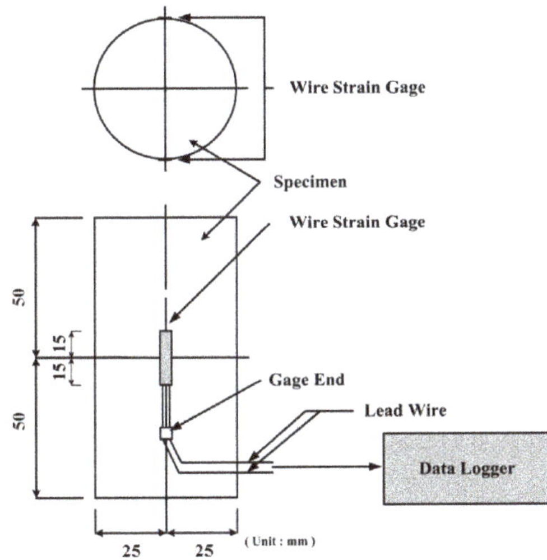

Figure 6. Conceptual drawing of the strain measurement system.

3.2.4. Test of Drying Shrinkage

Drying shrinkage tests were carried out with respect to ASTM C596-01: Standard Test Method for Drying Shrinkage of Mortar Containing Hydraulic Cement [21], using an environmental chamber held at a temperature of 23 ± 2 °C and relative humidity of 65% ± 5%. The specimen dimensions were 70 mm × 70 mm × 320 mm. The longitudinal strain was measured at the center of the cross-section using an embedded-type strain gauge (Tokyo Sokki, PMFL-series) and data logger (Tokyo Sokki, TDS-602) since placement. After 24 h curing in the molds with an air-tight plastic sheet placed on the top surface, the specimens were demolded for external drying; the shrinkage up to the first 24 h was also monitored to measure autogenous shrinkage. The procedure for installing the strain gauge is described in Figure 7.

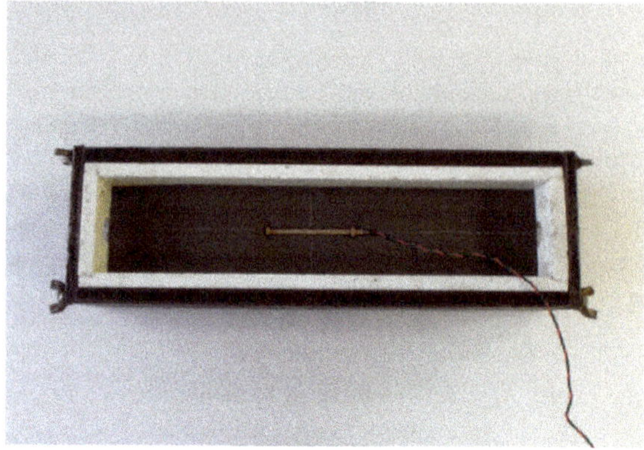

Figure 7. Test setup for measuring drying shrinkage.

3.2.5. Test of Thermal Expansion

Thermal expansion tests were carried out with respect to ASTM C531-18: Standard Test Method for Linear Shrinkage and Coefficient of Thermal Expansion of Chemical Resistance Mortars, Grouts, Monolitic Surfacings, and Polymer Concretes [22]. The temperature was increased from 25 °C to 80 °C once the thermal equilibrium was achieved throughout the specimen. The specimen dimensions and procedure for instrumentation were the same as in the drying shrinkage tests, shown in Figure 7. To determine the true thermal strain, thermal calibration was performed.

4. Results and Discussion

4.1. Stress–Strain Relationship and Modulus of Elasticity

Stress–strain curves provide basic data for estimating the modulus of elasticity. Given that concrete is not completely elastic, however, estimating the elastic modulus based on these curves results in various issues. Strictly speaking, the modulus of elasticity only applies to the linear elastic section of the stress–strain curve. In cases where it is difficult to judge the linear elastic section from the curved part, as is the case with concrete, the secant modulus of elasticity, also known as the chord modulus, is alternatively used [23]. In extrusion-based additive construction, the modulus of elasticity is a critical factor that is used to estimate both the stress caused by drying shrinkage in layered EVA-modified cementitious mortar ($\sigma = \varepsilon_{sh}E$) and the stress caused by thermal expansion ($\sigma = \alpha E\Delta T$).

In concrete, stress and strain typically have a non-linear relationship, but the relationship can be considered the linear elastic section at a lower stress level. This stress range can reach up to 40% of the ultimate strength, within which concrete can be considered an elastic material.

The stress–strain diagrams of the developed EVA-modified cementitious mortars are presented in Figure 8. The shapes of these curves resemble those obtained for cementitious paste [23]. Based on these diagrams, the obtained compressive strengths were 48.3 MPa, 41.8 MPa, 38.2 MPa, 35.7 MPa, and 33.5 MPa when the EVA/cement ratios were 0, 0.05, 0.10, 0.15, and 0.20, respectively. The compressive strength decreased as the EVA/cement ratio increased. The secant modulus was calculated using Equation (1). The results were 21.6 GPa, 19.1 GPa, 18.1 GPa, 17.6 GPa, and 16.7 GPa when the EVA/cement ratios were 0, 0.05, 0.10, 0.15, and 0.20, respectively. The relationship between these elastic modulus measurements and the applied EVA/cement ratios was analyzed, as presented in Figure 9. A high correlation was found (i.e., the modulus of elasticity tended to decrease as the EVA/cement ratio was increased).

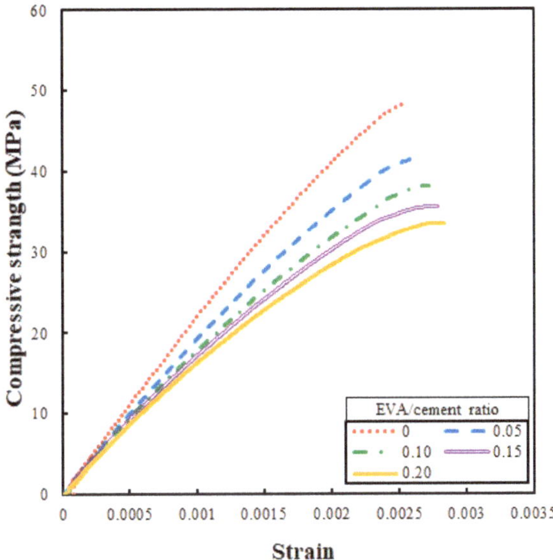

Figure 8. Stress–strain diagrams for different EVA/cement ratios.

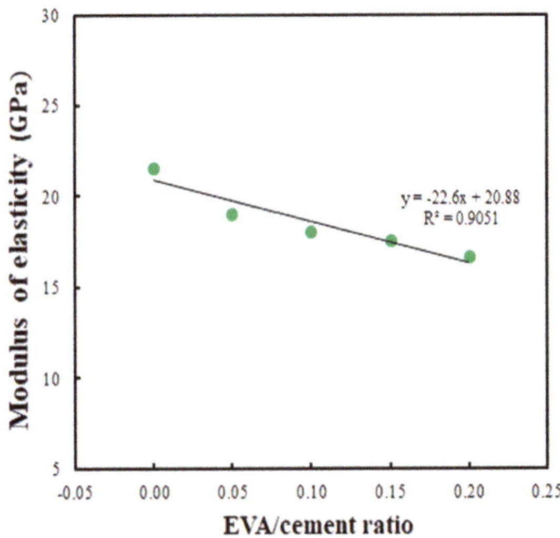

Figure 9. Relationship between the EVA/cement ratio and modulus of elasticity.

Earlier studies showed that the modulus of elasticity in the compression of unmodified cementitious concrete was 21.1 GPa, while the value ranges were 22.4 GPa to 23.6 GPa, 20.2 GPa to 24.3 GPa, and 10.0 GPa to 19.0 GPa for polyacrylic ester (PAE)–modified cementitious concrete, styrene butadiene rubber (SBR)–modified cementitious concrete, and polyvinyl acetate (PVAC)–modified cementitious concrete, respectively [18]. The results varied depending on the polymer type and polymer/cement ratio. Overall, the modulus of elasticity tended to decrease as the polymer/cement ratio increased, and the degree of reduction increased when the polymer content was excessive. A decrease in the elastic modulus reduces the stiffness of EVA-modified cementitious mortar but also decreases the level of stress caused by drying shrinkage and temperature change.

It is also known that the modulus of elasticity of polymer-modified cementitious mortar is relatively low, 10 GPa to 30 GPa, because it contains polymer (elastic modulus: 0.1 GPa to 10 GPa) [24]; the elastic modulus of cementitious concrete in compression generally falls within the range of 14 GPa to 40 GPa [17]. In the present research, the maximum compressive strains were 0.00256, 0.00267, 0.00272, 0.00278, and 0.00283 when the EVA/cement ratios were 0, 0.05, 0.10, 0.15, and 0.20, respectively. These results indicate that the maximum compressive strain tended to increase as the EVA/cement ratio increased. Given that the figures were 0.002 and 0.0027 for 30 MPa cement concrete and 34 MPa cement paste [23], respectively, the measured maximum compressive strains were found to be comparable to that of cement paste. Concrete in compression shows some inelastic strain before failure. It is worth noting that the typical level of strain at failure is 0.002 [17] and the strain of 100 MPa concrete is typically 0.003 to 0.004, while the strain of 20 MPa concrete is 0.002. Each stress corresponds to the ultimate strength. However, under the same stress, regardless of strength, stronger concrete exhibits a lower strain [23].

In addition, the relationship between the compressive strength and estimated modulus of elasticity was analyzed, as presented in Figure 10. Here, the modulus of elasticity tended to increase with increases in compressive strength. It is highly certain that in concrete, the compressive strength and modulus of elasticity have a proportional relationship, but an agreement has not been reached on the precise form of that relationship [23]. As a result, thus far the American Concrete Institute (ACI) Building Code and the Comit Euro-International du B ton and the F d ration International de la Pr contrainte (CEB-FIP) Model Code have proposed different equations [17]. In the present study, the relationship between the modulus of elasticity and compressive strength of each EVA-modified cementitious mortar was derived as shown in Equation (2):

$$E_c = 1.41(f'_c)^{0.7}, \tag{2}$$

where E_c is the modulus of elasticity (GPa) and f'_c is the compressive strength in MPa.

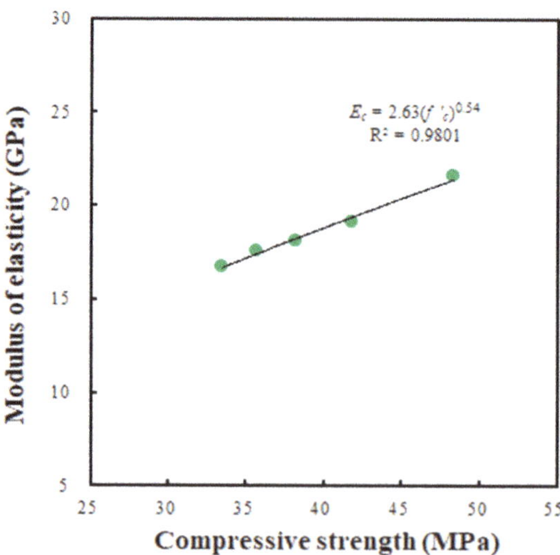

Figure 10. Relationship between the compressive strength and modulus of elasticity.

4.2. Drying Shrinkage

Drying shrinkage is one of the major causes of cracking in concrete. Once exposed to air, concrete starts to dry out and the dried surface contracts. The internal moisture, however, suppresses shrinkage in the outer part. Accordingly, the surface regions undergo tensile stresses, and when the stress-induced drying shrinkage exceeds the direct tensile strength of the concrete, cracking occurs.

One advantage of 3DCP is that formwork is not necessary. This removes a barrier between the curing concrete and ambient environment. Printed layers often have a greater exposed surface area than cast concrete. However, lower water/cement ratios than those seen in casting concrete are typical in 3DCP mortars. Hence, the likelihood of cracking resulting from autogenous shrinkage is increased. Therefore, mix designs must minimize dimensional changes due to dry and autogenous shrinkage and greater care should be taken when curing [25].

In typical cementitious mortar and concrete, the degree of drying shrinkage ranges from 200×10^{-6} to 1200×10^{-6}, depending on the aggregate/cement ratio [26]. Figure 11 presents the drying shrinkage test results for up to 28 days in relation to the EVA/cement ratio. Figure 11a shows the results for up to 24 h, while Figure 11b indicates the results for up to 28 days. Here, the results are presented in two separate figures to make the initial-stage strain caused by drying shrinkage more distinct.

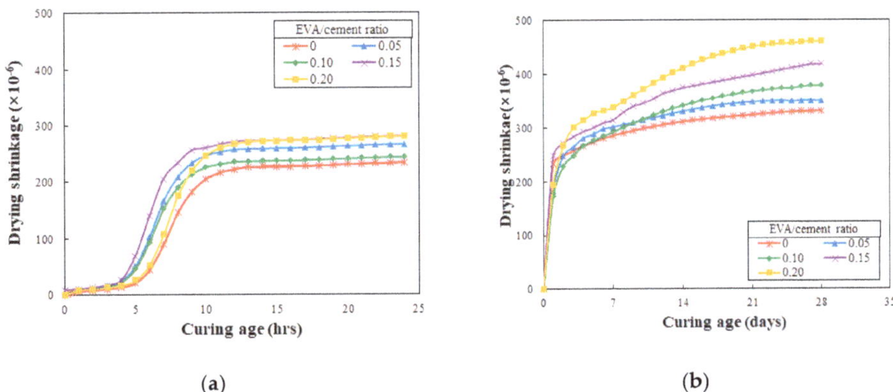

Figure 11. Drying shrinkages for different EVA/cement ratios at (**a**) 24 h and (**b**) 28 days.

One hour after placement, autogenous shrinkage was initiated; substantial shrinkage occurred from 4 to 11 h. After demolding, drying shrinkage continued to increase until 28 days. At 28 days, the drying shrinkage was 331×10^{-6}, 349×10^{-6}, 379×10^{-6}, 418×10^{-6}, and 461×10^{-6} when the EVA/cement ratios were 0, 0.05, 0.10, 0.15, and 0.20, respectively, indicating that the shrinkage increased as the EVA/cement ratio increased (see Figure 12). Based on these results, an increased rate was calculated when the EVA/cement ratio of zero was set as a reference. The rates were 5%, 14%, 26%, and 39% when the EVA/cement ratios were 0.05, 0.10, 0.15, and 0.20, respectively. This drying shrinkage development trend was related to an increase in the water/cement ratio from 0.45 to 0.55 with an increase in the EVA/cement ratio, as shown in Table 8. This is considered a disadvantage of the redispersible EVA powder.

In a previous study by Weng et al. [27], at 28 days and a water/cement ratio of 0.5, the drying shrinkages were 0.0128%, 0.0217%, 0.0222%, and 0.0224% when the EVA/cement ratios were 0, 0.03, 0.05, and 0.08, respectively. When the water/cement ratio was 0.6, the drying shrinkages were 0.0380%, 0.0527%, 0.0538%, and 0.0546% and the EVA/cement ratios were 0, 0.03, 0.05, and 0.08, respectively. This indicates that the drying shrinkage increased as both the EVA/cement ratio and water/cement ratio increased.

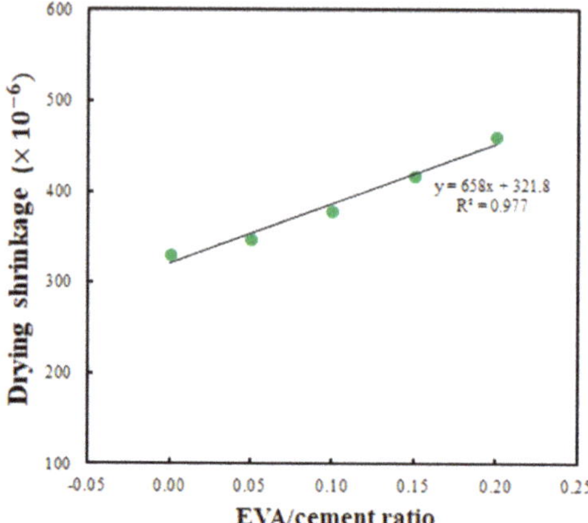

Figure 12. Relationship between EVA/cement ratio and drying shrinkage.

In contrast, in SBR latex-modified cementitious mixtures, drying shrinkage was reported to decrease with an increase in polymer content [19,24,28]. Likewise, when modified with latex, cementitious mortar exhibited less drying shrinkage; this is ascribed to the effects of the surfactants and antifoamers contained in the latex [29]. The significant drying shrinkage seen in EVA-modified cementitious mixtures can be significantly reduced by using shrinkage reducing agents such as polyethylene glycol [30] and ethylene [18], but attention must be paid when following this course because adverse effects, such as strength degradation, may occur.

4.3. Coefficient of Thermal Expansion

The coefficient of thermal expansion is defined as the change in the unit length of a material for a unit change in temperature. Thermal shrinkage strain is determined by the degree of temperature drop in concrete and its coefficient of linear thermal expansion [17]. The coefficient of the thermal expansion of concrete is determined by the combined values of the dissimilar thermal coefficients of its two main constituents (i.e., cement paste and aggregates) [23]. Concrete structures are deformed by temperature variations resulting from the hydration reaction of cement or atmospheric temperature changes. When this temperature variation causes the tensile stress of concrete to exceed its tensile stress, cracking is initiated. Layered cementitious materials built through extrusion-based additive construction are expected to undergo cracking and delamination due to temperature change.

In the present study, thermal strain tests were conducted, and the results are presented in Figure 13. The thermal strains were 437×10^{-6}, 530×10^{-6}, 643×10^{-6}, 812×10^{-6}, and 997×10^{-6} when the EVA/cement ratios were 0, 0.05, 0.10, 0.015, and 0.020, respectively. It was found that in all cases, the thermal strain was reached within about two hours. The coefficient of thermal expansion was estimated by dividing each thermal strain measurement by the corresponding temperature rise. Once calculated, its relationship with the EVA/cement ratio was analyzed, as shown in Figure 14. Here, the coefficients of thermal expansion were $7.9 \times 10^{-6}/°C$, $9.6 \times 10^{-6}/°C$, $11.7 \times 10^{-6}/°C$, $14.8 \times 10^{-6}/°C$, and $18.1 \times 10^{-6}/°C$ when the EVA/cement ratios were 0, 0.05, 0.10, 0.15, and 0.20, respectively. This indicates that the thermal expansion coefficient tended to increase as the EVA/cement ratio increased, and the correlation between the two factors was high.

Previous studies have reported that the coefficient of thermal expansion of polymer-modified cementitious mortar ranged from $9 \times 10^{-6}/°C$ to $10 \times 10^{-6}/°C$ [24]. Notably, however, when the polymer/cement ratio ranged from 0.10 to 0.20, the coefficient of thermal expansion of SBR latex-modified cementitious mortar was reported to be between $7.7 \times 10^{-6}/°C$ and $8.6 \times 10^{-6}/°C$, which was not significantly different from that of unmodified mortar at $7.9 \times 10^{-6}/°C$ [18]. For ordinary cementitious concrete, Neville [23] reported that the coefficient of linear thermal expansion of hydrated cementitious paste varied from $11 \times 10^{-6}/°C$ to $20 \times 10^{-6}/°C$. Mehta et al. [17] presented that the coefficient of linear thermal expansion of cementitious mortars was approximately $18 \times 10^{-6}/°C$ for cementitious paste, $12 \times 10^{-6}/°C$ for mortar, and between $6 \times 10^{-6}/°C$ and $12 \times 10^{-6}/°C$ for concrete. As explained above, the measured coefficient of thermal expansion was determined to be significantly higher in the present study when compared to the results reported by previous studies on polymer-modified cementitious mortars. At the same time, the coefficient of thermal expansion of ordinary cementitious concrete and mortars measured in the present study was largely comparable to those reported in similar previous research.

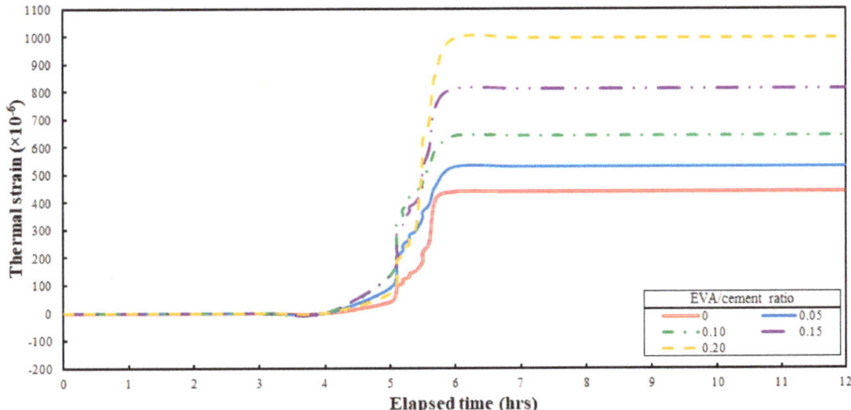

Figure 13. Thermal strains for different EVA/cement ratios.

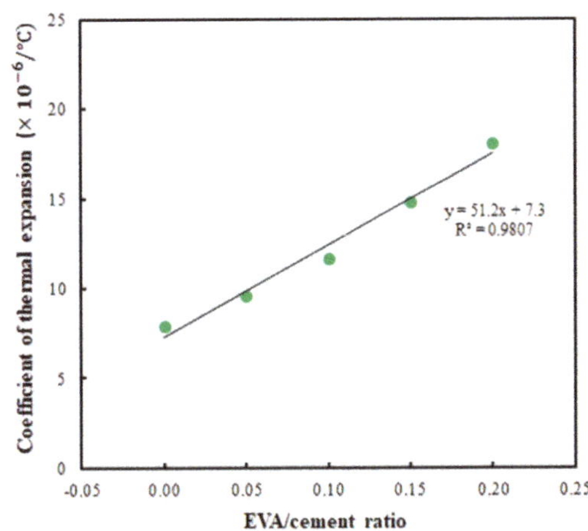

Figure 14. Relationship between the EVA/cement ratio and coefficient of thermal expansion.

5. Conclusions

EVA powder is easier to handle onsite than liquid polymer (i.e., latex or emulsion types) with regards to pre-mixing. Hence, EVA was selected as the admixture used to produce the 3D-printable EVA-modified cementitious mortars examined here. This research is a follow-up study to previous research published by the author, entitled Fresh Properties of EVA-Modified Cementitious Mixtures for use in Additive Construction by Extrusion. The present study experimentally examined the short-term deformability of EVA-modified cementitious mortar for extrusion-based additive construction, especially with regards to the elastic modulus, drying shrinkage, and thermal expansion. These three factors are critical for determining the properties of hardened materials used in additive construction. The major findings of this study are as follows.

As the EVA/cement ratio was increased, the compressive strength and elastic modulus tended to decrease but the maximum compressive strain increased. Drying shrinkage began one hour after the mortar was placed, was most significant in the 4–11 h range, and continued to increase until 28 days of aging. At 28 days, the drying shrinkage tended to increase as the EVA/cement ratio was increased. The thermal expansion coefficient also tended to increase as the EVA/cement ratio was increased. A correlation analysis indicated that each of the three factors (i.e., elastic modulus, drying shrinkage, and thermal expansion) had a high determination coefficient (r^2) with regards to the EVA/cement ratio.

Of the three factors described above, drying shrinkage and thermal expansion increased as the EVA/cement ratio was increased, indicating that the materials concerned were disadvantageous in terms of dimensional stability. At the same time, the elastic modulus decreased; thus, the level of stress caused by the other two factors may also have been reduced, thereby improving the degree of extensibility and offsetting any adverse impact. That being said, in order to increase the elastic modulus while reducing the degree of drying shrinkage and thermal expansion, it is still desirable to reduce the water/cement ratio by reducing the EVA/cement ratio.

Funding: This research received no external funding.

Acknowledgments: The author is very grateful to former collaborators Kwan-Kyu Kim and Hee Jun Lee from Korea Conformity Laboratories and Kangwon National University for the assistance in conducting the experiment.

Conflicts of Interest: The author declares no conflicts interest.

References

1. Chinese Company Assembles 10 3D-Printed Concrete Houses in a Day for Less Than $5,000 Each. Available online: https://inhabitat.com/chinese-company-assembles-ten-3d-printed-concrete-houses-in-one-day-for-less-than-5000-each/ (accessed on 1 September 2019).
2. Apis Cor and Gerdau to Print Homes Together on Earth and Beyond. Available online: https://www.apis-cor.com/apiscor-and-gerdau (accessed on 1 September 2019).
3. TU/e and BAM Infra Get to Work on 3D Printed Concrete Bicycle Bridge. Available online: https://3dprint.com/178462/eindhoven-3d-printed-bridge/ (accessed on 1 September 2019).
4. Nerella, V.N.; Mechtcherine, V. Studying the printability of fresh concrete for formwork-free concrete onsite 3D printing technology (CONPrint3D). In *3D Concrete Printing Technology*, 1st ed.; Sanjayan, J.G., Nazari, A., Nematollahi, B., Eds.; Elsevier: Amsterdam, The Netherlands, 2019; Volume 16, pp. 333–347.
5. Rushing, T.S.; Stynoski, P.B.; Barna, L.A.; Al-Chaar, G.K.; Burroughs, J.F.; Shannon, J.D.; Kreiger, M.A.; Case, M.P. Investigation of concrete mixtures for additive construction. 3D concrete printing technology. In *3D Concrete Printing Technology*, 1st ed.; Sanjayan, J.G., Nazari, A., Nematollahi, B., Eds.; Elsevier: Amsterdam, The Netherlands, 2019; Volume 7, pp. 137–160.
6. Khoshnevis, B. Automated Construction by Contour Crafting—Related Robotics and Information Technologies. *Autom. Constr.* **2004**, *13*, 5–19. [CrossRef]
7. Tay, Y.W.D.; Panda, B.; Paul, S.C.; Noor Mohamed, N.A.; Tan, M.J.; Leong, K.F. 3D Printing Trends in Building and Construction Industry: A Review. *Virtual Phys. Prot.* **2017**, *12*, 261–276. [CrossRef]
8. Hager, I.; Golonka, A.; Putanowicz, R. 3D Printing of Buildings and Building Components as the Future of Sustainable Construction? *Proc. Eng.* **2016**, *151*, 292–299. [CrossRef]
9. Yeon, K.S.; Kim, K.K.; Yeon, J. Feasibility study of the use of polymer-modified cement composites as 3D concrete printing material. In Proceedings of the International Congress in Polymers in Concrete, Washington, DC, USA, 29 April–1 May 2018; Taha, M.R., Ed.; Springer: Berlin, Germany, 2018.
10. Bentz, D.P.; Jones, S.Z.; Bentz, I.R.; Peltz, M.A. Towards the formulation of robust and sustainable cementitious binders for 3D additive construction by extrusion. In *3D Concrete Printing Technology*, 1st ed.; Sanjayan, J.G., Nazari, A., Nematollahi, B., Eds.; Elsevier: Amsterdam, The Netherlands, 2019; Volume 15, pp. 307–331.

11. Kazmian, A.; Yuan, X.; Meier, R.; Khoshnevis, B. Performance-based testing portland cement concrete for construction-scale 3D printing. In *3D Concrete Printing Technology*, 1st ed.; Sanjayan, J.G., Nazari, A., Nematollahi, B., Eds.; Elsevier: Amsterdam, The Netherlands, 2019; Volume 2, pp. 13–35.

12. Li, Z.; Wang, L.; Ma, G. Method for the enhancement of buildability and bending resistance of three-dimensional-printable tailing mortar. In *3D Concrete Printing Technology*, 1st ed.; Sanjayan, J.G., Nazari, A., Nematollahi, B., Eds.; Elsevier: Amsterdam, The Netherlands, 2019; Volume 8, pp. 161–180.

13. Weng, Y.; Li, M.; Tan, M.J.; Qian, S. Design 3D printing cementitious materials via Fuller Thompson Theory and Marson-Percy Model. In *3D Concrete Printing Technology*, 1st ed.; Sanjayan, J.G., Nazari, A., Nematollahi, B., Eds.; Elsevier: Amsterdam, The Netherlands, 2019; Volume 14, pp. 281–306.

14. Wu, Y.Y.; Ma, B.G.; Wang, J.; Zhang, F.C.; Jian, S.W. Study on Interface Properties of EVA-Modified Cement Mortar. *Adv. Mater. Res.* **2011**, *250–253*, 875–880. [CrossRef]

15. Yeon, K.; Kim, K.K.; Yeon, J.; Lee, H.J. Fresh Properties of EVA-Modified Cementitious Mixtures for Use in Additive Construction via Extrusion. *Materials* **2019**, *12*, 2292. [CrossRef] [PubMed]

16. Ohama, Y. *Handbook of Polymer-Modified Concrete and Mortars*, 1st ed.; Noyes Publications: Park Ridge, NJ, USA, 1995.

17. Mehta, P.K.; Monteiro, P.J.M. *Concrete: Microstructure, Properties, and Materials*, 3rd ed.; McGraw-Hill: New York, NY, USA, 2006.

18. ASTM C109/C109M-02. Standard Test Method for Compressive Strength of Hydraulic Cement Mortars. Available online: https://www.astm.org/DATABASE.CART/HISTORICAL/C109C109M-02.htm (accessed on 1 September 2019).

19. Ohama, Y.; Kan, S. Effects of Specimen Size on Strength and Drying Shrinkage of Polymer-Modified Concrete. *Int. J. Cem. Compos. Lightweight Concr.* **1982**, *4*, 229–233. [CrossRef]

20. ASTM C469/C469M-14. Standard Test Method for Static Modulus of Elasticity and Poisson's Ratio of Concrete in Compression. Available online: https://www.astm.org/Standards/C469 (accessed on 1 September 2019).

21. ASTM C596-01. Standard Test Method for Drying Shrinkage of Mortar Containing Hydraulic Cement. Available online: https://www.astm.org/DATABASE.CART/HISTORICAL/C596-01.htm (accessed on 1 September 2019).

22. ASTM C531–18. Standard Test Method for Linear Shrinkage and Coefficient of Thermal Expansion of Chemical-Resistant Mortars, Grouts, Monolithic Surfacings, and Polymer Concretes. Available online: https://www.astm.org/Standards/C531.htm (accessed on 1 September 2019).

23. Neville, A.M. *Properties of Concrete*, 4th ed.; John Wiley & Sons Inc.: Hoboken, NJ, USA, 1996.

24. Chandra, S.; Ohama, Y. *Polymers in Concrete*, 1st ed.; CRC Press: Boca Raton, FL, USA, 1994.

25. Buswell, R.A.; Leal de Silva, W.R.; Jones, S.Z.; Dirrenberger, J. 3D Printing Using Concrete Extrusion: A Road Map for Research. *Cem. Concr. Res.* **2018**, *112*, 37–49. [CrossRef]

26. Lea, F.M. *The Chemistry of Cement and Concrete*, 3rd ed.; Chemical Publishing Company: New York, NY, USA, 1970.

27. Weng, T.L.; Lin, W.T.; Li, C.H. Properties Evaluation of Repair Mortars Containing EVA and VA/VeoVa Polymer Powders. *Polym. Polym. Compos.* **2017**, *25*, 77–86. [CrossRef]

28. Kardon, J.B. Polymer-Modified Concrete: Review. *J. Mater. Civ. Eng.* **1997**, *9*, 85–92. [CrossRef]

29. Kawano, T. Studies on the mechanism of reducing drying shrinkage of cement mortar modified by rubber latex. In Proceedings of the Third International Congress on Polymers in Concrete, Koriyama, Japan, 13–15 May 1981; Transport Research Laboratory: Wokingham, UK, 1981.

30. Kim, W.; Ohama, Y.; Demura, K. Drying Shrinkage Reduction of Polymer-Modified Mortars Using Redispersible Polymer Powder by Use of Shrinkage-Reducing Agents. *J. Soc. Mater. Sci. Jap.* **1997**, *46*, 84–88. [CrossRef]

Article

Mechanical Properties and Microstructure of Polyvinyl Alcohol (PVA) Modified Cement Mortar

Jie Fan [1], Gengying Li [2,*], Sijie Deng [1] and Zhongkun Wang [2]

[1] School of Civil Engineering, Guizhou Institute of Technology, Guiyang 515063, China; jfan1988@163.com (J.F.); dengsijie1988@163.com (S.D.)

[2] College of Water Conservancy and Civil Engineering, South China Agricultural University, Guangzhou 510642, China; 13526578601@163.com

* Correspondence: ligengying@scau.edu.cn; Tel.: +86-136-2303-9690

Received: 1 May 2019; Accepted: 25 May 2019; Published: 28 May 2019

Abstract: The mechanical properties of cement mortars with 0~2.0% (by mass) polyvinyl alcohol (PVA) were experimentally studied, and the effects of PVA incorporation on the hydration products and microstructure of the cement mortar were determined with differential scanning calorimetry (DSC), Fourier transform infrared spectroscopy (FTIR) and scanning electron microscopy (SEM). The results show that the rational content of PVA formed evenly dispersed network-like thin films within the cement matrix, and these network-like films can bridge cracks in the cement matrix and improve the mechanical properties of the cement mortar. Over- incorporation of PVA may result in the formation of large piece polymer films that coat the cement particles, delay the hydration of the cement mortar and adversely affect its performance. The mechanical properties of the cement mortar show a significant increase and then decrease with a change in the PVA incorporation. When the PVA content was 0.6% and 1.0%, the mortar had the best compressive and flexural strengths, respectively. The compressive strength of the cement mortar increased by 12.15% for a PVA content of 0.6%, and the flexural strength of the cement mortar increased by 24.83% for a PVA content of 1.0%.

Keywords: polyvinyl alcohol; cement mortar; mechanical properties; microstructure

---

1. Introduction

Cement-based materials behave like typical porous, brittle materials [1–3]. Brittleness makes the cement material easy to crack under the action of external force, which adversely affects the bearing capacity and durability of cement concrete structures. Polymers have good elastic deformation properties, flexibility, acid and alkali corrosion resistance, and good compatibility with cement materials, which can effectively improve the mechanical properties, deformation properties and durability of cement-based materials [4–9]. Therefore, polymers are widely used in repairing roads, bridges, reservoirs and dams and in the bonding engineering of facing materials [10–12].

Polyvinyl alcohol (PVA) is a typical water-soluble nonionic synthetic polymer containing vinyl [13,14]. It is harmless and, therefore, considered safe and relatively environmentally friendly [15]. Commercial- grade PVA is usually divided into two categories based on its degree of polymerization and hydrolysis: (1) fully hydrolyzed groups (98% by mole of acetate or more) where the PVA group has been replaced by an alcohol group, and (2) hydrolyzed groups (approximately 87–89% by mole of acetate group) where the PVA group has been replaced by alcohol groups. Completely hydrolyzed PVA is soluble in hot water and has good film-forming properties (the formed film is insoluble in low-temperature water) and good adhesion properties [15,16]. The polyvinyl alcohol after film formation has good deformation properties, toughness, and wear resistance [17,18]. The physical properties of the polyvinyl alcohol film system are very stable. It not only has good heat (no obvious change occurs under 140 °C), light, and chemical stability but also has excellent gas and water barrier properties (the inside of polyvinyl alcohol can remain dry under high humidity) [19,20]. The good physical and chemical properties of polyvinyl alcohol and its formed films enable its wide use in the textile, construction, chemical, and papermaking industries, to name a few.

Owing to the excellent performance as described above, researchers have utilized PVA to modify the mechanical properties of a cement-based material. Singh et al. [21] found that the addition of 3% of PVA could increase the compressive strength of cement mortar about 12 % due to there exists chemical interaction between PVA and cement hydration. This chemical interaction is effective in improving the interfacial bond between cement hydration and aggregates. Moreover, the chemical products could fill the pores of the cement matrix, resulting in the mechanical properties of the cement matrix increase greatly. Similarly, Kapen et al. [22] demonstrated that the flexural strengths of cement mortar increased by 21% after loading 1% PVA when cured in dry condition, respectively. They stated that this improvement was mainly attributed to the formation of PVA film within the cement matrix. Kim et al. [23] found that the improvement in bond strength after loading PVA seems to arise from suppression of the porous interfacial transition zone and inhibition of calcium hydroxide nucleation on the aggregate surface.

Using PVA to enhance the properties of cement composites has attracted extensive attention. However, the previous research mainly focused on the hydration and mechanical properties after loading a mono content of PVA, there is limited information on the effect of PVA with different content on the fresh properties and strength development at the early ages. Moreover, the morphology and microstructure of cement composite with different PVA content lack systematic investigation. The goal of this study was to study the effect of PVA content on the compressive and flexural strength of mortar after 3 days, 7 days, 14 days and 28 days curing, in which PVA/binder weight ratios of 0% 0.2%, 0.6%, 1.0%, and 2.0% were adopted. Moreover, the fresh properties and water absorption of mortars were also investigated. Additionally, the morphology and microstructure of mortars after 28 days curing were also analyzed by using differential scanning calorimetry (DSC), Fourier transform infrared spectroscopy (FTIR), and SEM.

2. The Experiment

2.1. Materials and Mix Design

The cement used was Type P.O 42.5 R Portland cement, produced by Guangdong Tapai Cement Co., Ltd. (Figure 1a). The chemical composition is listed in Table 1. Polyvinyl alcohol (PVA-124 AR) from Xilong-Chem was used. The molecular weight (Mw) was 105,000, and the degree of hydrolysis was 97% (Figure 1b). The physicochemical characteristics of this polyvinyl alcohol are given in Table 2. The fine aggregate was China ISO standard sand with the execution criterion of GSB08-1337-2001, from Xiamen ISO Standard Sand Co., Ltd. (Figure 1c). The chemical composition and particle size of standard sand are listed in Table 3. The defoaming agent was produced by Guangdong Defeng Chemical Industry Co., Ltd.

(a)   (b)   (c)

Figure 1. The raw materials: (**a**) Portland cement; (**b**) polyvinyl alcohol (PVA) power, (**c**) standard sand.

The water/cement/sand ratio by weight of the specimens was selected as 0.4:1:1.5, and the PVA was incorporated at concentrations of 0%, 0.2%, 0.6%, 1.0%, and 2.0% by weight of cement. The defoamer was added at the amount of 0.14% of cement to eliminate the entrained air bubble during the mixing process. The detailed mix design is shown in Table 4.

Table 1. Chemical composition of Portland cement.

P.O 42.5 R	Raw Material (%)								
	SiO_2	Al_2O_3	CaO	MgO	Na_2O	K_2O	Fe_2O_3	SO_3	Loss
	18.3	4.5	62.4	2.1	0.3	1.5	2.3	3.5	2.6

Table 2. Physicochemical characteristics of PVA.

PVA-124 AR	Molecular Weight (Mw)	Degree of Hydrolysis (mole %)	PH	Volatile Content (%)	Ash Content (%)
	105,000	97	5~7	5.0	0.7

Table 3. Chemical composition and particle size of standard sand.

Standard Sand	SiO_2 Content(%)	Mud Content(%)	Ignition Loss(%)	Particle Size (mm)
	>96	<0.2	<0.4	0.08~2

Table 4. Mix proportions of cement mortar.

Code	Ratio of Material Mass to Cement Mass (%)				
	Cement	Sand	Water	PVA	Defoamer
PCM0	100	150	40.2	0	0.14
PCM1	100	150	40.24	0.2	0.14
PCM2	100	150	40.32	0.6	0.14
PCM3	100	150	40.4	1.0	0.14
PCM4	100	150	40.5	2.0	0.14

2.2. Sample Preparation

PVA powder was dispersed in cold water at room temperature for 10 min, and this mixture was heated to 95 °C with stirring to completely dissolve the PVA. Using a cement mortar mixer, cement and fine sand were first dry-mixed thoroughly for approximately three minutes. The prepared dispersed PVA solution and the remaining water were then poured into the cement mortar and mixed for two minutes with a mid speed stirring. The defoaming agent was then slowly added into the mix, and the mixing continued for three minutes with a high-speed stirring until the relatively homogeneous cement slurry was formed. The prepared mixture was poured into oiled molds and vibrated to remove air bubbles, then pouring the mixes into oiled molds (40 × 40 × 160 mm^3). Afterwards, the specimens were removed from their molds after 24 h and cured in a moist room (relative humidity ≥97%) for 28 days. The detailed process of material preparation method is shown in Figure 2.

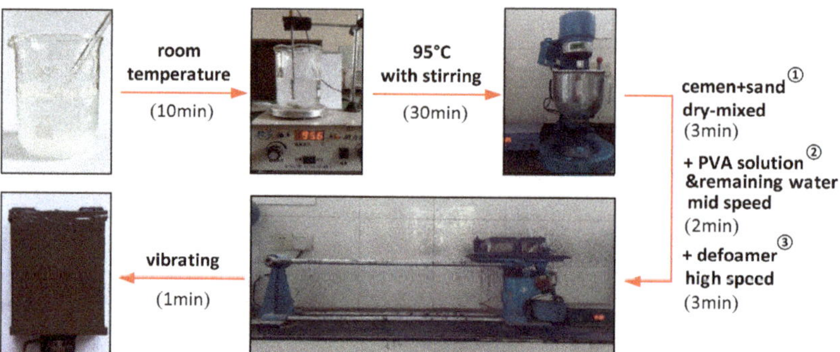

Figure 2. Illustration of the PVA modified cement mortar fabrication process.

2.3. Testing Methods

The flow table test was carried out on the freshly prepared mortars as described by GB/T2419-2005 [24]. The jumping table beats 25 times in 15 s and measures the average diameter of the fresh mortar on two mutually perpendicular sides.

The hardened density of the specimens was determined by using a water displacement method according to GB/T 11970-1997 [25].

The capillary water absorption test on cement mortar was conducted after 28-day standard curing and determined according to DIN 52617 [26]. Fifteen samples of size $40 \times 40 \times 80$ mm^3 were adopted in this test, and the surface of samples was wax-sealed, and then the molded surface was immersed in water, the distance between the height of the water surface and the bottom edge of the test piece was 5 mm, the schematic diagram is shown in Figure 3. The results are the average values of three specimens.

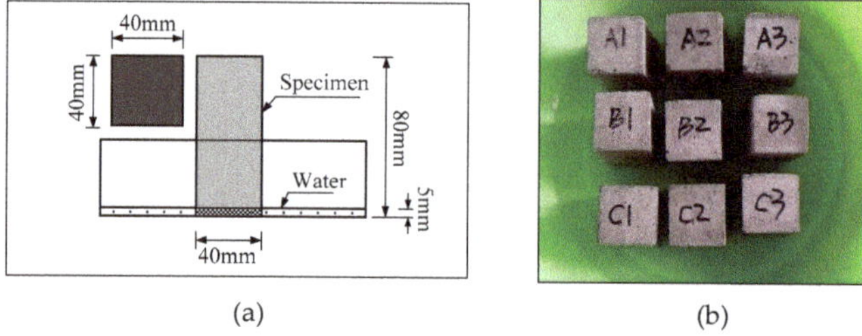

Figure 3. Capillary water absorption test schematic diagram (**a**) and measured diagram (**b**).

An electronic universal testing machine (CMT-5105, China) was used to measure flexural strength and compression strength of the composites (three samples of size $40 \times 40 \times 160$ mm^3 were adopted in this test). The flexural strength was tested at 3 days, 7 days, 14 days, and 28 days using a three-point method in accordance with GB/T 17671-1999 [27]. The broken pieces (portions of the prisms broken in the flexural strength test) were used to determine the cube compressive strength of the mortar samples in accordance with GB/T 17671-1999 [27].

Differential thermal analysis was using a Shimadzu DSC 50 thermal analyzer at a heating rate of 20 °C/min. The samples chamber was purged with nitrogen at a flow rate of 40 cc/min.

FT-IR spectra were carried out using a MAUNA-IR 750 spectrometer manufactured by Nikolai, USA. Powdered samples were mixed with KBr and pressed into pellets. The analyses were carried out in the frequency range of 500–4000 cm^{-1}.

The microstructures of the cement mortar were observed using a scanning electron microscope (SEM, SU3500) at an accelerating voltage of 10 kV. The sample was coated with a thin layer of gold before observation.

3. Results and Analysis

3.1. Flow Properties

Fluidity is an important index that reflects the working performance of a cement mortar. The flow properties of cement mortar can be evaluated by using a flow table test. Figure 4 shows the test results of the slump value of all mortars, in which the w/c ratio was fixed at 0.4. Obviously, the incorporation of PVA greatly reduced the flow properties of mortars, as shown in Figure 4, the slump value of fresh mortars decreased from 219.4 mm to 108.5 mm with the PVA content increased from 0% to 2.0%. The incorporation of 0.2%, 0.6%, 1.0% and 2.0% PVA reduced the slump value by 13.9%, 31.5%, 41.1%, and 50.5%, as compared to the control one, respectively. This reduction possibly attributed to the strong, cohesive force of PVA, which limits the movement of cement particles during the mixing process. Moreover, the -OH bonds in the PVA matrix may react with the -OH bonds of water, which reduce the water participating in dispersing of cement particles. The working performance impacts the mechanical properties of cement mortar, and generally, a higher slump value leads to a better uniformity for the fresh mortar, which further results in a higher mechanical properties. Thus, the content of PVA for actual engineering applications should be as small as possible.

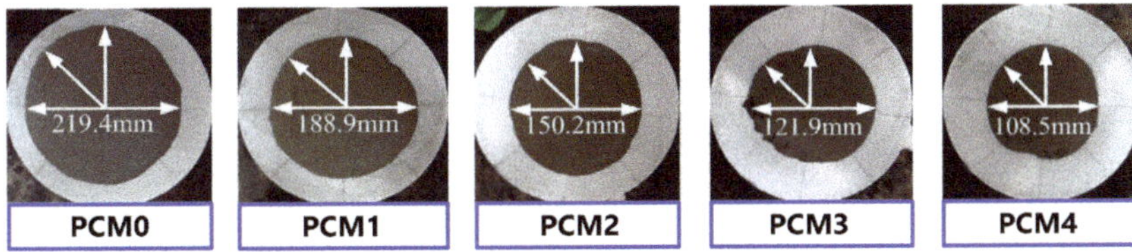

Figure 4. Effect of PVA content on the flow spread diameter of fresh mortar mixes.

3.2. Flexural Strength

Figure 5 shows the flexural strength of cement mortars with 0.2~2.0% PVA after 3~28 days of curing. Clearly, the flexural strength of mortars increased at first and then decreased with the increase of PVA content. PCM3 containing 1.0% PVA shows the highest flexural strength. The 7-day, 14-day and 28-day flexural strengths were about 31.9%, 28.8% and 24.8% higher than those of the control mixture without PVA. In addition, the curing ages also impacted the flexural strength development, which generally increased with the increase of curing age. However, the increasing rate was greatly influenced by the PVA content. When the PVA content was 0~0.6%, the increasing rate of flexural strength for different samples was very similar. However, when incorporating 1.0~2.0 wt. % PVA, the flexural strength slightly increased with the increase of curing ages. The 28-day flexural strength of PCM4 containing 2.0% PVA was just 12.5% higher than the three-day flexural strength, while for PCM2, the flexural strength after 28 days of curing was about 29.0% higher than that after three days of curing.

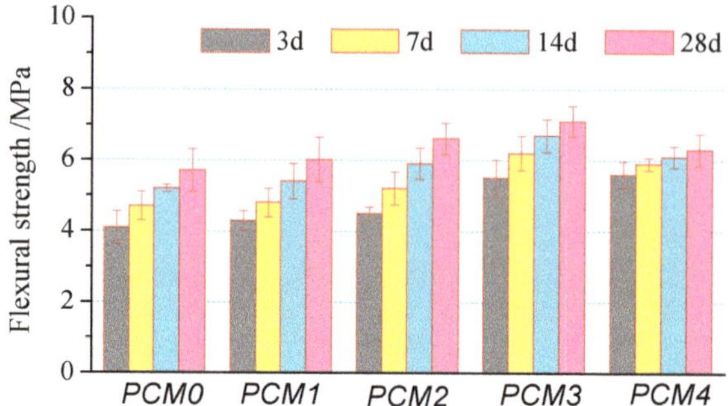

Figure 5. Effects of PVA content on the flexural strength of cement mortar.

3.3. Compressive Strength

Figure 6 shows the compressive strength development of cement mortars with 0~2.0% PVA at the early age. Clearly, the growth trend of compressive strength of all mortars was quite similar to that of flexural strength, in which the strength increased at first and then decreased with the increase of PVA content. However, Figure 6 shows that the optimal use of PVA was 0.6% of cement content by mass, and the compressive strength increased by 12.15% as compared to PCM0 after 28 days of curing. Moreover, it is worth noting that the curing age impacted the improvement effect of PVA, and the reinforcement effect was more remarkable at the early curing age. For example, the three-day compressive strength of PCM2 was about 17.6% higher than that of PCM0, while its 14-day strength just increased by 14.9%, as compared to PCM0. Additionally, the excessive adding of PVA might reduce the compressive strength. As shown in Figure 6, the 28-day compressive strengths of PCM4 were about 5.5% lower than those of PCM0.

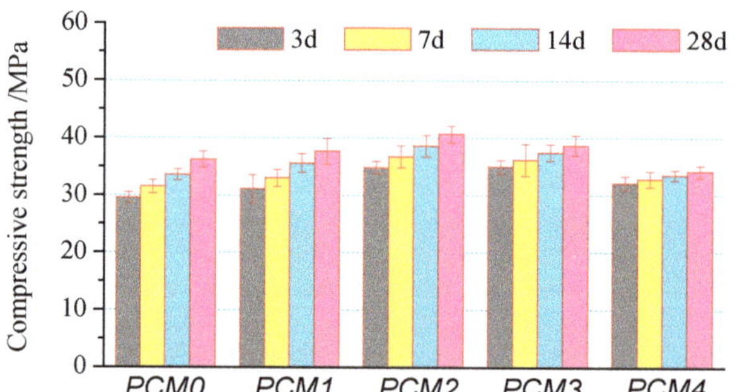

Figure 6. Effects of PVA content on the compressive strength of cement mortar.

3.4. Volume Density

The mass per unit volume of a material that contains open and closed pores is referred to as the volume density, and the volume density can reflect the degree of internal compactness of materials. Figure 7 shows the effect of the PVA content on the volume density of the cement mortar. For the measured value, the volume density of the mortar increased with increasing PVA content until it reached an optimal amount of approximately 0.6 wt. %, and then started to decrease. This indicates that the incorporation of a proper amount of PVA can improve the internal pore structure of the mortar and improve the compactness of the mortar. The maximum density of 2148.8 kg/m^3 corresponded to 0.6% PVA and compared with the mortar without PVA (2028.5 kg/m^3), the increase in the density was determined to be 5.9%. When the PVA content continued to increase, it was difficult to uniformly mix the mortar during the stirring and shaking process due to its poor working performance, so the internal pores increased, and the volume density decreased. The minimum density of 1992.1 kg/m^3 corresponded to 2.0% PVA content.

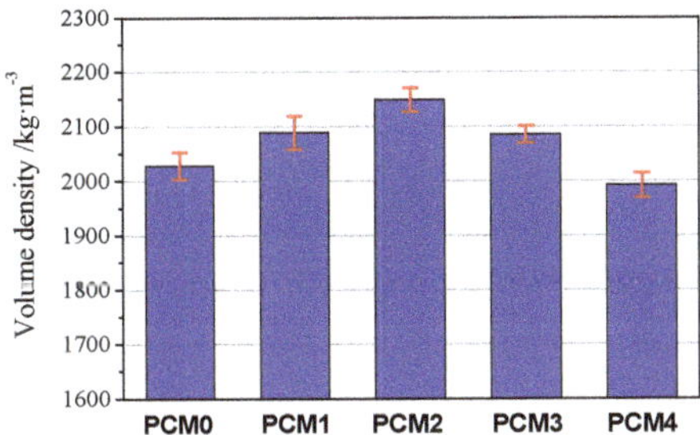

Figure 7. Effects of PVA content on the volume density of cement mortar.

3.5. Capillary Water Absorption

The water absorption of a material depends not only on its hydrophily but also on its porosity and pore size distribution. Figure 8 shows the capillary water absorption of cement mortars containing 0.2~2.0% PVA. The capillary water absorption rate is plotted on the ordinate, and the square root of time $t^{0.5}$ is the abscissa. By analyzing the relationship between the capillary water absorption and the square root of the time, the pore characteristics of the mortar can be reflected. According to the test results in the figure, it can be found that the curve shape of water absorption of cement mortars with or without PVA was quite similar, and which can be divided into two stages: A rapid ascending phase (stage I), and a steady phase (stage II). At stage I, water diffuse along the micro-cracks or interconnecting holes and then fill the pores of surrounding mortar. Thus, the more micro-cracks and interconnecting holes, the faster this stage will ascend. Clearly, according to Figure 8, PCM0 had the highest micro-cracks or interconnecting holes, followed by PCM4, and PCM2 had the lowest micro-cracks and interconnecting holes as its ascending rate was the lowest. In stage II, the pores were filled by water to a certain level, the equilibrium between the rates of transporting moisture and filling pores is built. At this stage, the capillary water absorption value is related to the porosity, and a higher water absorption value means a higher porosity. The effect of PVA on the water absorption of mortar may be because (1) PVA reduces the hydrophily, and (2) PVA film fills the pores of the cement matrix. For this reason, the samples containing PVA had a much lower absorption over the control one. Especially for PCM2, the capillary water absorption of cement mortar was the lowest value of 3.21 kg/m^2, which was 60% lower than that of PCM0. However, it should be noted that the addition of PVA will reduce the flow properties of cement mortar, as shown in Figure 4, the higher the PVA content, the lower the slump value was. Particularly for PCM4, the slump value reduced by about 50% as compared to PCM0. The poor flow behavior might cause the uneven dispersion of cement mortar, consequently, the interconnecting holes probably increased with the increase of PVA.

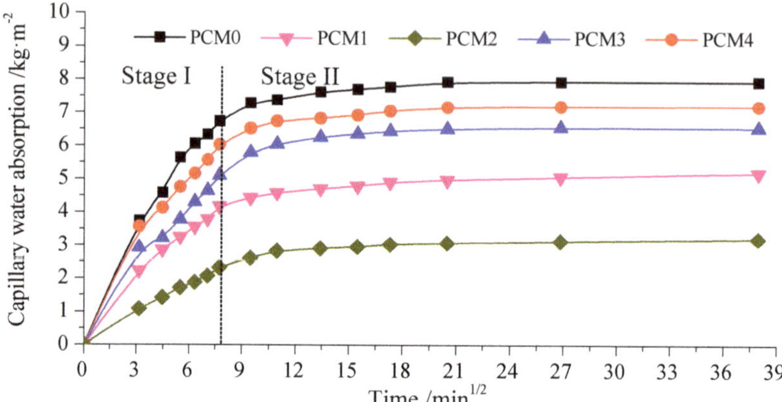

Figure 8. Capillary water absorption of cement mortar with different PVA contents (0~24 h).

3.6. Thermal Analysis

The above-mentioned test results indicate that the adding of 0.6% PVA had a considerable effect on increasing the mechanical property and reducing the water absorption of cement mortar, while the excessive incorporation of 2.0% PVA hardly affected the properties of cement mortar. To understand the mechanism behind these opposite levels of effectiveness of PVA on cement mortar, the microstructure of cement mortar incorporating of 0.6% and 2.0% PVA was studied, for comparison, the property of cement mortar without PVA was also investigated.

Firstly, the thermal decomposition performance of cement mortars containing 0%, 0.6%, and 2.0% PVA was investigated by using differential scanning thermal analysis (DSC), and the results are shown in Figure 9a. There are two major endothermic peaks in the three curves, as shown in Figure 9a. The first endoscopic endothermic peak is located at 95 °C, which is caused by the evaporation of adsorbed water. The second endoscopic endothermic peak is located at 430~450 °C, which is caused by the decomposition of $Ca(OH)_2$. The two absorption peaks of the three curves are integrated in Figure 9b to assess the effect of PVA on the thermal decomposition of cement mortar. Comparing the first endothermic peaks of all DSC curves, it can be found that PCM0 without PVA is 18.245 J/g, and the incorporating 0.6% PVA increased the endothermic enthalpy values by 13.8%. However, the adding of 2.0% PVA reduced the first absorption peak by 9.3% as compared to PCM0 without PVA. This result indicates that the incorporation of 0.6% PVA might accelerate the hydration rate of the cement matrix, while the addition of 2.0% of PVA delay the hydration rate. Comparing the second absorption peak of the three samples, it can be found that PCM0 had the highest value of 22.145 J/g, followed by PCM2 having a peak value of 19.51J/g, and PCM4 had the lowest value of 17.243 J/g. Figure 9b shows that the first endothermic peak corresponds to the evaporation of adsorbed and interlayer water of PCM2 was much higher than that of PCM0, while the second peak corresponds to the decomposition of $Ca(OH)_2$. This conflicting result may be due to the -OH bonds in PVA chain react with the -OH bonds of $Ca(OH)_2$, leading to the reduction of the content of $Ca(OH)_2$, which is further studied by using FTIR and reported in the next section.

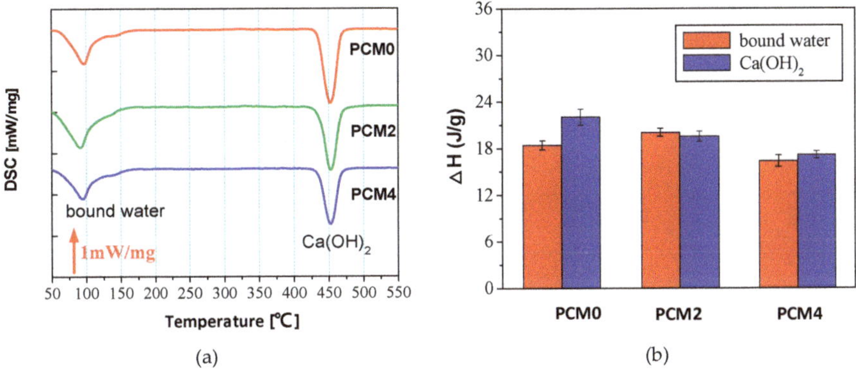

Figure 9. Differential scanning calorimetry (DSC) analysis of cement mortars at 28 days. (**a**) DSC curves and (**b**) bound water and Ca(OH)$_2$ content.

3.7. FTIR Spectroscopy

The FTIR analyses of cement mortar containing 0.6% and 2.0% PVA were conducted after 28 days of curing, and the spectra are shown in Figure 10 for comparison. As shown in Figure 10, all curves have obvious characteristic peaks at 3643, 1643, 1452, and 973 cm^{-1}. The absorption peak near 3643 cm^{-1} corresponds to the characteristic peak of free hydroxyl in the Ca(OH)$_2$ molecule produced by cement hydration. The characteristic absorption peak of Si-O of C-S-H of cement hydration products is at near 973 cm^{-1}. The peak at 1452 cm^{-1} corresponds to the stretching of CaCO$_3$. The FTIR spectra of PCM2 containing 0.6% PVA shows more intense absorption peaks at 973 cm^{-1} as compared to CM, indicating the positive effect of PVA on accelerating the hydration rate of the cement matrix. However, this peak of PCM4 was much lower than that of PCM0, indicating the excessive addition of PVA impacted the hydration of the cement matrix in a negative way. Moreover, Figure 10 shows that the addition of PVA reduced the spectral data at 3643 cm^{-1}, which corresponds to the -OH band of Ca(OH)$_2$. This result coincides well with the results of DSC.

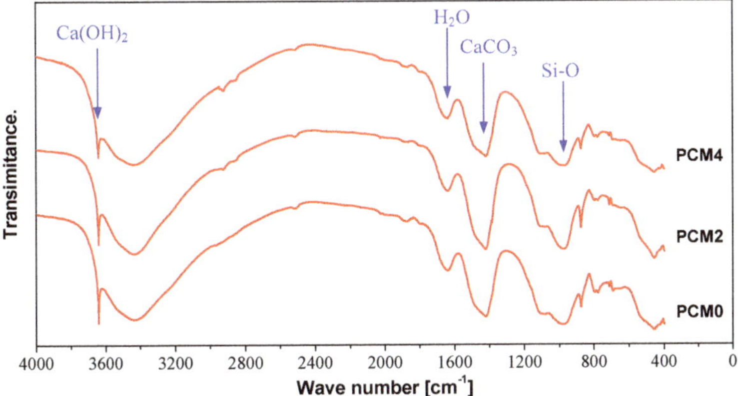

Figure 10. Fourier transform infrared spectroscopy (FTIR) analysis of cement mortar with different PVA content.

3.8. SEM Investigation

Scanning electron microscopy is an important means of analyzing the internal microstructure of materials. Scanning electron microscopy images can be used to determine the composition of the pore structure inside the cement matrix, the internal shape of the cement matrix and the type of hydration products. Figure 11 shows the microstructure of the modified mortar with a PVA content of 0.6% for 28 days of curing. As shown in Figure 11, the internal structure, pore distribution, and pore size composition of the modified mortar changed with the incorporation of PVA. The polymer was evenly distributed in the cement matrix while the cement mortar was being agitated and intertwined with the cement hydration product. As the cement hydration reaction progressed, the polymer continuously precipitated in the cement to form a polymer film. The presence of these films can play the following roles: (1) The films can combine with the hydration product to enhance the connection between the hydration products, thereby improving the mechanical properties of the cement mortar. (2) Part of the film can become interspersed with the cement base. Between the pores, it can fill the pores and form bridges in the cement plate. Under a load, it can absorb the energy generated by the external force so that the mechanical properties of the cement matrix are improved [23]. (3) The presence of the polymer can refine the pores, making a large number of harmful pores decrease into medium-sized or even harmless pores, thereby enhancing the mechanical properties of the cement mortar. In addition, it can be clearly observed in Figure 11 that the polymer film spans the crack between the cement plates to form a load transfer bridge. Under the action of an external force, the film is torn, and the energy generated by the external load is absorbed so that the crack generation and development process of the cement plate are delayed.

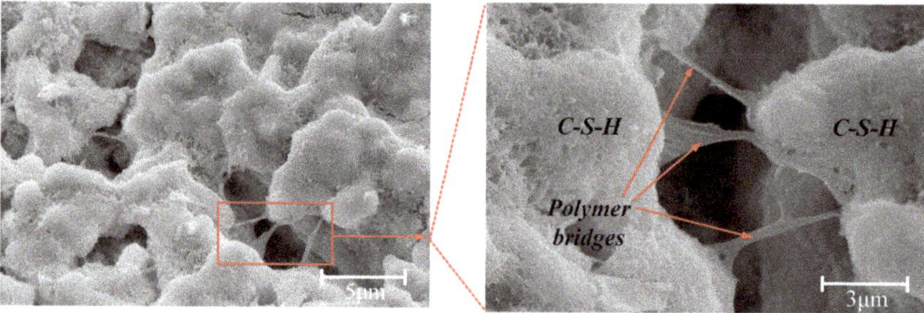

Figure 11. Polymer film stretched in open space in 0.6% PVA modified mortar at 28 days.

Figure 12 shows the microstructure of the modified mortar with a PVA content of 1.0%. As shown in Figure 12, as the amount of PVA increased, polymer film combines with the hydration product in the cement matrix to form a stable continuous network-like structure. This structure improves the mechanical properties of the mortar, especially the bending performance.

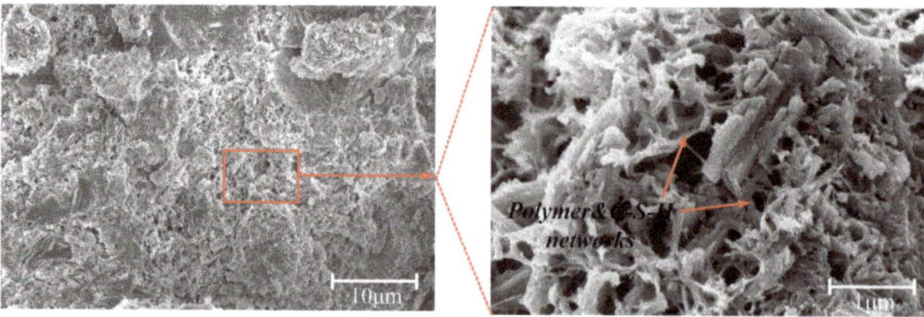

Figure 12. Network-like structure in cement mortar with the PVA content of 1.0% at 28 days.

Figure 13 is the SEM image of PCM4 containing 2.0% PVA. Figure 13 shows that the morphology and microstructure of PCM4 are quite different from PCM2. The excessive incorporation of PVA led to the formation of heave polymer sheets within the cement matrix. These heavy PVA film might coat cement particles and prevent them from having contact with water. This coated effect of PVA might delay the hydration rate of cement, as demonstrated by DSC and FTIR test results as shown in Figures 9 and 10. The lower compressive strength of PCM4 over PCM0 might be due to its lower hydration rate caused by the coated effect of PVA. In addition, Figure 13 shows the positive effect of PVA on bridging cracks, which lead to the increase of flexural strength of mortar as shown in Figure 6, where the three-day flexural strength of PCM4 was about 36.6% and 1.8% higher than that of PCM0 and PCM2, respectively.

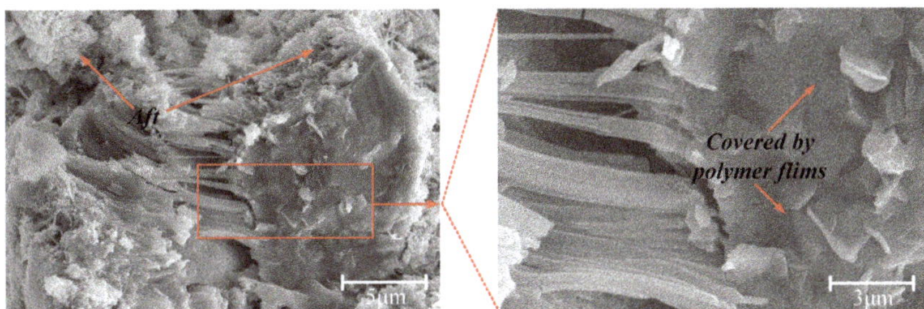

Figure 13. SEM image of cement mortar with 2.0% PVA content at 28 days.

The results of the scanning electron microscopy analysis show the following: (a) The proper amount of PVA incorporation (0.6%~1.0%) can be evenly distributed inside the matrix with an appropriate aggregate stirring process. A uniformly distributed polymer film can be formed that has a certain tensile strength. The polymer film bridging between the cement plates can delay or inhibit crack development of the cement mortar. Additionally, the film also can combine with hydration products to form a network-like structure. The structural system improves the mechanical properties of the cement mortar, especially for the flexural strength. (b) When the PVA content is too high (2.0%), the mortar has a poor working performance, and it is difficult to stir evenly. The incorporation of a large amount of PVA forms an uneven distribution inside the matrix. The cement curing process leads to the formation of a large and thick polymer film that coats the surface of the cement particles and hinders the hydration of the cement, thereby reducing the mechanical properties of the cement mortar.

4. Conclusions

This study investigated the strength development of cement mortars at the early age (3~28 days) after loading various polyvinyl alcohol (PVA) contents to obtain an optimum proportion. Then, the microstructure and morphology of PVA-modified cement mortar were studied by using DSC and FTIR to reveal its working mechanism. According to the test results in this study, the main achievements were:

(1) PVA has a strong cohesiveness and water retention, and its incorporation reduces the fluidity of cement mortar. The incorporation of 0.2%, 0.6%, 1.0%, and 2.0% PVA reduced the slump value by 13.9%, 31.5%, 41.1%, and 50.5% as compared to the control one, respectively.

(2) The mechanical properties of cement mortar show a significant increase at first and then decrease with the increase of PVA content. Cement mortar containing 0.6% PVA has the highest compressive strength, its 28-day compressive strength is 12.1% higher than that of the control one. The incorporating 1.0% PVA increased the flexural strength greatly, its 28-day flexural strength is about 24.8% higher than that of the control one.

(3)   The bulk density and water absorption test results show that the incorporation of 0.6% PVA increases the bulk density of cement mortar by 5.90%, and the corresponding capillary water absorption decreases by 60% as compared to the control mortar, respectively.

(4)   SEM tests show that three-dimensional PVA networks are formed in cement mortar. The crack-bridging effect of PVA film can be observed in the SEM images. When 2.0% PVA was incorporated, PVA formed heavy films, coating cement particles and preventing them from hydration.

(5)   DCS and FTIR analyses results manifest the adding of 0.6% PVA accelerates the hydration of the cement matrix, while the incorporation of 2.0% PVA impacts the hydration rate of cement in a negative way.

Author Contributions: Conceptualization, J.F. and G.L.; data curation, S.D. and Z.W.; writing of the original draft, J.F. and Z.W; writing of review and editing, G.L.

Funding: This work is supported by the National Natural Science Foundation of PR China (No. 51708145, No. 51878299), the project of Natural Science Research Foundation of Guizhou Province (No. [2018]1064), and the Science and Technology Plan of Guangdong Province (2015A010105029).

Acknowledgments: This work is supported by the National Natural Science Foundation of PR China (No. 51708145, No. 51878299), the project of Natural Science Research Foundation of Guizhou Province (No. [2018]1064), and the Science and Technology Plan of Guangdong Province (2015A010105029). The authors are grateful for the above supporters.

Conflicts of Interest: The authors declare no conflict of interest.

References

1.   Luo, Z.; Li, W.; Wang, K.; Shah, S.P. Research progress in advanced nanomechanical characterization of cement-based materials. *Cem. Concr. Compos.* **2018**, *94*, 277–295. [CrossRef]

2.   Sang, G.; Zhu, Y.; Yang, G. Mechanical properties of high porosity cement-based foam materials modified by EVA. *Constr. Build. Mater.* **2016**, *112*, 648–653. [CrossRef]

3.   Yang, H.; Cui, H.; Tang, W.; Li, Z.; Han, N.; Xing, F. A critical review on research progress of graphene/cement-based composites. *Compos. Part A Appl. Sci. Manuf.* **2017**, *102*, 273–296. [CrossRef]

4.   Ohama, Y. *Handbook of Polymer-Modified Concrete and Mortars*; Noyes Publications: Park Ridge, NJ, USA, 1995.

5.   Sakai, E.; Sugita, J. Composite mechanism of polymer modified cement. *Cem. Concr. Res.* **1995**, *25*, 127–135. [CrossRef]

6.   Park, D.; Ahn, J.; Oh, S.; Song, H.; Noguchi, T. Drying effect of polymer-modified cement for patch-repaired mortar on constraint stress. *Constr. Build. Mater.* **2009**, *23*, 434–447. [CrossRef]

7.   Li, G.; Zhao, X.; Rong, C.; Wang, Z. Properties of polymer modified steel fiber-reinforced cement concretes. *Constr. Build. Mater.* **2010**, *24*, 1201–1206. [CrossRef]

8.   Mirza, J.; Mirza, M.S.; Lapointe, R. Laboratory and field performance of polymer-modified cement-based repair mortars in cold climates. *Constr. Build. Mater.* **2002**, *16*, 365–374. [CrossRef]

9.   Assaad, J.J. Development and use of polymer-modified cement for adhesive and repair applications. *Constr. Build. Mater.* **2018**, *163*, 139–148. [CrossRef]

10.   Shaker, F.A.; El-Dieb, A.S.; Reda, M.M. Durability of Styrene Butadiene latexmodified concrete. *Cem. Concr. Res.* **1997**, *27*, 711–720. [CrossRef]

11.   Almeida, A.E.D.S.; Sichieri, E.P. Mineralogical study of polymer modified mortar with silica fume. *Constr. Build. Mater.* **2006**, *20*, 882–887. [CrossRef]

12.   Xu, F.; Zhou, M.; Chen, J.; Ruan, S. Mechanical performance evaluation of polyester fiber and SBR latex compound-modified cement concrete road overlay material. *Constr. Build. Mater.* **2014**, *63*, 142–149. [CrossRef]

13.   Yahya, G.O.; Ali, S.A.; Al-Naafa, M.A.; Hamad, E.Z. Preparation and viscosity behavior of hydrophobically modified poly(vinyl alcohol) (PVA). *J. Appl. Polym. Sci.* **2010**, *57*, 343–352. [CrossRef]

14.   Rim, I.W.; Robertson, R.E.; Robert, Z. Effects of Some Nonionic Polymeric Additives on the Crystallization of Calcium Carbonate. *Cryst. Growth Des.* **2005**, *5*, 513–522.

15. Wang, S.; Ren, J.; Li, W.; Sun, R.; Liu, S. Properties of polyvinyl alcohol/xylan composite films with citric acid. *Carbohydr. Polym.* **2014**, *103*, 94–99. [CrossRef]

16. Sahoo, S.K.; Panyam, J.; Prabha, S.; Labhasetwar, V. Residual polyvinyl alcohol associated with poly (D,L-lactide-co-glycolide) nanoparticles affects their physical properties and cellular uptake. *J. Control. Release* **2002**, *82*, 105–114. [CrossRef]

17. Zhang, Y.; Ye, L.; Cui, M.; Yang, B.; Li, J.; Sun, H.; Yao, F. Physically crosslinked poly(vinyl alcohol)–carrageenan composite hydrogels: Pore structure stability and cell adhesive ability. *RSC Adv.* **2015**, *5*, 78180–78191. [CrossRef]

18. Chopra, P.; Nayak, D.; Nanda, A.; Ashe, S.; Rauta, P.R.; Nayak, B. Fabrication of poly(vinyl alcohol)-Carrageenan scaffolds for cryopreservation: Effect of composition on cell viability. *Carbohydr. Polym.* **2016**, *147*, 509–516. [CrossRef]

19. Ma, Y.; Bai, T.; Wang, F. The physical and chemical properties of the polyvinylalcohol/ polyvinyl-pyrrolidone/hydroxyapatite composite hydrogel. *Mater. Sci. Eng. C Mater. Biol. Appl.* **2016**, *59*, 948. [CrossRef]

20. Zheng, Y.; Huang, X.; Wang, Y.; Xu, H.; Chen, X. Performance and characterization of irradiated poly(vinyl alcohol)/polyvinylpyrrolidone composite hydrogels used as cartilages replacement. *J. Appl. Polym. Sci.* **2009**, *113*, 736–741. [CrossRef]

21. Singh, N.B.; Rai, S. Effect of polyvinyl alcohol on the hydration of cement with rice husk ash. *Cem. Concr. Res.* **2001**, *31*, 239–243. [CrossRef]

22. Knapen, E.; Van Gemert, D. Polymer film formation in cement mortars modified with water-soluble polymers. *Cem. Concr. Compos.* **2015**, *58*, 23–28. [CrossRef]

23. Kim, J.H.; Robertson, R.E. Effects of Polyvinyl Alcohol on Aggregate-Paste Bond Strength and the Interfacial Transition Zone. Adv. *Cem. Based Mater.* **1998**, *8*, 66–76. [CrossRef]

24. *Test Method for Fluidity of Cement Mortar, China*; GB/T 2419-2005; Architecture & Building Press: Beijing, China, 2005. (In Chinese)

25. *Test Methods for Bulk Density, Moisture and Water Absorption of Aerated Concrete, China*; GB/T 11970-1997; Architecture & Building Press: Beijing, China, 1997. (In Chinese)

26. *Determination of The Water Absorption Coefficient of Construction Materials*; DIN 52617; German Institute for Standardisation: Berlin, Germany, 1987.

27. *Method of Testing Cements-Determination of Strength, China*; GB/T 17671-1999; Architecture and Building Press: Beijing, China, 2003. (In Chinese)

 *applied sciences*

Article

Experimental Study on Flexural Behavior of TRM-Strengthened RC Beam: Various Types of Textile-Reinforced Mortar with Non-Impregnated Textile

Jongho Park [1], Sungnam Hong [2] and Sun-Kyu Park [1,*

[1] Department of Civil, Architectural and Environmental System Engineering, Sungkyunkwan University, Suwon 16419, Gyeonggi-do, Korea; rhapsode@skku.edu
[2] Department of Ocean Civil Engineering, Gyeongsang University, Tongyeong 53064, Gyeongsangnam-do, Korea; snhong@gnu.ac.kr
* Correspondence: skpark@skku.edu; Tel.: +82-031-290-7530

Received: 10 April 2019; Accepted: 9 May 2019; Published: 15 May 2019

Abstract: In this study, to compare strengthening efficiency and flexural behaviors of textile- reinforced mortar (TRM) according to various types of strengthening methods without the textile being impregnated, ten specimens were tested. The results showed that TRM was beneficial for uniform distribution of cracks and increased the strengthening efficiency and load-bearing capacity, as textile reinforcement ratio and textile lamination increased and the mesh size of the textile decreased and mechanical end anchorage applied. However, the strengthening effect was shown obviously until the yield load considering structural safety and serviceability.

Keywords: textile reinforced mortar; non-impregnated textile; textile reinforcement ratio; textile lamination; mesh size; end anchorage; strengthening efficiency

---

1. Introduction

Due to aging infrastructures, the need for improvement in repair, strengthening, and maintenance processes has been continuously increasing. To improve the performance of aging infrastructures, astronomical funds have been invested [1]. Moreover, with the development of design code, structures are required to be strengthened to meet structural safety [2]. To keep up with these requirements, fiber-reinforced polymer (FRP) strengthening method has been studied and applied for many years [3–5]. Strengthening with FRP uses organic materials such as epoxy resin for bonding with concrete substrate. However, organic materials have several disadvantages, including the impossibility of applying them to wet surfaces, low glass transition temperature, low fire resistance, and low permeability.

To overcome the disadvantages of the use of organic materials, textile-reinforced mortar (TRM) method using inorganic materials such as cementitious matrix has been actively researched recently [2,6–8]. TRM is a structure-strengthening method for attaching textiles composed of fiber materials, such as carbon and glass fibers with excellent chemical resistance, to the surface of masonry and reinforced concrete (RC) structures generally using mortar. As it uses mortar, TRM is highly resistant to temperatures and applicable to wet surfaces, thus allowing efficient construction in various environments. However, since the textile consists of numerous filaments, the mortar cannot penetrate into the textile and the bonding with mortar is divided into the inner and the outer.

Many researchers investigated the bonding behavior between textile and mortar and between TRM layer and concrete substrate to explain the bonding characteristics of textile composite materials [9–14]. Furthermore, to improve the bonding between textile and mortar, textile has been impregnated with resin or attached small sand grains [15–20]. This reinforcement of textile enables its tensile strength to be used more efficiently.

Papanicolaou et al. [15] coated a textile using a polymeric resin. Compared with the uncoated specimens, the load of specimen reinforced with the textile alone was increased by 26–112% in terms of the first-crack, yield and ultimate load. The load of specimen reinforced with the textile and steel rebar increased by 11–16%. The effect of strengthening using coated textile was increased when reinforcing by textile alone, or when using a smaller amount of textile. Kamani et al. [19] applied a partially impregnated epoxy at the interaction point of weft and warp roving in textile. Experimental results show that the maximum load of specimens with partially impregnated epoxy was increased by 49–73%. In Raoof et al. [21], the flexural strength of the specimens with coated textile was increased by 16% at crack and by 5% at yield and ultimate load. The failure mode of specimens was changed from premature failure due to local slippage of uncoated textile to interlaminar shearing of mortar of coated textile. Jesse et al. [22] also showed that the ultimate load of the specimen with coated textile was increased by 23% compared with the uncoated specimen.

Commonly, impregnation or coating of textile improves the bond condition between textile and mortar. In addition, fiber alignment, prevention of relative slippage, strain compatibility of filaments, and the ease of load transferring due to strong binding of the intersection point of textile was studied [15,19]. The process of impregnation makes it possible to use the textile more efficiently, but the efficiency differs depending on the type of the resin and textile, impregnation method, and the shape of the structure. In addition, according to Gutierrez et al. [23], the additional cost of resin and impregnation process accounts for about 38% of the fiber in terms of raw materials.

Therefore, in this study, various types of TRM were investigated to solve the problem of cost due to textile impregnation and to find a method of using textile efficiently without the impregnation process. According to previous studies, the non-impregnated textile will cause the bond of textile and mortar to be divided into inner and outer filaments so that the tensile strength of TRM composite is decreased. Hence, to prevent the slippage and damage of textiles, various methods of textile lamination, textile geometry, and end anchorage were applied. Ten specimens were fabricated and compared with load and deflection relationship, cracking, strain, and ductility index to investigate the strengthening efficiency and flexural behavior of TRM without textile impregnation.

2. Experimental Program

2.1. Textile

The textile used in this study was made by alkali resistant (AR) glass fibers containing 16.5% zircon in 8 × 8 mm intervals as shown in Figure 1. Properties of the textiles provided by the manufacturer (Taishan fiberglass Inc., Tai´an, China) are listed in Table 1. The diameter of the filament was 14 μm, and the area of roving was 0.246 mm^2. Its tensile strength, modulus of elasticity, and maximum strain were 1789 MPa, 68 GPa, and 0.0262, respectively.

Figure 1. Alkali resistant (AR) glass textile.

Table 1. Detailed specifications of the AR glass provided by the manufacturer; AR: Alkali Resistant.

Properties and Geometric Parameters [1]	AR Glass
Number of filaments per roving	1600
Tensile strength of warp (N/50 mm)	2142
Tensile strength of weft (N/50 mm)	1833
Rupture elongation ratio of warp (%)	2.85
Rupture elongation ratio of weft (%)	2.36
Tex of yarn (g/km)	640
Diameter or filament (μm)	14

[1] Tested by Chinese standard: GB/T 7689.5 idt ISO 4606.

2.2. Matrix

Ready-mixed concrete with specified concrete strength up to 35 MPa was used. The average compressive strength of the concrete using three cylindrical specimens with a 100 mm diameter and a 200 mm height was 42.19 MPa. For the matrix of the TRM section, polymer mortar was used for effective bonding with the textile and concrete substrate. The concrete and polymer mortar mix proportions are listed in Tables 2 and 3. Specifications of polymer mortar provided by the manufacturer (SsangYong Cement, Seoul, South Korea) are listed in Table 4. The average compressive strength of the polymer mortar using three cylindrical specimens was 48.91 MPa.

Table 2. Mix proportions of concrete.

W/B (%)	Unit Weight (kg/m^3)						
	Cement	Water	Fine Aggregate	Coarse Aggregate	Fly Ash	Blast Furnace Slag	Water Reducer
35.8	319	163	780	898	68	68	4.1

Table 3. Mix proportions of polymer mortar.

W/M [1] (%)	Content Per 1 Bag of 25 kg (%)					
	Cement	Fine Aggregate	PVA Fiber [2]	Acrylate Copolymer	CSA [3]	Water Reducer
19	<50	35~40	>1	3>	>5	<1

[1] Water/mortar. [2] Polyvinyl alcohol fiber. [3] Expansive admixture.

Table 4. Detailed specifications of the polymer mortar provided by the manufacturer.

Strength Type	Experimental Value (MPa)	Standard of KS (MPa)
Flexural	8	6.0 >
Compressive	45	20.0 >
Bond [1]	1.5	1.0 >

[1] The bond strength is improved 20% with primer.

2.3. Specimen

Nine TRM-strengthened reinforced-concrete beams were fabricated as TRM specimens, and one RC beam was fabricated as a control specimen. TRM specimens are composed of the RC and TRM. Total length was 1500 mm, width was 120 mm, and height was 150 mm, which consist of a height of 135 mm of RC and a height of 15 mm of TRM. The control specimen had a height of 135 mm without TRM strengthening layer. RC was reinforced with two D10 bars with diameter 9.53 mm in tension. To prevent the shear failure at support section, D6 shear stirrups with diameter 6.35 mm were placed at 80 mm and 100 mm intervals except for the pure bending section. Yield strength and modulus of elasticity of the reinforcement were 400 MPa and 200 GPa for rebar D6 and D10, respectively. Specifications of these study specimens are shown in Figure 2.

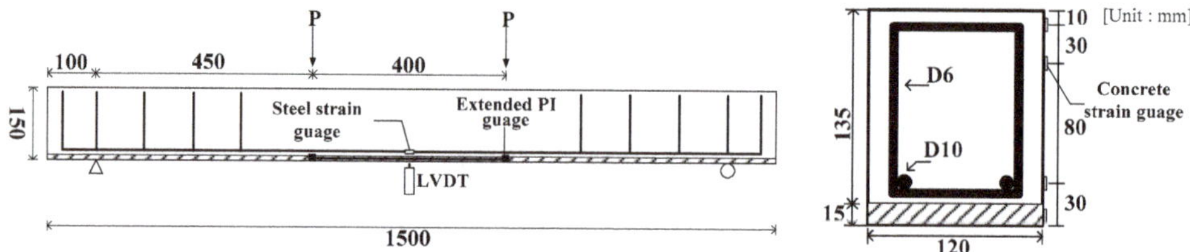

Figure 2. Specifications of study specimens.

Experimental parameters of the TRM specimen and the experimental comparison group were as follows: (1) Textile reinforcement ratio (RC, T1, T2, T3); (2) textile lamination and geometry (T1, T1O, T1N, T1M); and (3) end anchorage (T2, T2E, T2A, T2AP).

The procedures for preparing the TRM specimen were as follows: (1) Grinding the bottom surface of RC beam with grid of grooves with depth of 2–3 mm; (2) setting up a RC beam to a strengthening device; (3) applying a primer; (4) pouring polymer mortar with thickness of 4 mm by using a trowel; (5) fixing the first textile on the clamp device at each side as shown in Figure 3; (6) stretching the textile to parallel to reinforcing axis; (7) pouring second polymer mortar with thickness of 2–3 mm and applying hand pressure to penetrate mortar into the textile; (8) repeat step (5)–(7) until all designed textile layers were placed; (9) after last textile was placed, mortar poured with 4 mm thickness. The average thickness of the TRM strengthening layer of all specimens was 19.91 mm, which was larger than the design as shown in Figure 2. This is because the amount of mortar required for TRM was increased due to the grooves 2–3 mm deep and the thicker textile layer due to lamination. Detailed specifications of these study specimens are listed in Table 5.

Figure 3. Textile-reinforced mortar (TRM)-strengthening device with textile stretching.

Table 5. Detailed specifications of study specimens.

Name	Strengthening Configuration	Textile		End Anchorage
		Layer	Lamination [2]	
Control	Reinforced concrete	-	-	-
T1 [1]	Textile reinforcement ratio was 0.049%	3	-	×
T1N	Textile applied manually (non-stretching)	3	-	×
T1O	Three textiles lamination in one layer	1	3	×
T1M [2]	Textile mesh size changed from 8 mm to 24 mm	3	(1/3) × 3	×
T2 [1]	Textile reinforcement ratio was 0.098%	3	2	×
T2E [3]	Fixed to both ends of the beam by TRM	3	2	○
T2A [3]	L-shaped angle + bolt anchor	3	2	○
T2AP [3]	Steel plate + bolt anchor	3	2	○
T3 [1]	Textile reinforcement ratio was 0.15%	3	3	×

[1] A reference specimen for each experimental group. [2] The number of roving T1M per one layer was reduced to one third due to changed mesh size. [3] TRM specimens were applied with end anchorage.

T denotes the beam reinforced with TRM, and the number denotes the ratio of textile reinforcement. Textile reinforcement ratio of T1 was 0.049% (8.86 mm^2). T2 and T3 were two and three times that of T1, respectively. T1N applied textile manually without textile stretching. All specimens except T1N were fabricated using a TRM-strengthening device as shown in Figure 3, which was fixing textiles on both sides of the specimen to keep the textile parallel to reinforcing axis when mortar pouring [24]. Textile stretching was applied with a slight tension so as not to damage the fibers [17–19]. T1O was laminated with three textiles to one layer. The mesh size of T1M was changed from 8 mm to 24 mm. T2E was fixed to both ends of the beam by TRM. T2A and T2AP were applied mechanical end anchorage with L-shaped angle and steel plate, respectively. The schematic of differences of each specimen is shown in Figure 4.

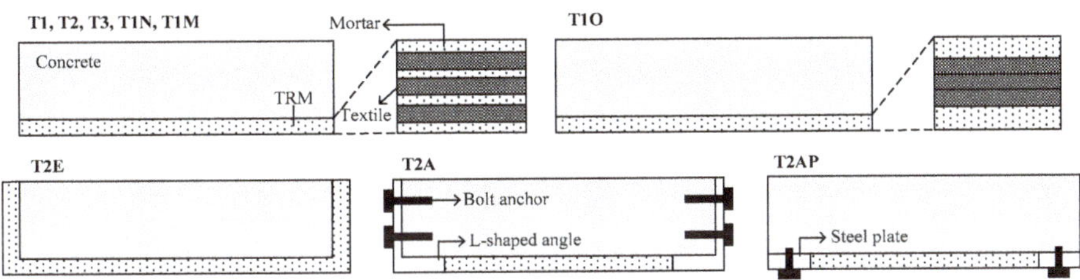

Figure 4. Schematic of TRM specimens.

2.4. Test Set-Up

The 4-point loading method with 3000 kN UTM was used at a rate of 0.1 mm/s under displacement control. As shown in Figure 2, the span of the specimen was 1300 mm, and the loading point was 400 mm in the mid-span. A linear variable differential transformer (LVDT) was installed at the mid-span to measure the deflection. Four concrete strain gauges and two steel strain gauges were installed at the mid-span. Extended PI gauge which consisted of two PI gauges with a 50 mm length and thin steel plate with a 300 mm length were attached at the mid-span of the TRM to measure the longitudinal deformation in the pure bending section.

3. Experimental Results and Discussion

3.1. Crack and Failure

Crack was observed with visual inspection. Similar flexural cracks appeared in all specimens as shown in Figure 5. The first cracks appeared in TRM at the pure bending section and developed towards both sides. After the occurrence of the crack of the TRM, crack of RC propagated at the same location where the TRM crack appeared. This is because the flexural stress was concentrated on the crack of the TRM and the concrete stress easily reached the tensile strength.

The number of cracks in RC was shown in Figure 6. The number of cracks in the pure bending section of the TRM specimen appeared more than that of control specimen. Therefore, TRM reinforcement appears beneficial for uniform distribution of cracks [17,20,25].

When the tensile force is applied to the textile and released in the matrix, the compressive force is applied to the matrix, and the initial crack strength of the TRM is increased as much as compressive stress. For a prestressed TRM, for the new crack to occur, the bond stress between textile and matrix must be increased as much as the sum of the initial compressive stress and tensile strength of the matrix. That is, the occurrence of a large crack is essential, and the number of cracks is decreased [20,26,27]. The textile of this study did not introduce enough force to cause sufficient elastic deformation of TRM. However, a slight tension of textile was maintained by applying mechanical anchors to T2A and T2AP. Therefore, it can be seen that the crack width and spacing was increased in the pure bending section as shown in Figure 6.

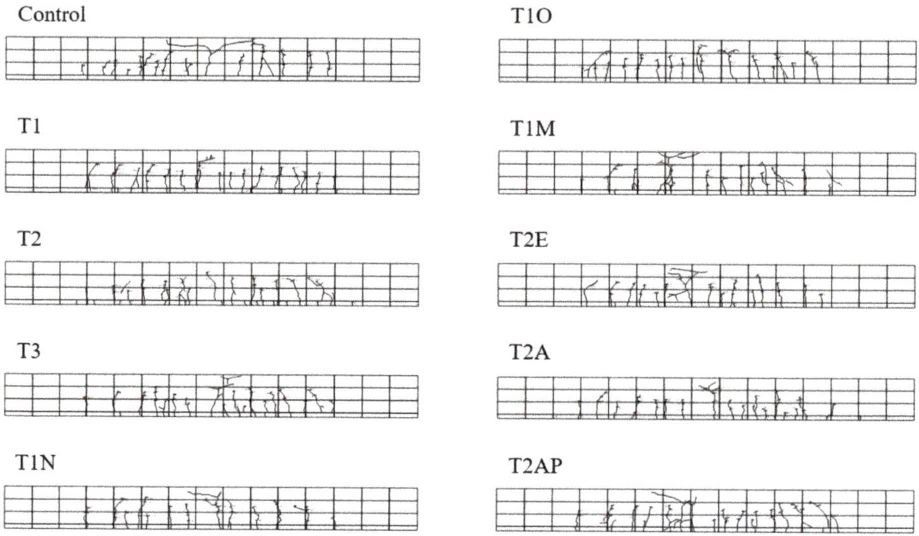

Figure 5. Crack patterns for all specimens at concrete crushing.

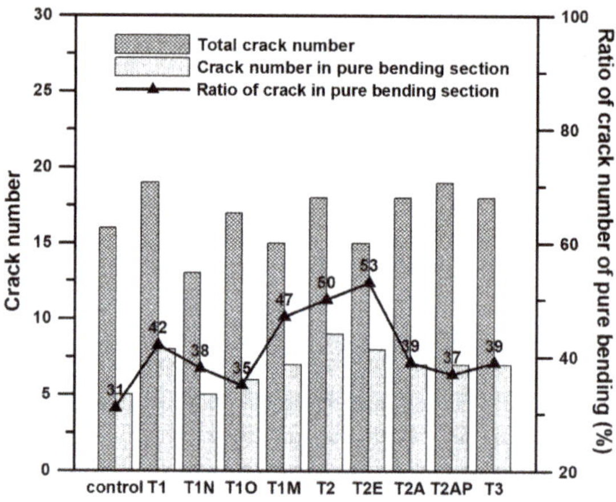

Figure 6. Crack number of all specimens.

The T1O applied one layer with lamination of three textile and T3 applied three layers with lamination of three textile. Thus, the bond area between textile and mortar of T1O and T3 was smaller than the other TRM specimens so that ratio of crack numbers of pure bending was low, as shown in Figure 6.

The failure modes of the TRM were shown as follows: (1) Textile rupture; (2) debonding between concrete substrate and TRM and textile rupture; and (3) textile slippage. After the failure of TRM, nearly the same plastic behavior of the control specimen was shown and, finally, crushing of concrete occurred. The failure modes of all specimens, listed in Table 6 and Figure 7, show three types of failure of the TRM. Figure 7a shows the textile rupture after the test of the T2AP. Figure 7b shows the debonding of the TRM strengthening layer of T2 after the ultimate load was reached. This debonding was also observed in T1O. Figure 7c shows the textile slippage of the T1M. The slippage of the textile occurred from yield load to crushing of concrete. After the end of the test, the textile had not failed, but showed crimp shape.

Table 6. Test results.

Name	Experiment Results					Failure Mode
	P_y (kN)	δ_y (mm)	P_u (kN)	δ_u (mm)	μ (δ_u/δ_y)	
RC	32.06	6.75	35.39	18.92	2.8	Concrete crushing
T1	33.66	7.48	-	-	-	Textile rupture
T1N	38.96	7.85	-	-	-	Textile rupture
T1O	38.22	8.18	39.21	10.79	1.32	Debonding and textile rupture
T1M	35.38	8.42	37.11	17.48	2.08	Textile slippage
T2	34.89	10.02	36.49	15.86	1.58	Debonding and textile rupture
T2E	40.69	8.85	42.29	12.45	1.41	Textile rupture
T2A	40.19	7.31	-	-	-	Textile rupture
T2AP	38.34	7.41	40.19	9.03	1.22	Textile rupture
T3	38.71	8.53	41.92	12.79	1.50	Textile rupture

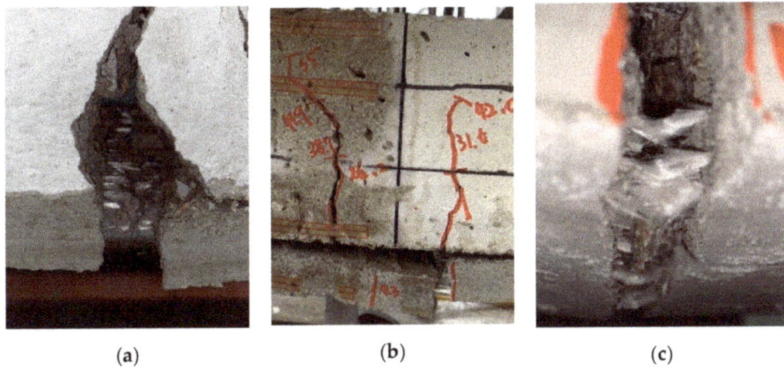

(a)   (b)   (c)

Figure 7. Failure modes of TRM: (**a**) Textile rupture of T2AP; (**b**) debonding and textile rupture of T2; (**c**) textile slippage of T1M.

3.2. Load and Deflection Relationship

The load and deflection curves for each comparison group are shown in Figures 8–10, and the results are listed in Table 6. The stiffness of TRM and control specimens were similar, except for T2, T2A, and T2AP. The decrease of the stiffness of the T2 was observed due to damage of the concrete beam during the grinding process. Increasing stiffness of T2A and T2AP was affected by applying a mechanical end anchorage.

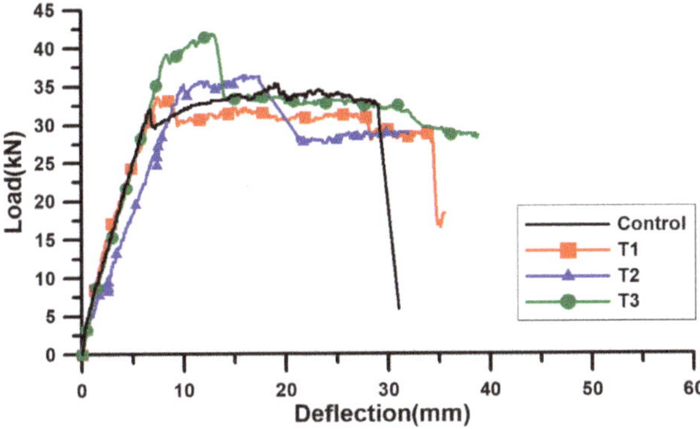

Figure 8. Load and deflection curve: Textile reinforced ratio.

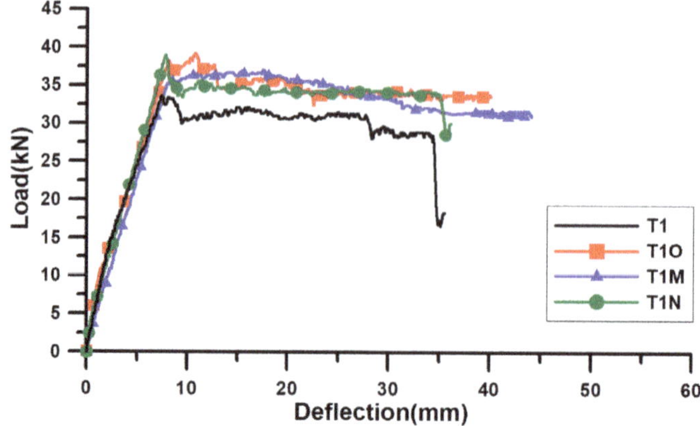

Figure 9. Load and deflection curve: Textile lamination and geometry.

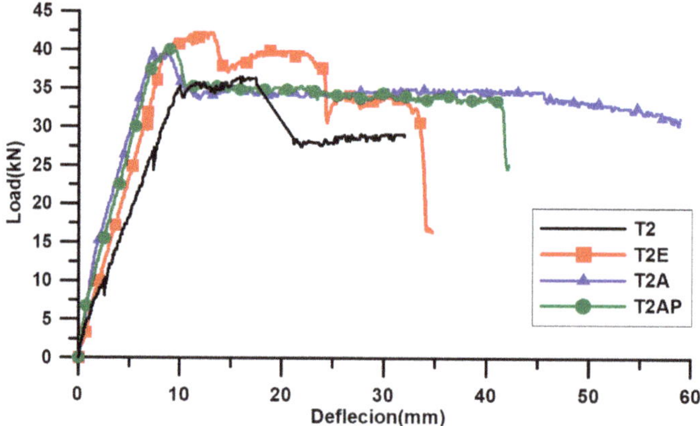

Figure 10. Load and deflection curve: End anchorage.

3.2.1. Textile Reinforcement Ratio

Figure 8 shows flexural behavior of the control, T1, T2, and T3 specimens according to the textile reinforcement ratio. The yield and ultimate loads of the control specimen were 32.06 kN and 35.39 kN, respectively. The yield load of T1 was 33.66 kN, which was 5% higher than that of the control specimen. As for T2, its yield and ultimate loads were 34.89 kN and 36.49 kN, respectively, which were 9% and 3% higher than that of the control specimen, respectively. For T3, its yield and ultimate loads were 38.71 kN and 41.92 kN, respectively, which were 21% and 18% higher than the control specimen, respectively. The strengthening effect of T1 and T2 was negligible because the ratio of increasing strength was less than 10%. In particular, the textile reinforcement ratio of T2 was twice that of T1, but the ratio of increase of the ultimate load was 3%.

Hegger et al. [28] reported that when the transversal action and the bending is applied to textile, the filament is damaged, and the load-bearing capacity is decreased. In addition, the bond performance of inner filaments is better, but inner filaments exhibit lower strain than outer filaments. To determine the complex behavior, inclined roving model in the crack was introduced, and the textile effectiveness was reduced about 50% for an angle of 45°. The load of T1 was decreased after the yielding of the steel rebar. T2 exhibited a yield load similar to T1 and showed a negligible increase in the ultimate load compared with the control specimen. This behavior of T1 and T2 indicates that the damage of the non-impregnated textile in the crack increased with the increase of the load, and the strengthening efficiency decreased sharply.

The T3 showed sufficient improvement with higher ratio of increase of strength compared with T1 and T2. As shown in Table 6, T3 had similar load-bearing capacity of T2E, T2A, and T2AP with end anchor because there was enough textile to bear the load even if some of textile of the T3 was damaged. T2E, T2A, and T2AP were considered to be free of textile damage due to anchor. This is discussed in detail in Section 3.2.3. The higher the ratio of textile reinforcement, the more sufficient load-bearing capacity appears even if damage to the non-impregnated textile occurred.

3.2.2. Textile Lamination and Geometry

Figure 9 shows flexural behaviors of T1, T1N, T1O, and T1M according to textile lamination and geometry. The yield load of T1N was 38.96 kN, which was 16% higher than that of the T1 specimen. The yield load of T1O was 38.22 kN, which was 14% higher than that of T1. The ultimate load of T1O was 39.21 kN, which was a little higher than its yield load. T1O and T1N improved the load-bearing capacity with nearly the same increment. The textile of T1 was vulnerable to damage, as described in the previous section, Section 3.2.1. However, in case of T1O, because the textile applied at the same reinforcing axis by lamination, all filaments in the textile resist the loads acting on the TRM layer simultaneously, even if the inner filaments have a higher ratio in the total cross-section area of the textile. Therefore, T1O increased load-bearing capacity higher than that of T1. In the case of T1N with the process of pressing the mortar with a trowel after the application of the textile without fixing both ends of the textile, the spacing of the textile layer of T1N was decreased compared to that of T1, so that same effect of textile lamination of T1O was observed. However, unlike other specimens, the uncertainty in the fabrication process of non-stretching caused unstable behavior in which the load suddenly decreased due to the textile rupture at the yield load.

The yielding and ultimate load of T1M was 35.38 kN and 37.11 kN, respectively. The textile mesh size of T1M was changed from 8 mm to 24 mm. The yield load of T1M was 5% higher than that of T1, but the yield and ultimate load were decreased compared with T1N and T1O. Because the weft (transverse direction) roving of the textile applied to beams or one-way slabs is perpendicular to the longitudinal axis, there is no load-bearing capacity but it performs a mechanical interlock such as an anchor of warp (longitudinal direction) roving between interaction points [25]. Therefore, T1M had wider spacing of anchors that fix warp roving than other TRM specimens, so that the possibility of textile slippages increased, and load-bearing capacity decreased [16].

3.2.3. End Anchorage

Figure 10 shows flexural performances of T2, T2E, T2A, and T2AP according to the type of end anchorage. The yield and ultimate load of T2E were 40.69 kN and 42.29kN, respectively, which were 17% and 16% higher than that of T2, respectively. The yield load of T2A was 40.19 kN, which was 15% higher than that of T2. The yield and ultimate load of T2AP were 38.34 kN and 40.19kN, respectively, which were 10% higher than that of T2. In the case of T2E, the load dropped two times after the ultimate load, unlike other TRM specimens. The first load drop was caused by the rupture of the outer filaments due to increase in the stress in the crack. In general, the slippage of textile inside the matrix occurred if there was no clamping device with sufficient compressive force such as an anchor [25]. The textile of T2E was fixed to both ends of the beam, hence the inner filaments in the textile were fixed by anchor and the load was once again increased slightly. The slippage of inner filaments had occurred before the ultimate load was reached but was limited by bonding of the end anchorage until the second load drop occurred. Therefore, T2E improved the load-bearing capacity more than the non-end anchorage specimens but did not completely prevent slippage. The T2A and T2AP with bolt anchor prevented all slippage of the textile, and all the filaments of textile resisted the load simultaneously so that the stiffness and load-bearing capacity increased significantly.

3.2.4. Ductility

Figure 11 compares the ductility index, which is defined as the ultimate deflection, δ_u, divided by deflection at yield load, δ_y. The ductility of T1O, T1M, T2, T2E, T2AP, and T3 were 1.32, 2.08, 1.58, 1.41, 1.22 and 1.5, respectively. T1M showed higher ductility compared with other TRM specimens. Contrary to T1M, T2AP with mechanical end anchorage had the lowest ductility while suppressing slippage. The ductility of the control specimen was 2.8, and all TRM specimens were less ductile than that of the control specimen, and thus failed to ensure sufficient ductility for structural safety.

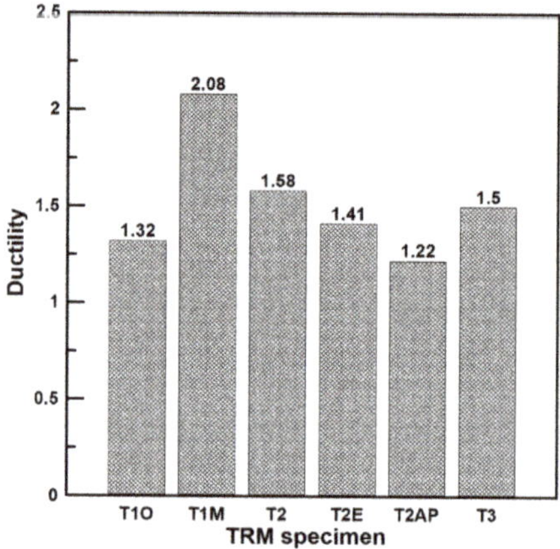

Figure 11. Comparison of ductility index of TRM specimen.

3.3. Strain

Strain changes according to height of specimen are shown in Figure 12. Strains of all specimens before yield were proportional to the distance from the neutral axis. As shown in Figure 12a, T1M clearly shows a sudden increase of TRM strain due to textile slippage between yield and ultimate load. As shown in Figure 12b, T3 shows the load was relatively proportional to the distance of the neutral axis until the ultimate load, although a significant change occurred from yield to ultimate load. Figure 12c shows a strain change of T2 at ultimate load and debonding of TRM. At this point, there is no proportional strain change along the neutral axis. In contrast, as shown in Figure 12d,e, T2AP and T2E showed proportional strain change of the TRM. T2AP had a constant strain variation until the ultimate load and perfectly prevented the slippage of textile. T2E showed almost the same behavior as the T3, but it can be seen that the strain variation was smaller than that of T3 by partially limited slippage of textile.

3.4. Comparison of Yield Loads of Studies Specimens

Experiment results showed that yield and ultimate load of TRM specimens were higher than those of the control specimen except T1, T1N, and T2A in ultimate load. The ratio of increase of yield load was especially higher than that of the ultimate load. After the yielding point of the TRM specimen, textile rupture occurred in T1, T1N, and T2A, and ductility of all specimens was lower than that of the control specimen. Therefore, considering the structural safety and serviceability, the strengthening effect of reinforced concrete with TRM using non-impregnated textile could be compared within yield load. Figure 13 shows yield load of all TRM specimens. Loads increased in proportion to the reinforcement ratio of T1, T2, and T3. The other specimens except T1M showed similar yield loads to that of the T3 specimen, which had the largest reinforcement ratio. Therefore, TRM strengthening using non-impregnated textiles could improve load-bearing capacity until yielding by using small mesh size, textile lamination, and fixing the textile with mechanical end anchorage.

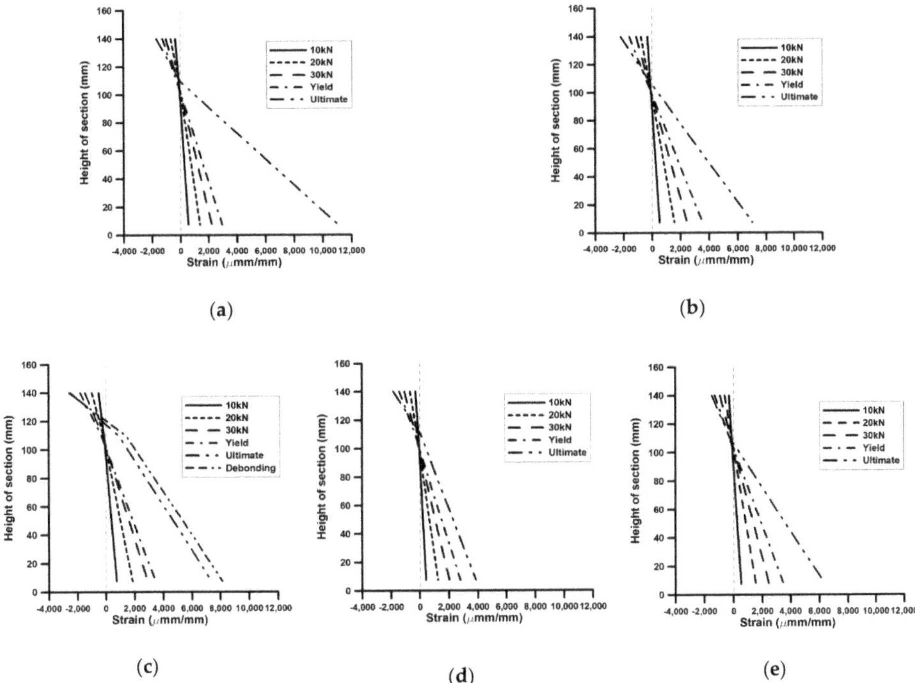

Figure 12. Strain distribution of different load levels: (**a**) T1M; (**b**) T3; (**c**) T2; (**d**) T2AP; (**e**) T2E.

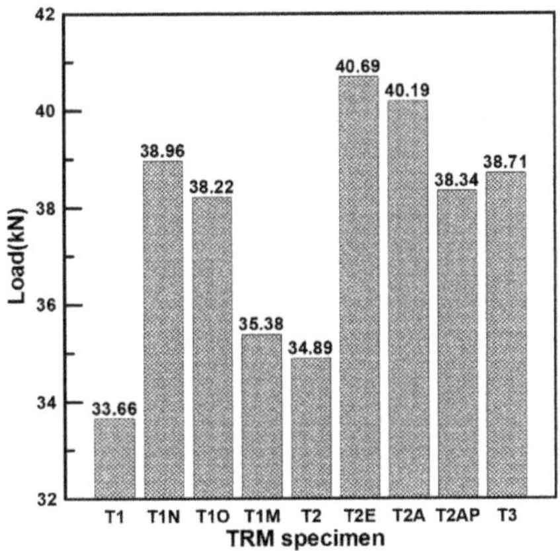

Figure 13. Comparison of yield loads of TRM specimen.

4. Conclusions

This study investigated the strengthening efficiency and flexural behavior of reinforced concrete (RC) beams strengthened by textile-reinforced mortar (TRM) with non-impregnated textile. For this purpose, the control specimen, an RC beam, was compared with nine TRM specimens.

1. There were multiple cracks in the pure bending of the TRM specimen compared with control specimen. However, in the case of using textile lamination, the ratio of bonding area with textile and mortar decreased compared to the total cross-sectional area of the textile, and in the case of

applying mechanical end anchorage (T2AP, T2A), the initial crack strength increased. In both cases, the cracking ratio of the pure bending section was lowered. Nevertheless, TRM strengthening was beneficial for uniform distribution of cracks.

2. Non-impregnated textile is easily damaged and slipped. Experimental results show that the case of lamination of textile can reduce the damage of the textile, but it does not prevent slippage. In the case of applying mechanical end anchorage, the slippage of the textile was expected to be completely prevented.

3. Considering the low ductility of all TRM specimens compared with control specimens and that no ultimate load was observed in some specimens, the behavior of TRM specimens after yield load was considered to be unstable. Therefore, the TRM strengthening effect of non-impregnated textiles could be compared with the behavior before the yield load considering structural safety and serviceability.

4. TRM strengthening using non-impregnated textiles increased the efficiency as the textile reinforcement ratio and textile lamination increased and the mesh size of the textile decreased and mechanical end anchorage applied. In this study, it was possible to have a load-bearing capacity similar to T3 with the largest textile reinforcement ratio by applying the above method.

Note that the conclusion is based on the experimental results. To generalize experimental results, a more experimental and analytical study is needed.

Author Contributions: Investigation, methodology, and writing—original draft, J.P.; writing—review and editing, S.H.; project administration and supervision, S.-K.P.

Funding: This research was funded by National Research Foundation of Korea, grant number NRF-2017R1A2B2012535.

Acknowledgments: This work was supported by the National Research Foundation of Korea (NRF) grant funded by the Korea government (MSIP) (No. NRF-2017R1A2B2012535) and by the Basic Science Research Program through the National Research Foundation of Korea (NRF) and funded by the Ministry of Education (2016R1A6A3A11931804).

Conflicts of Interest: The authors declare no conflict of interest.

References

1. American Society of Civil Engineers. *2017 Infrastructure Report Card*, 1st ed.; American Society of Civil Engineers: Reston, VA, USA, 2017; pp. 7–8.

2. Koutas, L.N.; Bournas, D.A. Flexural Strengthening of Two-Way RC Slabs with Textile-Reinforced Mortar: Experimental Investigation and Design Equations. *J. Compos. Constr.* **2017**, *21*, 1–11. [CrossRef]

3. Arduini, M.; Nanni, A. Behavior of Precracked RC Beams Strengthened with Carbon FRP Sheets. *J. Compos. Constr.* **1997**, *1*, 63–70. [CrossRef]

4. Hong, S.; Park, S.K. Effect of prestress levels on flexural and debonding behavior of reinforced concrete beams strengthened with prestressed carbon fiber reinforced polymer plates. *J. Compos. Mater.* **2013**, *47*, 2097–2111. [CrossRef]

5. Hong, S.; Park, S.K. Prestressing effects on the performance of concrete beams with near-surface-mounted carbon-fiber-reinforced polymer bars. *Mech. Compos. Mater.* **2016**, *52*, 305–316. [CrossRef]

6. Brameshuber, W. *Textile Reinforced Concrete—State of the Art Report of RILEM TC 201-TRC*, 1st ed.; RILEM: Aachen, Germany, 2006; pp. 1–271.

7. Babaeidarabad, S.; Loreto, G.; Nanni, A. Flexural Strengthening of RC Beams with an Externally Bonded Fabric-Reinforced Cementitious Matrix. *J. Compos. Constr.* **2014**, *18*, 1–11. [CrossRef]

8. Baiee, A.; Rafiq, I.; Lampropoulos, A. Innovative technique of textile reinforced mortar (TRM) for flexural strengthening of reinforced concrete (RC) beams. In Proceedings of the 2nd International Conference on Structural Safety Under Fire and Blast Loading, London, UK, 10–12 September 2017.

9. Hartig, J.; Häußler-Combe, U.; Schicktanz, K. Influence of bond properties on the tensile behaviour of Textile Reinforced Concrete. *Cem. Concr. Compos.* **2008**, *30*, 898–906. [CrossRef]

10. Colombo, I.G.; Magri, A.; Zani, G.; Colombo, M.; di Prisco, M. Textile Reinforced Concrete: Experimental investigation on design parameters. *Mater. Struct.* **2013**, *46*, 1933–1951. [CrossRef]

11. Banholzer, B.; Brockmann, T.; Brameshuber, W. Material and bonding characteristics for dimensioning and modelling of textile reinforced concrete (TRC) elements. *Mater. Struct.* **2006**, *39*, 749–763. [CrossRef]

12. Williams Portal, N.; Lundgren, K.; Walter, A.M.; Frederiksen, J.O.; Thrane, L.N. Numerical modelling of textile reinforced concrete. In Proceedings of the 8th International Conference on Fracture Mechanics of Concrete and Concrete Structures, FraMCoS 2013, Toledo, Spain, 11–14 March 2013; pp. 886–897.

13. Zastrau, B.; Richter, M.; Lepenies, I. On the Analytical Solution of Pullout Phenomena in Textile Reinforced Concrete. *J. Eng. Mater. Technol.* **2003**, *125*, 38–43. [CrossRef]

14. Iorfida, A.; Verre, S.; Candamano, S.; Ombres, L. Tensile and Direct Shear Responses of Baslat-Fibre Reinforced Mortar Based Materials. In Proceedings of the Strain-Hardening Cement-Based Composites (SHCC4), Deresden, Germany, 18–20 September 2017; RILEM: Aachen, Germany, 2018; pp. 544–552.

15. Papanicolaou, C.G.; Papantoniou, I.C. Mechanical Behavior of Textile Reinforced Concrete (TRC) / Concrete Composite Elements. *J. Adv. Concr. Technol.* **2010**, *8*, 35–47. [CrossRef]

16. Yin, S.P.; Xu, S.L. An Experimental Study on Improved Mechanical Behavior of Textile-Reinforced Concrete. *Adv. Mater. Res.* **2011**, *168–170*, 1850–1853. [CrossRef]

17. Yin, S.; Xu, S.; Lv, H. Flexural Behavior of Reinforced Concrete Beams with TRC Tension Zone Cover. *J. Mater. Civ. Eng.* **2014**, *26*, 320–330. [CrossRef]

18. Yin, S.P.; Xu, S.L.; Wang, F. Investigation on the flexural behavior of concrete members reinforced with epoxy resin-impregnated textiles. *Mater. Struct.* **2015**, *48*, 153–166. [CrossRef]

19. Kamani, R.; Kamali Dolatabadi, M.; Jeddi, A.A.A. Flexural design of textile-reinforced concrete (TRC) using warp-knitted fabric with improving fiber performance index (FPI). *J. Text. Inst.* **2017**, *109*, 492–500. [CrossRef]

20. Liu, L.; Du, Y.; Zhou, F.; Pan, W.; Zhang, X.; Zhu, D. Flexural Behaviour of Carbon Textile-Reinforced Concrete with Prestress and Steel Fibres. *Polymers* **2018**, *10*, 98.

21. Raoof, S.M.; Koutas, L.N.; Bournas, D.A. Textile-reinforced mortar (TRM) versus fibre-reinforced polymers (FRP) in flexural strengthening of RC beams. *Constr. Build. Mater.* **2017**, *151*, 279–291. [CrossRef]

22. Jesse, F.; Weiland, S.; Curbach, M. Flexural strengthening of RC structures with textile-reinforced concrete. *ACI Spec. Publ.* **2008**, *250*, 49–58.

23. Gutierrez, E.; Bono, F. *Review of Industrial Manufacturing Capacity for Fibre-Reinforced Polymers as Prospective Structural Components in Shipping Containers*; Publications Office of the European Union: Luxembourg, 2013; pp. 5–12.

24. Peled, A. Pre-tensioning of fabrics in cement-based composites. *Cem. Concr. Res.* **2007**, *37*, 805–813. [CrossRef]

25. Koutas, L.N.; Tetta, Z.; Bournas, D.A.; Triantafillou, T.C. Strengthening of Concrete Structures with Textile Reinforced Mortars: State-of-the-Art Review. *J. Compos. Constr.* **2019**, *23*, 1–20. [CrossRef]

26. Reinhardt, H.W.; Krüger, M.; Große, C.U. Concrete Prestressed with Textile Fabric. *J. Adv. Concr. Technol.* **2003**, *1*, 231–239. [CrossRef]

27. Du, Y.; Zhang, M.; Zhou, F.; Zhu, D. Experimental study on basalt textile reinforced concrete under uniaxial tensile loading Experimental study on basalt textile reinforced concrete under uniaxial tensile loading. *Constr. Build. Mater.* **2017**, *138*, 88–100. [CrossRef]

28. Hegger, J.; Will, N.; Bruckermann, O.; Voss, S. Load-bearing behaviour and simulation of textile reinforced concrete. *Mater. Struct.* **2006**, *39*, 765–77628. [CrossRef]

Article

Experimental Study on Mechanical Properties and Fractal Dimension of Pore Structure of Basalt–Polypropylene Fiber-Reinforced Concrete

Ditao Niu [1,2,*], Daguan Huang [2], Hao Zheng [2], Li Su [2], Qiang Fu [2,*] and Daming Luo [2]

[1] State Key Laboratory of Green Building in Western China, Xi'an University of Architecture & Technology, Xi'an 710055, China
[2] College of Civil Engineering, Xi'an University of Architecture & Technology, Xi'an 710055, China; hdg0505@163.com (D.H.); niurougan522@163.com (H.Z.); suli9290@outlook.com (L.S.); dmluo@xauat.edu.cn (D.L.)
* Correspondence: niuditao@163.com (D.N.); fuqiangcsu@163.com (Q.F.)

Received: 16 March 2019; Accepted: 16 April 2019; Published: 17 April 2019

Abstract: This study investigates the effects of basalt–polypropylene fibers on the compressive strength and splitting tensile strength of concrete and calculates the fractal dimension of the pore structure of concrete by using a fractal model based on the optical method. Test results reveal that hybrid fibers can improve the compressive strength and splitting tensile strength of concrete, and the synergistic effect of the hybrid fibers is strongest when the contents of basalt fiber (BF) and polypropylene fiber (PF) are 0.05% each, and that the maximum increments in compressive strength and splitting tensile strength are 5.06% and 9.56%, respectively. The effect of hybrid fibers on splitting tensile strength is greater than on compressive strength. However, hybrid fibers have adverse effects on mechanical properties when the fiber content is too high. The pore structure of basalt–polypropylene fiber-reinforced concrete (BPFRC) exhibits obvious fractal characteristics, and the fractal dimension is calculated to be in the range of 2.297–2.482. The fractal dimension has a strong correlation with the air content and spacing factor: the air content decreases significantly whereas the spacing factor increases with increasing fractal dimension. In addition, the fractal dimension also has a strong positive correlation with compressive strength and splitting tensile strength. Therefore, the fractal dimension of the pore structure can be used to evaluate the microscopic pore structure of concrete and can also reflect the influence of the complexity of the pore structure on the macroscopic mechanical properties of concrete.

Keywords: basalt fiber; polypropylene fiber; hybrid fiber-reinforced concrete; mechanical properties; pore structure; fractal dimension

---

1. Introduction

Concrete is widely used in engineering structures because of its low cost, simplicity of preparation, and excellent strength [1]. However, concrete also has disadvantages such as a low tensile strength, poor toughness, and high brittleness, which adversely affect the safety, applicability, and durability of the concrete structure [2]. Many studies have shown that the mechanical properties and durability of concrete can be effectively improved by incorporating fibers into it, thus obtaining fiber-reinforced concrete for high strength, toughness, and durability [3,4].

Fibers can be incorporated into concrete in two ways: incorporation of a single type of fiber, and incorporation of fibers of different types or sizes. Incorporation of a single type of fiber provides limited improvement in concrete performance. When hybrid fibers obtained by mixing fibers of different types or sizes are incorporated into concrete, the hybrid fibers can induce their respective reinforcing effects in different layers and stress stages of concrete, and consequently, the improvement in the concrete performance is more remarkable [5]. Steel–polypropylene fiber-reinforced concrete is currently the most widely used hybrid fiber-reinforced concrete. The elastic modulus and strength of steel fiber (SF) are high, because of which its incorporation into concrete can effectively reduce the brittleness of concrete, and improve its mechanical properties and load-bearing capacity. Polypropylene fiber (PF) has a low elastic modulus and good ductility; as a result, its incorporation into concrete can effectively improve the splitting tensile strength and flexural strength of concrete [6,7]. Badogiannis et al. reported that SF and PF could significantly improve the mechanical properties of pumice concrete and that the maximum increments in the compressive strength and splitting tensile strength were 76% and 110%, respectively [8]. Pajak et al. studied the flexural properties of steel–polypropylene fiber-reinforced concrete and found that the hybrid fibers could effectively improve the mechanical properties and toughness of concrete [9]. Aslani et al. demonstrated that the compressive strength and elasticity modulus of steel–polypropylene fiber-reinforced concrete were higher than those of SF-reinforced concrete and PF-reinforced concrete [10]. However, the chemical composition of SF is the same as that of reinforcing bars; because of this, SF rusts easily in a corrosive environment, which limits the application scope of steel–polypropylene fiber-reinforced concrete [11].

Basalt fiber (BF) is made from basalt as raw material via melting and drawing at high temperatures. BF has advantages such as a high tensile strength, high-temperature resistance, corrosion resistance, and a high elastic modulus. Moreover, it is a type of a green, environmentally-friendly fiber and a good substitute for SF [12,13]. Therefore, basalt–polypropylene fiber-reinforced concrete (BPFRC)—prepared by mixing BF instead of SF with PF—has a wider application range. The mechanical properties of BPFRC have been preliminarily studied by some researchers. Ghazy et al. reported that compressive strength improved significantly when it was incorporated with BPFRC composed of 0.1% BF and 0.1% PF [4]. Wang et al. demonstrated that when the BF and PF contents were 0.15% and 0.033%, respectively, the compressive strength and splitting tensile strength of concrete increased by 14.1% and 48.6%, respectively [14]. Kong et al. investigated the mechanical properties of basalt–polypropylene hybrid fiber-reinforced recycled concrete under high-temperature conditions [15]. They showed that the compressive strength and splitting tensile strength of concrete reinforced with hybrid fibers were higher than those of plain concrete. However, several other studies showed that incorporation of hybrid fibers composed of BF and PF improved the splitting tensile strength of concrete, but lowered its compressive strength [16,17].

3D printing techniques have developed rapidly and been introduced to the field of civil engineering structures. Rapid development of the technique in the construction field depends on the development of high-performance cementitious materials compatible with 3D printers [18]. Some studies have shown that the addition of fibers in the preparation of 3D printing cementitious materials can improve structural performance [19]. Hambach et al. found that when fibers were aligned along the printing path, the flexural strength of the printed structure in the designated direction increases significantly [20]. Panda et al. reported that when the glass fiber content was 1%, the flexural and tensile strength of the printed specimens improved significantly and in an obvious directional dependency, but there was little effect on compressive strength [21]. Ma et al. demonstrated that a cementitious composite containing 0.5% basalt fibers had favorable printability and mechanical properties, and the 3D-printed samples performed obviously mechanical anisotropy [22]. The research on the anisotropic mechanics of 3D-printed fiber-reinforced materials is not adequately comprehensive. Therefore, more experiments should be carried out to promote the practical application of 3D printing techniques in the field of construction.

The structure of composites includes two characteristic scales: macroscopic and microscopic. The macroscopic mechanical behavior of composites is affected by the volume fraction, shape and distribution of the components. Researchers have proposed many mathematical techniques to solve this interaction problem. The common methods include the Self-Consistent Method [23], Generalized Self-Consistent Method [24], Mori–Tanaka Method [25], Asymptotic Homogenization Method [26], and so on. The Asymptotic Homogenization Theory has developed into the main method to solve the equivalent mechanical properties of composite materials in recent years. By establishing the asymptotic displacement field that depends on the change of two-scale coordinate variables, the governing equation reflecting the microstructure is derived, and the average material properties are obtained [27]. The pore structure of concrete is an important aspect of its microstructure, which directly affects mechanical properties, frost resistance, and other macroscopic properties of concrete. Because of the increasing complexity of high-performance concrete, conventional parameters cannot effectively describe the complexity of the pore structure quantitatively, which hinders the study of the relationship between the pore structure and the macroscopic properties of concrete [28]. Fractal theory, as a new approach for describing the complexity and irregularity of matter, has been introduced and gradually applied to the study of pore structure, wherein it quantifies the complexity of the micropore structure as a fractal dimension. The relationship between the microstructure and macroscopic properties of concrete can be studied by utilizing the fractal dimension. In recent years, with advances in research, fractal models of multiple pore structures based on various pore structure characterization methods have been established, and the relationship between fractal dimension and macroscopic performance has been discussed [29–31]. Jin et al. established a model of the relationship between the fractal dimension and compressive strength. Their results revealed that the fractal dimension could well-characterize the relationship between the micropore structure and the macroscopic mechanical properties [30]. Cui et al. revealed that the fractal dimension obtained by a thermodynamics method could describe the pore size distribution of concrete better than that obtained using the Menger sponge model, and that there was a strong positive correlation between the mechanical properties of concrete and the fractal dimension [31]. Zhao et al. calculated the fractal dimension of BF-reinforced concrete by using a fractal model based on an optical method; their results revealed that with an increasing fractal dimension, compressive strength increased and air content decreased [32]. Yu et al. tested the pore structure and mechanical properties of perlite cement stone, but their results revealed that with an increasing fractal dimension, compressive strength decreased and porosity increased [33]. This finding is consistent with the results reported in other literature [34].

From a review of the aforementioned studies, it is clear that research conducted on the mechanical properties of BPFRC is not comprehensive enough and that the research conclusions are insufficiently unified, which hinders the application and development of BPFRC to a certain extent. Because of the different testing methods of the pore structure and different interpretations of the physical meaning of the fractal dimension, there is a large gap in existing research results. Research on the pore structure of concrete and its mechanical properties needs to be improved. Therefore, it is necessary to further study the fractal characteristics of the pore structure. The main objective of this work is to investigate the mechanical properties of concrete reinforced with basalt–polypropylene hybrid fibers and to determine the optimal contents of BF and PF in order to significantly improve the compressive strength and splitting tensile strength of concrete. The fractal dimension of the pore structure is calculated using a fractal model based on an optical method. In addition, the relationship of the fractal dimension with the mechanical properties of concrete and other parameters of the pore structure is also investigated.

2. Materials and Test Methods

2.1. Materials

P.O 42.5R Portland cement (OPC), granulated blast furnace slag powder (BFS), silica fume (SF) and fly ash (FA) were used to prepare test specimens. The physical properties and chemical composition of the cementitious material are listed in Table 1. BF and PF are shown in Figure 1a,b, respectively, and their physical and mechanical properties are listed in Table 2. The coarse aggregate (CA), with 5–20 mm continuous gradation, consisted of gravel sourced from the Shaanxi Jingyang mountain. The fine aggregate (S) used was medium sand with a fineness modulus of 2.8. Tap water (W) was used for mixing. A polycarboxylate superplasticizer (PBS) was used to achieve a water-reducing rate of 30%.

(**a**)

(**b**)

Figure 1. (**a**) Basalt fiber. (**b**) Polypropylene fiber.

Table 1. Chemical composition and physical properties of cementitious materials.

Item	OPC	BFS	SF	FA
SiO_2 (%)	21.18	34.65	85.04	35.71
Al_2O_3 (%)	5.02	14.21	0.97	16.57
Fe_2O_3 (%)	3.14	0.49	1.04	8.92
CaO (%)	63.42	34.11	1.63	21.14
MgO (%)	3.12	11.15	0.32	1.41
SO_3 (%)	2.30	1.00	-	1.94
Other	1.82	3.74	10	12.49
Loss of ignition (%)	2.79	0.3	5.48	2.85
Density (g/cm^3)	3.10	2.86	2.1	2.35

Table 2. Physical and mechanical properties of fibers.

Item	Length (mm)	Diameter (μm)	Density (g/cm^3)	Elastic Modulus (GPa)	Tensile Strength (GPa)	Elongation (%)
BF	18	15	2.56	75	4.5	3.15
PF	19	30	0.91	3	0.27	40

2.2. Mix Proportions

The details of all eight mixtures used in this study are presented in Table 3. The eight mixtures had the following fiber contents: 0%, 0.10% (0.05% BF + 0.05% PF), 0.15% (0.05% BF + 0.10% PF), 0.15% (0.10% BF + 0.05% PF), 0.20% (0.05% BF + 0.15% PF), 0.20% (0. 15% BF + 0. 05% PF), 0.20% (0.10% BF+0.10% PF), and 0.30% (0.15% BF + 0.15% PF) (all contents by volume of concrete). These mixtures were termed BF0PF0, BF5PF5, BF5PF10, BF10PF5, BF5PF15, BF15PF5, BF10PF10, and BF15PF15, respectively. For all the mixtures, the water–binder ratio was 0.38 and the concrete constituents excluding the fibers were the same.

Table 3. Mix proportions of concrete (kg/m^3).

Specimen	OPC	SF	FA	BFS	PBS	Water	Sand	CA	BF Volume Fraction (%)	PF Volume Fraction (%)
BF0PF0	241.6	15.8	79.2	59.4	3.96	150.5	683.4	1163.6	0.0	0.0
BF5PF5	241.6	15.8	79.2	59.4	3.96	150.5	683.4	1163.6	0.05	0.05
BF5PF10	241.6	15.8	79.2	59.4	3.96	150.5	683.4	1163.6	0.05	0.10
BF5PF15	241.6	15.8	79.2	59.4	3.96	150.5	683.4	1163.6	0.05	0.15
BF10PF5	241.6	15.8	79.2	59.4	3.96	150.5	683.4	1163.6	0.10	0.05
BF15PF5	241.6	15.8	79.2	59.4	3.96	150.5	683.4	1163.6	0.15	0.05
BF10PF10	241.6	15.8	79.2	59.4	3.96	150.5	683.4	1163.6	0.10	0.10
BF15PF15	241.6	15.8	79.2	59.4	3.96	150.5	683.4	1163.6	0.15	0.15

2.3. Test Method

2.3.1. Mechanical Properties

The compressive strength and splitting tensile strength of all specimens (100 mm × 100 mm × 100 mm) were tested according to GB/T 50081-2002 [35]. The specimens were tested after 28 days of curing under standard curing conditions with a constant temperature of 20 ± 2 °C and a relative humidity greater than 95%. Three specimens of each mixture were tested, and their average value was taken as the final value of the strength.

2.3.2. Pore Structure

The linear traverse method was used to determine the pore characteristic parameters of concrete according to ASTM C 457-9. The pore size distribution of the specimen was measured with a RapidAir 457 air void analyzer (as shown in Figure 2a). Through image recognition, this instrument can measure pore characteristics in hardened concrete by itself, reducing the manual error in testing, and improving measurement speed and accuracy.

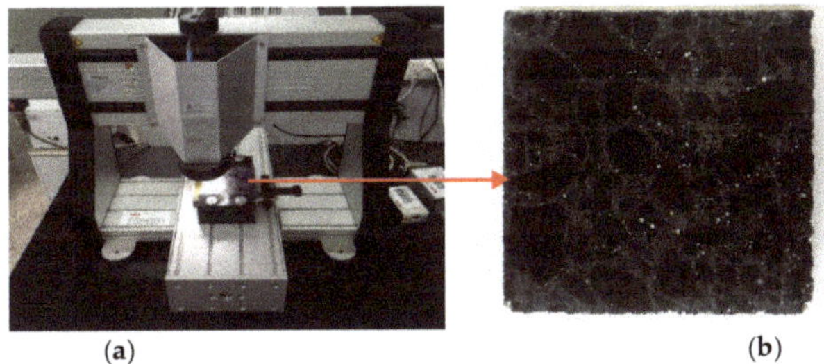

(a)   (b)

Figure 2. (**a**) Test instrument. (**b**) Test specimen.

The specimens were prepared according to the following procedure. First, all the original specimens were cut after 28 days of standard curing to obtain a specimen with dimensions of 100 mm × 100 mm × 20 mm, without any obvious saw marks on the specimen surface. Second, the specimen was ground with silicon carbide grinding fluid on a lapping machine. Three different grit sizes were used (in the listed order): 320 grit, 600 grit, and 800 grit. Third, the surface of the ground specimen was smeared with carbon black and then zinc paste, and the specimen was heated to 80 °C. When the zinc paste had enough fluidity, it was evenly smeared on the surface of the specimen. Finally, the zinc paste was removed from the specimen surface after cooling. The resultant specimen is shown in Figure 2b.

The test steps are as follows: (1) place the specimen on the sample holder of RapidAir 457 air void analyzer; (2) start the test software, input the basic information of the sample, and make sure that the size of the air voids on the "Analysis Image" have the same size as on the live "Raw Image" by adjusting the focus, lighting and threshold; (3) set the test area of the sample to 80 mm × 80 mm, and then start the test.

3. Fractal Model Based on Optical Method

The contour of the pore section of concrete is complicated and has obvious fractal features. Previous studies have shown that a fractal model based on an optical method is constructed mainly around the solution of the fractal dimension of the pore section contour, e.g., by the perimeter-area method [36]. The fractal model of the concrete pore section is established according to the principle of the perimeter-area method, and the calculation formula is given as follows:

$$\lg P = 0.5 D_p \lg A + C \tag{1}$$

where P is the perimeter, A is the area, D_p is the fractal dimension of pore section contour, C is a constant.

Hu and Tang used the above-described model to study the fractal dimension of concrete [37,38]. Their results demonstrated that the fractal dimension obtained by this model could effectively describe the roughness of concrete, but the correlation between the fractal dimension and the macro-performance of concrete was very low. Therefore, W. Hu improved the model and proposed a new fractal dimension for calculating the pore size distribution [26]; the calculation formula is given as follows:

$$\lg n = D_d \lg d + C \tag{2}$$

where d is the diameter of a pore, n is the number of pores with diameter larger than or equal to d, D_d is the fractal dimension of the pore size distribution, and C is a constant.

The fractal dimension can effectively describe the pore size distribution of concrete by analyzing the experimental data, and it has a good linear relationship with the compressive strength of concrete. According to the test method and the characteristics of the test data, Zhang et al. introduced the box dimension into the previous model. The box dimension is determined by the covering of the same shape set, and it is relatively simple to calculate [39]. The mathematical expression of the box dimension is given as follows: F is an arbitrary nonempty subset of R^n and $N_\delta(F)$ is the minimum number of sets that have a maximum diameter of δ and that can cover set F; then, the box dimension of F can be calculated by formula (3):

$$Dim_B F = \lim_{\delta \to 0} \frac{\lg N_\delta(F)}{\lg(1/\delta)} \tag{3}$$

The pore structure analyzer for hardened concrete measures the number of circular pores at a given circular degree under the assumption that the tested pore in concrete is a regular circular air-hole, and n circular box are selected for measurement according to the definition of the box dimensions. These boxes are used to cover pores whose diameter is greater than or equal to d_i. According to the principle of equal area, the pore with a diameter greater than d_i is transformed into that with the diameter of d_i, and the number of converted pores, N_{ci}, with diameter d_i is obtained. As a result, the data set (d_1, N_{c1}), (d_2, N_{c2}), (d_3, N_{c3}), ... , (d_i, N_{ci}) is obtained. Finally, data on the pore diameter and the number of converted pores plotted on the double-logarithmic axis are subjected to linear regression analysis, and the slope of the regression line is the box dimension. The corresponding mathematical expression is given in Equation (4):

$$\lg N_c = -D_d \cdot \lg d + C \tag{4}$$

where N_c is the number of converted pores, D_d is the fractal dimension of the pore size distribution, d is the pore diameter, and C is a constant.

4. Experimental Results and Discussion

4.1. Mechanical Properties

The measurement results of the compressive strength and splitting tensile strength of concrete incorporated with mixtures of different fiber contents are shown in Figure 3a,b, respectively. Figure 3a shows that compressive strength can be improved by the incorporation of an appropriate amount of hybrid fibers. The compressive strength of BF5PF5 is the highest, 44.43 MPa, which is 5.06% higher than that of the reference concrete (BF0PF0). However, with an increase in the fiber content, compressive strength becomes lower than that of the reference concrete; BF5PF15 shows the lowest compressive strength, which is 22.63% lower than that of the reference concrete. These findings are in agreement with the results reported by other researchers [40,41]. Sadrinejad et al. demonstrated that the compressive strength and splitting tensile strength of concrete are improved when the content of hybrid fibers was lower than 0.1%, but degraded when the content of hybrid fibers was higher than 0.1% [40]. The incorporation of BF and PF into concrete can reduce microcracks and prevent the expansion of macrocracks, which consequently improves the internal structure of the concrete.

The transverse deformation of concrete under compression is restrained, and the fibers can bond strongly to the matrix. When stress is transferred from the matrix to the fibers, the fibers consume energy because of the deformation, which consequently delays the destruction process of the concrete; as a result, the compressive strength of the concrete improves. However, when the fiber content is too high, the fibers are not uniformly dispersed, and they overlap and agglomerate in the matrix, as shown in Figure 4a; this consequently worsens the bonding effect between the fibers and the matrix [42]. As a result, zones with poor cohesion and weak structures are formed that ultimately nullify the reinforcing effect of the fibers on compressive strength, and this leads to a reduction in compressive strength. Therefore, accurate control of the fiber content of concrete is necessary.

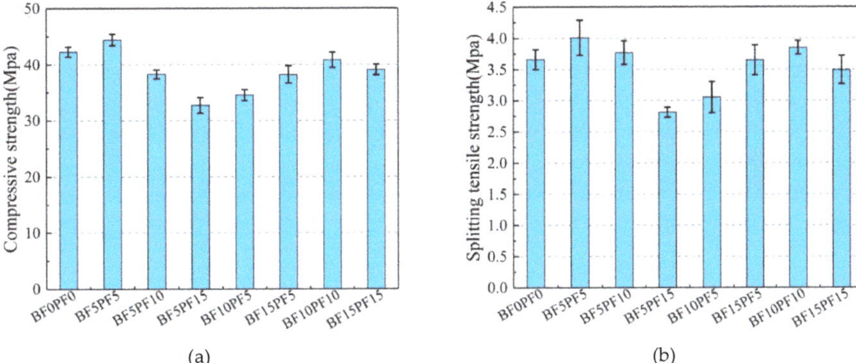

(a)   (b)

Figure 3. The compressive strength and splitting tensile strength of different mixtures. (**a**) Compressive strength. (**b**) Splitting tensile strength.

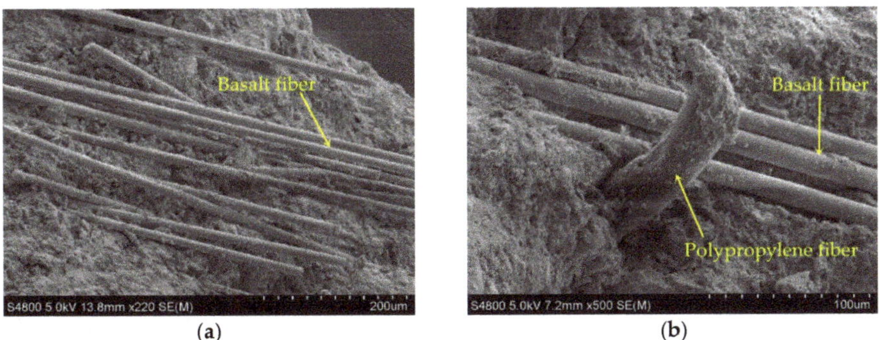

(a)   (b)

Figure 4. Scanning electron microscopy (SEM) images of fibers. (**a**) Distribution of BF in basalt–polypropylene fiber-reinforced concrete (BPFRC). (**b**) Distribution of BF and PF in BPFRC.

As observed in Figure 3b, the improvement in splitting tensile strength shows a trend similar to that of compressive strength. The test results show that hybrid fibers can effectively improve the splitting tensile strength. The splitting tensile strength of BF5PF5 is the highest, 4.01 MPa, which is 9.56% higher than that of the reference concrete. When the BF content is fixed at 0.05%, splitting tensile strength decreases rapidly with an increasing PF content. BF5PF15 has the lowest splitting tensile strength, which is 23.22% lower than that of the reference concrete. When the content ratio of BF to PF is 1:1, and the total fiber content is higher than 0.2%, the splitting tensile strength is lower than that of the reference concrete. This result is in agreement with previously reported results [4,43]. BF and PF are evenly dispersed or intertwined in the concrete matrix to form a three-dimensional support network, as shown in Figure 4b. When concrete is subjected to loading, the fibers intersect the cracks transfer the load to the upper and lower surfaces of the cracks, and therefore, the cracks can continue to bear the load. In addition, the fibers can also share part of the tensile force on the section, reduce the stress concentration factor at the micro-cracks in concrete, increase the ultimate tensile strain of concrete, and prevent the formation and propagation of cracks, all of which lead to an improvement in splitting tensile strength. However, when the fiber content is too high, the total surface area of the fibers increases, and a larger amount of cement paste needs to be applied, which adversely affects not only the bonding between the cement paste and the aggregate, but also the splitting tensile strength. The splitting tensile strengths of BF15PF5 and BF10PF10 were higher than that of BF5PF15 when the total fiber content is 0.2%. In general, the results illustrate that the effect of BF on the splitting tensile strength is greater than that of PF [12,14]. This is may be because the elastic modulus and tensile strength of BF are much higher than those of PF. When the microcracks expand further, PFs are pulled out or broken, and then, the tensile stress is borne mainly by the BFs. Therefore, BF plays a major role in the intermediate and later stages of fracture development.

Results of the analysis of variance are presented in Table 4, which indicate whether or not there are significant differences in the compressive strength and splitting tensile strength between the BPFRC and reference concrete. According to the significance level of 0.05, when the significance coefficient of BPFRC is less than or equal to 0.05, there is a significant difference in the strength between BPFRC and the reference concrete. Otherwise, there is no significant difference. Therefore, when the BF and PF contents are 0.05% each, the compressive strength and splitting tensile strength are not significantly improved. However, when the fiber contents are (0.05% BF + 0.15% PF) and (0.10% BF + 0.05% PF), the compressive strength and splitting tensile strength are significantly reduced and there is a significant difference between BPFRC and the reference concrete.

Table 4. Compressive strength and splitting tension strength results of different mixtures.

Specimen	Compressive Strength (MPa)			Splitting Tensile Strength (MPa)		
	Mean	SD.	Sig.	Mean	SD.	Sig.
BF0PF0	42.29	0.88	-	3.66	0.16	-
BF5PF5	44.43	1.02	0.066	4.01	0.28	0.199
BF5PF10	38.31	0.77	0.022	3.77	0.19	0.561
BF5PF15	32.72	1.37	0.003	2.81	0.08	0.002
BF10PF5	34.51	0.99	0.009	3.05	0.25	0.045
BF15PF5	38.21	1.57	0.021	3.65	0.24	1.000
BF10PF10	40.81	1.35	0.447	3.85	0.11	0.237
BF15PF15	39.07	0.93	0.113	3.49	0.23	0.441

Mean: the strength of concrete specimen (MPa); SD.: the standard deviations of the measured values; Sig.: the mean difference is significant at the 0.05 level.

4.2. Fractal Dimension of Hardened Concrete

The fractal dimension of the pore structure of BPFRC was calculated using the fractal dimension models of Hu [37] and Zhang [39]; the calculation results are shown in Figures 5 and 6, respectively. Figure 5 reveals a good correlation between the logarithmic values of the pore diameter and the number of pores, and the values of the correlation coefficient R^2 of BF5PF10 and BF5PF15 are 0.866 and 0.890, respectively. However, the graph shows an obvious curvilinear relationship, indicating that the model cannot accurately reflect the fractal characteristics of the pore size distribution of BPFRC. Figure 6 shows a double-logarithmic scatter plot of the number of converted pores versus the pore diameter; the figure reveals that the correlation between $\lg N_c$ and $\lg d$ is very high, and the values of the correlation coefficient R^2 of BF5PF10 and BF5PF15 are 0.990 and 0.984, respectively. The graph obtained using Zhang's fractal dimension model shows a more obvious linear relationship than that obtained using Hu's fractal dimension model, which indicates that the fractal features are more obvious when Zhang's model is used and that this model is more reliable. The advantage of Zhang's fractal model is that the definition of the box dimension is used, and the concept of "number of converted pores" is introduced in the process of constructing the fractal model. Therefore, the fractal dimension calculated using this model is more accurate and reliable, and it can more accurately reflect the fractal characteristics of the pore size distribution. Therefore, in this study, the fractal dimension of BPFRC was calculated using Zhang's fractal model.

The pore structure of concrete can be improved by incorporating fibers into concrete, which affects the fractal dimension of the pore structure [31]. The fractal dimensions of the BPFRC mixtures are listed in Table 5. It can be seen from this table that when the BF content is fixed, the fractal dimension decreases with an increasing PF content; the fractal dimension of BF5PF5 is the largest, 2.482. The fractal dimension of BF5PF15 is the smallest; 7.69% smaller than that of BF5PF5. When the PF content is fixed, the fractal dimension first decreases and then increases with an increasing BF content; the fractal dimension of BF10PF5 is the smallest, 6.16% smaller than that of BF5PF5. The variation in the fractal dimension is very small when the total fiber content is higher than 0.2%.

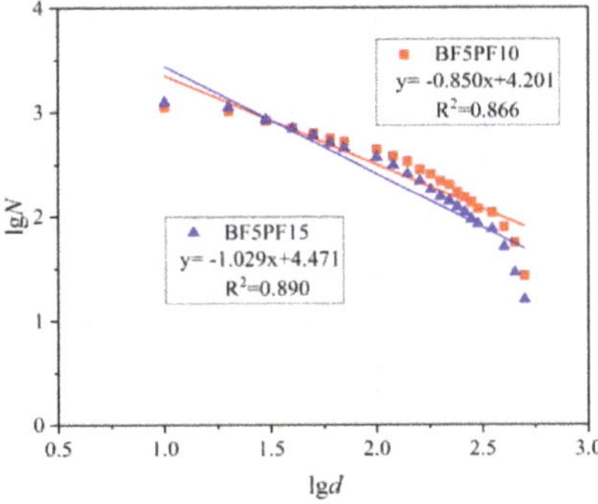

Figure 5. Double logarithmic scatter plot of cumulative number with pore and pore diameter.

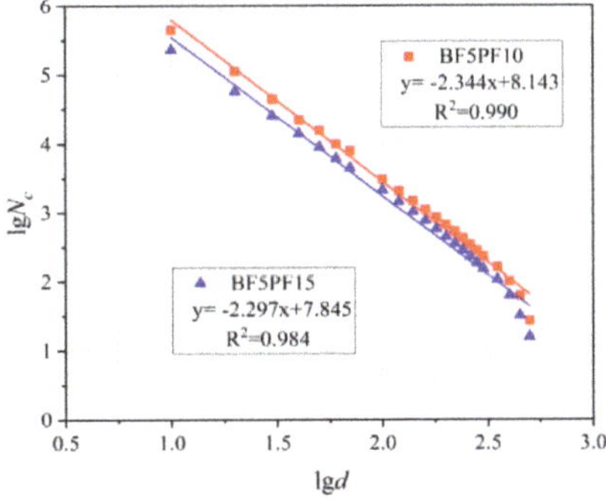

Figure 6. Double logarithmic scatter plot of conversions number with pore and diameter.

Table 5. Results of the fractal dimension.

Specimen	Fractal Dimension	R^2
BF0PF0	2.381	0.979
BF5PF5	2.482	0.977
BF5PF10	2.344	0.990
BF5PF15	2.297	0.984
BF10PF5	2.329	0.986
BF15PF5	2.375	0.980
BF10PF10	2.372	0.982
BF15PF15	2.379	0.977

4.3. Relationship Between Fractal Dimension and Mechanical Properties

The strength of concrete is one of its most important macroscopic performance indexes. The strength is closely related to pore structure parameters such as the air content, porosity and spacing factor [31]. The fractal dimension can synthetically characterize the micropore structure of concrete under certain conditions; therefore, the fractal dimension may have a certain relationship with the strength of concrete [30]. Figure 7 show results of the correlation analysis of the fractal dimension with compressive strength and splitting tensile strength. It can be seen that the fractal dimension has a good correlation with both compressive strength and splitting tensile strength, and the corresponding correlation coefficients are 0.883 and 0.777, respectively. This demonstrates that the fractal dimension as a pore size distribution parameter can well-describe the relationship between pore structure and the mechanical properties of BPFRC. The compressive strength and splitting tensile strength of BPFRC show the same changing trend with an increase in the fractal dimension. That is, compressive strength and splitting tensile strength show an obvious increasing trend with increasing fractal dimension. According to fractal theory, the larger the fractal dimension, the more complex the spatial distribution of pores in concrete and the stronger is the pores' ability to occupy space [44]. Therefore, when the specimens are subjected to stress, the internal stress is distributed evenly, which prevents concentration of premature stress and improves compressive strength and splitting tensile strength; this indicates that the complexity of the pore structure is an important factor affecting the macroscopic performance of concrete. The equation in Figure 7 expresses the relationship between the macroscopic mechanical properties and the micropore structure index and the mode of interaction. This finding is consistent with results reported in the literature [30,31,44].

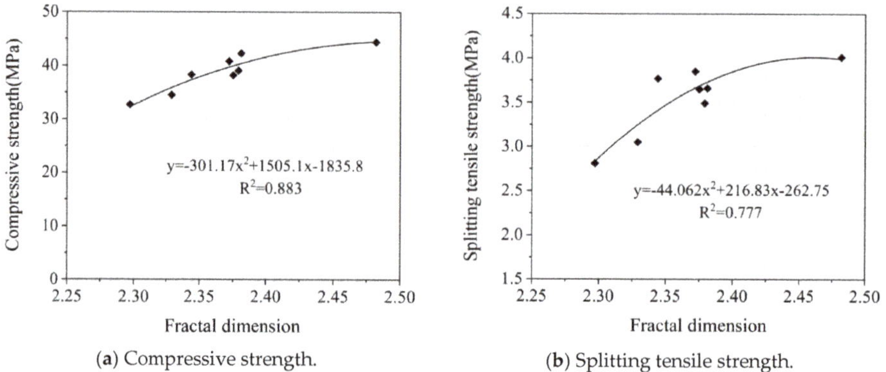

(**a**) Compressive strength.   (**b**) Splitting tensile strength.

Figure 7. Fitting of strength and fractal dimension.

4.4. Relationship Between Fractal Dimension and Pore Structure Parameters

The variations in the air content and spacing factor of BPFRC with the fractal dimension are shown in Figures 8 and 9, respectively. Air content refers to the ratio of the pore volume in mortar to the total volume of concrete. As can be seen from Figure 8, a good correlation exists between air content and the fractal dimension, and the correlation coefficient is 0.775; air content shows a decreasing trend with increasing fractal dimension. Therefore, under certain conditions, the relative air content can be deduced by a comparison of the fractal dimensions of concrete incorporated with different mixtures. Spacing factor refers to the maximum distance between any point in concrete and any adjacent pore sphere. Figure 9 illustrates the relationship between the fractal dimension and the spacing factor. It can be seen that the spacing factor and fractal dimension also have a good correlation ($R^2 = 0.635$); however, the spacing factor shows an increasing trend with increasing fractal dimension. These results are in agreement with those reported by Liu et al. [45]. It can be noted from our results that a hybrid mixture of BF and PF can improve the pore structure of concrete. When the fractal dimension increases, the air content decreases whereas the spacing factor increases, which means that both the pore size and the number of pores decrease; that is, the pore structure is refined and optimized. Results of previous works have shown that the air content and spacing factor are the main factors affecting the frost resistance of concrete [46]. Powers also estimated the relationship between the air content and the spacing factor and proposed the critical values of the air content and spacing factor to ensure the frost resistance of concrete [47]. In order to meet the requirement of high frost resistance of concrete, many countries have proposed a recommended value of the air content, which is generally 3–6%; furthermore, studies have recommended that the spacing factor be no larger than 0.25 mm [2]. From the above analysis, it can be seen that the fractal dimension has a strong correlation with the air content and spacing factor, and therefore, the fractal dimension can be used to evaluate pore structure characteristics comprehensively. Therefore, the fractal dimension can be used to predict frost resistance of concrete more effectively.

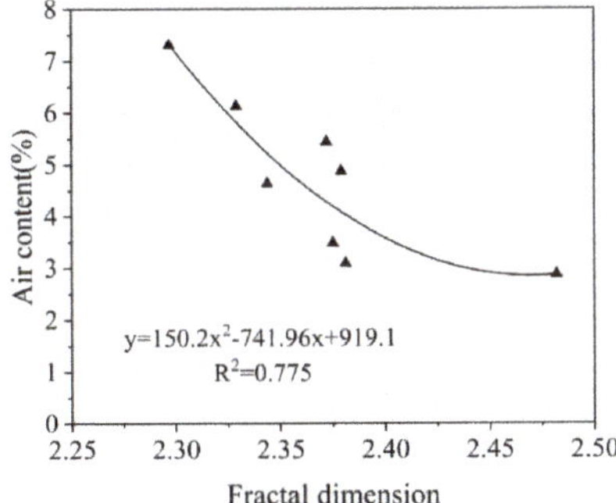

Figure 8. Fitting of air content and fractal dimension.

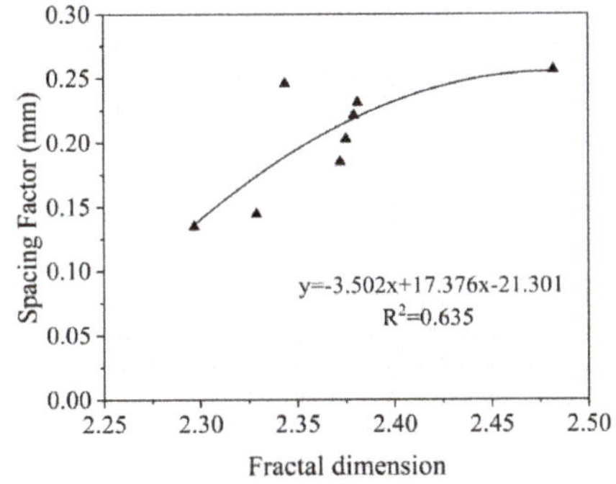

Figure 9. Fitting of spacing factor and fractal dimension.

5. Conclusions

In this study, the mechanical properties and pore structure of BPFRC are evaluated. The following conclusions can be drawn from the study.

1. Incorporation of a hybrid mixture of BF and PF into concrete has both positive and adverse effects on the mechanical properties of concrete. The synergistic effect of the hybrid fibers is greatest when the BF and PF contents are 0.05% each; the corresponding increments in the compressive strength and splitting tensile strength 5.06% and 9.56%, respectively. The effect of the hybrid fibers on the splitting tensile strength is greater than that on compressive strength. However, when the fiber content is too high, the hybrid fibers have adverse effects on the mechanical properties. Therefore, accurate control of the fiber content of concrete is necessary.

2. The pore structure of BPFRC exhibits fractal characteristics. The fractal dimension of the pore structure calculated using a fractal model based on an optical method is in the range of 2.297–2.482, with a high correlation coefficient ($R^2 > 0.977$); this indicates that the fractal dimension calculated using this model can well-characterize the pore size distribution characteristics of concrete.

3. The fractal dimension of BPFRC is closely related to the air content and spacing factor. As the fractal dimension increases, the air content decreases and the spacing factor increases. Therefore, the pore structure characteristics of BPFRC can be evaluated comprehensively using the fractal dimension. In addition, the fractal dimension has a strong positive correlation with the compressive strength and splitting tensile strength of concrete. That is, the larger the fractal dimension, the higher the compressive strength and splitting tensile strength. This indicates that the complexity of the pore structure is an important factor affecting the macroscopic mechanical properties of concrete.

Author Contributions: D.N. and D.H. conceived the main concept and contributed the analysis; D.H., L.S., and H.Z. conducted experiments; Q.F. and D.L. helped analyzing experimental data; D.H. wrote the paper.

Funding: This research was funded by National Natural Science Foundation of China (Grant Nos. 51590914, 51608432, 51808438), and the APC was funded by National Natural Science Foundation of China (Grant No. 51590914).

Acknowledgments: This study is financially supported by National Natural Science Foundation of China (No. 51590914, No. 51608432, and No. 51808438).

References

1. Jin, W.L.; Zhao, Y.X. *Durability of Concrete Structures*, 2nd ed.; Science Press: Beijing, China, 2014; pp. 13–17.
2. Li, Y.; Zhao, W. *Resistance Crack Toughening and Durability of Hybrid Fiber Concrete*, 1st ed.; Science Press: Beijing, China, 2012; pp. 1–6.
3. Xue, M.K. Research of Mechanical Properties for Basalt and Polypropylene Mixed Fiber Concrete. Master's Thesis, Anhui University of Science and Technology, Huainan, China, 2018.
4. Ghazy, A.; Bassuoni, M.T.; Maguire, E.; Oloan, M. Properties of Fiber-Reinforced Mortars Incorporating Nano-Silica. *Fibers* **2016**, *4*, 6. [CrossRef]
5. Afroughsabet, V.; Biolzi, L.; Ozbakkaloglu, T. High-Performance Fiber-Reinforced Concrete: A Review. *J. Mater. Sci.* **2016**, *51*, 6517–6551. [CrossRef]
6. Tadepalli, P.R.; Mo, Y.L.; Hsu, T.T.C. Mechanical Properties of Steel Fibre Concrete. *Mag. Concr. Res.* **2013**, *65*, 462–474. [CrossRef]
7. Sun, Z.Z.; Xu, Q.W. Microscopic, Physical and Mechanical Analysis of Polypropylene Fiber Reinforced Concrete. *Mater. Sci. Eng. A* **2009**, *527*, 198–204. [CrossRef]
8. Badogiannis, E.G.; Christidis, K.I.; Tzanetatos, G.E. Evaluation of the Mechanical Behavior of Pumice Lightweight Concrete Reinforced with Steel and Polypropylene Fibers. *Constr. Build. Mater.* **2019**, *196*, 443–456. [CrossRef]
9. Pajak, M. Investigation on Flexural Properties of Hybrid Fibre Reinforced Self-Compacting Concrete. *Procedia Eng.* **2016**, *161*, 121–126. [CrossRef]
10. Aslani, F.; Nejadi, S. Self-Compacting Concrete Incorporating Steel and Polypropylene Fibers: Compressive and Tensile Strengths, Moduli of Elasticity and Rupture, Compressive Stress-Strain Curve, and Energy Dissipated Under Compression. *Compos. Part B Eng.* **2013**, *53*, 121–133. [CrossRef]
11. Li, C.R.; Wang, X.Z.; Liu, H.X.; Hu, K.X.; Li, G. Research Progress of Hybrid Fiber Reinforced Concrete. *J. Mater. Sci Eng.* **2018**, *36*, 504–510.
12. Jiang, C.H.; Fan, K.; Wu, F.; Chen, D. Experimental Study on the Mechanical Properties and Microstructure of Chopped Basalt Fibre Reinforced Concrete. *Mater. Des.* **2014**, *58*, 187–193.
13. Fiore, V.; Scalici, T.; Bella, G.D.; Valenza, A. A Review on Basalt Fibre and its Composites. *Compos. Part B Eng.* **2015**, *74*, 74–94. [CrossRef]
14. Wang, D.H.; Ju, Y.Z.; Shen, H.; Xu, L.B. Mechanical Properties of High Performance Concrete Reinforced with Basalt Fiber and Polypropylene Fiber. *Constr. Build. Mater.* **2019**, *197*, 464–473. [CrossRef]

15. Kong, X.Q.; Yuan, S.L.; Dong, J.K.; Gang, J.M.; Zhang, W.J. Experimental Study on Performance of Polypropylene-Basalt Hybrid Fiber Reinforced Recycled Aggregate Concrete after Exposure to Elevated Temperatures. *Sci. Technol. Eng.* **2018**, *18*, 101–106.

16. Zhao, B.B.; He, J.J.; Wang, X.Z.; Zheng, S.W. Experimental Study on Mechanical Properties of Concrete with Basalt-Polypropylene Hybrid Fiber. *China Concr. Cem. Prod.* **2014**, *8*, 51–55.

17. He, J.J.; Shi, J.P.; Wang, X.Z.; Han, T.L. Effect of Hybrid Effect on the Mechanical Properties of Hybrid Fiber Reinforced Concrete. *Fiber Reinf Plast Compos.* **2016**, *9*, 26–31.

18. Wu, P.; Wang, J.; Wang, X.Y. A Critical Review of the Use of 3-D Printing in the Construction Industry. *Autom. Constr.* **2016**, *68*, 21–31. [CrossRef]

19. Christ, S.; Schnabel, M.; Vorndran, E.; Groll, J.; Gbureck, U. Fiber Reinforcement during 3D Printing. *Mater. Lett.* **2015**, *139*, 165–168. [CrossRef]

20. Hambach, M.; Volkmer, D. Properties of 3D-Printed Fiber-Reinforced Portland Cement Paste. *Cem. Concr. Compos.* **2017**, *79*, 62–70. [CrossRef]

21. Panda, B.; Paul, S.C.; Tan, M.J. Anisotropic Mechanical Performance of 3D Printed Fiber Reinforced Sustainable Construction Material. *Mater. Lett.* **2018**, *184*, 1005–1010. [CrossRef]

22. Ma, G.W.; Li, Z.J.; Wang, L.; Wang, F.; Sanjayan, J. Mechanical Anisotropy of Aligned Fiber Reinforced Composite for Extrusion-Based 3D Printing. *Constr. Build. Mater.* **2019**, *202*, 770–783. [CrossRef]

23. Yousfi, A.; Fréour, S.; Jacquemin, F. Eshelby-Kröner Viscoelastic Self-Consistent Model: Multi-Scale Behavior of Polymer Composites Under Creep Loading. *Adv. Mater. Res.* **2013**, *682*, 105–112. [CrossRef]

24. Xiao, J.H.; Xu, Y.L.; Zhang, F.C. A Generalized Self-Consistent Method for Nano Composites Accounting for Fiber Section Shape under Antiplane Shear. *Mech. Mater.* **2015**, *81*, 94–100. [CrossRef]

25. Mori, T.; Tanaka, K. Average Stress in Matrix and Average Elastic Energy of Materials with Misfitting Inclusions. *Acta Metall.* **1973**, *21*, 571–574. [CrossRef]

26. Lukkassen, D.; Persson, L.-E.; Wall, P. Some Engineering and Mathematical Aspects on the Homogenization Method. *Compos. Eng.* **1995**, *5*, 519–531. [CrossRef]

27. Cai, Y.W. Asymptotic Homogenization of Periodic Plate and Micro-Structural Optimization. Ph.D. Thesis, Dalian University of Technology, Dalian, China, 2014.

28. Lian, C.; Zhuge, Y.; Beecham, S. The Relationship between Porosity and Strength for Porous Concrete. *Constr. Build. Mater.* **2011**, *25*, 4294–4298. [CrossRef]

29. Arandigoyen, M.; Alvarze, J.I. Pore structure and Mechanical Properties of Cement–Lime Mortars. *Cem. Concr. Res.* **2007**, *37*, 767–775. [CrossRef]

30. Jin, S.S.; Zhang, J.X.; Han, S. Fractal Analysis of Relation between Strength and Pore Structure of Hardened Mortal. *Constr. Build. Mater.* **2017**, *135*, 1–7. [CrossRef]

31. Cui, S.A.; Liu, P.; Cui, E.Q.; Su, J.; Huang, B. Experimental Study on Mechanical Property and Pore Structure of Concrete for Shotcrete use in a Hot-Dry Environment of High Geothermal Tunnels. *Constr. Build. Mater.* **2018**, *173*, 124–135. [CrossRef]

32. Zhao, Y.R.; Guo, Z.L.; Fan, X.Q.; Shi, J.N.; Wang, L. Basalt Fiber Reinforced Concrete Stress-Strain Relationship and Pore Structure Analysis. *Bull. Chin. Ceram. Soc.* **2017**, *36*, 4142–4150.

33. Yu, L.H.; Ou, H.; Duan, Q.P. Research on Pore Volume Fractal Dimension and its Relation to Pore Structure and Strength in Cement Paste with Perlite Admixture. *J. Mater. Sci. Eng.* **2007**, *25*, 201–204.

34. Ji, X.; Chan, S.Y.N.; Feng, N. Fractal Model for Simulating the Space-Filling Process of Cement Hydrates and Fractal Dimensions of Pore Structure of Cement-Based Materials. *Cem. Concr. Res.* **1997**, *27*, 1691–1699. [CrossRef]

35. Chinese National Standards. *GB/T 50081-2002, Standard for Method of Mechanical Properties on Ordinary Concrete*, 1st ed.; China Architecture and Building Press: Beijing, China, 2003; pp. 12–24.

36. Chu, W.Y. *Fractals in Materials Science*, 1st ed.; Chemical Industry Press: Beijing, China, 2004; pp. 3–32.

37. Hu, W. Modeling the Influence of Composition and Pore Structure on Mechanical Properties of Autoclaved Cellular Concrete. Ph.D. Thesis, University of Pittsburgh, Pittsburgh, PA, USA, 1997.

38. Tang, M. Study on Fractal Characteristics and Application of Concrete Materials. Ph.D. Thesis, Harbin Institute of Technology, Harbin, China, 2003.

39. Zhang, J.X.; Jin, S.S. *Micropore Structure of Cement Concrete and Its Function*, 1st ed.; Science Press: Beijing, China, 2014; pp. 34–44.

40. Sadrinejad, I.; Madandoust, R.; Ranjbar, M.M. The Mechanical and Durability Properties of Concrete Containing Hybrid Synthetic Fibers. *Constr. Build. Mater.* **2018**, *178*, 72–82. [CrossRef]
41. Liu, D.C. Study on the Properties of Basalt and Polypropylene Mixed Fiber Concrete. Master's Thesis, Chongqing Jiaotong University, Chongqing, China, 2018.
42. Mydin, M.A.O.; Soleimanzadeh, S. Effect of Polypropylene Fiber Content on Flexural Strength of Lightweight Foamed Concrete at Ambient and Elevated Temperatures. *Adv. Appl. Sci. Res.* **2012**, *3*, 2837–2846.
43. Yap, S.P.; Alengaram, U.J.; Jumaat, M.Z. Enhancement of Mechanical Properties in Polypropylene-and Nylon-Fibre Reinforced Oil Palm Shell Concrete. *Mater. Des.* **2013**, *49*, 1034–1041. [CrossRef]
44. Li, Y.X.; Chen, Y.M.; He, X.Y. Pore Volume Fractal Dimension of Fly Ash-Cement Paste and its Relationship between the Pore Structure and Strength. *J. Chin. Ceram. Soc.* **2003**, *31*, 774–779.
45. Liu, H.Y.; Li, H.Y.; Zou, C.X. Effect of Fiber Elastic Modulus on Concrete Bubble Fractal Dimension and Frost Durability of Light-Weight Aggregate Concrete. *Bull. Chin. Ceram. Soc.* **2015**, *34*, 3039–3044.
46. Shi, Y.; Yang, H.Q.; Chen, X.; Li, X.; Zhou, S.H. Influence of Aggregate Variety on Pore Structure and Microscopic Interface of Concrete. *J. Build. Mater.* **2015**, *18*, 133–138.
47. Powers, T.C. Void Spacing as a Basis for Producing Air-Entrained Concrete. *J. Am. Concr. Inst.* **1954**, *50*, 741–759.

Review

Crossover Effect in Cement-Based Materials: A Review

Sumra Yousuf [1], Payam Shafigh [2,3,*], Zainah Ibrahim [1], Huzaifa Hashim [1] and Mohammad Panjehpour [4]

[1]   Department of Civil Engineering, Faculty of Engineering, University of Malaya, Kuala Lumpur 50603, Malaysia

[2]   Department of Building Surveying, Faculty of Built Environment, University of Malaya, Kuala Lumpur 50603, Malaysia

[3]   Center for Building, Construction & Tropical Architecture (BuCTA), Faculty of Built Environment, University of Malaya, Kuala Lumpur 50603, Malaysia

[4]   Centre for Advanced Concrete Technology (CACT), INTI International University, Nilai 71800, Malaysia

*   Correspondence: pshafigh@gmail.com

Received: 5 May 2019; Accepted: 21 June 2019; Published: 10 July 2019

Abstract: Cement-based materials (CBMs) such as pastes, mortars and concretes are the most frequently used building materials in the present construction industry. Cement hydration, along with the resulting compressive strength in these materials, is dependent on curing temperature, methods and duration. A concrete subjected to an initial higher curing temperature undergoes accelerated hydration by resulting in non-uniform scattering of the hydration products and consequently creating a great porosity at later ages. This phenomenon is called crossover effect (COE). The COE may occur even at early ages between seven to 10 days for Portland cements with various mineral compositions. Compressive strength and other mechanical properties are important for the long life of concrete structures, so any reduction in these properties is of great concern to engineers. This study aims to review existing information on COE phenomenon in CBMs and provide recommendations for future research.

Keywords: crossover effect; cementitious materials; compressive strength; accelerated curing; concrete; mortar

1. Introduction

Cement-based materials (CBMs) such as mortars and concretes are the most frequently used building materials in the construction industry [1]. Cement hydration is extremely exothermic in these materials [2]. The hydration process of cement continues for many years at a reducing rate, depending on the water quantity in mixture and appropriate curing temperature conditions [3]. For long-term durability and strength of concretes and mortars, their appropriate mixes should be placed and cured in a proper environment during early hardening by preventing water loss of evaporation from capillaries [4]. In addition, internal water loss by self-desiccation should be replaced by water from external sources [5]. The hydration process and resulting compressive strength of concrete depend on curing temperature, curing methods, curing duration and mix composition [6]. The compressive strength of concrete, along with other mechanical properties, has essential role in the design and construction of all types of concrete structures [7].

Concretes exposed to accelerated curing temperatures at the early ages gain high compression and splitting tensile strengths at early ages but low compression and splitting tensile strengths at later ages compared with those exposed to moderate normal curing temperatures. The modulus of elasticity of concrete has the same tendency, but the variation in its value with accelerated curing temperatures is not as prominent as that of compressive strength [8]. Fall et al. [9] stated that splitting tensile strength of the CBMs is more affected by the mechanism of crossover effect (COE) than compressive strength.

Researchers [5,10] described that increasing curing temperature of CBMs, such as cement pastes, mortars and concretes, increases rate of early-age strength gain but lowers the long-term strength (see Figure 1). Garcia and Sharp [11] also reported that concrete exposed to high curing temperatures exhibits accelerated hydration and non-uniform spreading of the hydration products. Such an exposure creates less porosity at early-ages and great porosity at later-ages. It results in increased compressive strength at early-ages and reduced compressive strength at later-ages [8].

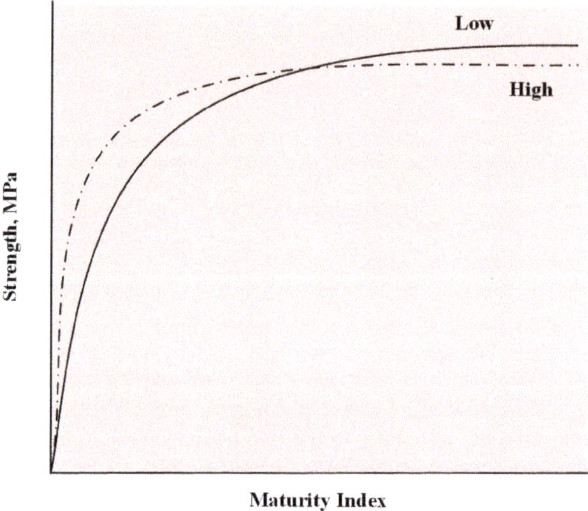

Figure 1. Crossover effect (COE) with maturity and strength at low and high temperatures.

Ogirigbo and Black [12] described that concrete cured at elevated temperature of 38 °C had less porosity and more compressive strength at seven days than those cured at 20 °C. However, at 28-days, the porosity of elevated temperature cured concrete increased and compressive strength decreased than that of normally cured concrete. Luz and Hooton [13] studied the effect of elevated curing temperatures up to 50 °C on supersulfated cements incorporating low and high alumina slags. They observed lower porosity and higher compressive strength due to elevated temperatures (38 °C and 50 °C) than the samples cured at normal temperature (23 °C) at the age of seven days. Ezziane et al. [14] emphasised that under elevated temperatures, cement mortars and concretes have more porous and poorer physical structures than those developed at normal room temperature. They also stated that high temperatures increase strength gain at the early ages, whereas substantial amount of the formed hydrates have insufficient time for proper arrangement at the later ages. This process results in COE, a decrease in ultimate compressive strength of CBMs.

According to the research reports [15,16], COE may occur between seven and 10 days in Portland cements of various mineralogical compositions. Weerdt et al. [17] stated that the initial fast hydration of Portland cement can produce a lesser homogenous microstructure that causes the COE in cements and concretes. Castellano et al. [18] also reported that such a process of Portland cement can decrease hydration at later ages. This decrease is due to the postponing or ending of the hydration of anhydrous cement. Moreover, the hydration products at lesser temperatures have adequate time to precipitate and diffuse homogeneously in the whole cement matrix. According to Kim et al. [8], the COE of type V Portland cement concrete is not very clear compared with that of type I Portland cement concrete because of varying rates of hydration.

The COE in CBMs due to accelerated initial and prolonged curing is of great concern to researchers because of the utmost importance of concrete structures in the present construction industry. Some researchers [15,17,19] explained that the loss in compressive strengths of CBMs at later ages is because of different initial curing temperatures. Yang et al. [19] studied the strength development of concrete due to different initial curing temperatures with ageing (see Figure 2). They found a significant difference in compressive strengths of concretes at later ages because of various initial curing temperatures. At the later ages, compressive strength of materials cured at lower initial curing temperatures were greater than those cured at higher initial curing temperatures. They termed this phenomenon as COE.

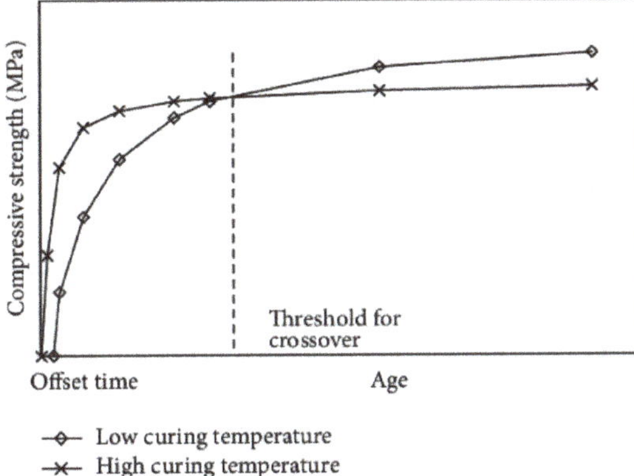

Figure 2. COE during strength development of concrete due to different initial curing temperatures. Adapted with permission from [19], Hindawi, 2015.

According to Lothenbach et al. [20], the COE in compressive strength occurs in Portland cements, but its occurrence for blended cements having slag [11,21], natural pozzolan [22] and fly ash [23] is debated. The term COE has recently been used by researchers, but information is insufficient for a detailed discussion of this phenomenon in the CBMs. However, given the importance of mechanical properties and durability of concrete for long life of concrete structures, even any small reduction in the properties greatly concerns engineers.

A comprehensive review of the existing information should be available to address the lack of knowledge on the COE. The authors believe that resources such as the ACS (American Chemical Society) Publications, Emerald, Informs, Elsevier, Springer, Taylor and Francis and the Scopus database suffice to prepare a critical review and extend the knowledge. Finally, the identified gaps will be presented as suggestions for future research. This study aims to review the existing information about the COE in CBMs and provide recommendations for future research.

2. Methods of Accelerating Strength Gain

The strength-gaining rate and hardening of CBMs such as mortars and concretes in the normal curing conditions are slow. Therefore, heat treatment is commonly used in the precast concrete industry to speed up the strength-gaining rate of these CBMs. The application of heat treatment depends on the acceleration of hydration reactions [24]. Under heat treatment application, cement fineness, Portland cement composition, types and amounts of the additives used in blended cements are the main factors to determine behaviour of cements [25]. The degree and speed of cement hydration, the consequent strength of CBMs and other mechanical properties are also directly influenced by the curing process [26].

2.1. Effect of Elevated Temperature Curing on Strength Gain

Kjellsen and Detwiler [27] found that the 28-day compressive strength of samples cast and cured at 5 °C is around 80% of that cast and cured at 21 °C to 46 °C. The later-age compressive strength will be lower with higher casting and curing temperatures. According to Newman and Choo [28], the later-age compressive strength of heat-treated CBMs is generally lower than that of standard cured specimens. It is due to the lesser rate of reaction at lower temperatures, means that mortars and concretes should be cured for a longer time to gain the desired degree of strength. Khan and Abbas [29] reported that the fast reaction at elevated temperatures provides comparatively high early-age strengths, but long-term strength and durability are commonly decreased.

Lothenbach et al. [20] reported that cement hydration is highly affected by the temperature. The higher temperature leads the hydration towards high early-age strength. At later ages, the strength of the hydrated cements at high temperature is decreased compared with those hydrated at room temperature. Sajedi and Razak [30] stated that variations in temperature caused by either heat of hydration or changes in outer environment greatly influence mechanical properties, especially the compressive strength of concrete and mortar at the early ages. Kim et al. [8] proved by experimental results that concretes and mortars subjected to high curing temperatures gain greater early-age compressive and splitting tensile strengths but lower later-age compressive and splitting tensile strengths compared with those exposed to room curing temperatures.

Namarak et al. [31] reported that an elevated curing temperature can speed up both pozzolanic and hydration reactions and achieve greater compressive strength at the early ages. However, at later ages, mortars and concretes cured at high temperature achieve compressive strength less than that of a normal curing temperature. The similar behaviour was also stated by Ezziane et al. [14] with the reduction in compressive strength when mortars were subjected to a high curing temperature with replacement of cement by natural pozzolan.

2.2. Effect of Steam Curing on Strength Gain

In steam curing, the temperature ranges from 40 °C to 100 °C, but the temperature from 65 °C to 85 °C is commonly applied [5]. Cakır and Akoz [4] stated that the steam curing has been frequently used in prefabrication because the rate of strength gain increases by increasing range of curing temperature. Yazıcı et al. [32] observed in their study that the steam curing improved 1-day compressive strength from about 10–20 MPa of concrete mixtures containing fly ash up to 40%, 50% and 60% in replacement of ordinary Portland cement (OPC) by volume. They stated that this compressive strength is sufficient to enable formwork to be removed and significantly supports the precast concrete industry. Aldea et al. [33] emphasised that air curing at room temperature and 100% RH in a controlled chamber until 28 days is the best for improved durability of concrete. However, atmospheric pressure steam curing is better than autoclaving for fast strength development.

The researchers [34,35] investigated that steam cured concrete has less compressive strength than moist cured concrete at 28 days, indicating that ordinary concrete has lower strength gaining rate after steam curing. However, the steam curing concrete having SCMs has high rate of strength development. They also reported that the hydration rate of the cement increases rapidly at elevated curing temperatures. As a result of this fast hydration, a gel is formed around each cement particle in the cement matrix. The cement hydration rate and distribution rate of hydration products are fast, but the dissolving rate of hydration products is slow, so the gel layer becomes dense and thick progressively. The diffusion of water into this thick and dense gel layer is lessened, and the hydration of unhydrated cement particles is consequently delayed. Later-age compressive strength has lower increment. Therefore, the later strength-gaining ratio will be lower if higher temperature and longer steam curing time are used.

2.3. Effect of Autoclaving on Strength Gain

Autoclaving is high-pressure steam curing in which the temperature is from 160 °C to 220 °C, and the pressure of steam is from 6 to 10 atmospheres [4]. Autoclaving is used to accelerate strength gaining of CBMs to achieve an adequate high early-age strength [5]. Aldea et al. [33] stated that under autoclave curing conditions, addition of the slag up to 50% of cement replacement in concrete does not affect the compressive strength, whereas 75% slag replacement decreases compressive strength. In their research, autoclave curing induced the greatest reduction in compressive strength of the tested materials because of less homogeneous scattering of hydration products in the mix. The fast initial hydration resulted in the huge capillary pore distribution [36].

2.4. Effect of Hot Water Curing on Strength Gain

According to Ozkul [37], the water acts as an insulator to conserve the heat of cement hydration in hot water curing method. Thus, the hot water curing method is more effective for rapid hardening cements (e.g., Portland cement type III) than low heat cements (e.g., pozzolan cement). Chithra and Dhinakaran [38] found that hot water curing is more effective than hot air oven curing due to the more uniform and fast distribution of temperature into the specimens in hot water curing. Their test results revealed that higher strength of concrete and mortar can be achieved from few hours of thermal curing before normal curing. Moreover, the replacement of cement with 40% of ground granulated blast furnace slag (GGBFS) under hot water curing at 60 °C produced maximum compressive and tensile strengths of concrete. Sajedi and Razak [30] reported that each specified cementitious material has an optimal temperature to achieve high early compressive strength. They found that 60 °C is the optimal temperature for the slag. Authors [8,14,39,40] also reported that a constant increase in early-age compressive strength can be achieved by enhancing curing temperature up to 60 °C.

Sajedi and Razak [41] stated that among different curing methods for mortars with or without cementitious materials, only curing in the air of all specimens cannot be recommended practically. However, they recommended curing under air in room temperature for specimens initially heated in the bath water. They also observed that heating samples in water followed by air curing and heating samples in oven followed by water curing have greater compressive strengths compared with curing in water after water heating and curing in air after oven heating, respectively. Aprianti et al. [42] also investigated that the short-time initial hot water curing is a very effective technique to achieve 1-day compressive strength for the CBMs. Their experimental results showed that the hot water curing for duration of 2.5 h considerably increases compressive strength of the mortar with 50% GGBFS. However, increasing the hot water curing duration from 2.5–5 h will not affect the gain of compressive strength.

2.5. Effect of Curing Period on Strength Gain

A minimum of seven days is recommended for moist curing for concretes containing normal Portland cement, whereas a longer curing period is required to ensure the strength contribution from pozzolanic reaction for concretes comprising slow hardening cements or mineral admixtures [43]. Sajedi and Razak [30] concluded that time period of heating is crucial for gaining high early-age strength. They reported that 20-h heating time is the optimum duration for slag used in their research. Moreover, compressive strength will increase as heating time duration increases towards the optimum, and heating of mortar more than optimum heating time will not increase the early-age strength of mortar.

Sajedi and Razak [41] concluded from their research that the duration of curing must be prolonged to 14 days when the cement has SCMs such as silica fume, fly ash and slag because of the slow hydration reactions between calcium hydroxide and SCMs. Kim et al. [8] found that for Portland cement type I concrete, the compressive strengths increases at one and three days with increasing curing temperatures from 10–50 °C. However, the seven-day compressive strength decreases up to 5.5% with increase in curing temperatures from 23–35 °C, and the 28-day compressive strength reduces up to 5% with increase in curing temperatures from 10–50 °C.

3. COE and Mechanism

CBMs exposed to elevated curing temperatures at the early ages attain higher early-age compressive strength, but a loss of ultimate strength may occur at later ages because of the unsuitable arrangement of hydrates during fast hydration. This phenomenon is called COE [44–46]. The scanning electron microscope (SEM) images of cement paste cured at room temperature and elevated temperature of

60 °C at the ages of three and 28 days were studied by Teixeira et al. [47] as shown in Figures 3 and 4, respectively. From the images, they observed more uniformly distributed hydration products in the cement pastes cured at room temperature than in the cement pastes cured at 60 °C. The observed that cement paste cured at room temperature had porous microstructure with uniformly distributed capillary pores at early age of three days. This porous microstructure densified due to curing at elevated temperature and an increase in compressive strength of cement paste was observed. A dense rim of hydration product, generally composed of calcium silicate hydrate (C-S-H), was formed around cement particles cured at elevated temperature of 60 °C at the ages of three and 28 days. The concluded that increase in heterogeneous phases in microstructure of cement paste cured at elevated temperature might be the reason of compressive strength reduction at the age of 28 days.

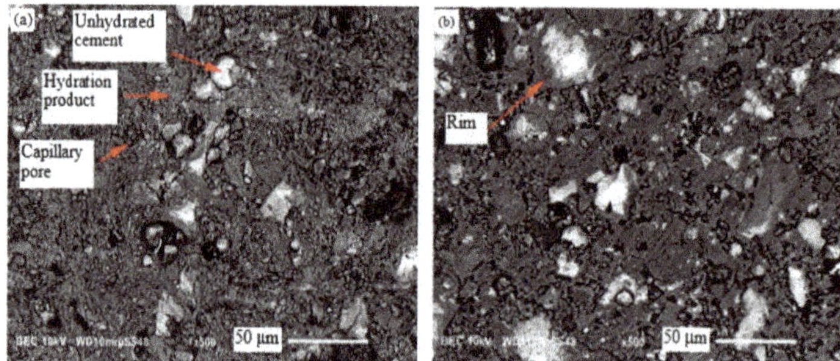

Figure 3. SEM images of cement (ordinary Portland cement (OPC)) paste at the age of three days, (**a**) cured at room temperature, and (**b**) cured at 60 °C. Adapted with permission from [47], MDPI, 2016.

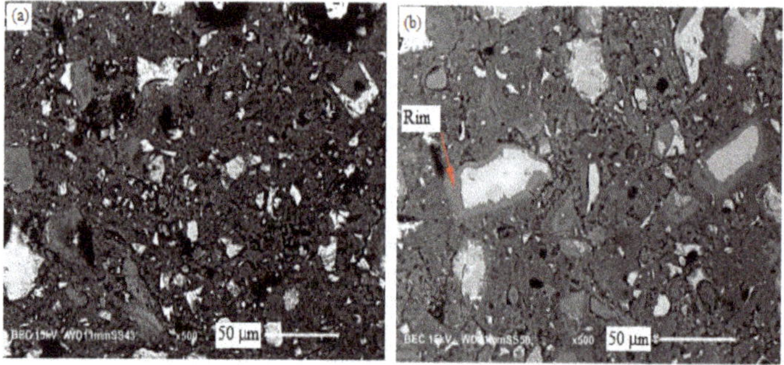

Figure 4. SEM images of cement (OPC) paste at the age of 28-days, (**a**) cured at room temperature, and (**b**) cured at 60 °C. Adapted with permission from [47], MDPI, 2016.

According to the research reports [18,48], the detection of COE depends on the range of temperature, duration of testing, type of cement used, proportions of the mineral additions and mineralogical composition of the cement used. Reports from the researchers [49,50] showed that an elevated temperature speeds up the binder hydration process and increases the quantity of formed hydration products (for example calcium-silicate-hydrate). Moreover, accelerated curing temperature increases strength gain at initial ages by faster precipitation of hydration products and refining pore structure of CBMs. According to several reports [49,51–54] COE is caused by harder and thicker formation of hydrated shells, low porosity hydration products round cement grains, slow diffusion of hydration products in secondary cement hydration reaction and non-uniform scattering of reaction products in pores of the hardening cement paste at elevated temperature. Kim et al. [55] reported that COE can appear even after seven days. Kjellsen et al. [56] reported that the non-uniform distribution of hydration products results in porosity of bulk cement paste. Additionally, speedy temperature increase during early ages can increase inside stresses that surpass tensile strength of immature concrete, leading to increased porosity, cracking and decreased potential strength.

According to the reports [11,49,52], the rate of hydration and concrete strength increase rapidly by increasing early curing temperature, but porosity increases and micro cracks appear because of non-homogeneous distribution of hydration products and variance in coefficients of thermal expansion of the concrete's ingredients, which finally leads to reduced compressive strength at the later ages. Aldea et al. [33] stated that the hydration products do not have sufficient time for their uniform distribution within the pores of hardening CBMs at elevated temperature. Consequently, hydrated shells with low porosity are formed near the cement particles by ending further hydration. This non-uniform scattering of the hydration products causes large capillary pores by decreasing later-age compressive strength. Mindess and Young [57] also reported that steam curing makes the COE by causing great reduction in compressive strength because of less homogeneous scattering of hydration products in the paste due to fast initial hydration.

Researchers [41,58] reported that the major cause of the strength reduction at later ages is the shortage of the inner water present in samples to complete hydration. In a common water to cement ratio, internal water is present and sufficient for hydration for the duration of 1–28 days. However, after 28 days, this amount of water may be reduced and become insufficient for further hydration, which will result in strength loss at later ages. Reports by the authors [27,49,50] stated that the lesser hydration reaction of anhydrous cement is also due to delaying effect of the shell on distribution of hydrates in the secondary cement hydration reaction.

4. Effect of Different Types of Portland Cements on COE Phenomenon

According to Neville [26], all Portland cements have the same constituents. However, types of Portland cement differ only because of the differences in the proportions of these constituents. Constituents of C_3S and C_2S form 70–80% of all Portland cements. Strunge et al. [59] stated that Portland cements with high alite (C_3S) and tricalcium aluminate (C_3A) contents such as Portland cement type III develop high early-age strength along with high early heat. Cements with low alite (C_3S) and tricalcium aluminate (C_3A) contents such as Portland cement types IV and V have slow strength and heat development. Alkali oxides (Na_2O and K_2O) of Portland cement also increase the compressive strength to the period of seven days but reduce it at the later ages.

All types of Portland cements as well as blended cements can be used in concretes cured by steam curing at atmospheric pressure or by the accelerating heat methods [60]. According to Mehta and Monteiro [43], Portland cement type III has higher fineness and hydrates more quickly compared with other Portland cement types at ordinary temperature. Therefore, concretes containing Portland cement type III will have a higher strength at any given water to cement ratio and early ages of hydration as one, three and seven days. Popovics [61] described that the early-age strengths can be intensely increased during heat curing using cements with SO_3 contents somewhat higher than usual. Portland cement type IV is used for minimum rate and quantity of heat generated from hydration. It gives strength at lower rate compared with other types of Portland cements. Portland cement type V gains strength more slowly than Portland cement type I.

5. Effect of Cementitious Materials on COE Phenomenon for Prolonged Curing

5.1. Ordinary Portland Cement (OPC)

Sajedi and Razak [41] used various mixes of OPC mortars with different curing conditions in their research. They observed COE in compressive strengths of tested mortars at 56 and 90 days when temperature of water curing was 60 °C for a period of 20 h. They used specific names for their mortar mixes and curing regimes. The detail of mix names, curing conditions, proportions of the mortars and range of COE detected are given in Table 1. The age at which COE occurred and the percentage of COE were determined from the graphical representation of their obtained results.

Table 1. Mix proportions of ordinary Portland cement (OPC) mortars in one batch, curing conditions and crossover effect (COE) records.

Number	Name of Mix	Curing Condition	Age at Which COE Occurred	COE (%)
1	OM-WH-ac	Water bath heated followed by air curing under room temperature	90 days	7.14
2	OM-WH-wc	Curing at water bath heated and then water curing	90 days	5.13
3	OM-OH-ac	Oven heated curing and then air curing under room temperature	56 days	4.30
4	OM-OH-wc	Oven heated curing and then water curing	90 days	4.70
5	OM-G6/WH-ac	Water bath heated followed by air curing under room temperature	56 days	5.0

The major cause of the strength reduction at later ages is shortage of inner water in samples to complete hydration reaction. For 1–28 days, the internal water is present and sufficient for hydration. However, after 28 days, the water is insufficient for hydration due to consumption, which causes strength reduction in mortar specimens [62]. The loss in strength of OPC mortars also depends on fineness of cement, chemical composition of products, temperature and regime of curing. The early curing temperature has a vital effect on compressive strength and may decrease or increase it at long curing periods [41].

Garcia and Sharp [16] used two samples of OPC in their research. They observed loss in compressive strengths in the range of 4–13% of these samples at 360 days as a result of hot water curing at 40–60 °C for 24 h. Kim et al. [8] observed that concrete exposed to accelerated curing temperature at initial age gains higher initial age but lesser later-age compressive strength. They found a loss of 4.90% in compressive strength at seven days as compared to two days in Portland type V cement concrete due to hot water curing at 50 °C for 24 h.

5.2. Blast Furnace Slag, Fly Ash and Silica Fume

Sajedi and Razak [30] used different OPC-slag mortars in their research with 0%, 40% and 50% slag replacement with OPC by mass. They studied the influence of temperatures of 50 °C, 60 °C and 70 °C on compressive strengths of these mortars for up to 90 days. They found that hot water curing at a temperature of 60 °C for 20 h period is the most effective to enhance the early age strengths of all the mortars. Therefore, they selected 60 °C as the optimum temperature. They observed COE of 2.2% in OPC-slag mortar with 50% replacement of slag with OPC by mass at 56 and 28 days. This COE appeared as a result of 60 °C hot water curing for 20 h. However, they did not observe any COE in other two Portland cement slag mortars with 0% and 40% replacement of slag with OPC by mass for up to 90 days.

Bougara et al. [63] used Portland cement–slag mortars with various percentages of slag as 0%, 30% and 50% in their study. They observed a COE of about 3.7% in Portland cement slag mortar with 30% slag at 28 days when samples were initially water cured at 60 °C for 20 h. Schindler [64] suggested that the addition of higher percentages of slag as replacement with OPC in mortars tends to reduce the occurrence of COE. Verbeck [46] explained that curing at elevated temperature resulted in a non-homogeneous scattering of the hydration products in microstructure. However, at lower temperature, the hydration products have adequate time period to precipitate and diffuse more homogeneously throughout the bulk cement matrix.

Wade et al. [65] used 13 concrete mixtures to examine the effect of OPC type I and III, water to CMs ratios of 0.37, 0.41, 0.44 and 0.48 and various types and dosage of SCMs. They used class F and C fly ashes at replacement values of 20% and 30%, respectively. They used GGBFS at replacement levels of 30% and 50% and silica fume at a replacement level of 10% in a ternary mix with 20% class F fly ash.

They made three batches of each mix as cold batch (4 °C and 13 °C for 24 h), hot batch (32 °C and 41 °C for 24 h) and control batch (20 °C and 23 °C). They measured compressive strengths at 24 and 48 h and seven, 14 and 28 days. They observed that hot batches achieved strength rapidly, but they had lower later-age strengths compared with control and cold batches. The cold batches achieved strength slowly but sustained to achieve strength at later ages. They concluded that all control mixtures exhibited COE ranging from 7–12% at seven and 16 days. No COE was observed in case of Portland cement type I mixtures with 20% and 30% class F fly ash having w/b of 0.41. The COE for mixtures with 20% class C fly ash for Portland cement type I mixture with w/b ratio of 0.41 was delayed but not completely removed. The replacement of 30% class C fly ash for the type I cement mixture with 0.41 w/b ratio excellently eliminated COE. The replacement of cement up to 30% or 50% with GGBFS for cement type I mixture with 0.41 w/b ratio showed strength losses of 6–17% for hot batches in some cases. Changing the cement type I to cement type III for cement type I mixture with 0.44 w/b ratio marginally increased COE but significantly reduced the time at which crossover occurred from 16 to 4 days. The cement replacement with 20% class F fly ash and 10% silica fume for type I cement mixture with w/b ratio of 0.44 increased strength loss from 7–31% and reduced the time at which crossover take place from 16 to 5 days. The type III cement mixture with 0.37 w/b ratio had crossover of 17% which occurred within less than two days after mixing for hot batches. The ternary blended prestressed concrete mixture had crossover of 23%.

6. Effect of Cementitious Materials on COE Phenomenon Due to Different Initial Curing Temperatures

6.1. Blast Furnace Slag

Castellano et al. [15] used Portland cement (OPC) with GGBFS in their investigation. They studied the effect of curing temperature on hydration process and the phenomenon of COE. Two blast-furnace cements, BFS40 and BFS80, were made by replacement of OPC with 40% and 80% slag by mass, respectively. They found that for all test mixtures (OPC, BFS40 and BFS80) compressive strength decreased when w/b ratio increased, as shown in Figure 5A–C, respectively. All w/b ratios had 20–60 °C increase in the water curing temperature, and loss in compressive strengths was observed in OPC and BFS40 pastes at 7–365 days. Moreover, compressive strengths at early ages are lower at lower water curing temperatures for all w/b ratios.

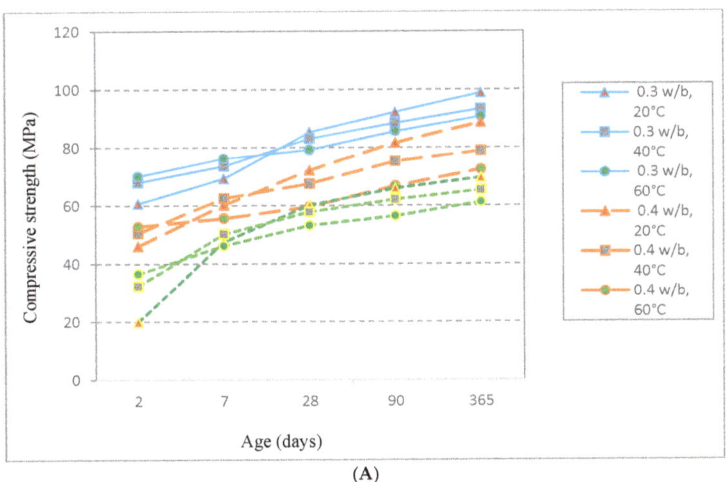

(A)

Figure 5. *Cont.*

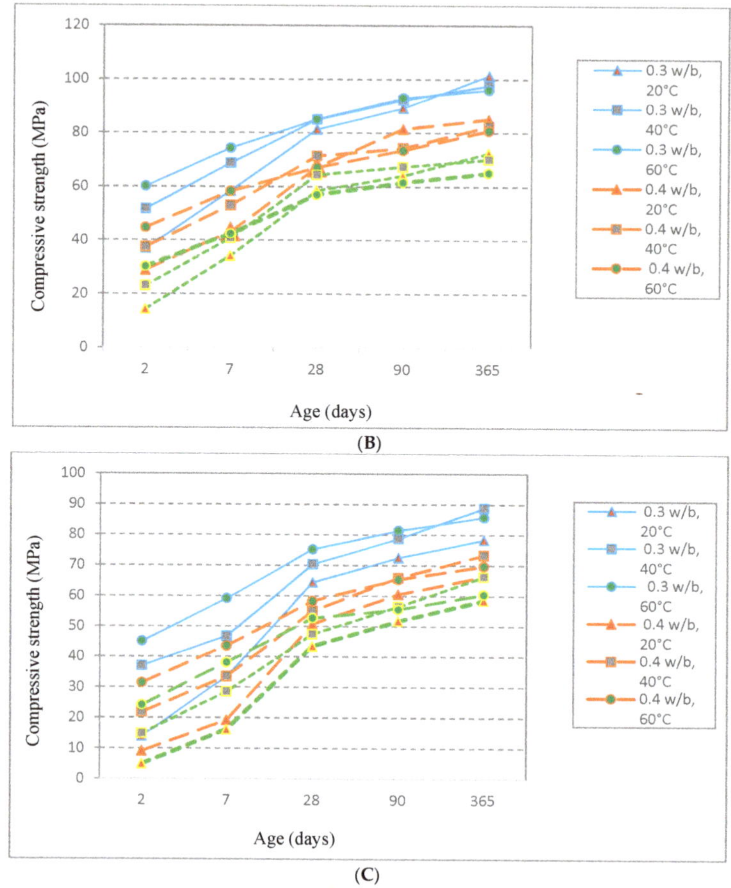

Figure 5. Compressive strength (MPa) at various water curing temperatures and ages for (**A**) OPC; (**B**) BFS40; (**C**) BFS80 pastes.

The losses in compressive strengths (COE %) and age of COE for OPC and BFS40 pastes were calculated from the original results provided by the researchers (Table 2). However, no COE was observed for BFS80 at all w/b ratios with 20–60 °C increase in water curing temperature up to 365 days.

Ezziane et al. [66] described the effect of increasing curing temperature from 20–60 °C on the long-term strength of OPC and two slag blended cement mortars (BFS30 and BFS50). They used 30% and 50% slag in replacement of OPC by mass. They found that the long-term strengths of slag cements are mostly greater than those of OPC for all replacement levels. Moreover, the long-term strengths of OPC and slag blended cement mortars were decreased with increasing curing temperature from 20–60 °C.

Figure 6 shows the loss in compressive strengths at later ages in OPC and slag blended cement mortars with 20–60 °C increase in water curing temperature. These losses in compressive strengths of Portland cement and slag blended cement mortars were calculated from the actual data obtained by researchers (Table 3). From the results in Table 3, it can be concluded that the amount of COE increases with age and it will be higher at later ages.

Table 2. COE records with 20–60 °C increase in water curing temperature at ages up to 365 days for ordinary Portland cement (OPC), BFS40 and BFS80 pastes.

Paste Name	w/b Ratio	Age at Which COE Occurred	COE (%)
OPC paste	0.3	28	7.2
		90	7.2
		365	8.2
	0.4	7	7.6
		28	17.3
		90	18.3
		365	18.3
	0.5	7	2.9
		28	11.5
		90	14.4
		365	12
BFS40	0.3	365	5.2
	0.4	28	0.1
		90	9.8
		365	5
	0.5	28	2.6
		90	3.9
		365	10.2
BFS80	No COE was observed at all w/b ratios up to 365 days		

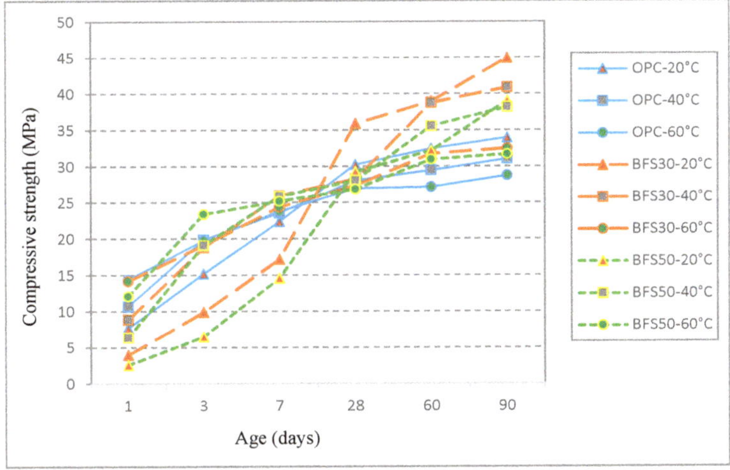

Figure 6. Compressive strength (MPa) at various water curing temperatures and ages for OPC and slag blended cement mortars (BFS30 and BFS50).

Table 3. COE records with 20–60 °C increase in water curing temperature at ages up to 90 days for OPC and slag blended cement mortars.

Paste Name	% Age Replacement	Age at Which COE Occurred	% Age of COE
OPC	0	28	10.9
		60	16.3
		90	15.3
BFS30	30	28	23.7
		60	18.5
		90	27.8
BFS50	50	28	18.2
		60	3.7
		90	18.9

6.2. Ternary Blended Cements

Weerdt et al. [17] used total five pastes, one OPC paste and four composite cement pastes (OPC95-L5, OPC70-FA30, OPC65-FA35 and OPC65-FA30-L5) with limestone powder (L) and fly ash (FA) in their study. They monitored the hydration of these cement pastes at three various water curing temperatures of 5 °C, 20 °C and 40 °C for a period up to 180 days. They described that loss in the compressive strength occurs at later ages because of increasing values of initial curing temperatures. Figure 7A,B display that increasing initial curing temperature from 5–40 °C resulted in loss in compressive strength of about 5.3%, 20.4% and 26.1% for OPC paste and 8.6%, 22.9% and 26.5% for composite cement paste containing OPC and lime stone powder at 7, 28 and 90 days, respectively. However, loss in compressive strengths was not observed for mixtures containing fly ash.

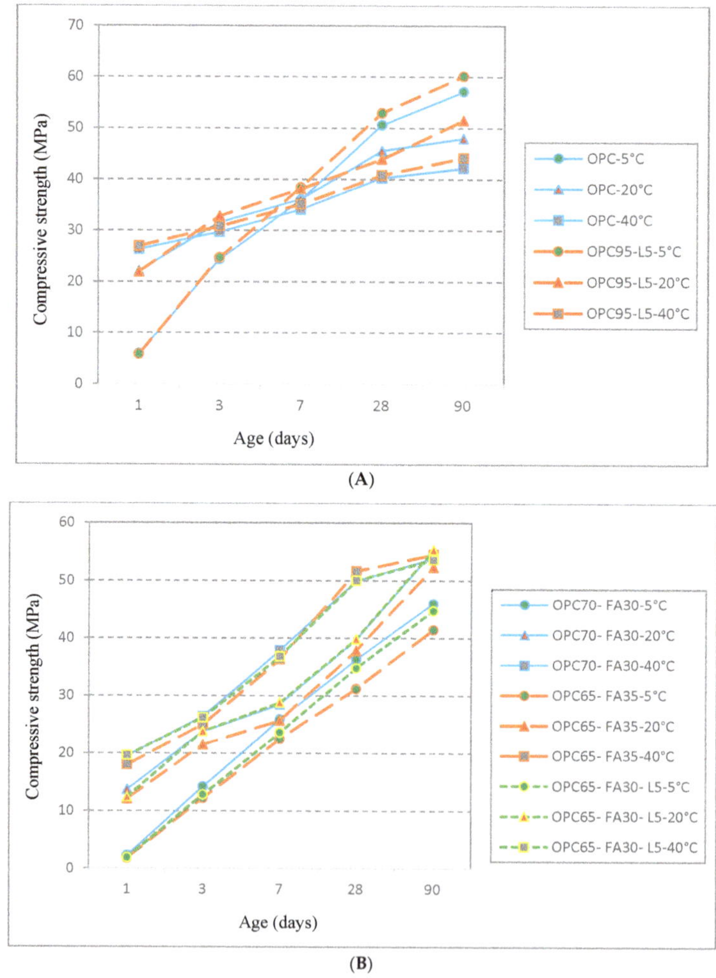

Figure 7. Compressive strengths (MPa) of (**A**) OPC and OPC95-L5; (**B**) OPC70-FA30, OPC65-FA35 and OPC65-FA30-L5 for water curing temperatures of 5 °C, 20 °C and 40 °C.

6.3. Natural Pozzolan and Limestone Powder

Ezziane et al. [66] stated the effect of increasing curing temperature from 20–60 °C on the long-term strength of OPC, natural pozzolan and limestone powder blended cement mortars. They used 100%

OPC, 10% (OPC-NP10), 20% (OPC-NP20), 30% (OPC-NP30) and 40% (OPC-NP40) natural pozzolan and 5% (OPC-L5), 15% (OPC-L15) and 25% (OPC-L25) limestone powder in replacement of OPC by mass. They observed that at replacement levels of 10%, 20% and 30%, the long-term strengths of pozzolan blended cement mortars decreased with increasing curing temperature from 20–60 °C and were greater than those of OPC mortar. For limestone powder blended cement mortars, the long-term strengths for 25% replacement level were less than OPC mortar and showed reduction in strength only at 90 days.

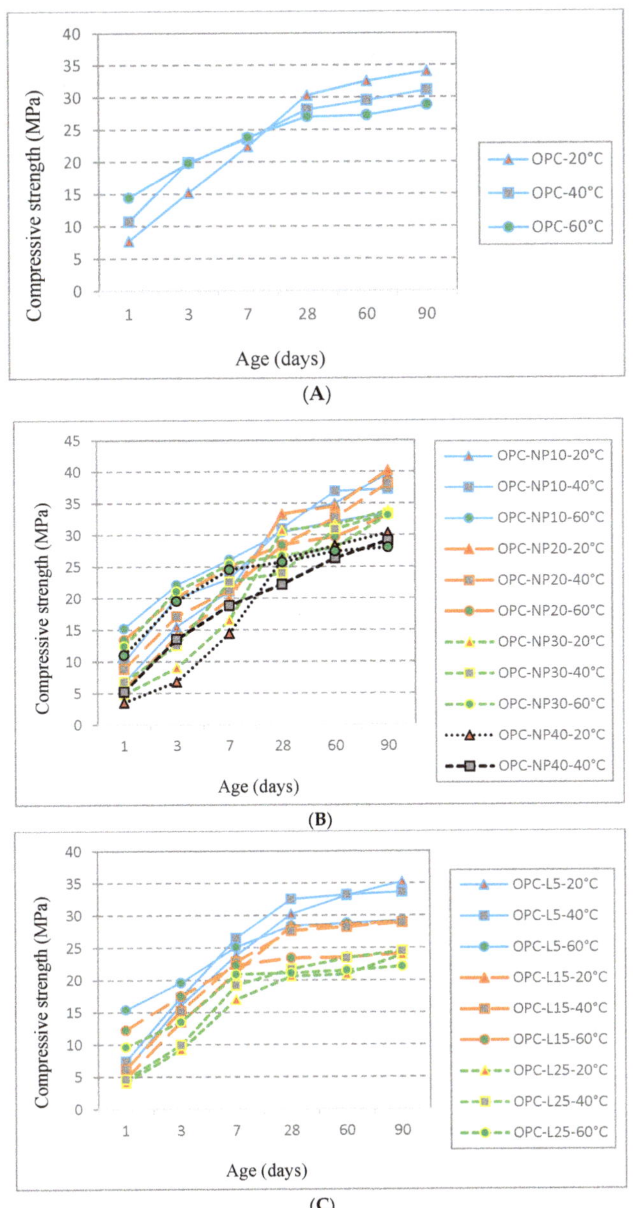

Figure 8. Compressive strength (MPa) at various water curing temperatures and ages for (**A**) OPC; (**B**) natural pozzolan blended cement; (**C**) limestone powder blended cement mortars.

Figure 8A–C show the loss in compressive strengths at later ages in Portland cement, natural pozzolan and limestone powder blended cement mortars with 20–60 °C increase in the water curing temperature, respectively. These losses in compressive strengths of Portland cement, natural pozzolan and limestone powder blended cement mortars are shown in Table 4.

Table 4. COE records with 20–60 °C increase in water curing temperature at ages up to 90 days for OPC, natural pozzolan and limestone powder blended cement mortars.

Paste Name	Age at Which COE Occurred	COE (%)
OPC	28	10.9
	60	16.3
	90	15.3
OPC-NP10	28	8.2
	60	8.3
	90	15.7
OPC-NP20	28	14.7
	60	13.9
	90	17.7
OPC-NP30	28	13
	60	11.4
	90	2.1
OPC-NP40	28	1.5
	60	3.2
	90	17.6
OPC-L5	28	6
	60	13
	90	17.1
OPC-L15	28	17.3
	60	17.5
	90	17.2
OPC-L25	90	7.5

7. Materials Controlling COE Phenomenon

The existing literature does not contain sufficient information about COE phenomenon in CBMs. Therefore, few materials have been mentioned to decrease or control the loss in compressive strengths at later ages at high curing temperatures. Castellano et al. [15] found that COE in compressive strength was not observed for BFS80 (by replacing Portland cement with 80% GGBS by mass) in temperature range of 20–60 °C up to 365 days.

Wang [67] stated that the compressive strength will increase by replacement of slag powder after 28 days. Moreover, slag powder will significantly increase the later-age compressive strength. Aldea et al. [33] suggested that 25% replacement of slag is optimum for the strength development with steam and normal curing. Autoclaving has no effect on the strength developed by replacing 50% slag with Portland cement, whereas 75% slag replacement with Portland cement reduces compressive strength in all types of curing methods. Normal curing methods provide higher compressive strength than the accelerated curing methods up to 50% replacement of slag with OPC.

Kim et al. [8] stated that Portland type V cement concretes do not undergo COE compared with Portland type I cement concretes up to curing temperature of 50 °C for 28 day because the rate of hydration of Portland type V cement concrete is lower than that of Portland type I cement concrete. Thus, the COE of Portland type V cement concrete is postponed. Ezziane et al. [14] used blended cement mortars with different percentages of natural pozzolan replacement with OPC. They studied their behaviour under water curing temperatures of 20 °C, 40 °C and 60 °C for up to 90 days. They detected no COE in compressive strength of blended cement mortars at curing temperatures of 20 °C, 40 °C and 60 °C due to pozzolan in mixtures. Its presence decreases negative influence of rise in temperature. The optimal replacement rate of OPC by natural pozzolan is approximately 15% for normal room temperature curing and 20% for accelerated temperature curing of 40 °C and above.

8. Conclusions

This paper presents a detailed literature review about the COE phenomenon in CBMs due to accelerated initial curing. The basic concept and mechanism of COE, various methods of accelerating for strength gain at the early ages and reasons for strength decrease at the later ages of CBMs, effect of cementitious materials on COE and materials controlling COE have been discussed in detail. The following conclusions are drawn from this study.

(1) Pure and blended cement products subjected to accelerated curing temperatures show faster hydration and non-homogeneous scattering of hydration products that may result in the COE of their compressive strengths.

(2) The COE in compressive strengths of pure and blended cement products depends on the initial curing temperatures from lower to higher values and the amount of internal water present for hydration.

(3) The detection of the COE depends on the duration of testing, loss of water loss, temperature and techniques of curing, porosity, type of cement used, proportions of mineral additions, mineralogical characteristics of mineral additions and specific surface and composition of cement used.

(4) The COE in pure and blended cement products can appear even after 7 days onwards.

(5) The compressive strength loss at later ages in accelerated cured specimens ranges from 2.2% to 31% at 40–60 °C for accelerated prolong curing.

(6) The COE due to accelerated curing can occur at curing temperature as low as 40 °C.

(7) The COE can be controlled using slag blended cement with 80% replacement of GGBFS by OPC, 25–50% addition of slag powder in blended cement mortars, type V Portland cement concrete under an accelerated curing temperature of up to 50 °C as well as blended mortars with different percentages of natural pozzolan.

9. Recommendations for Future Studies

The COE in CBMs has a crucial role due to importance of the mechanical properties and durability of concrete for long life of concrete structures. Because of the lack of detailed investigation on COE, further studies are required to understand this phenomenon in CBMs. The following are some recommendations for better understanding of the COE phenomenon.

1. COE phenomenon was studied mainly on CBMs containing pure OPC or OPC-slag blended cements. Further studies on the CBMs incorporating other types of SCMs are required.

2. The mechanism of COE phenomenon is still unclear. A quantitative study in terms of microstructure and chemistry of materials exhibiting COE may help to understand this phenomenon.

3. The existing data of the accelerated curing is mainly for short time. While, COE may happen at later ages. In addition, the most elevated temperatures reported in the literature for curing of CBMs are in the range of 40–60 °C. Therefore, studies are required to investigate the effect of elevated temperatures more than 60 °C for long terms.

Author Contributions: All authors have equal contribution.

Funding: The authors gratefully acknowledge the support given through the University of Malaya RU—Faculty Research Grant (Project No: GPF015A-2018), University of Malaya Bario Program Research Grant (Project No: RG561-18HTM) and University of Malaya Postgraduate Research Grant (Project No: PG151-2015B).

Conflicts of Interest: The authors declare no conflict of interest.

Abbreviations

CBMs	cement-based materials
COE	crossover effect
SCMs	supplementary cementitious materials

References

1. Hosseini, P.; Booshehrian, A.; Delkash, M.; Ghavami, S.; Zanjani, M. Use of nano-SiO_2 to improve microstructure and compressive strength of recycled aggregate concretes. *Nanotechnol. Constr.* **2009**, *3*, 215–221.
2. Cervera, M.; Faria, R.; Oliver, J.; Prato, T. Numerical modelling of concrete curing, regarding hydration and temperature phenomena. *Comput. Struct.* **2002**, *80*, 1511–1521. [CrossRef]
3. Al-Gahtani, A. Effect of curing methods on the properties of plain and blended cement concretes. *Constr. Build. Mater.* **2010**, *24*, 308–314. [CrossRef]
4. Çakır, Ö.; Aköz, F. Effect of curing conditions on the mortars with and without GGBFS. *Constr. Build. Mater.* **2008**, *22*, 308–314. [CrossRef]
5. Neville, A.; Brooks, J. *Concrete Technology Revised Edition—2001 Standards Update*; Pearson Education Limited: Paris, France, 1990.
6. Li, B.; Mao, J.; Lv, J.; Zhou, L.J.E.J.o.E.; Engineering, C. Effects of micropore structure on hydration degree and mechanical properties of concrete in later curing age. *Eur. J. Environ. Civ. Eng.* **2016**, *20*, 544–559. [CrossRef]
7. Silva, R.; De Brito, J.; Dhir, R.J.E.J.o.E.; Engineering, C. The influence of the use of recycled aggregates on the compressive strength of concrete: A review. *Eur. J. Environ. Civ. Eng.* **2015**, *19*, 825–849. [CrossRef]
8. Kim, J.-K.; Han, S.H.; Song, Y.C. Effect of temperature and aging on the mechanical properties of concrete: Part I. Experimental results. *Cem. Concr. Res.* **2002**, *32*, 1087–1094. [CrossRef]
9. Fall, M.; Célestin, J.; Pokharel, M.; Touré, M. A contribution to understanding the effects of curing temperature on the mechanical properties of mine cemented tailings backfill. *Eng. Geol.* **2010**, *114*, 397–413. [CrossRef]
10. Carino, N.J.; Malhotra, V. *Handbook on Nondestructive Testing of Concrete*, 2nd ed.; CRC Press: Boca Raton, FL, USA, 2003.
11. Escalante-Garcıa, J.; Sharp, J. The microstructure and mechanical properties of blended cements hydrated at various temperatures. *Cem. Concr. Res.* **2001**, *31*, 695–702. [CrossRef]
12. Ogirigbo, O.R.; Black, L. Influence of slag composition and temperature on the hydration and microstructure of slag blended cements. *Constr. Build. Mater.* **2016**, *126*, 496–507. [CrossRef]
13. Da Luz, C.A.; Hooton, R.D. Influence of curing temperature on the process of hydration of supersulfated cements at early age. *Cem. Concr. Res.* **2015**, *77*, 69–75. [CrossRef]
14. Ezziane, K.; Bougara, A.; Kadri, A.; Khelafi, H.; Kadri, E. Compressive strength of mortar containing natural pozzolan under various curing temperature. *Cem. Concr. Compos.* **2007**, *29*, 587–593. [CrossRef]
15. Castellano, C.C.; Bonavetti, V.L.; Donza, H.A.; Irassar, E.F. The effect of w/b and temperature on the hydration and strength of blastfurnace slag cements. *Constr. Build. Mater.* **2016**, *111*, 679–688. [CrossRef]
16. Escalante-García, J.I.; Sharp, J.H. Effect of temperature on the hydration of the main clinker phases in portland cements: Part i, neat cements. *Cem. Concr. Res.* **1998**, *28*, 1245–1257. [CrossRef]
17. De Weerdt, K.; Haha, M.B.; Le Saout, G.; Kjellsen, K.; Justnes, H.; Lothenbach, B. The effect of temperature on the hydration of composite cements containing limestone powder and fly ash. *Mater. Struct.* **2012**, *45*, 1101–1114. [CrossRef]
18. Castellano, C.; Bonavetti, V.; Irassar, E. Effect of curing temperature on hydration and strength of cement paste with granulated blast-furnace slag. *Rev. Constr.* **2007**, *6*, 4–15.
19. Yang, K.-H.; Mun, J.-S.; Cho, M.-S. Effect of Curing Temperature Histories on the Compressive Strength Development of High-Strength Concrete. *Adv. Mater. Sci. Eng.* **2015**, *2015*, 12. [CrossRef]
20. Lothenbach, B.; Winnefeld, F.; Alder, C.; Wieland, E.; Lunk, P. Effect of temperature on the pore solution, microstructure and hydration products of Portland cement pastes. *Cem. Concr. Res.* **2007**, *37*, 483–491. [CrossRef]
21. Lee, H.-S.; Wang, X.-Y.; Zhang, L.-N.; Koh, K.-T.J.M. Analysis of the optimum usage of slag for the compressive strength of concrete. *Materials* **2015**, *8*, 1213–1229. [CrossRef]
22. Escalante-Garcia, J.-I.; Sharp, J. The chemical composition and microstructure of hydration products in blended cements. *Cem. Concr. Compos.* **2004**, *26*, 967–976. [CrossRef]

23. Paya, J.; Monzo, J.; Borrachero, M.; Peris-Mora, E. Mechanical treatment of fly ashes. Part I: Physico-chemical characterization of ground fly ashes. *Cem. Concr. Res.* **1995**, *25*, 1469–1479. [CrossRef]
24. Erdoğdu, S.; Kurbetci, S. Optimum heat treatment cycle for cements of different type and composition. *Cem. Concr. Res.* **1998**, *28*, 1595–1604.
25. Türkel, S.; Alabas, V. The effect of excessive steam curing on Portland composite cement concrete. *Cem. Concr. Res.* **2005**, *35*, 405–411. [CrossRef]
26. Neville, A.M. *Properties of Concrete*; Longman: London, UK, 1995; Volume 4.
27. Kjellsen, K.; Detwiler, R. Later-age strength prediction by a modified maturity model. *ACI Mater. J.* **1993**, *90*, 220–227.
28. Newman, J.; Choo, B.S. *Advanced Concrete Technology 3: Processes*; Butterworth-Heinemann: Oxford, UK, 2003.
29. Khan, M.S.; Abbas, H. Performance of concrete subjected to elevated temperature. *Eur. J. Environ. Civ. Eng.* **2016**, *20*, 532–543. [CrossRef]
30. Sajedi, F.; Razak, H.A. Thermal activation of ordinary Portland cement–slag mortars. *Mater. Des.* **2010**, *31*, 4522–4527. [CrossRef]
31. Namarak, C.; Satching, P.; Tangchirapat, W.; Jaturapitakkul, C. Improving the compressive strength of mortar from a binder of fly ash-calcium carbide residue. *Constr. Build. Mater.* **2017**, *147*, 713–719. [CrossRef]
32. Yazıcı, H.; Aydın, S.; Yiğiter, H.; Baradan, B. Effect of steam curing on class C high-volume fly ash concrete mixtures. *Cem. Concr. Res.* **2005**, *35*, 1122–1127. [CrossRef]
33. Aldea, C.-M.; Young, F.; Wang, K.; Shah, S.P. Effects of curing conditions on properties of concrete using slag replacement. *Cem. Concr. Res.* **2000**, *30*, 465–472. [CrossRef]
34. Liu, B.; Xie, Y.; Li, J. Influence of steam curing on the compressive strength of concrete containing supplementary cementing materials. *Cem. Concr. Res.* **2005**, *35*, 994–998. [CrossRef]
35. Hanif, A.; Kim, Y.; Usman, M.; Park, C.J.M. Optimization of Steam-Curing Regime for Recycled Aggregate Concrete Incorporating High Early Strength Cement—A Parametric Study. *Materials* **2018**, *11*, 2487. [CrossRef] [PubMed]
36. Mindess, S.; Young, J.F. *Concrete Prentice-Hall*; Pearson College: Victoria Englewood, BC, Canada, 1981; p. 481.
37. Ozkul, M.H. Efficiency of accelerated curing in concrete. *Cem. Concr. Res.* **2001**, *31*, 1351–1357. [CrossRef]
38. Chithra, S.; Dhinakaran, G. Effect of hot water curing and hot air oven curing on admixed concrete. *Int. J. Chemtech Res.* **2014**, *6*, 1516–1523.
39. Agarwal, S. Pozzolanic activity of various siliceous materials. *Cem. Concr. Res.* **2006**, *36*, 1735–1739. [CrossRef]
40. Carino, N.J. The maturity method: Theory and application. *Cem. Concr. Aggreg.* **1984**, *6*, 61–73.
41. Sajedi, F.; Razak, H.A. Effects of curing regimes and cement fineness on the compressive strength of ordinary Portland cement mortars. *Constr. Build. Mater.* **2011**, *25*, 2036–2045. [CrossRef]
42. Aprianti, E.; Shafigh, P.; Zawawi, R.; Hassan, Z.F.A. Introducing an effective curing method for mortar containing high volume cementitious materials. *Constr. Build. Mater.* **2016**, *107*, 365–377. [CrossRef]
43. Mehta, P.K. *Microstructure and properties of hardened concrete*, 3rd ed.; McGraw-Hill: New York, NY, USA, 2006; pp. 41–80.
44. Brooks, J.; Al-Kaisi, A. Early strength development of Portland and slag cement concretes cured at elevated temperatures. *Mater. J.* **1990**, *87*, 503–507.
45. Ma, W.; Sample, D.; Martin, R.; Brown, P.W. Calorimetric study of cement blends containing fly ash, silica fume, and slag at elevated temperatures. *Cem. Concr. Aggreg.* **1994**, *16*, 93–99.
46. Verbeck, G.J. Structures and physical properties of cement paste, Invited Paper. In Proceedings of the 5th International Symposium on the Chemistry of Cement, Tokyo, Japan, 7–11 October 1968; pp. 1–32.
47. Pimenta Teixeira, K.; Perdigão Rocha, I.; De Sá Carneiro, L.; Flores, J.; Dauer, E.A.; Ghahremaninezhad, A. The Effect of Curing Temperature on the Properties of Cement Pastes Modified with TiO_2 Nanoparticles. *Materials* **2016**, *9*, 952. [CrossRef] [PubMed]
48. Tank, R.C.; Carino, N.J. Rate constant functions for strength development of concrete. *Mater. J.* **1991**, *88*, 74–83.
49. Alexander, K.M. Concrete strength, cement hydration and the maturity rule. *Aust. J. Appl. Sci.* **1962**, *13*, 277–284.
50. Carino, N. *Temperature Effects on the Strength–Maturity Relation of Mortar, Report No. NBSIR 81-2244*; National Bureau of Standards: Washington, DC, USA, 1981.
51. Carino, N.J.; Lew, H. Temperature effects on strength-maturity relations of mortar. *J. Proc.* **1983**, *80*, 177–182.

52. Kjellsen, K.O.; Detwiler, R.J.; Gjørv, O.E. Pore structure of plain cement pastes hydrated at different temperatures. *Cem. Concr. Res.* **1990**, *20*, 927–933. [CrossRef]
53. Lee, C.; Lee, S.; Nguyen, N. Modeling of Compressive Strength Development of High-Early-Strength-Concrete at Different Curing Temperatures. *Int. J. Concr. Struct. Mater.* **2016**, *10*, 205–219. [CrossRef]
54. Yi, S.-T.; Moon, Y.-H.; Kim, J.-K. Long-term strength prediction of concrete with curing temperature. *Cem. Concr. Res.* **2005**, *35*, 1961–1969. [CrossRef]
55. Kim, J.-K.; Moon, Y.-H.; Eo, S.-H. Compressive strength development of concrete with different curing time and temperature. *Cem. Concr. Res.* **1998**, *28*, 1761–1773. [CrossRef]
56. Kjellsen, K.O.; Detwiler, R.J. Reaction kinetics of Portland cement mortars hydrated at different temperatures. *Cem. Concr. Res.* **1992**, *22*, 112–120. [CrossRef]
57. Mindess, S.; Young, J. *Concrete*; Prentice Hall: Upper Saddle River, NJ, USA, 1981.
58. Mohamed, O.A.J.S. Effect of Mix Constituents and Curing Conditions on Compressive Strength of Sustainable Self-Consolidating Concrete. *Sustainability* **2019**, *11*, 2094. [CrossRef]
59. Strunge, J.; Knoefel, D.; Dreizler, I. Influence of alkalis and sulfur on the properties of cement: I, Effect of the SO sub 3 content on the cement properties. *ZKG* **1985**, *38*, 150–158.
60. Voglis, N.; Kakali, G.; Chaniotakis, E.; Tsivilis, S. Portland-limestone cements. Their properties and hydration compared to those of other composite cements. *Cem. Concr. Compos.* **2005**, *27*, 191–196. [CrossRef]
61. Popovics, S. *Concrete Materials: Properties, Specifications, and Testing*; William Andrew: Amsterdam, The Netherlands, 1992.
62. Aïtcin, P.C. 2-Phenomenology of cement hydration. In *Science and Technology of Concrete Admixtures*; Aïtcin, P.-C., Flatt, R.J., Eds.; Woodhead Publishing: Cambridge, UK, 2016; pp. 15–25.
63. Bougara, A.; Lynsdale, C.; Ezziane, K. Activation of Algerian slag in mortars. *Constr. Build. Mater.* **2009**, *23*, 542–547. [CrossRef]
64. Schindler, A.K. Effect of temperature on hydration of cementitious materials. *Mater. J.* **2004**, *101*, 72–81.
65. Wade, S.A.; Nixon, J.M.; Schindler, A.K.; Barnes, R.W. Effect of temperature on the setting behavior of concrete. *J. Mater. Civ. Eng.* **2010**, *22*, 214–222. [CrossRef]
66. Ezziane, K.; Kadri, E.-H.; Bougara, A.; Bennacer, R. Analysis of Mortar Long-Term Strength with Supplementary Cementitious Materials Cured at Different Temperatures. *ACI Mater. J.* **2010**, *107*, M37.
67. Wang, C.C. Modelling of the compressive strength development of cement mortar with furnace slag and desulfurization slag from the early strength. *Constr. Build. Mater.* **2016**, *128*, 108–117. [CrossRef]

Article

Limestone and Calcined Clay-Based Sustainable Cementitious Materials for 3D Concrete Printing: A Fundamental Study of Extrudability and Early-Age Strength Development

Yu Chen [1,*], Zhenming Li [1], Stefan Chaves Figueiredo [1], Oğuzhan Çopuroğlu [1], Fred Veer [2] and Erik Schlangen [1]

[1]   Faculty of Civil Engineering and Geosciences, Delft University of Technology,
      2628 CN Delft, The Netherlands; Z.Li-2@tudelft.nl (Z.L.); S.ChavesFigueiredo@tudelft.nl (S.C.F.);
      O.Copuroglu@tudelft.nl (O.C.); Erik.Schlangen@tudelft.nl (E.S.)
[2]   Faculty of Architecture and the Built Environment, Delft University of Technology,
      2628 BL Delft, The Netherlands; F.A.Veer@tudelft.nl
*   Correspondence: Y.Chen-6@tudelft.nl; Tel.: +31-644015749

Received: 3 April 2019; Accepted: 25 April 2019; Published: 30 April 2019

Abstract: The goal of this study is to investigate the effects of different grades of calcined clay on the extrudability and early-age strength development under ambient conditions. Four mix designs were proposed. Three of them contained high, medium, and low grades of calcined clay, respectively, and one was the reference without calcined clay. In terms of extrudability, an extrusion test method based on the ram extruder was introduced to observe the quality of extruded material filaments, and to determine the extrusion pressure of tested materials at different ages. For evaluating the very early-age strength development, the penetration resistance test, the green strength test, and the ultrasonic pulse velocity test were applied. Furthermore, the mechanical properties of the developed mix designs were determined by the compressive strength test at 1, 7 and 28 days. Finally, the main finding of this study was that increasing the metakaolin content in calcined clay could significantly increase the extrusion pressures and green strength, shorten the initial setting time and enhance the compressive strength at 1, 7, and 28 days.

Keywords: sustainable; extrudability; early-age strength; limestone and calcined clay; 3D concrete printing

---

1. Introduction

3D concrete printing (3DCP), also referred to as digital fabrication or additive manufacturing of concrete, has been under development by both the academic research and construction industries in the past decade [1–3]. The unique advantages of 3DCP over conventional concrete construction may include saving labors and costs, eliminating formwork, increasing flexibility in architecture design, as well as reducing the construction time [3–6]. Until now, there are two main approaches in 3DCP: the extrusion-based method and the powder-based method [2,7]. This study only focuses on the extrusion-based 3DCP, which generally employs an additive, layer-based, formwork-free extrusion and deposition process [7]. Considering conventional concrete technology principles, the extrusion-based 3DCP cannot be regarded as a sustainable construction method. As reported by Panda et al. [8], the amount of ordinary Portland cement (OPC) forms more than 70% of the binder in most of the available cementitious materials for 3DCP. To satisfy the required fresh properties of 3D printable concrete, such as pumpability and extrudability, the content of aggregate is significantly reduced, which leads to a further increased quantity of OPC in the mix design [9]. Producing OPC consumes massive energy and emits large amounts of greenhouse gases [10]. According to Dong et al. [11], the cement industry is responsible for nearly 7% of global total CO_2 emission. Partially replacing OPC by supplementary cementitious materials (SCMs) might be a proper way to start making the printable concrete sustainable. Recently, many researchers started to look for suitable SCMs for decreasing the use of OPC in 3D printable concrete. The mix designs for 3DCP proposed by Nerella et al. [2] contain 55% of OPC (391 kg/m^3), 30% of fly ash and 15% of microsilica in the binder. It is a good example to show the feasibility of using comparable content of fly ash and microsilica/silica fume to substitute OPC. The attempt of using a high volume of fly ash or slag to replace OPC can be found in the recent works by Panda et al. [12] and Panda and Tan [13], respectively. A large amount of limestone is also utilized by Bentz et al. [14] for developing a sustainable cementitious binder. However, many limitations may appear by using the common SCMs like fly ash, silica fume, slag, and limestone as the main ingredients of 3D printable cementitious materials for a long-term application. The main problem for silica fume and slag is the limited quantity of their production worldwide [15,16]. The amount of fly ash is relatively higher, whereas more than 66% of available fly ash is not suitable for blending with cement due to quality reasons [17]. Using the above 10% of limestone alone as the substitution of OPC in the binder would result in higher porosity and weaker strength [16,18].

Chen et al. [19] pointed out that it is worthwhile to use limestone and calcined clay as the OPC substitutions for developing sustainable and printable cementitious materials. Reasons include the following: (1) kaolinitic clay and limestone exist in abundance; (2) the production of calcined clay requires much lower heating temperature (700–850 °C). Manufacturing 1 ton of calcined clay may emit only 0.3 ton of CO_2 [20] which is much less than producing the same mass of OPC (emitting 0.8–0.9 ton of CO_2) [21]; (3) a mixture using limestone and calcined clay with 1:2 mass ratio not only could replace higher contents of OPC but generate carbo-aluminate hydrates which can fill the capillary pores [16]. The porosity refinement could contribute to increasing the compressive strength at an older age; (4) the lower grades of calcined clay (metakaolin content: 40–50%) also shows the comparable pozzolanic reactivity which has been demonstrated by Avet et al. [22]. Metakaolin (MK) or high grades of calcined kaolinitic clay with high purity and price is generally used by other industries, for instance, paper, ceramics and refractory [16]. In the concrete industry, it appears to be an economical choice to use lower grades of calcined clay, which are widely available compared to MK. However, the calcined clay sourced from different suppliers may contain various percentages of MK. Different calcined clay may have different compositions, secondary phases, fineness, and specific surfaces [22] which will influence the rheological behaviors of fresh cementitious materials. To our knowledge, at this moment few researchers are focusing on developing limestone and calcined clay-based cementitious materials for 3DCP. The effects of different grades of calcined clay (different MK content) on fresh behavior and early-age strength development is to be further explored.

Extrudability and early-age strength should be emphasized as two major constraints of developing printable cementitious materials. As cement is a time-dependent material, the workability of a given printable mix design is not constant and shows loss of consistency with time. Printability window/open time was used to describe the valid time for extruding the fresh material with acceptable quality [8,23]. Thus, the printability window/open time of a printable mix design could be understood as the timespan of extrudability. As explained by Roussel [24], extrusion behavior of fresh cementitious materials are dependent on several rheological parameters, such as yield stress, plastic viscosity and shear strength. Many experimental methods have been proposed from the literature to quantify how these rheological parameters change over time. Le et al. [7] used a shear vane apparatus to measure the change of shear strength of fresh mortar with time. More recently, rheological behavior of fresh cementitious materials were precisely determined by using a coaxial-cylinder rheometer (equipped with a unit cell and vane rotor) [4,25]. Besides, the ram extruder was also one method to characterize the extrudability of cement-based material [2]. The ram extruder could be regarded as a model to simulate the process of material extrusion and deposition in the 3DCP process. The extrusion technology as a cost-efficient and environmental-friendly fabrication method has been used in the cementitious materials industry for a long time [26] before applied in 3DCP. According to Perrot et al. [27], it is feasible to use a ram extruder as a rheo-tribometer to calculate the extrusion factors of the high yield fluids, such as bulk and shear yield stresses, wall friction yield stress, and others. The extrusion pressure could reflect the combined impacts of rheological and tribological properties of cement-based materials. Since a cement-based material is heterogeneous and time-dependent, the extrusion behavior should be changing with time. In this study, a new extrusion test based on the ram extruder is proposed for characterizing the extrudability of the developed limestone and calcined clay-based cementitious materials at different ages.

On the other hand, elimination of formwork in 3DCP introduces many challenges for developing printable cementitious materials. Without the supports from formwork, the extruded fresh mortar layer should emerge with high yield stress immediately to maintain the shape and sustain the load from subsequent layers once they are deposited [28]. That is the main reason why it is essential to monitor the mechanical performance of the printable mix design at an early age. In this study, the early age strength contains two sections: the green strength, as well as the compressive strength at 1, 7, and 28 days. The phase of green strength which was defined by Voigt et al. [29] is used to describe the uniaxial compressive strength of mixtures at the fresh state. According to the simulated building rates (1.1-6.2 m/h) from Perrot et al. [28], it will take at most 3h to print a large concrete component with a height of 3m. Taking the potential mixing and transporting time of fresh mortar into account, 4h is sufficient to indicate green strength development which plays a dominant factor for the layer stacking process. However, it is difficult to measure the green strength of mortar specimens in the fresh state. Wolfs et al. [30] gave a good example for determining the green strength of 3D printable mortar at different fresh material ages of t = 0, 15, 30, 60, and 90 min (The time zero was recorded after mixing, demolding, and placing the specimens). In another paper, Wolfs et al. [31] attempted to build a correlation between the results of uniaxial unconfined compression test and ultrasonic pulse velocity test for the fresh printable mortar. According to their findings, ultrasonic pulse wave transmission tests probably can be used as a non-destructive method to indicate the early-age mechanical performance of their 3D printable mortar. Besides, Ma et al. [32] believed that penetration resistance could be regarded as an essential factor to quantify the stiffness and strength development of cement-based materials at rest and it is also feasible to reflect the structural build-up behavior. In this study, the penetration resistance test, green strength test, and ultrasonic pulse velocity test were performed to evaluate the developed mix designs.

This paper provides a lab-scale methodology for developing limestone and calcined clay based sustainable cementitious material to be used in the 3DCP process. This research also aims to investigate the effects of different grades (different percentages of MK) of calcined clay on the extrudability and early-age strength development under ambient conditions. In terms of extrudability, an extrusion test method based on the ram extruder was introduced to determine the extrusion pressure of different mix designs at different ages. The shape quality of extruded filaments was also observed. To evaluate the early-age strength development, a series of tests were applied, including the penetration resistance test, the green strength test and the ultrasonic pulse velocity test. Furthermore, the mechanical performance of the developed mix designs was indicated by the compressive strength test at 1, 7, and 28 days.

2. Materials and Methods

2.1. Raw Materials

A rapid hardening CEM I 52.5R type Portland cement (PC), a limestone filler (LF) and two types of calcined clay were used as the binder. The low-grade calcined clay (LCC) that contained 40–50% of MK and the high-grade calcined clay (HPCC) which contained near 90% of MK was produced by flash calcination of 95% kaolinite clay from Burgess (US). Different grades of calcined clay could be achieved by blending LCC and HPCC in different proportions. According to Chen et al. [9], most of the 3D concrete printers have limited nozzle sizes and up to 2 mm of sand was used in many studies [2,8,33].

Therefore, fine aggregate with a maximum diameter size of 2 mm was also selected in this study. Figure 1 shows the cumulative particle size distribution of all dry components. HPCC had the smallest average particle size amongst all of the dry components in this study. LCC contained a relatively higher content of coarse particles. The chemical composition of dry components was analyzed by X-ray fluorescence spectrometry (XRF) shown in Table 1. For figuring out the contents of the reactive SiO_2 and Al_2O_3 in LCC and HPCC, a chemical dissolution method was utilized [34]. A proper amount of calcined clay was dissolved in dilute hydrochloric acid solution and afterward treated with boiling sodium carbonate solution [35]. The insoluble residue (I.R.) was rinsed, heated to around 950 °C, and cooled to the room's temperature (20 °C). Finally, the residue part was weighed by a scale and tested by XRF. The mass loss due to chemical dissolution can be regarded as the amorphous phase content. The amount of reactive SiO_2 and Al_2O_3 is given in Table 2. The molar mass ratios of Si/Al in the reactive part of LCC and HPCC were 0.33 and 0.74 respectively.

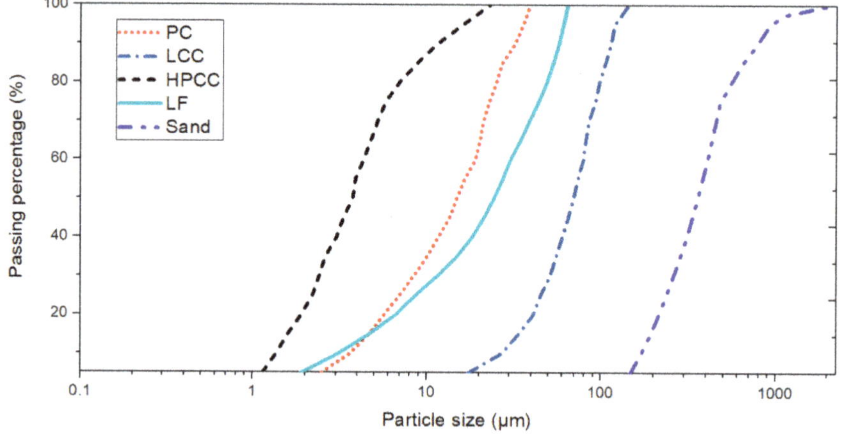

Figure 1. Particle size distribution of dry components.

Table 1. The chemical composition of the main components in the binder (wt. %).

Oxide (wt. %)	SiO_2	Al_2O_3	CaO	Fe_2O_3	K_2O	TiO_2	ZrO_2	Other	Total
PC	17.4	4.1	68.7	2.8	0.6	0.3	0	6.1	100.0
LF	0.2	0	39.6	0.1	0	0	0	60.1	100.0
LCC	55.1	38.4	0.6	2.6	0.2	1.1	0.1	1.9	100.0
HPCC	47.3	50.6	0	0.5	0.2	1.3	0	0.1	100.0

Table 2. Reactivity of low-grade calcined clay (LCC) and high-grade calcined clay (HPCC) (wt. %).

Components (wt. %)	I.R.	Reactive Content	Reactive SiO_2	Reactive Al_2O_3	Other Reactive Phases
LCC	51.2	48.8	12.3	32.0	4.5
HPCC	24.9	75.1	34.6	39.9	0.6

Besides, a Polycarboxylate Ether (PCE) based superplasticizer (MasterGlenium®51, BASF) was adopted in this research to achieve the required flowability, pumpability, and extrudability of mixtures. As described by Marchon et al. [36], superplasticizer could be used in the cementitious material to reduce yield stress and viscosity for enhancing the material workability at a constant solids volume. On the other hand, decreasing the water content at constant yield stress is also beneficial to reduce the porosity of the hardened cementitious material that contributes to improving the mechanical properties and durability. Currently, PCE is reported as the most effective superplasticizer for commercial use [35,36]. While superplasticizer is only used for adjusting printing properties such as pumpability and extrudability, it still requires other admixtures to retain the shape of the deposited material upon extrusion. Based on the research of Figueiredo et al. [37], Ma and Wang [38] and Kazemian et al. [23], adding a small portion of viscosity modifying admixture (VMA) is a method to improve the shape stability and buildability of the printed layers. In this study, Hydroxy Propyl Methyl Cellulose (HPMC) was selected as the VMA for increasing cohesion of the printable mixture.

2.2. Mix Designs and Preparation of Fresh Materials

The suitable binder-to-sand mass ratio (B/S) was 1:1.5 and water-to-binder ratio (W/B) was selected as 0.3. The B/S ratio adopted for this research was the same as the optimal mix design suggested by Le et al. [7] and Panda and Tan [33]. There were two reasons for keeping the W/B as 0.3. As previously reported in the literature [3,8], a printable cementitious mortar must have no slump but should offer sufficient workability. For most of our mix designs, 0.3 was a suitable W/B to ensure the sufficient workability of extrusion without the slump. Second, the smaller value of the W/B could contribute to a dense matrix with a lower capillary porosity and better mechanical performance. Lower W/B was also found in the 3D printable mortar proposals of Le et al. [7] (W/B = 0.26) and Ma et al. [32] (W/B = 0.27). As shown in Table 3, 60% of Portland cement was replaced by calcined clay and limestone in the binder mix. Based on the study of Antoni et al. [39], the limestone-to-MK weight proportion of 1:2 showed the highest compressive strength during the first 28 days of curing. Avet et al. [22] demonstrated that the calcined clay with more than 40% of MK in the ternary blend cement (30% of calcined clay, 15% of limestone, 50% of Portland cement and 5% of gypsum) were obtained the comparable compressive strength with plain Portland cement after 7 days. Therefore, in this paper, the mass ratio of limestone-to-calcined clay was also kept as 1:2 in the mix of MIX-L, MIX-M and MIX-H. L, M and H represented the calcined clay with a low (40–50%), medium (about 62.5%) and high (about 75%) dosage of MK. The HPCC was used to substitute 25% and 50% of LCC in mixtures MIX-M and MIX-H, respectively. According to Antoni et al. [39], up to 2 wt.% (of the binder mass) of PCE for mortar has no significant impact on the hydration kinetics of limestone and calcined clay cement. Thus, the maximum PCE content was regarded as approximately 17 kg/m^3. Without adding VMA, the fresh mixture showed similar properties as the self-compacting mortar. Based on test experiences [37], the VMA content was selected as 0.24 wt.% of the total binder mass to retain the initial printing shape. The mix design named MIX-R used LF to replace all calcined clay. MIX-R was not designed for printing, and its function was to determine the role of calcined clay on mechanical performance. Thus, MIX-R was only investigated by the penetration resistance test and compressive strength test (1, 7 and 28 days). For all tests in this paper, the fresh mortar samples were prepared following the mixing procedures of Table 4. Moreover, the time zero (t = 0) was defined as the time when the suspension (water and PCE) was added into the dry mixtures.

Table 3. Mix Designs.

Type	PC	Calcined Clay		LF	Sand	Water	PCE	VMA
	(kg/m^3)	LCC (kg/m^3)	HPCC (kg/m^3)	(kg/m^3)	(kg/m^3)	(kg/m^3)	(kg/m^3)	(kg/m^3)
MIX-R	331	0	0	497	1242	248	17	2
MIX-L	331	331	0	166	1242	248	17	2
MIX-M	331	248	83	166	1242	248	17	2
MIX-H	331	166	165	166	1242	248	17	2

2.3. Penetration Resistance Test

An automatic Vicat apparatus was used to determine the stiffness development and setting time of different mix designs. To investigate the structural build-up behaviors of cementitious materials, a penetration resistance test was used by Ma et al. [32]. In this paper, the entire test was performed under the specifications of NEN-EN 196-3 [40]. Before the initial set (penetration depth > 36.5 mm), the Vicat needle dropped automatically after every 10 min. Once the penetration depth was equal to or smaller than 36.5 mm, the next drop period was decreased to 5 min. In total there were 44 drops for each specimen.

Table 4. Timeline and procedures of fresh mortar preparation.

Time (min:s)	Steps Followed
-4:00	Homogenizing dry components, mixing with the low speed by a planetary mixer (HOBART).
0:00	While mixing with the low speed, add the blended liquid (water and superplasticizer).
4:00	Pause, scraping the walls and blade (A dough-like mixture is generated).
4:30	Mixing with the high speed. (until the dough-like material adhere to the walls of the bowl).
8:00	Stop, start to fill molds or do other tests

2.4. Extrusion Test

The primary objective of this test was to quantify the change of extrusion pressure by using different mixtures at different ages. As shown in Figure 2, the ram extruder contains four stainless steel components: a stand, a long die with the round opening (inner-diameter: 12.8 mm, length: 102.4 mm), a barrel (inner-diameter: 38.4 mm, length: 125 mm) and a piston. In this case, the ram extruder is built on the Instron universal testing machine that can apply the extrusion force and record the test results. For each experiment, 1 L of fresh mortar that was prepared as given in Table 4. After the mixing process, the required volume (about 0.16 L) of fresh material was filled into the barrel and die. The remaining material was collected into a sealed plastic bag. Before performing the extrusion test, a Fluon ring was inserted in the notch of the piston. The surface of the piston and Fluon ring were lubricated by Dow Corning®7 Release Compound to reduce the friction between the piston and barrel. The test was controlled by the actuator at a predefined speed for a specific displacement. As shown in Figure 3, the pre-test starts from the displacement of 0 mm to 56.5 mm. The piston moves into the barrel at 1 mm/s in the first stage (0 mm to 16 mm). Afterward, different speeds (2, 1, and 0.5 mm/s) were continuously performed to apply sufficient shear stress in the material. From the displacement of 56.5 mm to 70 mm, data readings were collected under 0.25 mm/s of loading rate. It has been found that 13.5 mm was long enough to get a steady-state regime of the extrusion process under the speed of 0.25 mm/s. Finally, the test ended at 2 mm/s to the maximum displacement of 83.5 mm. In total, the test duration was approximately 2 minutes. For each mix design, the material was tested at different ages, t = 10, 25, 45, 60, 90, 120, and 150 min. The test can be terminated earlier if the mixture has a relatively short initial setting time. Three repeated tests for every mix design were conducted to achieve the average results. The testing took place entirely under the same environmental condition (temperature: 20 ± 2 °C; relative humidity: 50 ± 5%). A high-resolution camera was used to take pictures of the final product of the extrusion test. Those filaments were used to assess the printing quality of the evaluated mixtures.

2.5. Green Strength Test

Plastic cylindrical molds with internal diameter (d) of 33.5 mm and 67.5 mm height (h) were used to make samples to investigate the green strength of the fresh mixtures. As Wolfs et al. [30] mentioned, nearly $h = 2d$ can satisfy the requirement of forming a diagonal shear failure plane. Before pouring the fresh mortar, a silicon spray was used to coat the interior surface of molds for easily demolding later. All of the fresh samples were compacted on a vibration table with 30 Hz for 6–8s to reduce air bubbles. The samples were covered by a plastic film and carefully demolded just 30s earlier before performing the uniaxial unconfined compression test. For each mix design proposal, the samples were tested at the ages of t = 30, 45, 60, 90, 150, and 240 min (4 h). Five specimens were prepared for each fresh mortar age. 30 specimens for testing per mixture. The same Instron universal testing machine, equipped with a 10 kN load cell and employed for the extrusion tests was also used for the green strength assessment. A double layer of plastic film covered both sides of a sample. The Polytetrafluoroethylene (PTFE) spray was applied between two plastic sheets to reduce the friction between the sample and base plates. The tests were conducted using displacement control at a rate of 0.2 mm/s. The maximum displacement was 20 mm. Each test took about 2min in total and strain values up to 30% were eventually achieved. A short test is preferable to preserve the fresh properties of the material under evaluation. The casting and test processes were performed under the same environmental condition (temperature: 20 ± 2 °C; relative humidity: 50 ± 5%). A high-resolution camera was used to acquire images during the mechanical test. Vertical and lateral deformations were obtained by image analysis using ImageJ. The cross-section of each sample was determined through a MATLAB-code.

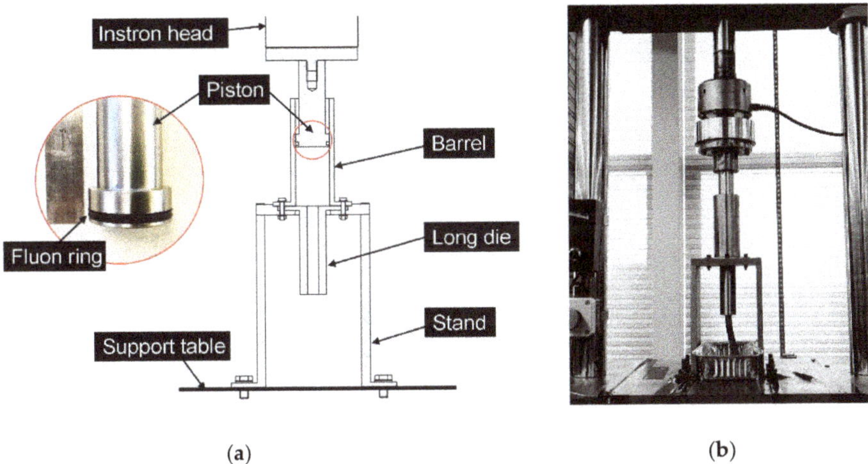

Figure 2. (**a**) Section drawing of the ram extruder; (**b**) Photograph of the ram extruder.

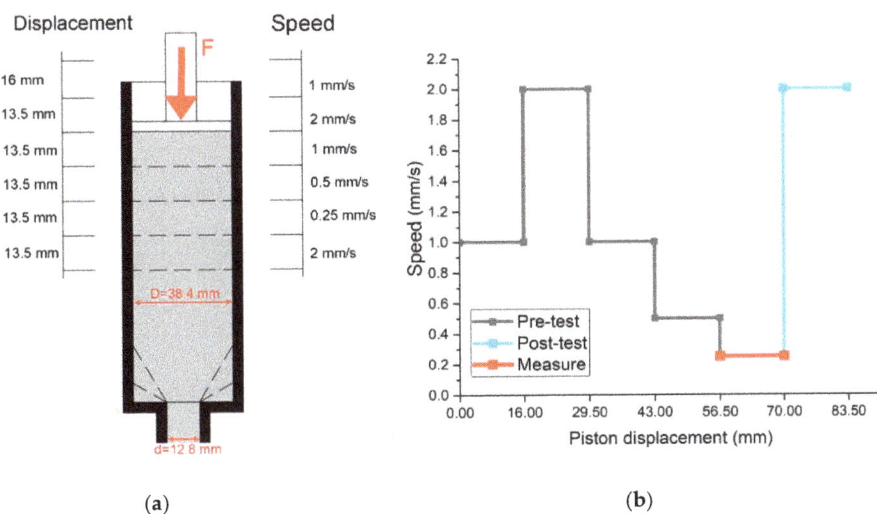

Figure 3. (**a**) Diagram of the extrusion process controlled by the actuator; (**b**) Predefined speed and displacement for the extrusion test.

2.6. Ultrasonic Pulse Velocity Test

This test was performed based on NEN-EN 12504-4 [41] and Wolfs et al. [31]. A Pundit ultrasonic pulse velocity test apparatus which could transmit and receive longitudinal wave (P-wave) was used. For keeping the constant test time of green strength test, the fresh mortar specimen was tested in the periods from 30 to 240 min. As shown in Figure 4, the ultrasonic pulse velocity test system contains a transmitter, a receiver and a steel rectangle mold with internal length (250 mm), width (50 mm) and height (160 mm). The transmitter and receiver were placed and fixed on both sides of the mold with a distance of 50 mm. After pouring the fresh mortar into the mold, the sample was compacted by jolting 5 times to attain the homogeneous state. A thin plastic sheet was used to cover the sample to avoid water loss during the test. Within the test periods (4 h), one P-wave was sent per second. With the hardening of mortar, the transmitting time of P-wave was getting shorter. The P-wave velocity (v) can be calculated by $v = D/t$, where D (50 mm) is the distance between the transmitter and receiver, t is the transmitting time. For each mix design, 3 times of repeated ultrasonic pulse velocity test were conducted to get the average result. All tests were performed in similar environmental conditions with Section 2.5.

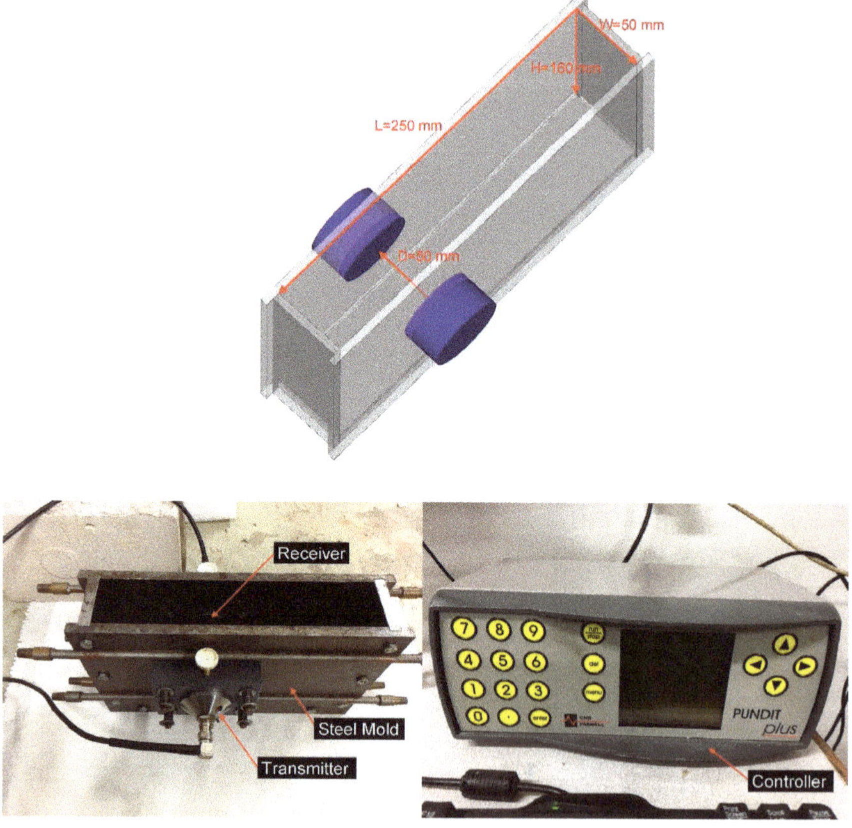

Figure 4. The ultrasonic pulse velocity test apparatus.

2.7. Compressive Strength Test

Compressive strength values were measured on mold-cast specimens. Mold-cast specimens were cured and stored at the fog room (20 ± 2 °C, above 99% RH) and tested at 1, 7 and 28 days to monitor the development of compressive strength with time. 6 samples were prepared for each mix design at the same age. All samples were tested in accordance with the specifications of NEN-EN 196-1 [42]. The loading rate was kept at 2.4 kN/s for each trial.

3. Results and Discussion

3.1. Penetration Resistance Test

The penetration resistance test results of all mix designs are shown in Figure 5. It was found that the transfer time between 40 mm and 0 mm of penetration depth is very short for all mix designs in this study. The sudden stiffening phase change may be attributed to the presence of VMA, which could prolong the initial setting time and has a small impact on the final set. Based on this test, it is not possible to measure the initial and final setting time for these mixtures precisely. The initial setting time of mixtures MIX-L, MIX-M and MIX-H are about 52, 78, and 147 min, respectively. The initial setting time was significantly reduced by increasing the content of MK in calcined clay. Without adding any calcined clay, the initial set was about 258 min in the case of mixture MIX-R.

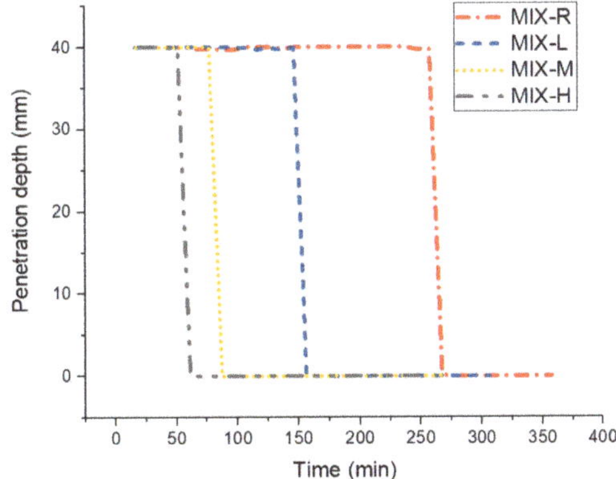

Figure 5. Results of the penetration resistance test by a Vicat apparatus.

3.2. Extrusion Test

An example of extrusion force (or load) with respect to piston displacement at the different material ages is presented in Figure 6a, which contains three zones: pre-testing (blue), measuring (red), and post-testing (white). The reasons why the pre-testing zone was designed in the beginning were explained as follows: (1) compacting the tested material. According to test experiences, pouring the dough-like material by hand made it difficult to fill the barrel and die. Thus, voids might get entrapped between the tested material and the extruder. On the other hand, the tested material might also contain many air voids as there was no vibrating process. The pre-testing procedures could compact the tested materials to reduce the effects of voids. (2) re-shearing the tested material. Due to the admixtures such as PCE and VMA, the mixture showed thixotropic properties. According to Marchon et al. [36], the fresh cementitious materials remained in the dynamic state and showed relatively small yield stress by under the constant shear stress in the printing system during mixing, pumping and extruding. After deposition, the static yield stress of fresh materials increased immediately to meet the requirements of buildability. In this test, the fresh material stayed at rest between two extrusion trials. Thus, it was essential to use pre-testing to 'activate' the fresh material from the static state to the dynamic state. From a microstructural point of view, after mixing with water, the fresh cementitious material first enters into the flocculation phase due to the colloidal attractive forces. Afterward, nucleation of hydrates happened at the pseudo-contact points between particles within the particles network. The hydrate bridges between flocculated cement particles contained higher energy at this stage. When the non-reversible hydrate bonds were generated between particles, the fresh material entered into the phase of structuration. The bonds between particles could be broken by higher forces from shearing or remixing and rebuilt again at rest with the reservoir of chemical species. From the macroscopic view, it was a reversible process to show the thixotropic behaviors of the material. With time passing, the energy required to break those hydrate bridges increased which could be determined as the increase of extrusion pressure with time in this test [24].

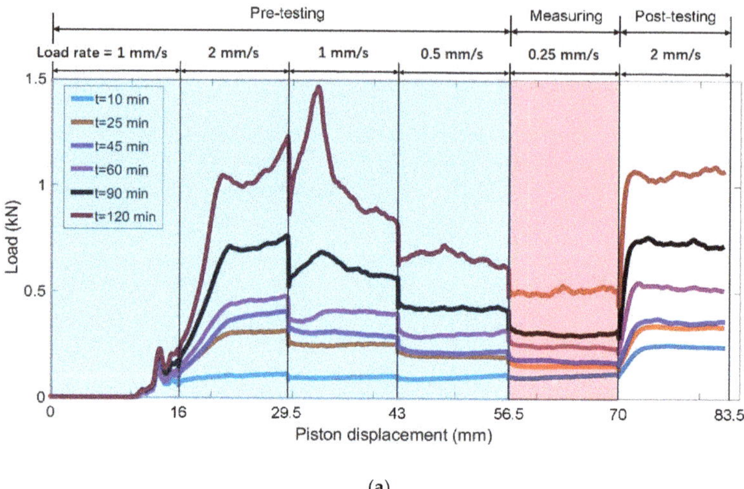

(a)

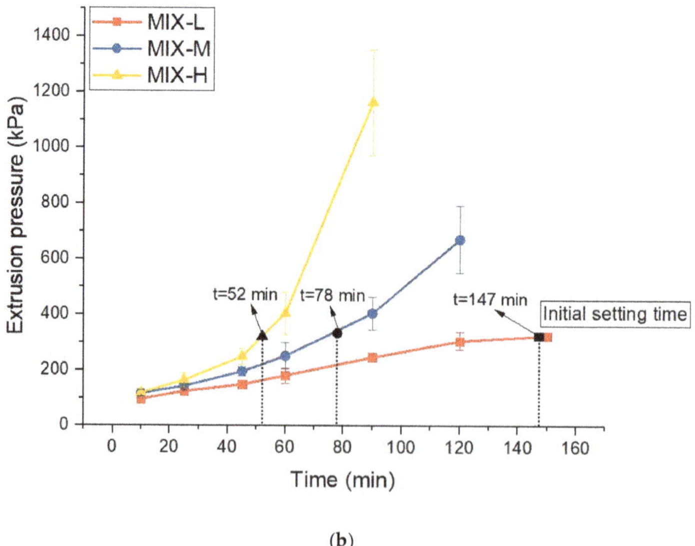

(b)

Figure 6. (**a**) Load and piston displacement test results of mixture MIX-L at different ages. Blue area – the pre-test process for reaching the dynamic status; red area – the test process for taking the average value. (**b**) Comparison of extrusion pressures (under 0.25 mm/s of piston rate) between mix designs at different ages.

As shown in Figure 6a, the extrusion force remained zero at the start until the piston touching the tested material in the barrel. In the second velocity protocol (2 mm/s) of the pre-testing zone, the load increased at the beginning of the test and then reached a steady state. The curves in this zone kept the relatively linear development under the following two constant velocities except for the tested material at the ages of 90 and 120 min. The irrational growth may be attributed to parts of the tested mortar being already hardened in the barrel. Finally, the extrusion force nearly reached the even state for all curves in the measuring zone. The driving velocity increases again from 0.25 to 2 mm/s and the extrusion process is terminated (the curves were cut-off before the end of the test). According to Perrot et al. [27] and Nerella et al. [2], the dramatic load increase with 2 mm/s of piston rate in the pre-testing zone might be partially due to the compacting of the tested material. In their cases, when the material was continuously extruded, the curves showed a slight and gradual decrease since the reduction of inner wall friction. Because of the limited extrusion distance (about 13.5 mm) by each driving velocity in our test, the slight reduction of frictional resistance was hard to be clarified. The decreasing trend was also not clear in the results of Zhou et al. [26]. Thus, we assumed the inner wall friction was consistently the same in the measuring zone. The extrusion pressure (P) in the measuring zone could be calculated by $P = 4F/\pi D^2$, where F is related to the average extrusion force under 0.25 mm/s of piston rate in the measuring zone and D represents the inner diameter of the barrel. The extrusion pressures of mixtures MIX-L, MIX-M and MIX-H at different ages are demonstrated in Figure 6b. The mix design with a higher content of MK showed the higher extrusion pressure with time.

As explained by Lothenbach et al. [43], most SCMs, including MK, could be regarded as fillers at the early-age, due to the pozzolanic reaction was mainly dependent on the hydrated cement products which were generated after alite hydration. One of the principal mechanisms about filler effect was illustrated as the finer SCMs could provide extra surface as nucleation sites for the hydration products. Mixtures MIX-H and MIX-M had a much higher growth rate of extrusion pressures with time to compare with that of mixture MIX-L. The extra MK in the matrix might accelerate the phase change of cement particles from flocculation to structuration. That could be one reason to explain the decreased initial setting time with an increased amount of MK addition. Besides, the HPCC used in this study had very small average particle diameter size (d50 = 3.75 μm, see Figure 1) which might physically increase the inner friction between particles and the number of available nucleation surfaces in the system. Figure 7 gives the extruded filaments collected after each extrusion test. Even at the age of 10 min, the extruded filaments of mixtures MIX-L, MIX-M and MIX-H showed stable shapes. The curves of mixtures MIX-H and MIX-M showed significantly high extrusion pressures as well as substantial standard deviations after the initial setting time. The high load was potentially harmful to the experimental setup.

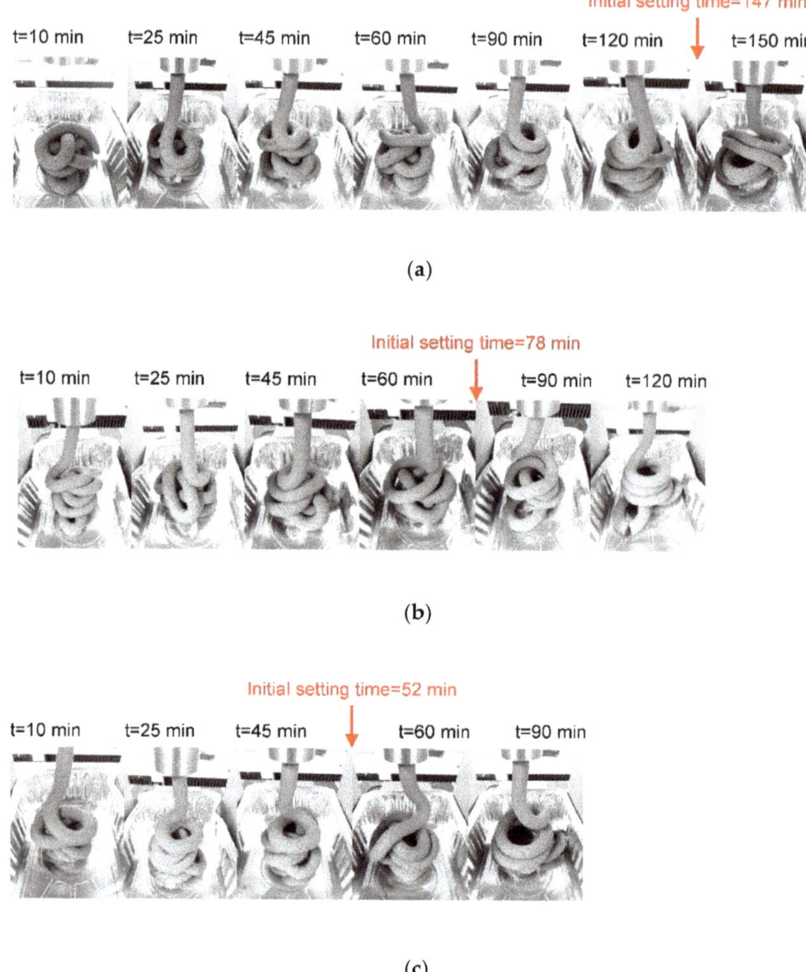

(**a**)

(**b**)

(**c**)

Figure 7. (**a**) Extruded filaments of mixture MIX-L; (**b**) Extruded filaments of mixture MIX-M; (**c**) Extruded filaments of mixture MIX-H.

3.3. Green Strength Test

The load and displacement results of mixtures MIX-L, MIX-M, and MIX-H are seen in Figure 8a–c. The changes in the mechanical performance of each mix design at a very early age (t = 30, 45, 60, 90, 150 and 240 min) were quite evident. For each mix design, the older samples (t = 90, 150 and 240 min) had a higher peak force than the samples with younger ages (t = 30, 45 and 60 min). As shown in Figure 8a–c, for each specimen, the load increased near linearly with the vertical displacement in the initial path. After that stage, the increasing rate of the load was reduced as deformation grew. The specimens with the younger ages, such as t = 30, 45, and 60 min, had a rising load as the deformation increased until reaching a certain height. Whereas the older specimens (t = 90, 150 and 240 min) showed a load decrease after the peak force. Wolfs et al. [30] and Panda et al. [44] explained this phenomenon. They mentioned that the difference between older and young ages samples was the occurrence of lateral deformations and failure mode. The specimens with the age of t = 30, 45, and 60 min had relatively small stiffness which led to an expansion in the horizontal direction with significant deformation.

As shown in Figure 9, the younger specimens fail by barreling instead to generate a distinct crack pattern. The area of cross-section grows under the increases of vertical displacement. The squeezing process resulted in force increases. However, the older specimens showed a typical failure pattern. After the peak force, the load decreased causes the formation of fractures. The older had higher brittleness and lower ductility than the younger specimens.

A high-resolution camera recorded the lateral displacement of each sample. Through using ImageJ (see Figure 8d) and MATLAB, the change of cross section area was determined. For each sample at the very early ages (t = 30, 45, 60 and 90 min), true stresses were calculated using the updated areas of the cross-section. Thus, the load-displacement curves in Figure 8a–c were transferred into the stress and strain curves with the average results demonstrated in Figure 10a–c. All curves were cut-off after the peak stress which is regarded as the ultimate green strength. The self-weight of the sample was excluded in the calculation, since the failure plane of each sample may happen at a different height [30]. In this study, the green strength of mixtures MIX-L, MIX-M, and MIX-H at the different ages are illustrated in Figure 10d. Increasing MK content in calcined clay could significantly increase the green strength. Mixture MIX-H showed the highest green strength and strength development rate from 30 min to 240 min. The test results in this section indicated that increasing MK content in calcined clay could benefit the very early-age strength development which may be used to improve the buildability of the fresh mixtures in 3DCP process.

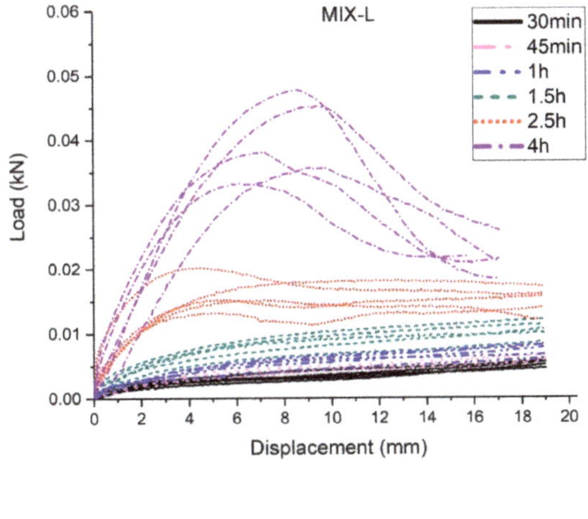

(**a**)

Figure 8. *Cont.*

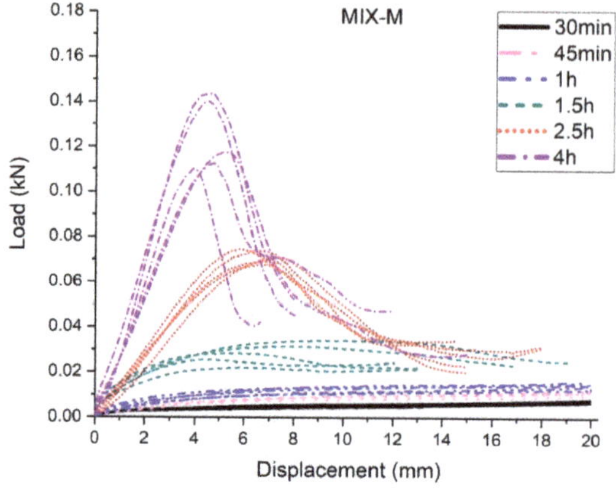

(**b**)

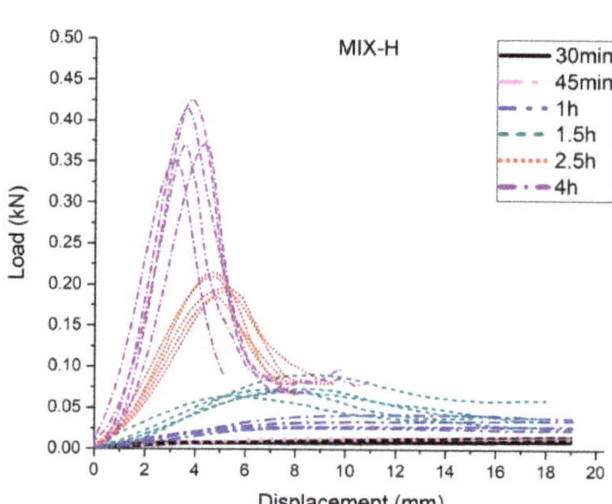

(**c**)

Figure 8. *Cont.*

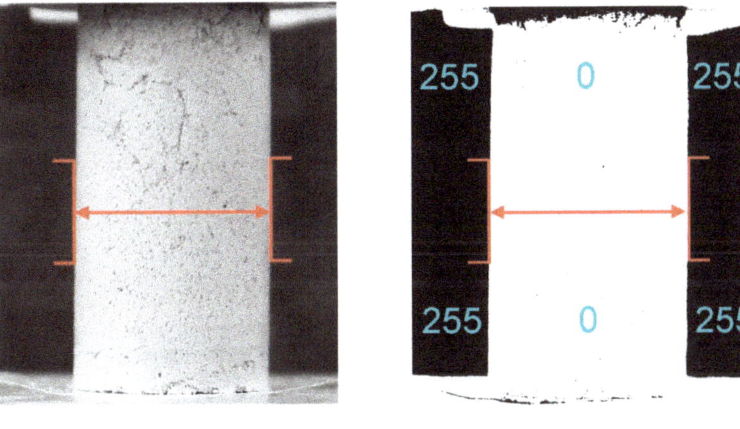

(d)

Figure 8. (**a**) The load and displacement curve of mixture MIX-L; (**b**) The load and displacement curve of mixture MIX-M; (**c**) The load and displacement curve of mixture MIX-H. (**d**) An example of a post-processed image obtained with the help of ImageJ.

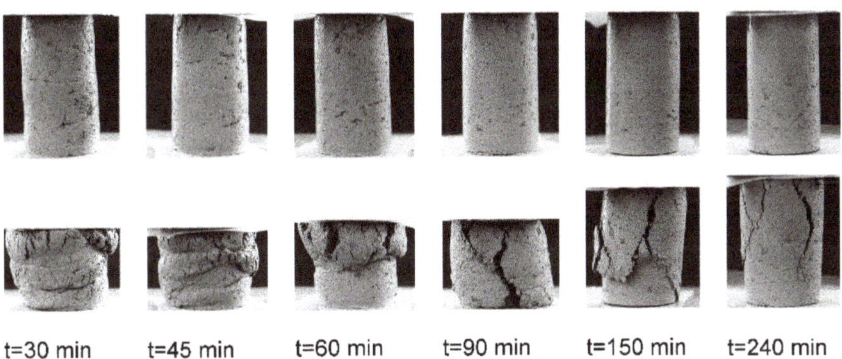

Figure 9. Mixture MIX-H samples in the green strength test. Photographs of demolded samples at different ages (first row); Photographs of damaged samples at different ages (second row).

3.4. Ultrasonic Pulse Velocity Test

The ultrasonic wave transmission measurement might be used to investigate the development of solid phases in the microstructure of cementitious materials. The results of the ultrasonic pulse velocity tests are presented in Figure 11. In the first hour, the measured values were stabilized at about 52 m/s. It might be because the specimens are somewhat fluid in this period and their stiffness is too low to be measured by the machine. After about 75 min, the p-wave velocity in the sample of mixture MIX-H showed nearly linear development with time in the studied time frame. The curves of mixtures MIX-L and MIX-M had a similar developing trend approximately 10 min later. Based on the linear fit of measuring results from 100 min to the end in Figure 11, mixture MIX-H demonstrates the highest gradient (about 3.424). Mixture MIX-M has a slightly higher slope than mixture MIX-L. The increase of p-wave velocity could reflect the growth of internal microstructure for cementitious materials. Therefore, this test could indicate that increasing the content of MK in calcined clay could accelerate the internal structure development in the first 4h.

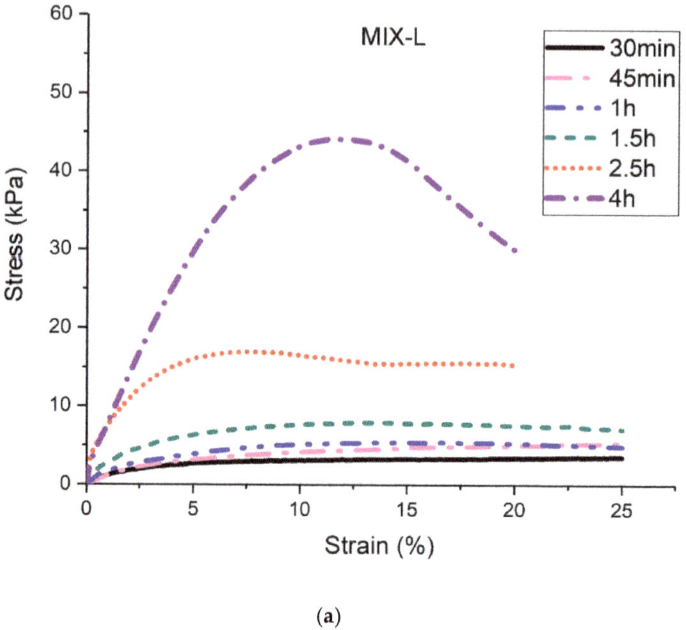

(**a**)

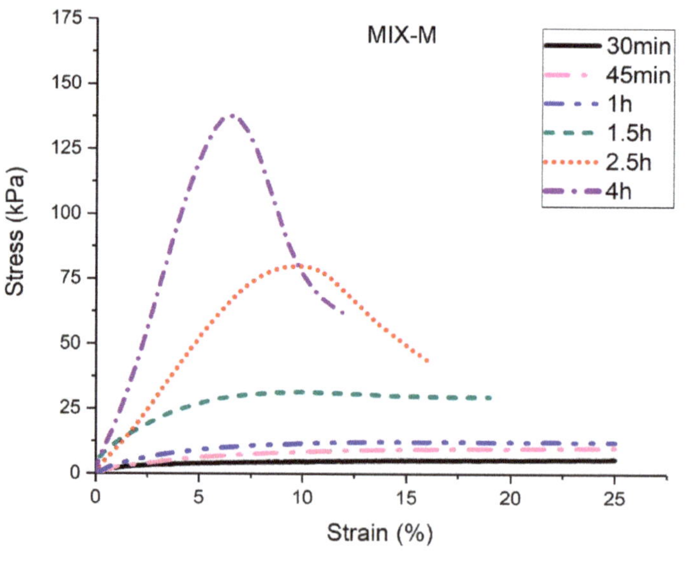

(**b**)

Figure 10. *Cont.*

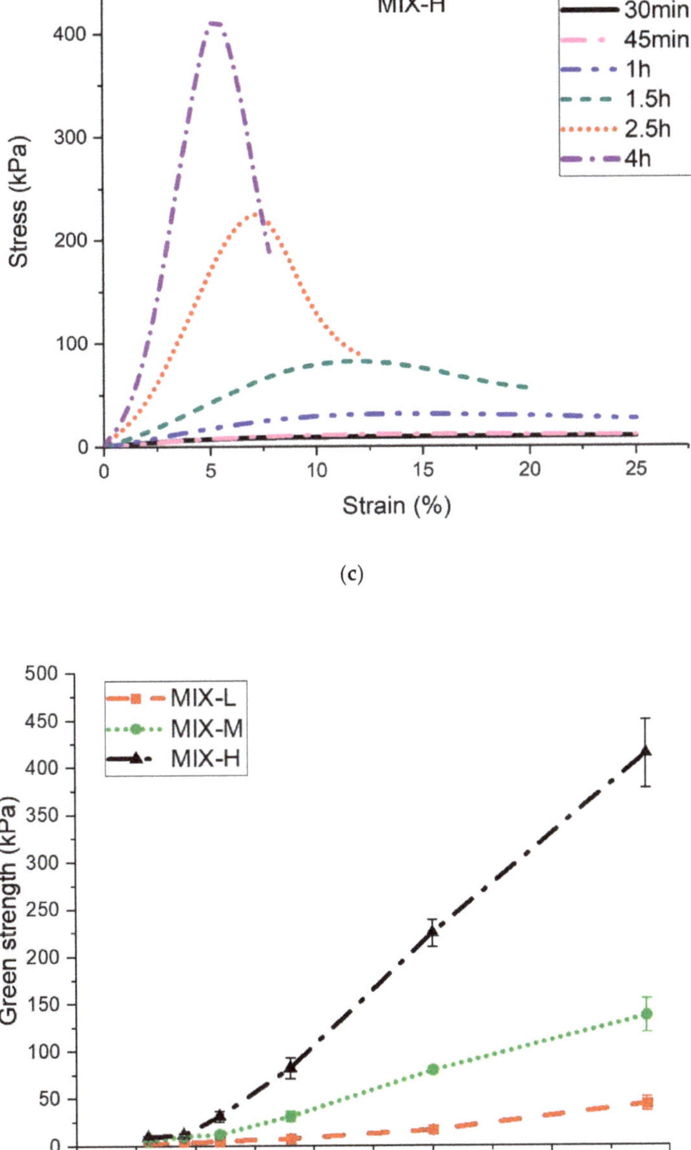

(c)

(d)

Figure 10. (**a**) The average stress and strain curve of mixture MIX-L; (**b**) The average stress and strain curve of mixture MIX-M; (**c**) The average stress and strain curve of mixture MIX-H; (**d**) Comparison of green strength of mix designs at different ages. The relative standard deviation (RSD) of peak stresses is under 20% which is similar to the findings of Wolfs et al. [30].

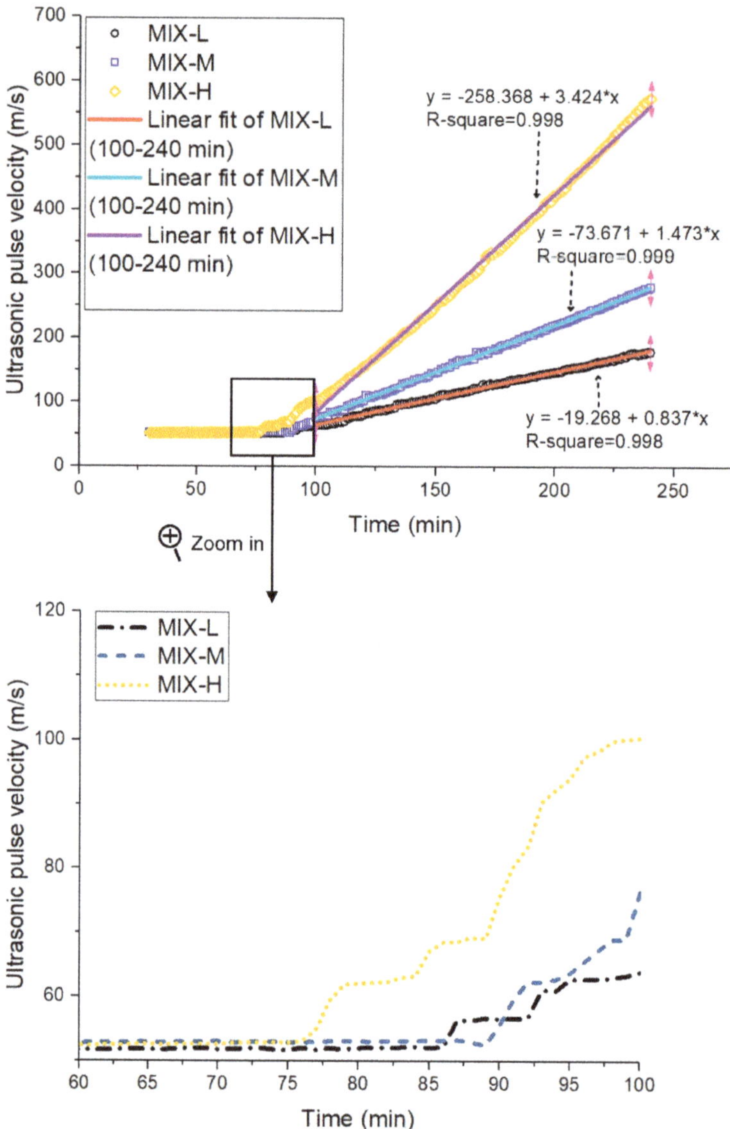

Figure 11. Measuring results of ultrasonic pulse velocity for mixtures MIX-L, MIX-M and MIX-H. The linear fit based on measuring results from 100 min to 240 min.

3.5. Compressive Strength at 1, 7 and 28 Days

The compressive strength values at 1, 7, and 28 days are given in Figure 12. Mix designs contained calcined clay showed higher strength than mixture MIX-R at all ages. For mixtures MIX-L, MIX-M, and MIX-H, the higher content of MK in calcined clay led to the higher compressive strength. Mixture MIX-H showed the highest compressive strength within 28 days. Avet et al. [22] mentioned that the mortar compressive strength of different grades of calcined clay based cementitious materials seemed only depended on the calcined kaolinite clay (MK) content regardless of other compositions, secondary phases, fineness and specific surfaces of the calcined clay. That also agreed with our test results.

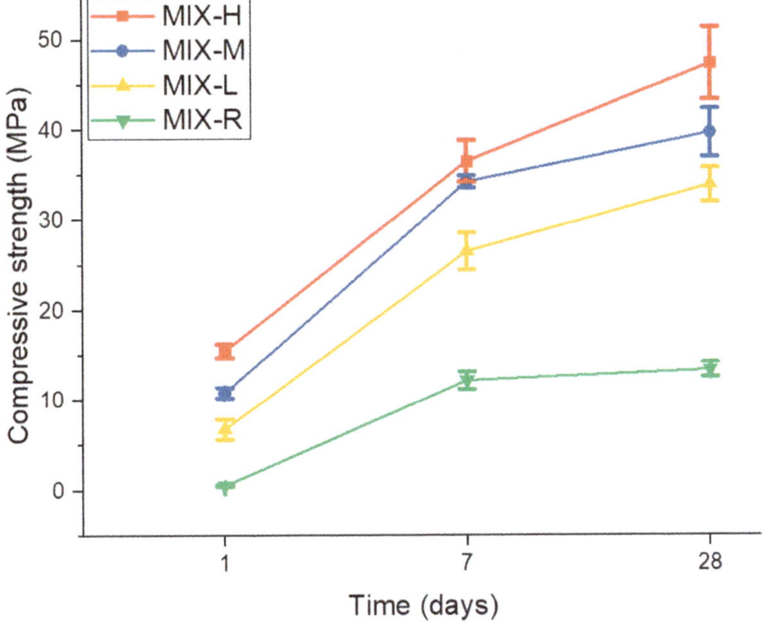

Figure 12. Measuring results of ultrasonic pulse velocity for mixtures MIX-L, MIX-M, and MIX-H. The linear fit based on measuring results from 100 min to 240 min.

4. Conclusions

This paper was a fundamental study to develop the limestone and calcined clay based sustainable cementitious materials for the extrusion-based 3DCP. Four mix designs with different binder compositions were proposed in this study. Mixtures MIX-L, MIX-M, and MIX-H contained 40% of Portland cement, 20% of limestone and 40% of calcined clay in the binder. While the different grades of calcined clay were used, respectively. Mixture MIX-R as the reference mix design had 40% of Portland cement and 60% of limestone as the binder. All mix designs were investigated through penetration resistance test and compressive strength test at 1, 7, and 28 days. The most important goal of this paper was to investigate the effects of different grades (MK content) of calcined clay on extrudability and the early-age strength development. Thus, the mixtures were also evaluated by the extrusion test, green strength test and ultrasonic pulse velocity test. The main findings of this study are summarized as follow.

- A lab-scale extrusion test method based on the ram extruder was proposed in this study to efficiently observe the extruded filaments of the fresh mixture and quantify the required extrusion pressures at different ages. Mixtures MIX-L, MIX-M, and MIX-H showed good shape stability at the ages before their initial setting time. The extrusion pressures and the growth rate of pressure with time were significantly increased by increasing the MK content in calcined clay. Thus, using a higher grade of calcined clay may increase the extrusion shear strength of fresh mixtures which may bring difficulties for extruding and reduce the printability window/open time.
- According to all results found during this study, increasing the MK dosage in calcined clay could significantly accelerate the initial cement hydration. Consequently, higher mechanical performance in the early stages (green strength) were achieved. Moreover, the buildability and structural build-up behavior of mix designs for 3DCP could be enhanced by using higher grades of calcined clay.

- Compared with mixture MIX-R, mix designs with calcined clay showed shorter initial setting time and higher compressive strength at 1, 7, and 28 days.

- Increasing the content of MK in calcined clay could significantly reduce the initial setting time and increase the compressive strength at 1, 7, and 28 days.

- Within the first 4h, the filler effect of the calcined clay may play a dominant role to promote the phase change of flocculation, nucleation and structuration of cement particles from the microstructure view. Increasing the percentage of MK in calcined clay could increase the content of very small particles in the binder system, which may enhance nucleation and facilitate the microstructure development.

- Overall, it is essential to find the balance between extrudability and early-age strength to apply the discussed mix designs in 3DCP. Results of this research showed that the MK content of calcined clay plays a key role in the performance of printable cementitious mortars.

Author Contributions: Conceptualization, Y.C.; Formal analysis, Y.C. and Z.L.; Investigation, Y.C. and Z.L.; Methodology, Y.C. and S.C.F.; Software, Y.C.; Supervision, O.C., F.V. and E.S.; Visualization, Y.C.; Writing – original draft, Y.C.; Writing – review & editing, Z.L., S.C.F., O.C., F.V. and E.S.

Funding: Yu Chen and Zhenming Li would like to acknowledge the funding supported by the China Scholarship Council (CSC) under grant No. 201807720005 and No. 201506120072 respectively. In addition, Stefan Chaves Figueiredo would like to acknowledge the funding from Science Without Borders from the National Council for Scientific and Technological Development of Brazil (201620/2014-6).

Acknowledgments: Burgess Pigment Company is thanked for the donation of the Optipozz®Burgess metakaolin. The authors appreciate Dr. Caglar Yalcinkaya and Mr. Maiko van Leeuwen for their supports in tests.

References

1. Buswell, R.A.; Leal de Silva, W.R.; Jones, S.Z.; Dirrenberger, J. 3D printing using concrete extrusion: A roadmap for research. *Cem. Concr. Res.* **2018**, *112*, 37–49. [CrossRef]

2. Nerella, V.N.; Näther, M.; Iqbal, A.; Butler, M.; Mechtcherine, V. Inline quantification of extrudability of cementitious materials for digital construction. *Cem. Concr. Compos.* **2019**, *95*, 260–270. [CrossRef]

3. Bos, F.; Wolfs, R.; Ahmed, Z.; Salet, T. Additive manufacturing of concrete in construction: Potentials and challenges of 3D concrete printing. *Virtual Phys. Prototyp.* **2016**, *11*, 209–225. [CrossRef]

4. Paul, S.C.; Tay, Y.W.D.; Panda, B.; Tan, M.J. Fresh and hardened properties of 3D printable cementitious materials for building and construction. *Arch. Civ. Mech. Eng.* **2018**, *18*, 311–319. [CrossRef]

5. Reiter, L.; Wangler, T.; Roussel, N.; Flatt, R.J. The role of early age structural build-up in digital fabrication with concrete. *Cem. Concr. Res.* **2018**, *112*, 86–95. [CrossRef]

6. Panda, B.; Paul, S.C.; Hui, L.J.; Tay, Y.W.D.; Tan, M.J. Additive manufacturing of geopolymer for sustainable built environment. *J. Clean. Prod.* **2017**, *167*, 281–288. [CrossRef]

7. Le, T.T.; Austin, S.A.; Lim, S.; Buswell, R.A.; Gibb, A.G.F.; Thorpe, T. Mix design and fresh properties for high-performance printing concrete. *Mater. Struct.* **2012**, *45*, 1221–1232. [CrossRef]

8. Panda, B.; Unluer, C.; Tan, M.J. Investigation of the rheology and strength of geopolymer mixtures for extrusion-based 3D printing. *Cem. Concr. Compos.* **2018**, *94*, 307–314. [CrossRef]

9. Chen, Y.; Veer, F.; Copuroglu, O. A Critical Review of 3D Concrete Printing as a Low CO2 Concrete Approach. *Heron* **2017**, *62*, 167–194.

10. Meyer, C. The greening of the concrete industry. *Cem. Concr. Compos.* **2009**, *31*, 601–605. [CrossRef]

11. Dong, Y.H.; Ng, S.T.; Kwan, A.H.K.; Wu, S.K. Substituting local data for overseas life cycle inventories—A case study of concrete products in Hong Kong. *J. Clean. Prod.* **2015**, *87*, 414–422. [CrossRef]

12. Panda, B.; Ruan, S.; Unluer, C.; Jen, M. Improving the 3D printability of high volume fly ash mixtures via the use of nano attapulgite clay. *Compos. Part B* **2019**, *165*, 75–83. [CrossRef]

13. Panda, B.; Tan, M.J. Material properties of 3D printable high-volume slag cement. In Proceedings of the First International Conference on 3D Construction Printing (3DcP) in Conjunction with the 6th International Conference on Innovative Production and Construction (IPC 2018), Melbourne, Australia, 26–28 November 2018.

14. Bentz, D.P.; Jones, S.Z.; Bentz, I.R.; Peltz, M.A. Towards the formulation of robust and sustainable cementitious binders for 3-D additive construction by extrusion. *Constr. Build. Mater.* **2018**, *175*, 215–224. [CrossRef]

15. Glavind, M. *Sustainability of Cement, Concrete and Cement Replacement Materials in Construction*; Woodhead Publishing Limited: Cambridge, UK, 2009; ISBN 978-1-84569-349-7.

16. Scrivener, K.; Martirena, F.; Bishnoi, S.; Maity, S. Calcined clay limestone cements (LC3). *Cem. Concr. Res.* **2018**, *114*, 49–56. [CrossRef]

17. Snellings, R. Assessing, Understanding and Unlocking Supplementary Cementitious Materials. *Rilem Tech. Lett.* **2016**, *1*, 50–55. [CrossRef]

18. Matschei, T.; Lothenbach, B.; Glasser, F.P. The role of calcium carbonate in cement hydration. *Cem. Concr. Res.* **2007**, *37*, 551–558. [CrossRef]

19. Chen, Y.; Veer, F.; Copuroglu, O.; Schlangen, E. Feasibility of Using Low CO2 Concrete Alternatives in Extrusion-Based 3D Concrete Printing. In Proceedings of the RILEM International Conference on Concrete and Digital Fabrication, Zurich, Switzerland, 10–12 September 2018; pp. 269–276.

20. Huang, W.; Kazemi-Kamyab, H.; Sun, W.; Scrivener, K. Effect of replacement of silica fume with calcined clay on the hydration and microstructural development of eco-UHPFRC. *Mater. Des.* **2017**, *121*, 36–46. [CrossRef]

21. Arbi, K.; Nedeljković, M.; Zuo, Y.; Ye, G. A Review on the Durability of Alkali-Activated Fly Ash/Slag Systems: Advances, Issues, and Perspectives. *Ind. Eng. Chem. Res.* **2016**, *55*, 5439–5453. [CrossRef]

22. Avet, F.; Snellings, R.; Alujas Diaz, A.; Ben Haha, M.; Scrivener, K. Development of a new rapid, relevant and reliable (R3) test method to evaluate the pozzolanic reactivity of calcined kaolinitic clays. *Cem. Concr. Res.* **2016**, *85*, 1–11. [CrossRef]

23. Kazemian, A.; Yuan, X.; Cochran, E.; Khoshnevis, B. Cementitious materials for construction-scale 3D printing: Laboratory testing of fresh printing mixture. *Constr. Build. Mater.* **2017**, *145*, 639–647. [CrossRef]

24. Roussel, N. Rheological requirements for printable concretes. *Cem. Concr. Res.* **2018**, *112*, 76–85. [CrossRef]

25. Nerella, V.N.; Beigh, M.A.B.; Fataei, S.; Mechtcherine, V. Strain-based approach for measuring structural build-up of cement pastes in the context of digital construction. *Cem. Concr. Res.* **2018**, *115*, 530–544. [CrossRef]

26. Zhou, X.; Li, Z.; Fan, M.; Chen, H. Rheology of semi-solid fresh cement pastes and mortars in orifice extrusion. *Cem. Concr. Compos.* **2013**, *37*, 304–311. [CrossRef]

27. Perrot, A.; Mélinge, Y.; Rangeard, D.; Micaelli, F.; Estellé, P.; Lanos, C. Use of ram extruder as a combined rheo-tribometer to study the behaviour of high yield stress fluids at low strain rate. *Rheol. Acta* **2012**, *51*, 743–754. [CrossRef]

28. Perrot, A.; Rangeard, D.; Pierre, A. Structural built-up of cement-based materials used for 3D-printing extrusion techniques. *Mater. Struct.* **2016**, *49*, 1213–1220. [CrossRef]

29. Voigt, T.; Malonn, T.; Shah, S.P. Green and early age compressive strength of extruded cement mortar monitored with compression tests and ultrasonic techniques. *Cem. Concr. Res.* **2006**, *36*, 858–867. [CrossRef]

30. Wolfs, R.J.M.; Bos, F.P.; Salet, T.A.M. Early age mechanical behaviour of 3D printed concrete: Numerical modelling and experimental testing. *Cem. Concr. Res.* **2018**, *106*, 103–116. [CrossRef]

31. Wolfs, R.J.M.; Bos, F.P.; Salet, T.A.M. Correlation between destructive compression tests and non-destructive ultrasonic measurements on early age 3D printed concrete. *Constr. Build. Mater.* **2018**, *181*, 447–454. [CrossRef]

32. Ma, G.; Li, Z.; Wang, L. Printable properties of cementitious material containing copper tailings for extrusion based 3D printing. *Constr. Build. Mater.* **2018**, *162*, 613–627. [CrossRef]

33. Panda, B.; Tan, M.J. Experimental study on mix proportion and fresh properties of fly ash based geopolymer for 3D concrete printing. *Ceram. Int.* **2018**, *44*, 10258–10265. [CrossRef]

34. Li, Z.; Zhang, S.; Zuo, Y.; Chen, W.; Ye, G. Chemical deformation of metakaolin based geopolymer. *Cem. Concr. Res.* **2019**, *120*, 108–118. [CrossRef]

35. *NEN-EN 196-2 Method of Testing Cement—Part 2: Chemical Analysis of Cement*; NEN: Delft, The Netherlands, 2013.

36. Marchon, D.; Kawashima, S.; Bessaies-Bey, H.; Mantellato, S.; Ng, S. Hydration and rheology control of concrete for digital fabrication: Potential admixtures and cement chemistry. *Cem. Concr. Res.* **2018**, *112*, 96–110. [CrossRef]

37. Figueiredo, S.C.; Rodrıguez, C.R.; Ahmed, Z.Y.; Bos, D.H.; Xu, Y.; Salet, T.A.M.; Copuroglu, O.; Schlangen, E.; Bos, F.P. An approach to develop printable strain hardening cementitious composites. *Mater. Des.* **2019**, *169*, 107651. [CrossRef]

38. Ma, G.; Wang, L. A critical review of preparation design and workability measurement of concrete material for largescale 3D printing. *Front. Struct. Civ. Eng.* **2017**, *12*, 1–19. [CrossRef]

39. Antoni, M.; Rossen, J.; Martirena, F.; Scrivener, K. Cement substitution by a combination of metakaolin and limestone. *Cem. Concr. Res.* **2012**, *42*, 1579–1589. [CrossRef]

40. *NEN-EN 196-3—Methods of Testing Cement—Part 3: Determination of Setting Times and Soundness*; NEN: Delft, The Netherlands, 2017.

41. *NEN-EN 12504-4:2005—Testing Concrete—Part 4: Determination of Ultrasonic Pulse Velocity*; NEN: Delft, The Netherlands, 2005.

42. *NEN-EN 196-1—Methods of Testing Cement—Part 1: Determination of Strength*; NEN: Delft, The Netherlands, 2016.

43. Lothenbach, B.; Scrivener, K.; Hooton, R.D. Supplementary cementitious materials. *Cem. Concr. Res.* **2011**, *41*, 1244–1256. [CrossRef]

44. Panda, B.; Hui, L.J.; Tan, M.J. Mechanical properties and deformation behaviour of early age concrete in the context of digital construction. *Compos. Part B Eng.* **2019**, *165*, 563–571. [CrossRef]

 *applied sciences*

Article

Advances in the Study of the Behavior of Full-Depth Reclamation (FDR) with Cement

Hernán Gonzalo-Orden [1], Alaitz Linares-Unamunzaga [1,*], Heriberto Pérez-Acebo [2] and Jesús Díaz-Minguela [3]

[1] Department of Civil Engineering, University of Burgos, c/Villadiego, s/n, 09001 Burgos, Spain
[2] Mechanical Engineering Department, University of the Basque Country UPV/EHU, Pº Rafael Moreno Pitxitxi, 2, 48013 Bilbao, Spain
[3] Spanish Institute of Cement and Its Applications (IECA), c/José Abascal, 53, 1º, 28003 Madrid, Spain
* Correspondence: alinares@ubu.es; Tel.: +34-94-725-9066

Received: 28 June 2019; Accepted: 24 July 2019; Published: 29 July 2019

Abstract: Road maintenance and rehabilitation are expected to meet modern society's demands for sustainable development. Full-depth reclamation with cement as a binder is closely linked to the concept of sustainability. In addition to the environmental benefits of reusing the existing pavement as aggregate, this practice entails significant technical and economic advantages. In Spain, in the absence of tests specifically designed to determine the behavior of recycled pavements stabilized with cement, these materials are treated as soil-cement or cement-bound granular material. This assumption is not entirely accurate, because this recycled pavement contains some bituminous elements that reduce its stiffness. This study aimed to obtain the relationships between flexural strength (FS) and the parameters that describe the pavement behavior (long-term unconfined compressive strength (UCS) and indirect tensile strength (ITS)) and compare the findings with the relationships between these parameters in soil-cement and cement-bound granular materials. The results showed that the similar behavior hypothesis is not entirely accurate for recycled pavements stabilized with cement, because they have lower strength values—although, this is not necessarily an indication of poorer performance.

Keywords: full-depth reclamation; recycling; pavement rehabilitation; cement-treated materials; base materials; unconfined compressive strength; flexural strength; splitting tensile strength; indirect tensile strength

---

1. Introduction

Pavement recycling is a road-rehabilitation technique in which a deteriorated pavement is transformed into a new course. Depending on the recycling processes and mixing temperature, pavement recycling technique can be classified as hot recycling (HR) and cold recycling (CR). HR methodology involves two techniques: Hot in-place recycling and hot central-plant recycling. On the other hand, there are three techniques for CR according to the processing place, the construction technology, and the reclamation depth: Cold in-place recycling, cold central-plant recycling, and full-depth reclamation [1,2].

Full depth reclamation (FDR) is a recycling technique in which all the asphalt pavement section and a previously quantified amount of underlying base material are treated. This mixture is pulverized, either mixed with a stabilizing agent or not, and compacted to produce a stabilized base course [3]. The usual depth that is reclaimed varies from 100 to 300 mm [4–6]. Sometimes, due to the structural capabilities of the mixture, it is not necessary to add any stabilizing additive, and therefore, the compacted material can be the base for a new surface layer. Nevertheless, if the obtained material does not provide enough structural strength, possible stabilizers are classified as chemical additives (Portland cement, hydrated lime, calcium chloride, and fly ash) and bitumen additives (bitumen emulsions). The most employed stabilizers worldwide are bituminous emulsions and Portland cement [3,6–9].

In this case, full depth reclamation with Portland cement (FDR-PC) yields a base course with significant structural capacity, for which the existing road is used as a "quarry" or source of aggregate. With this technique, the materials in the road are reused by pulverizing them, and adding cement, water, and sometimes a small percentage of aggregate or even an additive, in the proportions established through preliminary testing. This mixture is compacted and cured to form the course with the greatest structural strength in the new pavement [6,10–14].

This procedure is, without a doubt, more effective for ensuring user comfort and safety than reinforcing or rebuilding heavily cracked or deteriorated pavements.

FDR with cement has a number of technical, economic, and environmental advantages [15]. It results in a longer-lasting, less erosive, and water-resistant pavements, able to withstand the stress from traffic loads that reach the subgrade more efficiently. This high-performance technique requires no manufacturing plant or transportation of materials. It is also environmentally friendly; since the materials are reused in their present location, new aggregate deposits need not be found, nor existing quarries over-mined. The elimination of transport reduces CO_2 emissions and the associated impact on the road and the traffic [3,16]. Furthermore, the reclaimed pavement can be regarded as solid waste generated from deteriorated roadways [5]. Finally, a life cycle cost analysis of construction and maintenance practices indicates that the maintenance and rehabilitation strategy based on pavement in situ recycling is the least costly, providing savings to the overall economic performance of the road pavement over the life cycle [17].

In recent years, around one million square metres of pavement were recycled in Spain annually, covering a total surface area of nearly 30 million square metres between 1998 and 2018 [18]. This is a proven method that has been widely used and has shown exceptional results to date. Based on the many advantages described, it has a promising future.

Pavement recycling as a rehabilitation method is a technique that was first used in the United Kingdom in the 1940s to repair secondary roads damaged during World War II [19]. However, FDR did not make a comeback until the mid 1980s, when a better understanding of the characteristics of semi-rigid pavements and the development of advanced machinery led to the inclusion of cement in the mix [1,20].

In Spain, cold-recycling techniques first came into use in 1991 after several unsuccessful hot-recycling experiments in the early 1990s [19,21]. The first cold-recycling trial was in Huelva, where the Ministry of Public Works recycled a 12 km stretch of road, N-431, to a depth of 30 cm [22]. Other regions subsequently undertook experiments of their own, and in some cases, such as in the region of Castilla y León, standardized the technique through widespread use. The number of recycled roads quickly grew, ultimately reaching the aforementioned 30 million square metres.

Although FDR with cement is, in principle, more widely used on roads with traffic of low intensity [3,16,23], good results have also been obtained in some of the trials conducted on roads with high-intensity flows of heavy vehicles [8,19].

However, it must be taken into account that these pavements are being designed based on the assumption that FDR with cement and soil-cement exhibit similar behavior, but it differs due to the inclusion of reclaimed asphalt pavement aggregates in their blend, which reduces the stiffness and strength of the mixture [8,14,24]. Therefore, to verify this hypothesis, the long-term characteristics of the recycled material must be determined, as they have only been estimated to date [4].

In 2001, Kolias et al. [25] reported the results of an analysis of the mechanical properties of recycled pavements with different granular and bituminous mix percentages and 3% and 5% cement. The aim was to determine the effect of the bituminous mix percentage and temperature on the strength of the recycled pavement with cement. Compressive, tensile, flexural, and fatigue strength values were found for a small number of one and 60 day specimens. The authors determined that both compressive strength and the modulus of elasticity declined with rising bituminous content. At the same time, they concluded that flexural and tensile strength did not fall at low proportions of bituminous mix, but did so very quickly at higher contents.

In 2008, Díaz et al. [26] published a preview of the unconfined compressive strength (UCS), indirect tensile strength (ITS), and flexural strength (FS) results that form part of the first phase of this study. Since then, the number of trials conducted has grown significantly.

In FDR, the most used test to verify that material was manufactured correctly is the unconfined compressive strength (UCS) test at short-term [10,27–30]. But, in order for a better long-term characterization, it is necessary to perform flexural strength tests, and, more specifically, the four-point flexural beam test [8,31–38].

This FS test is carried out using prismatic specimens and manufacturing them requires a high level of qualification and experience within the testing team [10,39]. This is the main reason for usually estimating their behavior from standardized tests, such as the unconfined compressive strength and the indirect tensile strength (ITS) tests [31,40–42]. For this reason, the method used in this research is the one proposed by the University of Burgos [33].

This research aimed to fill the void in the understanding of the relationships among flexural strength, unconfined compressive strength, and indirect tensile strength based on the results of the tests conducted. To this end, the methods used for other materials mixed with hydraulic binders [33,34,40,43] and the tests described by Kolias et al. [25] were taken as a starting point. Here, however, the applicable European (EN) or Spanish (UNE or NLT, as appropriate) standards were used to characterize the behavior of an FDR with cement. The accuracy of the initial hypothesis of similarity with soil-cement and cement-bound granular material was also evaluated.

2. Materials and Methods

2.1. Material

While the number of possible granular material/bituminous material combinations is virtually countless, the proportion consisting of one-third mix asphalt and two-thirds granular material is the one most commonly used in roads [16], and was consequently chosen for this study (10 cm of mix asphalt and 20 cm of granular material), as can be seen in Figure 1. The bituminous layer has approximately 4.5% of bitumen.

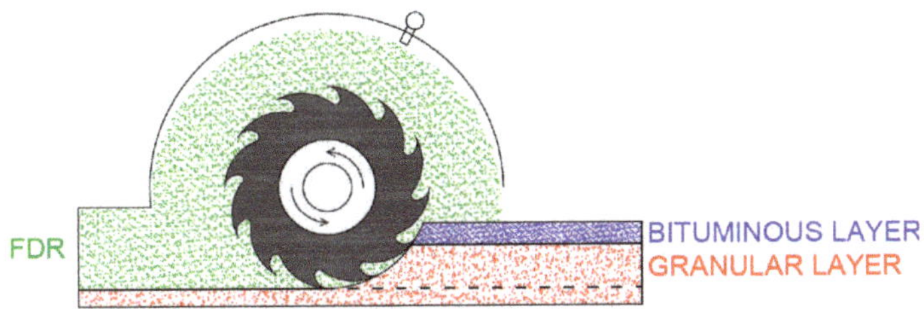

Figure 1. Full-depth reclamation section.

The granular material used in the laboratory trials was recycled pavement taken from road SA-801 (Peñaranda de Bracamonte to Campo de Peñaranda) from the west of Spain, with a maximum aggregate size of 40 mm. Figure 2 shows the granulometry of the material, which is inside the range of the SC40 (soil–cement with a maximum aggregate size of 40 mm) according to the Spanish standards [28]. It not was necessary to add any aggregate to improve the grading. The recycled material exhibited no plasticity and was free of organic matter and other substances that might prevent the cement setting.

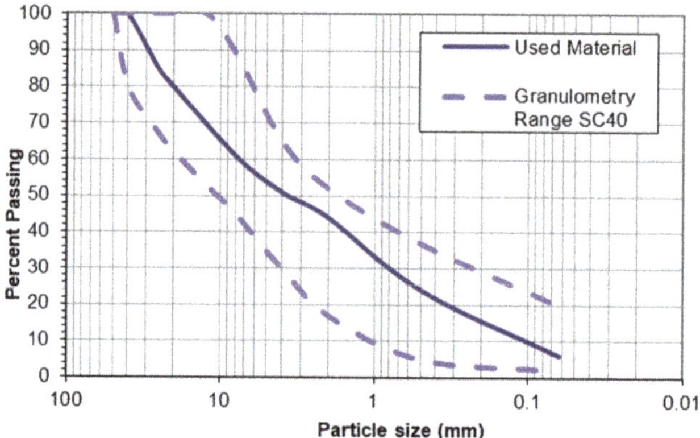

Figure 2. Used material granulometry and granulometry range for SC40.

The cement used was ESP VI-1 32.5 N [44]. This is a widely used cement type for recycled pavements stabilized with cement in roads, because of its low thermal shrinkage and long period workability due to the low quantity of clinker (<50%), high quantity of additives, and moderate strength, mainly short-term [45].

The characteristics of this type of cement are showed in Table 1.

Table 1. Cement ESP VI-1 32.5 N properties [44].

Main Standardized Component	Value	Cement Standardized Specifications	Value
Clinker (K)	25–55%	Sulfate	≤3.5%
Silica fumes (D) [1]		Initial setting time	≥60 min
Natural pozzolans (P) [1]		Final setting time	≤720 min
Calcined natural pozzolans (Q) [1]	45–75%	Expansion	≤10 mm
Siliceous fly ash (V) [1]		UCS at 28 days	22.5 ≤ R ≤ 42.5 MPa
Calcareous fly ash (W) [1]		UCS at 90 days [2]	≥32.5 MPa
Minority components	0–5%	Puzzolanicity	8 to 15 days
Chlorides	≤0.10%	-	-

[1] The natural pozzolans (P) content for Cements ESP VI-1 must be lower than 40%. [2] The code for special cements it is given by its UCS at 90 days.

2.2. Mix Design

The determination of maximum dry density and optimum moisture content was conducted following the UNE 103-501-94 [46] for cylindrical samples, whose prescriptions are analogous to the ASTM D1557-12 [47]. The density to be achieved in the test specimens was 2.10 g/cm^3 with an optimum modified Proctor moisture content of 7.61% [46] (Figure 3).

Further to the results of the proportioning study, 3.5% ESP VI-1 32.5 N cement [44] was used to ensure a 7-day compressive strength [48,49] of at least 2.5 MPa, the minimum value required by the Spanish Ministry of Public Works [29] and the Council of Castilla y León [50] (Table 2).

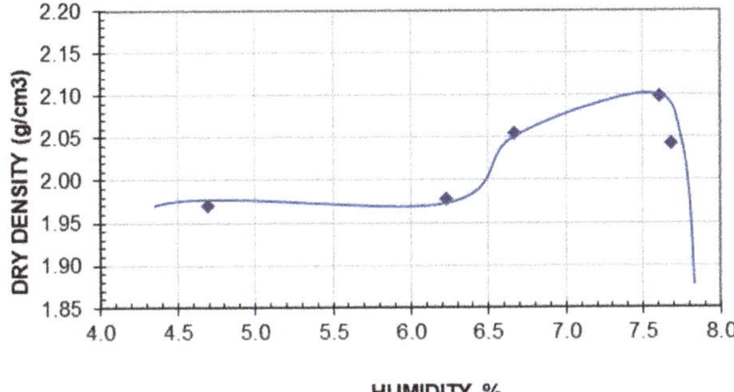

Figure 3. Modified Proctor density of the material.

Table 2. Unconfined compressive strength (UCS) at seven days for different cement content.

Sample	% Cement	Dry Density (g/cm^3)	UCS at 7 Days (MPa)	Average UCS (MPa)
P1.1	3.0	2.151	1.757	
P1.2	3.0	2.108	1.465	2.071
P1.3	3.0	2.122	2.991	
P2.1	3.5	2.070	2.259	
P2.2	3.5	2.143	2.560	2.637
P2.3	3.5	2.151	3.092	

2.3. Testing Program

Twenty-four prismatic specimens were prepared for flexural strength testing to characterize the recycled pavement in accordance with standard UNE-EN 12390-5, "Testing hardened concrete. Flexural strength of test specimens" [51], which is analogous to the ASTM D1635/D1635M-12 [52]. The mould dimensions where 15 cm × 15 cm × 60 cm. Samples were stored in a curing room at 20 ± 2 °C and 95% relative humidity [53]. At a curing age of at least 90 days, the four-point flexural beam test was conducted. This method ensures that the specimens break at the weakest section (uniformity of the bending moment between the two points where the load is applied).

The rollers over the specimen were placed at a distance of 15 cm (the height of the specimen), and the rollers bellow the specimen at a distance of 45 cm (three times the height of the specimen).

The applied load was transmitted by means of a plate between the specimen and the rollers over it. An increasing tension of 0.04 MPa was selected in the slowest way of the standard range of 0.04–0.06 MPa/s [51].

After each specimen of 15 × 15 × 60 cm was tested for flexural strength, specimens are broken approximately in the middle. The two resulting halves were also tested without being trimmed, one for the unconfined compressive strength (UCS) test and the other for the indirect tensile strength (ITS) test, to find the relationship between these values and the FS of the initial test specimen.

For simulating the behavior of a cubic sample in the UCS test, an auxiliary metal sheet (15 cm × 15 cm) was introduced between the lower plate and the lower side of the sample (a half from the prismatic sample), and between the top plate and the top side of the sample. This way, a uniform tensile distribution in a 15 cm cube is obtained (Figure 4a). In the case of the ITS test, the load was applied perpendicularly to the axle of the specimen with a modified metal sheet. Hence, the load was applied with a width of 15 cm (Figure 4b).

(a)   (b)

Figure 4. (**a**) UCS test; (**b**) indirect tensile strength (ITS) test.

UCS tests were conducted following the standard UNE-EN 13286-41 [49], with a load speed in the range interval of 0.1 ± 0.1 MPa/s [54]. ITS strength tests were performed in accordance with UNE-EN 12390-6, "Testing hardened concrete. Tensile splitting strength of test specimens" [55].

3. Results and Discussion

Obtained results from the 72 tests conducted on the 24 prismatic specimens are shown in Table 3.

Table 3. Long-term results obtained for flexural strength, unconfined compression strength, and indirect tensile strength tests.

Sample	FS (MPa)	UCS (MPa)	ITS (MPa)
S1	0.806	3.766	0.520
S2	0.598	3.344	0.313
S3	0.580	3.203	0.402
S4	0.775	3.947	0.447
S5	0.787	3.580	0.423
S6	0.610	3.317	0.398
S7	0.361	2.273	0.198
S8	0.350	2.896	0.273
S9	0.599	3.649	0.393
S10	0.366	2.169	0.209
S11	0.538	4.199	0.488
S12	0.667	4.340	0.483
S13	0.556	3.918	0.457
S14	0.221	2.313	0.155
S15	0.638	3.827	0.394
S16	0.427	2.919	0.174
S17	0.420	2.465	0.128
S18	0.585	5.103	0.314
S19	0.673	4.651	0.345
S20	0.609	4.559	0.427
S21	0.561	3.663	0.380
S22	0.516	3.660	0.379
S23	0.667	4.423	0.386
S24	0.501	3.600	0.310

3.1. Relationship Between Flexural and Unconfined Compressive Strength

The correlation between the values of the UCS at long-term and FS at long-term is shown in Figure 5.

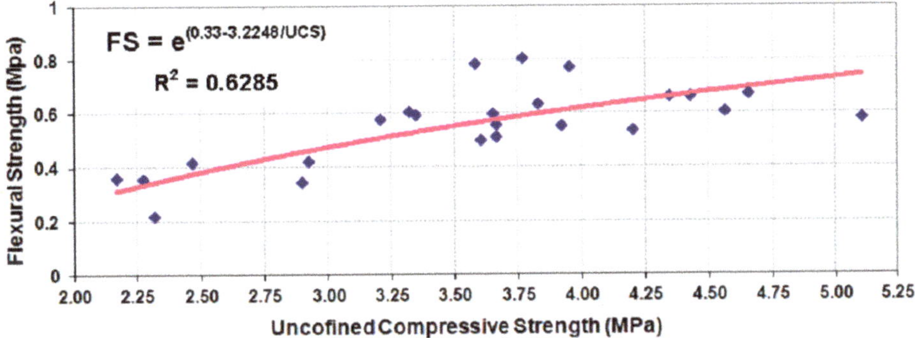

Figure 5. Relationship between unconfined compressive strength at long-term (UCS_{LT}) and flexural strength at long-term (FS_{LT}).

After examining various possible functions for correlating these two variables, the best correlation was obtained with an *S* shape function, with natural logarithm of the FS as dependent variable and 1/UCS as the independent variable. The developed relationship is shown in Equation (1).

$$Ln(FS_{LT\text{-}UCS}) = 0.33 - 0.3225/UCS_{LT} \tag{1}$$

where $FS_{LT\text{-}UCS}$ is the estimated value of the flexural strength at long-term by means of UCS_{LT}, and UCS_{LT} is the unconfined compressive strength at long-term, both expressed in MPa.

The coefficient of determination (R^2) has a value of 0.629, which indicates that the model can explain more than the 62% of the variability of the model.

The average values obtained in the tests for these two parameters for the recycled material were compared to the usual values for soil-cement and cement-bound granular material [26,42,45,56–58] in Table 4.

Table 4. Comparison between average UCS values at long-term and flexural strength (FS) values at long-term for soil-cement, cement-bound granular material, and obtained values for the full-depth reclamation (FDR) with cement of the study.

Materials	UCS_{LT} (MPa)	FS_{LT} (MPa)	UCS_{LT}/FS_{LT}
Soil-cement	4	0.9	4–5
Cement-bound granular material and compacted concrete	8	1.6	5–6
FDR with cement	3.73 (2.9 to 4.9)	0.60 (0.53 to 0.69)	6.21

As seen in Table 4, while the values obtained in the analysis were lower than soil-cement strength due to the bituminous matrix, the relationship between the two parameters was closer to the cement-bound granular material.

3.2. Relationship Between Flexural Strength at Long-Term and Indirect Tensile Strength at Long-Term

The correlation between FS and ITS values at long-term is shown in Figure 6, indicating a linear relationship between these parameters.

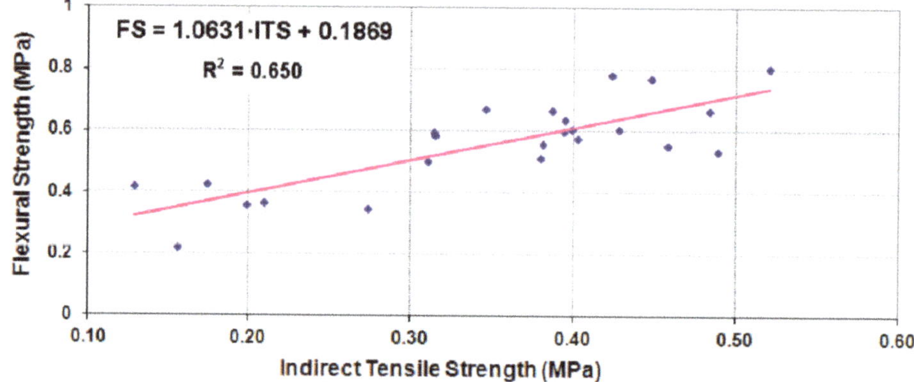

Figure 6. Relationship between indirect tensile strength and flexural strength.

A statistical analysis was performed and it was observed that the best relationship was obtained by means of a simple linear regression, expressed in Equation (2).

$$FS_{LT\text{-}ITS} = 0.187 + 1.063\ ITS_{LT} \tag{2}$$

where $FS_{LT\text{-}ITS}$ is the estimated value of the flexural strength at long-term obtained by means of ITS_{LT}, and ITS_{LT} is the indirect tensile strength at long-term, both in MPa.

The regression has a R^2 value of 0.65 and the F of Fisher-Snedecor test indicated that the relationship was true with a significance level over 99%. The Student's t-tests for the coefficients indicated that they were true, different from 0 with a significance level over 99%.

Once again, the average values obtained in the tests for these two parameters for the recycled material were compared to the usual values for soil-cement and cement-bound granular material [45] in Table 5. Although a direct relationship between ITS and FS is not established for cement-bound granular materials, it is indicated that the UCS value is approximately 10 times the ITS value [26,42,45,56–58]. This assumption is adopted for the analysis in Table 5.

Table 5. Comparison between averages of ITS values at long-term and FS values at long-term for soil-cement, cement-bound granular material, and obtained values for the FDR with cement of the study.

Materials	ITS_{LT} (MPa)	FS_{LT} (MPa)	ITS_{LT}/FS_{LT}
Soil-cement	0.4	0.9	0.4–0.5
Cement-bound granular material and compacted concrete	0.8	1.6	0.5–0.6
FDR with cement	0.40 (0.33 to 0.48)	0.60 (0.53 to 0.69)	0.67

It is observed that the ITS of the recycled material was similar to the value specified for soil-cement, while the relationship between ITS and FS was closer to a cement-bound granular material.

3.3. Relationship Between Indirect Tensile Strength and Unconfined Compressive Strength at Long-Term

The values of these two parameters (UCS and ITS) are compared in Figure 7.

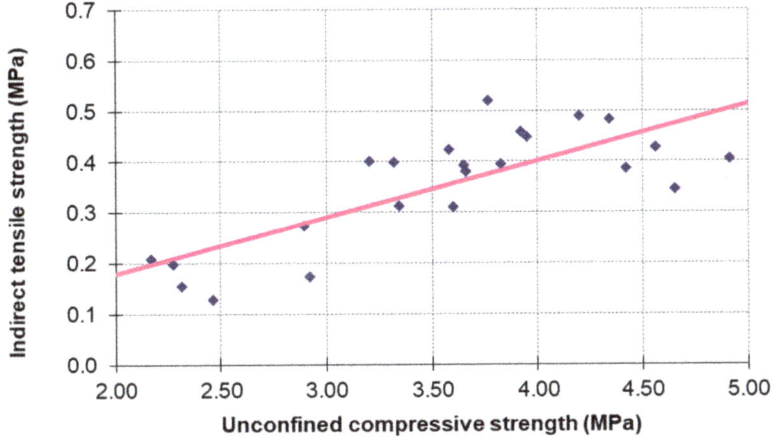

Figure 7. Relationship between ITS and UCS at long-term.

The correlation between both parameters was statistically analyzed and a linear correlation was proposed, as shown in Equation (3).

$$ITS_{LT} = 0.098\ UCS_{LT} \tag{3}$$

where ITS_{LT} and UCS_{LT} are as defined in Equations (1) and (2), respectively, both in MPa.

Equation (3) omitted the intercept because the *p*-value of the Student's *t*-test was over 0.99, indicating that it was not significant. The relationship has an R^2 value of 0.49. The *F* test indicated that the relationship was true with a significance level over 99%.

The relationship between these two parameters at long-term obtained for the recycled material and the usual values for cement-treated base courses [45] were found to be similar (Table 6).

Table 6. Comparison of the relationship between unconfined compressive strength and indirect tensile strength at long-term for soil-cement, cement-bound granular material, and obtained values for the FDR with cement of the study.

Materials	UCS_{LT}/ITS_{LT}
Soil-cement, cement-bound granular material, compacted concrete	8–10
FDR pavement with cement	10.20

3.4. Estimation of Flexural Strength at Long-Term Using the UCS and ITS Values

An additional equation for estimating the flexural strength at long-term of the FDR with cement was developed as a function of the unconfined compressive strength and the indirect tensile strength by means of a multiple linear regression, as shown in Equation (4).

$$FS_{LT\text{-}2} = 0.074\ UCS_{LT} + 0.826\ ITS_{LT} \tag{4}$$

where $FS_{LT\text{-}2}$ is the flexural strength at long-term by means of UCS_{LT} and ITS_{LT} simultaneously, and UCS_{LT} and STS_{LT} are as defined in Equations (1) and (2), respectively.

Equation (4) has a coefficient of determination (R^2) of 0.684. Including an intercept in Equation (4) made the coefficients of the intercept and UCS not significant. Without the intercept, both coefficients are different from 0, with a significance level over 99% (*p*-value of the Student's *t*-test >0.99).

Figure 8 shows the obtained values of FS and the values estimated by Equations (1), (2), and (4).

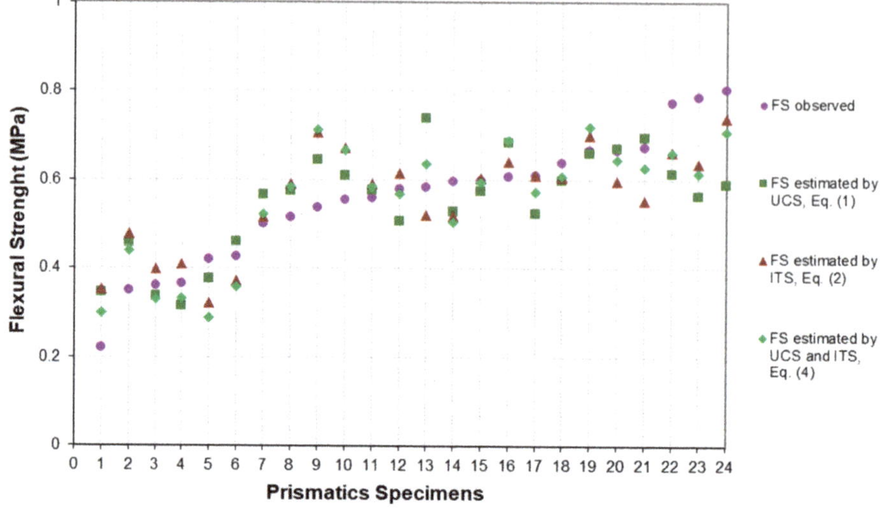

Figure 8. Relationship between observed flexural strength and predicted values by means of the proposed equations.

As seen in Figure 8, Equation (4) is suitable for calculating FS at long-term, especially with regard to the average values of the material, despite the disperse values obtained for some specimens of the recycled material. For the extreme values, the proposed model does not fit so accurately. For specimens with the lowest values in FS, higher values are predicted with all the developed equations. On the other hand, for the highest values of FS, lower values are predicted. This fact can be attributed to the heterogeneity of the material or flawed specimen preparation or testing.

From the point of view of sustainability, the advantages of FDR when compared with soil-cement and cement-bound granular mixture are considerable. When manufacturing FDR, it is avoided to transport -emove material to landfills; there is no need to use quarries and the quantity of material that must be transported is lower and, hence, CO_2 emissions are reduced. Moreover, the roads that are used for transporting the material are not so damaged.

With regard to the (expected) behavior, it can be said that the average UCS and ITS value at long-term are similar to soil-cement. In the case of the FS at long-term, the value is lower than usual for soil-cement. The fact that the FS values are lower could be regarded as a disadvantage, and perhaps the expected life of the pavement structure would not be as long as with soil-cement. However, if we compare the expected life of the new higher quality base that we are designing with the previous pavement structure, which was composed of unbound aggregates, an improvement is observed. The quality is not as high as with soil-cement, but it must be taken into account that there is a big increase in the quality of the new base compared to the previous one. With this technique, a material that is near to a standardized material is designed, which is cheaper and more sustainable. Consequently, the advantages overcome the disadvantages.

4. Conclusions

The study aimed to establish the long-term relationships among flexural, unconfined compressive, and indirect tensile strength in FDR with cement, and compared them to the strength relationships between soil-cement and cement-bound granular materials to verify the hypothesis that their behavior was similar.

The statistical analysis proved the existence of fairly close relationships among these three strength tests in the FDR, but with different behaviors to what it was expected. Flexural strength exhibited lower values in the recycled pavements than in soil-cements, whereas the indirect tensile strength and unconfined compressive strength values were similar. The relationships between unconfined compressive strength and flexural strength, and between indirect tensile strength and flexural strength, were even closer than in cement-bound granular material. In the analyzed recycled material, only the relationship between unconfined compressive strength and indirect tensile strength was similar to the relationship in cement-bound granular material and soil-cement.

With the research, in the case that only the unconfined compressive strength value is available, Equation (1) is recommended to calculate the flexural strength at long-term of the FDR. If only the indirect tensile strength is known, Equation (2) is then recommended to calculate the flexural strength of the FDR. If we have both the unconfined compressive strength and indirect tensile strength at long-term, Equation (4) is proposed to estimate the flexural strength of the FDR.

It is important to know the flexural strength, because the fatigue strength of the FDR material is calculated using this value. The hypothesis that the FDR with cement, soil-cement, and cement-bound granular material exhibit similar behaviors is not accurate and, therefore, there is a need to undertake a fatigue behavior study on this type of recycled base course to ensure the optimum design of this type of pavements.

Author Contributions: All the authors have contributed to this work in similar capacities.

Funding: The authors are grateful to the Education Council of Castilla y León for the funds received for project number BU009A06 and UB 07/03.

Conflicts of Interest: The authors declare no conflict of interest.

References

1. Xiao, F.; Yao, S.; Wang, J.; Li, X.; Amirkhanian, S. A literature review on cold recycling technology of asphalt pavement. *Constr. Build. Mater.* **2018**, *180*, 579–604. [CrossRef]
2. Yu, B.; Liu, Q.; Tian, X.; Zhou, L.; Lin, M. Empirical performance models of hot in-place recycling of asphalt pavements. *Int. J. Pavement Eng.* **2017**, *18*, 1081–1088. [CrossRef]
3. Jones, D.; Louw, S.; Wu, R. Full-Depth Reclamation: Cost-Effective Rehabilitation Strategy for Low-Volume Roads. *Transp. Res. Rec.* **2016**, *2591*, 1–10. [CrossRef]
4. Smith, S.; Braham, A. Comparing layer types for the use of PavementME for asphalt emulsion Full Depth Reclamation design. *Constr. Build. Mater.* **2018**, *158*, 481–489. [CrossRef]
5. Suebsuk, J.; Horpibulsuk, S.; Suksan, A.; Suksiripattanapong, C.; Phoo-ngernkham, T.; Arulrajah, A. Strength prediction of cement-stabilised reclaimed asphalt pavement and lateritic soil blends. *Int. J. Pavement Eng.* **2019**, *20*, 332–338. [CrossRef]
6. Ghanizadeh, A.R.; Rahrovan, M.; Bafghi, K.B. The effect of cement and reclaimed asphalt pavement on the mechanical properties of stabilized base via full-depth reclamation. *Constr. Build. Mater.* **2018**, *161*, 165–174. [CrossRef]
7. Hill, R.; Braham, A. Investigating the raveling test for full-depth reclamation. *Front. Struct. Civ. Eng.* **2018**, *12*, 222–226. [CrossRef]
8. Fedrigo, W.; Núñez, W.P.; Castañeda López, M.A.; Kleinert, T.R.; Ceratti, J.A.P. A study on the resilient modulus of cement-treated mixtures of RAP and aggregates using indirect tensile, triaxial and flexural tests. *Constr. Build. Mater.* **2018**, *171*, 161–169. [CrossRef]
9. Alizadeh, A.; Modarres, A. Mechanical and Microstructural Study of RAP–Clay Composites Containing Bitumen Emulsion and Lime. *J. Mater. Civ. Eng.* **2019**, *31*, 04018383. [CrossRef]
10. Reeder, G.D.; Harrington, D.; Ayers, M.E.; Adaska, W.S. *Guide to Full-Depth Reclamation (FDR) with Cement*; National Concrete Pavement Technology Center, Institute for Transportation of Iowa State University, Portland Cement Association: Ames, IA, USA, 2017.
11. Wirtgen. *Cold Recycling Manual*, 3rd ed.; Wirtgen GmbH: Windhagen, Germany, 2010.
12. Babashamsi, P.; Yusoff, N.I.M.; Ceylan, H.; Nor, N.G.M. Recycling toward sustainable pavement development: End-of-life considerations in asphalt pavement. *J. Teknol.* **2016**, *78*, 25–32.

13. Ghasemi, P.; Christopher Williams, R.; Jahren, C.; Ledtji, P.; Yu, J. *Field Investigation of Stabilized Full-Depth Reclamation (SFDR)*; Minnesota Department of Transportation Research Service & Library: St. Paul, MN, USA, 2018; No. MN/RC 2018-33.

14. Castañeda López, M.A.; Fedrigo, W.; Kleinert, T.R.; Matuella, M.F.; Núñez, W.P.; Ceratti, J.A.P. Flexural fatigue evaluation of cement-treated mixtures of reclaimed asphalt pavement and crushed aggregates. *Constr. Build. Mater.* **2018**, *158*, 320–325. [CrossRef]

15. Portland Cement Association. *Full Depth Reclamation: Recycling Roads Saves Money and Natural Resource*; PCA: Skokie, IL, USA, 2005; p. 6.

16. Godenzoni, C.; Graziani, A.; Bocci, E.; Bocci, M. The evolution of the mechanical behaviour of cold recycled mixtures stabilised with cement and bitumen: Field and laboratory study. *Road Mater. Pavement Des.* **2018**, *19*, 856–877. [CrossRef]

17. Santos, J.; Bryce, J.; Flintsch, G.; Ferreira, A. A comprehensive life cycle costs analysis of in-place recycling and conventional pavement construction and maintenance practices. *Int. J. Pavement Eng.* **2017**, *18*, 727–743. [CrossRef]

18. Díaz Minguela, J.; López Bachiller, M. *Reciclado de Firmes In Situ con Cemento (Full Depth Reclamation with Cement)*; Instituto Español del Cemento y sus Aplicaciones (IECA) & Asociación Nacional Técnica de Estabilizados de Suelos y Reciclado de Firmes (ANTER): Madrid, Spain, 2018; p. 270.

19. Jofré, C.; Kraemer, C.; Díaz, J. *Manual de Firmes Reciclados In Situ con Cemento (Guide of Full Depth Reclamation with Cement)*; Instituto Español del Cemento y sus Aplicaciones (IECA): Madrid, Spain, 1999.

20. Boz, I.; Solaimanian, M. Investigating the effect of rejuvenators on low-temperature properties of recycled asphalt using impact resonance test. *Int. J. Pavement Eng.* **2018**, *19*, 1007–1016. [CrossRef]

21. Miró, R.; Edmundo Pérez Jiménez, F.; Castillo Aguilar, S. Mixed recycling with emulsion and cement of asphalt pavements. Design procedure and improvements achieved. *Mater. Struct.* **2000**, *33*, 324–330. [CrossRef]

22. Díaz, J. *State of the Art of In Situ Subgrade Stabilisation and Pavement Recycling with Cement in Spain*; Asociación Española de la Carretera (AEC) e Instituto Español del Cemento y sus Aplicaciones (IECA): Madrid, Spain, 2001; pp. 133–161.

23. Ozarín, T.; Gonzalo-Orden, H. Reciclado "In Situ" de firmes con cemento en carreteras autonómicas de la provincia de Palencia (On-site pavement recycling with Portlant cement on regional roads in Palencia). *Carreteras* **2006**, *144*, 34–44.

24. Grilli, A.; Bocci, E.; Graziani, A. Influence of reclaimed asphalt content on the mechanical behaviour of cement-treated mixtures. *Road Mater. Pavement Des.* **2013**, *14*, 666–678. [CrossRef]

25. Kolias, S.; Katsakou, M.; Kaloidas, V. Mechanical properties of flexible pavement materials recycled with cement. In Proceedings of the First International Symposium on Subgrade Stabilisation and In Situ Pavement Recycling Using Cement, Salamanca, Spain, 1–4 October 2001; Asociación Española de la Carretera (AEC) e Instituto Español del Cemento y sus Aplicaciones (IECA): Madrid, Spain, 2001; pp. 659–674.

26. Díaz, J.; Murga, P.; Gonzalo-Orden, H.; González, D. *Estudio del Comportamiento de Firmes Reciclados In Situ con Cemento (Study of the Behaviour of Full Depth Reclamation with Cement)*; Asociación Española de la Carretera (AEC): Madrid, Spain, 2008; pp. 519–527.

27. BS. *BS-1924-1: Stabilized Materials for Civil Engineering Purposes. General Requirements, Sampling, Sample Preparation and Tests on Materials before Stabilization*; British Standards Institution: London, UK, 1990.

28. Ministerio De Fomento. *Pliego de Prescripciones Técnicas Generales para Obras de Carretera y Puentes (PG-3) (Statement of General Requirements for Construction of Roads and Bridges (PG-3))*; MFOM: Madrid, Spain, 2015.

29. Ministerio de Fomento. *Pliego de Prescripciones Técnicas Generales PG-4 (Statement of General Requirements for Road Maintenance Works (PG-4))*; MFOM: Madrid, Spain, 2001.

30. Mi, S. Material and mechanics performance of full depth asphalt pavement. *Fresenius Environ. Bull.* **2019**, *28*, 2063–2066.

31. Ismail, A.; Baghini, M.S.; Karim, M.R.; Shokri, F.; Al-Mansob, R.A.; Firoozi, A.A.; Firoozi, A.A. Laboratory Investigation on the Strength Characteristics of Cement-Treated Base. *Appl. Mech. Mater.* **2014**, *507*, 353–360. [CrossRef]

32. Otte, E. A Structural Design Procedure for Cement-Treated Layers in. Pavements.Sc.D. Thesis, University of Pretoria, Pretoria, South Africa, 1978.

33. Linares-Unamunzaga, A.; Gonzalo-Orden, H.; Minguela, J.; Pérez-Acebo, H. New Procedure for Compacting Prismatic Specimens of Cement-Treated Base Materials. *Appl. Sci.* **2018**, *8*, 970. [CrossRef]

34. Linares-Unamunzaga, A.; Pérez-Acebo, H.; Rojo, M.; Gonzalo-Orden, H. Flexural Strength Prediction Models for Soil–Cement from Unconfined Compressive Strength at Seven Days. *Materials* **2019**, *12*, 387. [CrossRef]

35. Kolias, S.; Williams, R.I.T. *Cement-Bound Road Materials: Strength and Elastic Properties Measured in the Laboratory*; Report SR 344; Transport and Road Research Laboratory: Crowthorne, UK, 1978.

36. Austroads. *Publication No. AP–T101/08. The Development and Evaluation of Protocols for the Laboratory Characterisation of Cemented Materials.*; Austroads: Sydney, Australia, 2008.

37. Mansoor, J.; Shah, S.; Khan, M.; Sadiq, A.; Anwar, M.; Siddiq, M.; Ahmad, H. Analysis of Mechanical Properties of Self Compacted Concrete by Partial Replacement of Cement with Industrial Wastes under Elevated Temperature. *Appl. Sci.* **2018**, *8*, 364. [CrossRef]

38. Díaz, J. El Estudio de Comportamiento de los Firmes Reciclados In Situ con Cemento (Study of the Behaviour of Pavements Recycled In Situ with Cement). Ph.D. Thesis, Universidad de Burgos, Burgos, Spain, 2011.

39. American Society for Testing and Materials International (ASTM). *D1632-17: Standard Practice for Making and Curing Soil-Cement Compression and Flexure Test Specimens in the Laboratory*; ASTM: West Conshohocken, PA, USA, 2017.

40. Xuan, D.X.; Houben, L.J.M.; Molenaar, A.A.A.; Shui, Z.H. Mechanical properties of cement-treated aggregate material—A review. *Mater. Des.* **2012**, *33*, 496–502. [CrossRef]

41. Lim, S.; Zollinger, D.G. Estimation of the compressive strength and modulus of elasticity of cement-treated aggregate base materials. *Transp. Res. Rec.* **2003**, *1837*, 30–38. [CrossRef]

42. Kersten, M.S. *Soil Stabilization with Portland Cement*; National Academy of Sciences-National Research Council: Washington, DC, USA, 1961.

43. Portland Cement Association (PCA). *Soil-Cement Laboratory Handbook*; PCA: Skokie, IL, USA, 1992.

44. Asociación Española de Normalización y Certificación (AENOR). *UNE-EN 197-1. Part 1: Composition, Specifications and Conformity Criteria for Common Cements*; AENOR: Madrid, Spain, 2011.

45. IECA-CEDEX. *Manual de Firmes con Capas Tratadas con Cemento (Guideline for Pavements with Cement Bound Materials)*, 2nd ed.; Centro de Estudios y Experimentación de Obras Públicas (CEDEX): Madrid, Spain, 2003; p. 265.

46. Asociación Española de Normalización y Certificación (AENOR). *UNE 103-501-94: Geotechnic Compactation Test. Modified Proctor*; AENOR: Madrid, Spain, 1994.

47. American Society for Testing and Materials International (ASTM). *D1557-12e1: Standard Test Methods for Laboratory Compaction Characteristics of Soil Using Modified Effort (56,000 ftlbf/ft3 (2,700 kN-m/m3))*; ASTM: West Conshohocken, PA, USA, 2012.

48. American Society for Testing and Materials International (ASTM). *D1634-00: Standard Test Method for Compressive Strength of Soil-Cement Using Portions of Beams Broken in Flexure (Modified Cube Method) (Withdrawn 2015)*; ASTM: West Conshohocken, PA, USA, 2006.

49. Asociación Española de Normalización y Certificación (AENOR). *Norma UNE-EN 13286-41. Unbound and Hydraulically Bound Mixtures—Part 41: Test Method for the Determination of the Compressive Strength of Hydraulically Bound Mixtures*; AENOR: Madrid, Spain, 2003.

50. JCyL. *Recomendaciones de Proyecto y Construcción de Firmes y Pavimentos (Recommendations for the Design and Construction of Road Pavements)*; Dirección General de Carreteras e Infraestructuras, Consejería de Fomento, Junta de Castilla y León: Valladolid, Spain, 2004.

51. Asociación Española de Normalización y Certificación (AENOR). *UNE-EN 12390-5. Testing Hardened Concrete—Part 5: Flexural Strength of Test Specimens*; AENOR: Madrid, Spain, 2009.

52. American Society for Testing and Materials International (ASTM). *D1635/D1635M-12: Standard Test Method for Flexural Strength of Soil-Cement Using Simple Beam with Third-Point Loading*; ASTM: West Conshohocken, PA, USA, 2012.

53. Asociación Española de Normalización y Certificación (AENOR). *UNE-EN 12390-2. Part 2: Making and Curing Specimens for Strength Tests*; AENOR: Madrid, Spain, 2009.

54. CEDEX. *Norma NLT-305/90. Resistencia a Compresión Simple de Materiales Tratados con Conglomerantes Hidráulicos (Unconfined Compressive Strength for Bound Materials)*; CEDEX, Dirección General de Carreteras, Ministerio de Fomento: Madrid, Spain, 1990.

55. Asociación Española de Normalización y Certificación (AENOR). *UNE-EN 12390-6. Testing Hardened Concrete—Part 6: Tensile Splitting Strength of Test Specimens*; AENOR: Madrid, Spain, 2010.
56. Thompson, M.R. *Mechanistic Design Concepts for Stabilized Base Pavements*; University of Illinois: Urbana, IL, USA, 1986; p. 52.
57. Solís Villa, L.A.; Díaz Minguela, J. Los firmes con suelocemento en la Red Autonómica de Castilla y León (Pavements with soil-cement in the regional road network of Castilla y León). *Cem. Hormig.* **2002**, *835*, 74–89.
58. Marshall, B.P.; Kennedy, T.W. *Tensile and Elastic Characteristics of Pavement Materials*; Research Report 183-1; Center for Highway Research, The University of Texas at Austin: Austin, TX, USA, 1974.

Article

Potential Activity of Recycled Clay Brick in Cement Stabilized Subbase

Chunyu Liang [1], Ying Wang [1], Wenzhu Song [2], Guojin Tan [1,*], Yanling Li [1] and Youmeng Guo [1]

[1] School of Communications, Jilin University, 5988 RenminRoad, Changchun 130000, China; liangcy@jlu.edu.cn (C.L.); wang_y17@mails.jlu.edu.cn (Y.W.); yanling@jlu.edu.cn (Y.L.); guoym16@mails.jlu.edu.cn (Y.G.)
[2] Jilin highway administration bureau, 2518 Jiefang Road, Changchun 130000, China; swz_glj@163.com
* Correspondence: tgj@jlu.edu.cn

Received: 28 September 2019; Accepted: 27 November 2019; Published: 29 November 2019

Abstract: Construction waste is one of the products in the process of urbanization. From the perspective of economy and environmental protection, this study used crushed construction waste clay brick to replace the fine aggregate of cement stabilized macadam subbase in certain proportions, and the optimum proportion was obtained according to the unconfined compressive strength of 7 days (d), 28 d, and 90 d. The "modified EDTA titration experiment" was also used to explain how the potential activity of construction waste clay brick works in cement stabilized macadam. The result obtained is that an optimal replacement ratio of 50% exists when using construction waste clay brick to replace the fine aggregate of cement stabilized macadam, and its unconfined compressive strength is higher than that of the 0% replacement ratio specimens; that is, the potential activity of the construction waste clay brick contributes the most to the unconfined compressive strength of the specimens at this proportion. According to the blending method and proportion obtained in this study, the application of construction waste clay bricks in a practical project can maximize environmental protection in road engineering and economic benefits simultaneously.

Keywords: road engineering; construction waste clay brick; cement stabilized macadam; unconfined compressive strength; potential activity

1. Introduction

Predictions show that a large amount of sand consumption in the future will not only create economic pressure on the construction industry but also cause the erosion and degradation of the world's major ecosystems. The global construction waste crisis will get worse; global waste production will increase by 70% by 2050 in comparison to 2016, and the global waste crisis will spur innovation in waste management. Due to the boom of the construction industry, the price of sand will increase significantly [1]. However, the large amount of construction waste produced by the process of demolition and reconstruction is one of the inevitable additional products in the process of urbanization. Among the construction waste includes a large number of waste concrete blocks, waste clay bricks, waste steel bars, waste wires and cables, waste plastic products, and so on. Among them, clay bricks and concrete blocks account for about 80% of the total [2]. Marzouk and Azabbuilt a system dynamics methodology of the construction and demolition waste management sector by developing a dynamic model that is capable of studying the behavior of landfill process on both the short and long run and its impacts on the environment and economy. In Egypt, this system dynamics methodology results show that the recycling of construction and demolition waste would reduce the costs required to mitigate air pollution by $16,161.35 billion over 20 years [3].

In order to reduce the environmental and economic pressure of construction waste, the reuse of it is an effective method. As for the reuse of construction waste in road engineering, it is usually used for filling roadbeds without secondary crushing, while after secondary crushing, it can be used as aggregate in asphalt pavement and as a cement stabilized base layer [4,5]. Waste concrete blocks are widely used in road engineering. The superiority of concrete materials for housing construction and considerable secondary hydration are the reasons why waste concrete blocks are widely used among several types of construction wastes. However, due to its high strength, the crushing process is complicated, the requirements for the crushing machine are higher, and the crushing process is difficult. There are also uncontrollable interference factors, such as cement that did not fully hydrate, aggregates from different sources with different strengths, and some waste bricks that cannot be completely removed [6]. In comparison, waste clay bricksare easy to be picked out and crushed and have a lower transportation burden. A stone-crushing machine can be used to crush waste clay bricks, which can save more expenditure.

In response to the management of construction waste, many developed countries began to take action to solve the problem in the 1970s and 1980s. Examples include the Solid Waste Treatment Act of 1965 in the United States and the Waste Treatment Act of 1970 in Japan [7]. Many other developed countries, such as Germany, Korea, Denmark, and Singapore also made initial efforts to solve this problem and had momentous success [8,9]. In China, regulations related to construction waste have been carried out since 1992. The Regulations on the Management of City Appearance and Environmental Hygiene have stipulated that the responsible persons for the disposal of construction waste must take timely clearance and transportation. A series of plans also have been implemented [10]. These actions above are the first step that has been taken in encouraging the recycling of construction waste and governing the unreasonable disposal of construction waste, which laid a foundation for improving the relevant laws and regulations for the recycling and reuse of construction waste.

In terms of construction waste recycling, most research studies have focused on cement concrete and asphalt concrete. Xue et al. considered that the optimum amount of construction waste composite powder in cement concrete is 30%. This amount of construction waste composite powder putsthe concrete in a more compact and low-alkali condition, thus enhancing the macroscopic frost resistance [11]. Peng et al. combine silicon powder, slag powder, and recycled construction waste regenerated powder in concrete; they comprehensively study the performance test of using construction waste regenerated powders to replace silicon powder and cement in concrete, obtaining the optimal water–binder ratio range of 0.18–0.20. The results show that the particle size of regenerated powders is mostly larger than that of silicon powder particles, and some of them are smaller than that of cement particles. Furthermore, the pozzolanic reaction of regenerated powders as active admixtures is weak, reflecting the macroscopical performance in which the resistance to chloride ion permeability and the resistance to shrinkage are enhanced with the increase of the amount of the regenerated powders, while simultaneously the unconfined compressive strength and flexural strength are weakened [12]. Cheng studied the application of fine aggregate partially replaced by fly ash and coarse aggregate partially replaced by waste brick in concrete. The results showed that when the replacement rate of both fly ash and waste brick was 30%, the compressive strength of recycled concrete reached the highest value [13]. The physical and mechanical properties of four kinds of coarse aggregate with different particle sizes in fresh concrete and hardened concrete were also studied. The results show that harder original brick directs the stronger compressive strength of the concrete, and the higher density directs the higher concrete strength. The flexural strength of concrete produced with a crushed brick aggregate was approximately 8% less than that of concrete made with granite aggregate [14]. Kristina et al. replaced natural macadam fine aggregate with recycled brick fine aggregate of 0%, 25%, and 50% in concrete, obtaining 25, 50, 75, and 100 routine freeze-thaw cycles at 28 and 60 days of curing age, respectively. Combining the subsequent compressive strength and flexural strength tests, it was concluded that the addition of recycled brick fine aggregate could improve the frost resistance and long-term strength of concrete to a certain extent [15]. Zhao et al. mix coarse and fine aggregates of construction waste into concrete, respectively. The hardening density, drying shrinkage coefficient, and water absorption rate were taken as reference, the optimum replacement ratio of coarse aggregate and fine aggregate is less than 25% and within the scope of 50%–75%, respectively [16]. Hu et al. mix recycled clay brick aggregate with 20%, 40%, 60%, 80%, and 100% of coarse aggregate and fine aggregate in cement stabilized macadam, respectively. The compressive strength, splitting strength, compressive modulus of resilience, shrinkage strain, and freeze resistance index were used as references in the study, and the maximum mixing ratio of coarse aggregate and fine aggregate obtained was 70% and 90%, respectively [17]. Turanli et al. used clay bricks as pozzolanic materials after fine grinding. An ASTM accelerated mortar bar test and scanning electron microscopy (SEM) showed that fine grinding clay bricks could effectively inhibit the expansion caused by the alkali–aggregate reaction [18]. Peng replaced recycled aggregate for coarse aggregate in asphalt concrete and obtained the optimum replacement ratio of 30%–40% with the high-temperature performance taken as a reference; their results also showed an optimum replacement ratio range of 60%–70%, using the residual stability and freeze–thaw splitting strength as references. When the replacement ratio of recycled aggregate is less than 60%, the adverse effects on low-temperature performance and fatigue resistance are controllable [19].

It can be seen that the research of recycled construction waste clay brick aggregate (RBA) in concrete and asphalt concrete is more indepth, and some referential conclusions are also drawn in terms of the replacement ratio and pozzolanic activity. In this study, the application of clay brick in cement stabilized macadam subbase will be studied, and the influence of the potential activity of RBA using it instead of fine aggregate partly on the unconfined compressive strength of this structure layer will be discussed. To study the microscopic mechanism of the tendency of strength, we also used the physical and chemical methods to verify the practicability and feasibility of RBA application in road subbase.

2. Material

2.1. Aggregate

Crushed basalt aggregate (CBA) produced from Tumen mountain, Jilin province in China, was crushed into four grades: CBA I (20–30 mm), CBA II (10–20 mm), CBA III (5–10 mm), and CBA IV (0–5 mm). The four grades of CBA will be screened into a single particle size, and the cement stabilized macadam test gradation is formed by a mix proportion design. RBA is taken from the demolition of the shanty renovation building in China, Jilin Province. It is crushed into two grades: RBA I (0–20 mm) and RBA II (0–5 mm). All these grades were applied to the experiment after sieving and then dispensed experiment gradation. The sieving results are shown in Figures 1 and 2.

The experimental gradation of cement stabilized macadam was designed [20], which satisfies the gradation range stipulated in the Technology Guidelines for the Construction of Highway Roadbases (JTG/T F20—2015) and is close to the lower bound. The experimental gradation curve is shown in Figure 3.

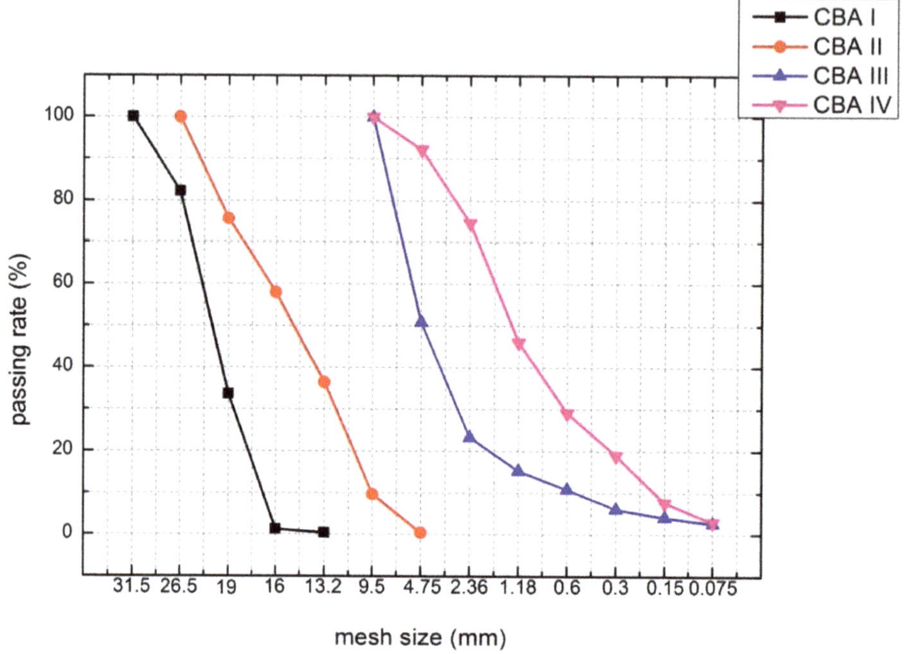

Figure 1. Grades sieving proportion.

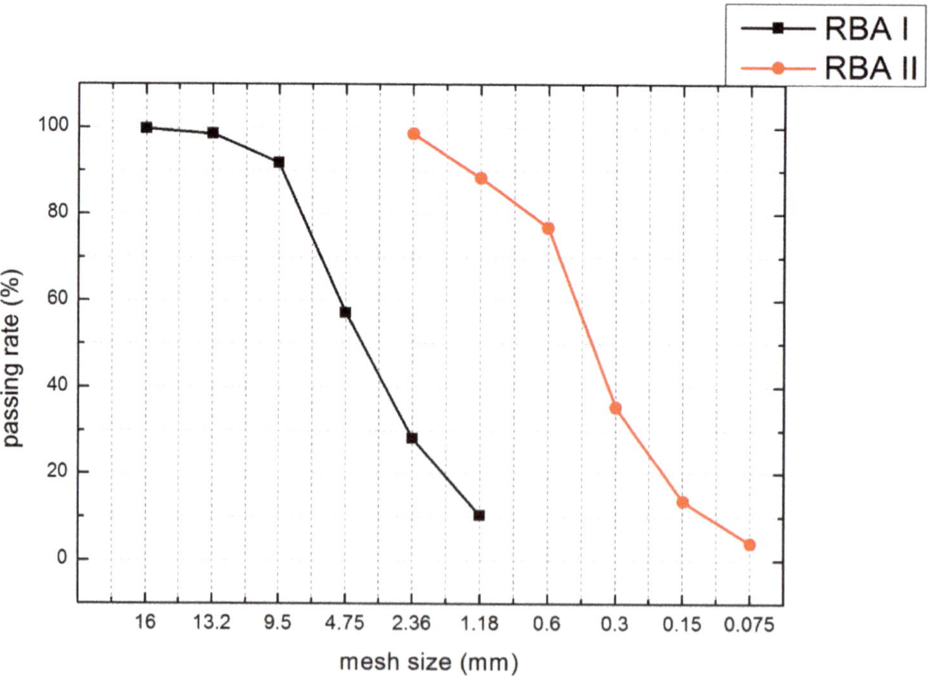

Figure 2. Grades sieving proportion.

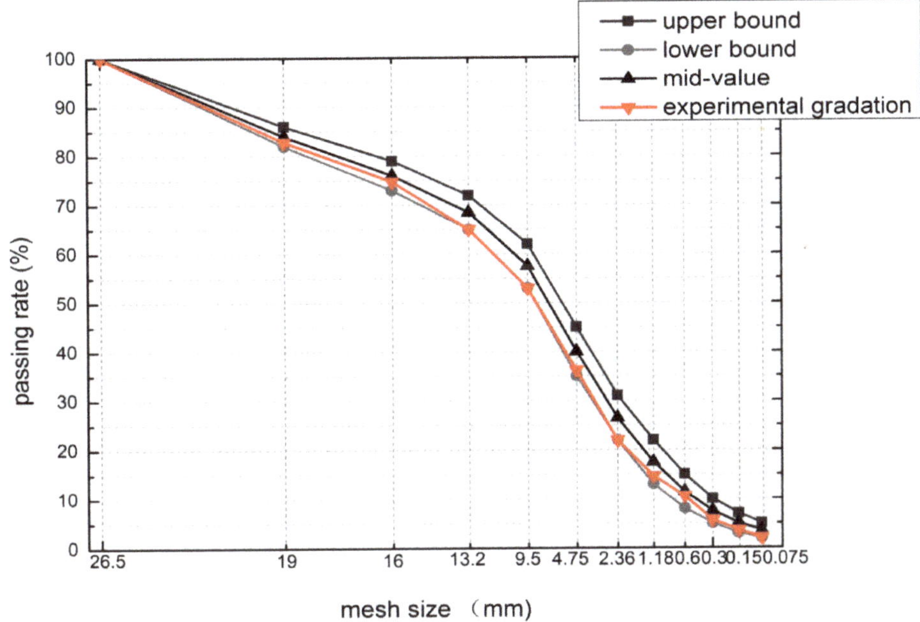

Figure 3. Gradation curve.

Several basic performance tests of aggregates, such as the crushed stone value, percentage of flat-elongated particle, water absorption, and apparent specific gravity were carried out on the basis of the Test Methods of Aggregate for Highway Engineering (JTG E42—2005). The plastic index test was carried out according to the Test Methods of Soils for Highway Engineering (JTG E40—2007). The results of the tests and the corresponding reference values in specification (JTG/T F20—2015) are shown in Table 1, and the coarse and fine aggregate were divided by a particle size of 4.75 mm. The CBA and RBA mixtures of different proportions were tested, whose proportions will be used in the experiment.

Table 1. Aggregate basic performance.

	Fine Aggregate			Coarse Aggregate			
	Plastic Index	Water Absorption (%)	Apparent Specific Gravity	Crushed Stone Value (%)	Flat-Elongated Particle (%)	Water Absorption (%)	Apparent Specific Gravity
0% [a]	4.34	1.73	2.68	21.63	9.57	1.24	2.73
20% [a]	6.16	3.89	2.59	22.85	9.53	3.75	2.69
40% [a]	9.84	8.21	2.50	25.74	9.72	7.42	2.58
50% [a]	13.97	10.32	2.41	28.45	9.89	9.98	2.51
60% [a]	24.53	13.02	2.33	31.22	9.59	11.86	2.45
80% [a]	40.64	15.53	2.24	38.41	9.63	14.97	2.39
100% [a]	60.50	17.60	2.11	41.53	9.97	17.36	2.33
Reference value	≤12	——	——	≤30	——	——	——

[a] Replace proportion of recycled construction waste clay brick aggregate (RBA) in crushed basalt aggregate (CBA) specimens.

According to the data in Table 1, as the proportion of RBA increase in specimens, the indexes such as the plastic index, water absorption, and crushed stone value also increase; however, the apparent specific gravity decreases at the same time, while the flat-elongated particlesare maintainedat a steady level. When only comparing two kinds of proportion, 0% and 100%, the apparent specific gravity of RBA (100%) is generally smaller than that of CBA (100%), and contraposing RBA, the fine aggregate is much smaller than the coarse aggregate. In terms of water absorption, RBA is 10 to14 times as much as CBA, which means that RBA can absorb more water than CBA under the same conditions. The crushed stone value of RBA is about twice that of CBA. This property leads to the strength of RBA being much weaker than that of CBA. The plastic index of CBA is much smaller than the reference value and less than that of RBA, which proves that the water absorption and water requirement of RBA are much larger than those of CBA, so it is not suitable to be used as aggregates alone. As an aspect of the percentage of flat-elongated particles, these two kinds of aggregates show similar test results, which prove that RBA will not produce more collapsed particles under the same crushing method.

2.2. Composition

Scanning electron microscopy (SEM) and quantitative energy dispersive X-ray diffraction (XRD) were used to scan the crushed powder of construction waste clay brick, and the main substances that exist in crystalline form and chemical elements were analyzed.

Construction waste clay brick powder was scanned by the above two micro detection methods; Figures 4 and 5 were obtained as scanning results. Figure 4 illustrates four points of the SEM scanning result, which showed that there are several major elements in construction waste clay brick such as O, Si, and C, with a relatively large mass percentage of each as shown in Table 2. Among the other elements, Al is the most abundant, and Ca, Fe, Mg, and Na all exist in a small mass percentage. In these scanning pictures, 1500x magnification was used to observe the details. Figure 5 is the XRD scanning result, which shows that quartz (SiO_2), calcite ($CaCO_3$), chalcopyrite ($CuFeS_2$), kyanite (Al_2SiO_5), and the residual graphite (C) from the process of firing are contained as the main crystalline compositions in construction waste clay brick.

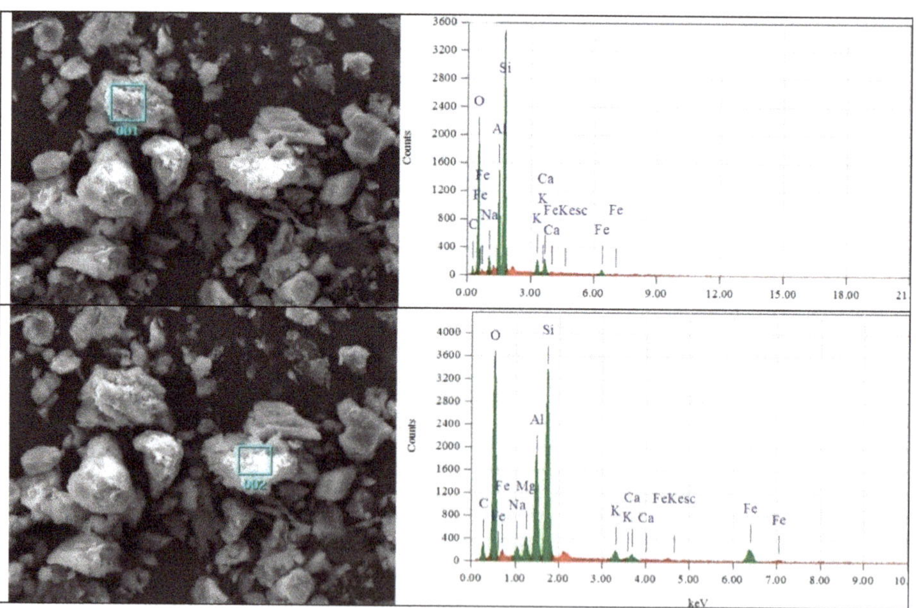

Figure 4. *Cont.*

Figure 4. Waste clay brick powder SEM scanning result.

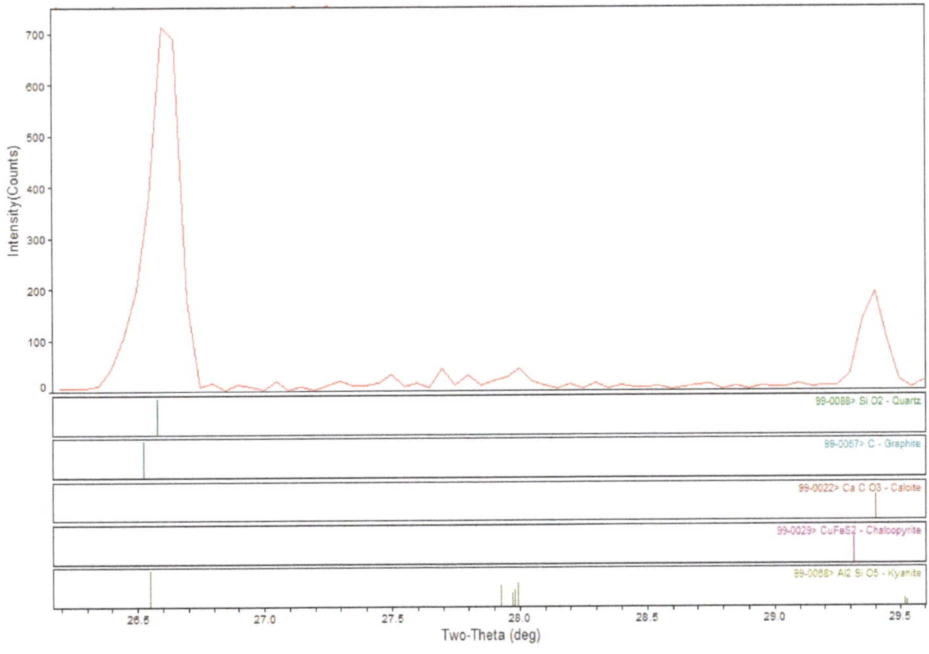

Figure 5. Construction waste clay brick powder XRD scanning.

Table 2. Scanning mean value of element composition.

Element	Mass (%)	Atom (%)	Sigma	Energy (keV)
C(Graphite)	16.69	24.57	0.19	0.277
O	49.08	54.60	0.29	0.525
Na	1.16	0.91	0.05	1.041
Mg	2.97	2.11	0.05	1.253
Al	6.53	4.33	0.08	1.486
Si	18.75	12.02	0.13	1.739
K	0.96	0.45	0.04	3.312
Ca	0.75	0.35	0.03	3.690
Fe	3.86	1.21	0.09	6.398

2.3. Cement

The cement used in this study is P.O 42.5R Portland cement produced by Jilin Yatai Company in China, which was stored in a cool, ventilated, desiccative room. The performance indexes of cement are shown in Table 3.

Table 3. Cementperformance index.

Type	Comparison	Physical Performances					
		Setting Time (h)		Strength (MPa)			
				Flexural Strength		Unconfined Compressive Strength	
		Initial set	Final set	3d	28d	3d	28d
P.O 42.5R	Reference Value	≥0.75	≤10	≥4.0	≥6.5	≥21.0	≥42.5
	Test result	2	3	5.2	8.5	25.0	53.0

The reference value in Table 3 is stipulated by specification (The Methods of Cement and Concrete for Highway Engineering JTG E30—2005).

3. Test Scheme

To achieve the purpose of applying RBA in the road structure layer with higher quality and efficiency, the mechanical properties and application requirements of the cement stabilized macadam structure in practical engineering are combined. The high water absorption and high crushed stone value of RBA in this study selected RBA to replace CBA below 4.75 mm by mass according to the experimental gradation. In addition, considering the reference values of the plastic index and crushed stone values in Table 1, six replacement rates of 0%, 20%, 40%, 50%, 60%, and 80% were selected in this study, and 60% and 80% were selected for research integrity. Cement stabilized macadam unconfined compressive strength was taken as the strength index, and the strength formation mechanism of the cement stabilized RBA and CBA mixture was analyzed comprehensively with the "modified EDTA titration experiment". By this method, the effect of RBA on strength depends on the interpreted potential activity activated in the cement stabilized macadam subbase.

3.1. Specimen Preparation

The cement stabilized macadam specimens were prepared under optimum moisture conditions to achieve maximum density. A heavy compaction test (T-0804) was used to get the optimum moisture content and the maximum density of each replacement, which are shown in Table 4. As specification (JTG E51-2009) stipulates, the heavy compaction test needs to choose the range of moisture content (5% by mass) and apply and change with a constant interval (0.5%) in each experiment to get the curve related to the moisture content and density. Cement, aggregate, and water were shovel-mixed and compacted into a 150-mm diameter by 150-mm high cylinder in three layers and finally pressed into a molding on the press machine. The experiments used 5% cement by mass of aggregate. According to the *Test Methods of Materials Stabilized with Inorganic Binders for Highway Engineering* (JTG E51-2009), the degree of compaction of the cylinder specimens was 96%.

Table 4. Moisture content and maximum density.

RBA Replacement Rate (%)	0	20	40	50	60	80
Maximum density (g·cm^{-3})	2.362	2.215	2.167	2.143	2.122	2.152
Optimum moisture content (%)	5.11	6.87	7.04	7.35	8.17	8.31

The specimens were maintained in standard curing room with a temperature of 20±2°C and a relative humidity of more than 95%.

3.2. Test Methods

3.2.1. Strength Test

The cement stabilized macadam specimens for unconfined compressive strength test followed the T0805-1994 of specification (JTG E51-2009), with curing ages of 7 days, 28 days, and 90 days.

3.2.2. Modified EDTA Titration Experiment

The "modified EDTA titration experiment" used the experimental principle of the EDTA cement titration test (T 0809-2009) in the specification (JTG E51-2009). An EDTA experiment is one method to test the effective cement content at the construction set, which mainly measures the Ca^{2+} content of samples. Meanwhile, the "modified EDTA titration experiment" changed the condition of specimens and contained them in a curing room for 28 days, also taking Ca^{2+} as the testing project to mainly measure the degree of the pozzolanic reaction of the admixture in the cement stabilized structure. The same chemical principle and reagents are applied in samples of different proportions and 1000 g quantity was stipulated and used in the "modified EDTA titration experiment". This method was used to measure the degree of pozzolanic activity from pozzolanic admixture in cement or lime stabilized structure at both the construction and experimental field. The quantity of $Ca(OH)_2$ consumed by RBA in specimens with different percent ages was measured in this study, and RBA is considered as a pozzolanic admixture.

Then, 1000g cement stabilized macadam mixture specimens were prepared with the replacement ratio in an optimum water contention mixed with cement and maintained for 28 days without compaction. Afterwards, 0.1mol·m^{-3} EDTA-2Na standard solution (ChemChina, Guangrao, Shandong Province, China), 10% NH_4Cl solution (ChemChina, Guangrao, Shandong Province, China), 1.8% NaOH solution (ChemChina, Guangrao, Shandong Province, China) with triethanolamine (Haihua Company, Weifang, Shandong Province, China), and calcium red indicator (ChemChina, Guangrao, Shandong Province, China) were also prepared and used according to the EDTA titration test (T 0809) in the specification(JTG E51-2009). In this experiment, the sample was soaked in a NaOH solution that contained twice the mass volume, which remained stationary after stirring for 5 minutes. The static time must be unified among all specimens in one test, and the specific duration is employed according to when the mixture becomes clear. Then, we took 10mL supernatant liquid and 50 mL NaOH into a 200 mL conical flask, mixed it up, and tested the PH value using test strips to make sure that it was between 12.5 and 13. Then, 0.2 g of calcium red indicator was added into the 200 mL conical flask, which was titrated with EDTA-2Na standard solution after being mixed up. In the comparison test, 300 g of RBA specimen was weighed and soaked in the NaOH solution of 600 mL, which was twice the mass volume. The PH value of the solution was 13 (the PH value range of the solution environment after cement hydration was 12–13 [21]). We stirred the solution for 5 minutes, put it in a 250ml conical flask after precipitation for 10 minutes, and added 0.2 g of Cal-Red as the EDTA experiment.

Given the various features of unconfined compressive strength, the activation condition of RBA mainly with its potential activity in cement stabilized macadam construction is discussed. Combined with the mechanism test results, the role of the potential activity of RBA in the strength formation process in cement stabilized macadam construction and the causes are emphatically analyzed. Many scholars have studied [15,22,23] and found that a large amount of Ca(OH)$_2$ was consumed in the process of the potential activity of clay bricks being activated to form strength, but these research studies lackeda specific chemical test as proof. These results prove that RBA can be activated by Ca(OH)$_2$ and produce hydration products. Through physical, mechanical, and chemical means, we obtain the conclusion that RBA has potential activity, and can improve the unconfined strength if blended it in a proper method.

4. Test Results and Analysis

4.1. Unconfined Compressive Strength Test

The unconfined compressive strength of 7 d is the only hard and fast rule regarding the design of cement stabilized macadam(JTG/T F20—2015). Specimens were used with curing ages of 7 days, 28 days, and 90 days, respectively, and the values of specimens with replacement amounts of RBA of 0%, 20%, 40%, 50%, 60%, and 80% are shown in Table 5.

Table 5. Representative value of unconfined compressive strength.

RBA Replacement Rate (%)	0	20	40	50	60	80
7d UCS (MPa)	5.37	3.88	4.22	5.46	4.22	3.77
28d UCS (MPa)	7.07	5.77	7.08	7.62	7.09	6.51
90d UCS (MPa)	8.05	8.61	8.29	9.61	8.20	9.78
7–28d UCSIV (MPa)	1.70	1.89	3.19	2.16	2.87	2.74
7–90d UCSIV (MPa)	2.68	4.73	4.07	4.15	3.98	6.01
7–28d UCSIP (%)	31.66	48.71	67.77	39.56	68.01	72.70
7–90d UCSIP (%)	49.91	121.91	96.45	76.01	94.31	159.42

7d UCS, 28d UCS, and 90d UCS represent the mean 7d unconfined compressive strength,28d unconfined compressive strength, and 90d unconfined compressive strength, respectively. 7–28d UCSIV(or7–90d UCSIV) represent the difference between 28d (or 90d) and 7d of unconfined compressive strength values. 7–28d UCSIP (or7–90d UCSIP) represents anunconfined compressive strength increase percentage of 28d (or 90d) compared with 7d. Unconfined compressive strength of 28d and 90d are both compared with 7d because the value of 7d is always used to indicate the practical engineering as the main index for whether the next step can be applied.

Combining Table 5, Figure 6, and Table 1, there are several features can be listed:

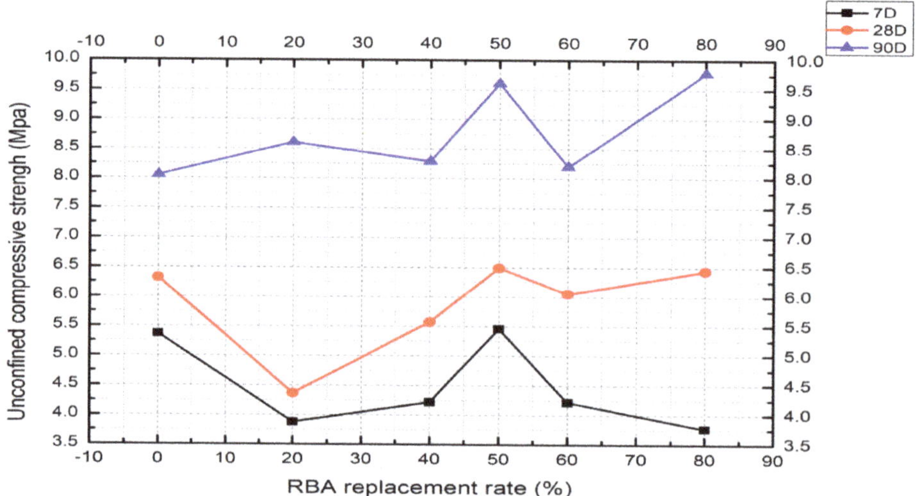

Figure 6. Compressive strength.

(1) Replacing part of CBA with RBA in cement stabilized macadam fine aggregate will have a great effect on its strength, and the change in the law of strength at various agesis slightly different from the change of RBA content.The strength of 7d and 28d decreases at first, andthen increases and decreases again with the increasing of RBA content. The 7d strength is slightly higher than that of the standard specimen (RBA 0%) when the RBA content is 50%, and the strength of other RBA contents is lower than that of the standard specimen. The 28d strength of 40%, 50%, and 60% RBA content is close to or slightly larger than the standard specimen, and the strength of other RBA contents is less than the standard specimen. However, the 90d strength of the specimens mixed with RBA is higher than that of the standard specimens.

(2) The property of aggregate has a non-negligible influence on strength at an early stage. Strengths of 7d and 28d show that 20% is smaller than 0% because the pozzolanic reaction is weak; as the replacement ratio increases, the pozzolanic reaction becomes stronger and counteracts the weakening caused by the RBA aggregate, so the strength increases gradually. At the same time, the crushed stone value and the plastic index also increase, exceeding the reference values until the 60% RBA replacement ratio directly caused either the strength to decrease again or be maintained at a relatively stable level.

(3) Since the increased percentage of strength was composed of the continuous hydration of cement and the activation of the potential activity of RBA, as the curing time increased, the strength of specimens containing RBA increased as well. The strength of the 90d kept increasing because the pozzolanic reaction started during the hydration reaction and enough hydration product was formed, leading to the higher strength of the RBA specimens.

(4) According to the various strength characteristics at 7d, 28d, and 90d, the 50% RBA replacement ratio has the optimal strength, and it can be the recommended as the replacement percentage of RBA in cement stabilized macadam.

4.2. Modified EDTA titration experiment

In the comparison test, the solution with Cal-Red was purple at first and turned blue after 0.40 mL of EDTA-2Na standard solution was consumed. This change (in Figure 7) proved that clay brick contains soluble calcium and can be dissolved in a solution environment after cement hydration. The titration results of specimens prepared in different replace proportions are shown in Table 6.

Figure 7. Diagram.

Table 6. EDTA titration value.

Replacement Rate (%)	0	20	40	50	60	80
EDTA-2Na standard solution consumption ΔV (mL)	6.05	4.15	3.20	2.35	2.40	2.55

Specimens with 50%, 60%, and 80% replacement ratios have the almost same content level of Ca^{2+}, of which 50% has the lowest content of Ca^{2+} among them. This indicated that more $Ca(OH)_2$

is consumed than other types of specimens and specimens with a 50% replacement proportion have the largest consumption of Ca(OH)$_2$. The increase of the RBA proportion indicated that more soluble calcium in the ammonium chloride solution makes the content of Ca^{2+} in the specimens with 80% RBA larger than that of the 60% specimens. The standard specimen (0%) has the maximum ΔV and 50% has the minimal ΔV, referring to the corresponding Ca(OH)$_2$ consumption.

4.3. Mechanism Analysis

From the results above, it can be seen that the unconfined compressive strength at 7d, 28d, and 90d have particular variation features as the proportion of RBA increases. The maximum value is obtained at 50% of the specimen at 7d. While at 28d it is kept in a stable range, after 50%, the proportion increases. At 90d, the strength increases as the proportion increases, and at 50%, it has the almost same value as that at 80%. It means that the potential activity of RBA activated and offset part of the weakness caused by the weak performances of it, which ultimately affected and enhanced the strength of the cement stabilized macadam layer. Combining with the plastic index and crushed stone value showed in Table 1, 50% is the most appropriate proportion in this study. Comparing with the reference value at 7d stipulated for the first-class highway by specification (JTG/T F20—2015), 3–5 MPa is stipulated, and the strength of the 50% specimens exceeds the upper limit. The "modified EDTA titration experiment "shows the titration results after 28 days' maintenance; the specimens with 50% of RBA have the minimum content of Ca^{2+}. Of this proportion, the potential activity of RBA was enormously stimulated and satisfied most of the specifications. When the replacement ratio increases to 80%, the CaO in clay brick increases because RBA in 80% consumes a similar Ca^{2+} amount of 50%, and the strength results showed that the strength generated by potential activation can offset the strength weakened by the weak performances of the part of RBA and improve the strength level.

It can be proved that the activation of RBA activity is based on the hydration reaction of cement. The potential activation of RBA from the strength makes use of the product of the cement hydration reaction. The consumption of this product makes the activated part of RBA form strength at a rate less than that of the cement hydration reaction. xCaO·SiO$_2$·yH$_2$O (calcium silicate hydrate, short for C–S–H) and Ca(OH)$_2$ are major products of the hydration reaction. After a period of time, a certain amount of Ca(OH)$_2$ is produced. Then, Ca(OH)$_2$ and SiO$_2$ in RBA react as shown in Equation (1).

$$Ca(OH)_2 + SiO_2 + H_2O \rightarrow xCaO \cdot SiO_3 \cdot yH_2O \tag{1}$$

In Equation (1), Ca(OH)$_2$ (form cement hydration), SiO$_2$ (the main substances in clay brick), and H$_2$O react and produce xCaO·SiO$_2$·yH$_2$O. This reaction to some extent increases the hydration product C–S–H of cement and increases the amount of the cement that participates in a hydration reaction. When the RBA content reaches a reasonable value, i.e., 50% in this paper, the SiO$_2$ in clay brick consumes a large amount of the Ca(OH)$_2$ produced by the cement hydration reaction, and produces xCaO·SiO$_3$·yH$_2$O with a cementitious effect and increases a part of the strength.

For 50%, the increased strength not only offsets the weakened strength due to the weak performances of RBA, but also achieves a higher strength level than the specimen without RBA replacing CBA, and satisfied most of the specifications.

5. Conclusions

In order to reuse a large amount of construction waste, clay bricks ought to be used in road engineering, as it reduces the environmental and economic pressure. Unconfined compressive strength of cement stabilized macadam subbase is the main strength index in the structure designing process; combining with SEM scanning, XRD scanning and "modified EDTA titration experiment", we studied the contribution of the potential activity of RBA to the strength of cement stabilized macadam subbase and the activation mechanism of its potential activity.

1. Component tests of SEM and XRD scanning showed that SiO_2 and $CaCO_3$ exist in clay brick as the dominant ingredient.
2. When construction waste clay brick is mixed into cement stabilized macadam in the form of aggregate below 4.75mm, there is an optimum content of 50% under the strength standard with the RBA in performance described in this article, and the unconfined compressive strength of 7 d, 28 d, and 90d are 5.46MPa, 7.62MPa, and 9.61MPa respectively, which are greater than the strength of the cement stabilized macadam standard specimen and the values stipulated in specification.
3. The titration results of the "modified EDTA titration test" show that 50% of specimens have the lowest content of Ca^{2+}, and the potential activity is maximized in this proportion.
4. All the experimental results show that the RBA used in this study has potential activation in cement stabilized structure, and the optimal replace proportion (50%) of RBA in cement stabilized macadam subbase is proven. In this proportion, RBA replaced CBA and enhanced the unconfined compressive strength; at this time, the potential activity is fully stimulated. Appropriate methods exist to digest construction waste in large quantities and solve the problem of funding for the disposal process. According to this blending method and proportion, it can be widely applied in the road structure subbase, which can solve the problems of environmental governance and economic problems.

Author Contributions: Conceptualization, C.L. and Y.W.; methodology, Y.W. and Y.G.; validation, C.L.; formal analysis, Y.L.; investigation, Y.W. and Y.G.; writing—original draft preparation, C.L.; writing—review and editing, Y.W. and G.T.; project administration, C.L.; funding acquisition, C.L. and W.S.

Funding: This research was funded by the Science Technology Development Program of Jilin Province grant number 2018ZDGC-9-1 and the National Natural Science Foundation of China grant number 51478203.

Acknowledgments: The authors would like to thank the anonymous reviewers for their constructive suggestions and comments to improve the quality of the paper. In addition, thanks to the Second Highway Bureau of China Communications Construction incorporated company for the supply of raw materials.

Conflicts of Interest: The authors declare no conflict of interest.

References

1. Ten Global Trends for 2019. Available online: https://www.atkearney.com/web/global-business-policy-council/article/?/a/year-ahead-predi (accessed on 12 March 2019).
2. Wang, J.Y. Study on the Comprehensive Utilization of Construction Waste in the Process of Urbanization. *Wall Mater. Innov. Energy Sav. Build.* **2013**, *6*, 38–41. (In Chinese)
3. Marzouk, M.; Azab, S. Environmental and economic impact assessment of construction and demolition waste disposal using system dynamics. *Resour. Conserv. Recycl.* **2014**, *82*, 41–49. [CrossRef]
4. Song, Y.Z.; He, Z.Q.; Shen, B.W. The application of the technology of dregs and granule filling in the subgrade construction of expressway. *Highway* **2016**, *61*, 86–89. (In Chinese)
5. Yang, J.P.; Zhang, M.X.; Li, X.B. Application of Recycled Construction Materials in Special Foundation Treatment of Road Engineering. *Road Mach. Constr. Mech.* **2016**, *33*, 95–98. (In Chinese)
6. Li, X.J. Study on the Pavement Structure of Light Traffic Highway in which Construction Waste Used as the Basic Course. Material. Master's Thesis, Southwest University of Science and Technology, Mianyang, China, 2016. (In Chinese).
7. Guo, Y.L. Comprehensive comparative analysis of construction waste treatment at home and abroad. *Manag. Res. Sci. Technol. Management and Research on Scientific & Technological Achievements.* **2015**, *11*, 28–30. (In Chinese)
8. Li, Y.; Xu, S.H. Study on the Current Situation of Construction Waste. *Constr. Technol.* **2007**, *36*, 480–484. (In Chinese)
9. Li, N.; Li, X.Z. Learn from the experience in recycling of construction waste in developed countries. *Renew. Resour. Circ. Econ.* **2009**, *6*, 41–44. (In Chinese)
10. Wang, L.C. Study on Legal Regulation of Construction Waste. Master's Thesis, Huaqiao University, Quanzhou, China, 2017.

11. Xue, C.Z.; Shen, A.Q.; Guo, Y.C.; Wan, C.G.; Zhang, J. Impact of Construction Waste Composite Powder Material on Concrete Anti-frost Performance. *Mater. Rev.* **2014**, *30*, 121–125. (In Chinese)

12. Zhu, P.; Mao, X.Q.; Qu, W.J.; Li, Z.Y.; Ma, Z.G. Investigation of using recycled powder from waste of clay bricks and cement solids in reactive powder concrete. *Constr. Build. Mater.* **2016**, *113*, 246–254. (In Chinese) [CrossRef]

13. Cheng, H.L. Experimental study on recycled concrete of fly ash and waste clay brick. *Concr. Cem. Prod.* **2005**, *5*, 48–50. (In Chinese)

14. Khalaf; Fouad, M. Using Crushed Clay Brick as Coarse Aggregate in Concrete. *J. Mater. Civ. Eng.* **2006**, *18*, 518–526. [CrossRef]

15. Fořtová, K.; Pavluring, T. The Performances of Fine Recycled Aggregate Concrete Containing Recycled Bricks from Construction and Demolition Waste. *Key Eng. Mater.* **2018**, *760*, 193–198. [CrossRef]

16. Xiao, Z.; Ling, T.C.; Kou, S.C.; Wang, Q.; Poon, C.S. Use of wastes derived from earthquakes for the production of concrete masonry partition wall blocks. *Waste Manag.* **2011**, *31*, 1859–1866. [CrossRef] [PubMed]

17. Hu, L.Q.; Sha, A.M. Performance Test of Cement Stabilized Crushed Clay Brick for Road Base Material. *China J. Highw. Transp.* **2012**, *3*, 73–79. (In Chinese)

18. Turanli, L.; Bektas, F.; Monteiro, P.J.M. Use of ground clay brick as a pozzolanic material to reduce the alkali–silica reaction. *Cem. Concr. Res.* **2003**, *33*, 1539–1542. [CrossRef]

19. Peng, C.J. Study on Performances of Recycled Coarse Aggregate Asphalt Concrete. Master's Thesis, South China University of Technology, Guangzhou, China, 2017. (In Chinese).

20. Li, L.H.; Zhang, N.L. *Road Building Materials*, 1st ed.; Tongji University Press: Shanghai, China, 1999; pp. 18–23. (In Chinese)

21. Hu, S.G. *Advanced Cement-based Composites*; Science Publishing & Media Ltd.: Beijing, China, 2017; pp. 81–85.

22. Pu, X.C.; Wang, Y.W. High-performance active mineral admixtures and concrete. *Concrete* **2002**, *2*, 3–7. (In Chinese)

23. Pu, X.C.; Wang, Y.W. High-performance active mineral admixtures and concrete (continued). *Concrete* **2002**, *3*, 21–24. (In Chinese)

Article

Mechanical Properties of Sandstone Cement-Stabilized Macadam

Qiang Du [1,2]**, Ting Pan** [3,*]**, Jing Lv** [3]**, Jie Zhou** [3]**, Qingwei Ma** [4] **and Qiang Sun** [3]

[1] School of Economics and Management, Chang'an University, Xi'an 710064, Shaanxi, China
[2] Center for Green Engineering and Sustainable Development, Xi'an 710064, Shaanxi, China
[3] School of Civil Engineering, Chang'an University, Xi'an 710061, Shaanxi, China
[4] Xi'an Highway Research Institute, Xi'an 710065, Shaanxi, China
* Correspondence: ting_pan@chd.edu.cn

Received: 25 June 2019; Accepted: 12 August 2019; Published: 22 August 2019

Abstract: Application of sandstone in cement-stabilized macadam (CSM) is an effective way to utilize sandstone. To determine the feasibility of using sandstone as a CSM aggregate, a series of experimental investigations, such as unconfined compressive strength (UCS) tests, Brazilian splitting tests and freeze-thaw cycle tests, were conducted on sandstone cement-stabilized macadam (SCSM). Three mixed variables, covering the cement content, aggregate type and curing period, were set as influence factors. The testing results indicated that the UCS, indirect tensile strength (ITS) and frost resistance property of the test-pieces increased with cement content and curing age. Considering the asphalt pavement design specifications for China, the UCS and ITS values of the SCSM complied with the requirements of light traffic road construction before freeze-thaw cycles. However, the SCSM subjected to freezing and thawing meets the requirements only when the cement content is 4.5%. Therefore, it is noteworthy that CSM containing sandstone aggregates should be applied with caution in cold region because of insufficient freeze resistance.

Keywords: cement stabilized macadam; sandstone; limestone; road performance; freeze-thaw cycles

---

1. Introduction

Cement-stabilized macadam (CSM) is a family of compacted blends containing aggregates with appropriate grading, cement of 3–8% by weight of aggregates and water at optimum moisture content levels [1,2]. CSM has been widely used in highway bases and sub-bases because of better bearing capacity and lower tensile stress or strain at the bottom of the bituminous layer [3]. With the rapid development of road construction, natural stone materials as aggregates of CSM are becoming depleted at an increasing rate, making them insufficient to meet the increasing construction demands [4]. The task of road construction in remote areas, in particular, is a challenge due to its very high demand for aggregate resources. However, the use of high-quality stone may cost more manpower and financial capital in most of these economically underdeveloped areas as a result of stone resource shortages. Hence, it is very significant to hunt for materials as aggregate replacements within a close range along highways to solve the problem of stone resource shortage.

Soft rock is widely distributed along the highways in the southwestern region, South Central China, central China, and Shaanxi-Gansu-Ningxia regions [5], including a sequence of sedimentary rocks, such as mudstone, sandstone, argillaceous sandstone, sandy mudstone, and siltstone [6]. If these soft rocks are abandoned and high-quality rock is purchased from a long distance away, large increases in the construction costs will result, together with extended construction periods and environmental pollution problems caused by the disposal impacts of the abandoned rock. For example, $14 \times 10^6 \ m^3$ of red sandstone was converted into eligible roadbed materials in the construction of the Heng-Zao Expressway in Hunan Province of China, which not only saved 842.5 hectares of farmland and 167.9

hectares of forest land, but also reduced millions related to the construction cost [7]. The use of widely distributed and not widely developed sedimentary rocks for CSM aggregates may be an effective means of addressing the shortages of high-quality stone, thereby simultaneously providing economic and environmental benefits.

Compared with natural aggregates, sedimentary rocks are considered undesirable road materials because of their high-water absorption and large crushing value [8]. Construction and demolition waste (CDW) with the same characteristics has been used as aggregates in existing studies, such as crushed clay brick and recycled concrete [9]. The successful research on the use of CDW for concrete gravel is mostly concentrated in the United States and European countries [10]. Poon and Chan [11] studied the properties of concrete blends with CDW used as an aggregate. Although the mixtures have a lower dry density and higher moisture content as a mixture of aggregates than natural materials, they can still be used in road sub-bases. Disfani et al. [12] studied the properties of CSM mixtures with crushed brick as aggregate through laboratory tests and found that the physical and strength properties of the mixture meet the road requirements. In addition, the researchers [13] conducted unconfined compression tests, split tensile tests and flexural strength tests; it was concluded that road base materials containing recycled concrete aggregates could be used for high-grade road construction. With the shortage of resources and worsening of the environment, recycling has gradually become a concern of academic experts in China. The feasibility of using recycled concrete aggregates as substitutes for natural aggregates was evaluated in lime-fly ash crushed stone bases [14]. The UCS values are better than those of natural aggregates and meet the requirements of road engineering. Based on the above facts, sedimentary rock may be feasible as a substitute for traditional aggregates in road construction.

At present, the research on sedimentary rocks used in highway engineering mainly focuses on the performance of red sandstone, and its uniaxial compressive strength and indirect tensile strength are superior to white sandstone and yellow sandstone [15]. Yao et al. [16] evaluated the physical and mechanical properties of red sandstone distributed in southern Anhui, with the aim of using this soft rock as a road construction material. The results demonstrated that the mixture consisting of sandstone can be directly applied into highway construction after particular preliminary steps are performed. The authors of [17,18] studied the improvement of red sandstone construction technology and applied it to the construction of some roadbeds in Hunan, a province in South Central China, and sufficient engineering results were achieved. In addition, Zhou [19] and Yang et al. [5] studied the performance of improved sandstones in the Yungui area and Gansu Province, respectively, and showed that the improved sandstones could fully meet the technical requirements. The mechanical properties of sandstones in different regions vary widely because of differences in their mineral composition. In addition to the above areas, there is also a large amount of sandstone in northern Shaanxi. However, there are few studies on whether the Cretaceous sandstone with weaker rock quality in the northern Shaanxi area can be used as a road base aggregate.

Therefore, the target of this paper is to assess the physical and mechanical properties of CSM consisting of sandstone through laboratory tests including UCS tests, splitting tensile strength tests and freeze-thaw stability tests. The research also assesses the feasibility of applying sandstone as a raw material of CSM by comparing the test results with the requirements of specifications. This research thus provides possible solutions for the lack of natural stone materials for infrastructure construction and more possibilities for material selection.

2. Materials and Testing Methods

2.1. Materials

2.1.1. Cement

Ordinary Portland cement is used in this study and its main mineral composition includes $3CaO \cdot SiO_2$, $2CaO \cdot SiO_2$, $3CaO \cdot Al_2O_3$ and $3CaO \cdot Al_2O_3 \cdot Fe_2O_3$. Ordinary Portland cement can better hydrate and harden when in contact with water as well as maintaining and developing its strength [20]. The chemical composition of the ordinary Portland cement employed in this research is summarized in Table 1, and its physical properties include a specific gravity of 3.14 and a fineness value of 329 m^2/kg.

Table 1. The chemical composition of the ordinary Portland cement used in this study.

Label	SiO_2	Al_2O_3	CaO	Fe_2O_3	MgO	SO_3
Cement	20.36	5.67	62.81	3.84	2.68	2.51

2.1.2. Aggregate

Four different coarse aggregates, including three types of sandstone marked A, B, C and one type of limestone marked D, were selected for the experiment. These sandstones were randomly obtained from three different production areas in northern Shaanxi, China. The main physical characteristics of the coarse aggregates are recapitulated in Table 2. The bulk density, porosity, water absorption and compressive strength values before and after the ruggedness testing were measured according to JTG E41-2005 [21]. The crushing value was determined following JTG E42-2005 [22].

Table 2. The properties of the coarse aggregates.

Label	A	B	C	D
Bulk density (g/cm^3)	2.6	2.6	2.7	2.7
Porosity (%)	10.1	10.3	9.8	1.2
Water absorption (%)	3.54	3.61	3.18	0.32
Compressive strength (MPa)	48	44.7	59.7	118.6
Compressive strength after ruggedness test (MPa)	26.3	22.4	30.8	107.7
Crushed value (%)	26.7	27.4	23.1	14.3

As shown in Table 2, the porosity, water absorption and crushing values of the sandstones (A, B, and C) are significantly higher than those of the limestone. Among them, the porosity and water absorption values of the sandstones are approximately 10 times those of the limestone, while the crushing values are approximately 2 times the limestone value. In contrast, the sandstones have lower compressive strength values compared with that of limestone. Moreover, after the robustness tests, the compressive strengths of the sandstones were significantly reduced to approximately half of the values before the tests, while the limestone showed little change.

2.2. Experimental Programme

2.2.1. Gradation Design

Under the premise of fully considering the residual porosity and other factors, the coarse and fine aggregates should be sandstone/limestone with a continuous grading, and the grading was artificially compounded through experiment. The accumulated screening rates are demonstrated in Figure 1, where the upper and lower limits refer to the technical specifications of JTG F30-2003 [23].

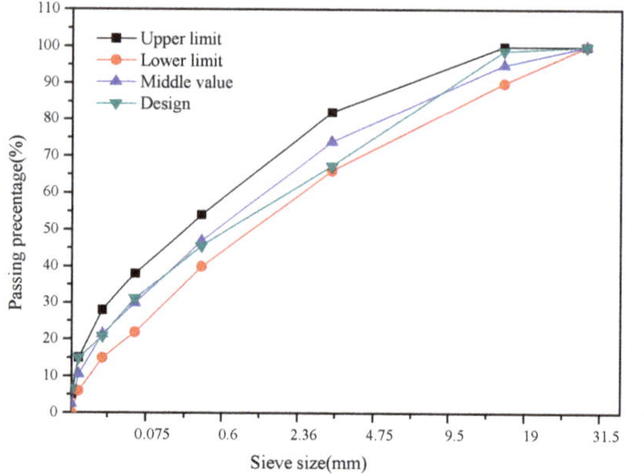

Figure 1. The gradation curve of the mixture.

2.2.2. Mixing Proportion Design

As demonstrated in Table 3, the specimens were assigned to four sets: three types of sandstone and one type of limestone. In China, the cement content of CSM may not exceed 6%, so ordinary Portland cement contents of 3.5%, 4.0%, and 4.5% were selected for the CSM of sandstone. Limestone with 4.0% cement content was chosen to analyse the difference between the sandstones and traditional aggregate materials.

Table 3. The cement content.

Aggregate Type	Code Number	Cement Content
Sandstone A	A1	3.5%
	A2	4.0%
	A3	4.5%
Sandstone B	B1	3.5%
	B2	4.0%
	B3	4.5%
Sandstone C	C1	3.5%
	C2	4.0%
	C3	4.5%
Limestone	D	4.0%

2.2.3. Unconfined Compressive Strength

According to the JTG E51-2009 [24], the SCSM and CSM mixtures were processed into standard test specimens of Φ150 mm × 150 mm by a compressor with a 98% degree of compaction and cured for 7 days, 28 days, 60 days, 90 days and 180 days under standard conditions (20 ± 2 °C and 90 ± 5% relative humidity (RH)). First, the specimens were placed on the pressure machine, and a flat ball base was placed on the lift platform. Then, the specimens were tested at an invariable loading velocity of 1 mm/min. The results are the average values of the three repetitive specimens from each specific combination.

2.2.4. Indirect Tensile Strength

Brazilian splitting tests were conducted in accordance with ASTM C496/C496 M-11 [25]. Concrete specimens were cast into Φ100 mm × 100 mm cylindrical mould for the Brazilian splitting tests and cured for 7 days, 28 days, 60 days, 90 days and 180 days under standard conditions. The concrete specimens were tested by applying force along the longitudinal axis of the cylinder utilizing an alignment fixture at a constant rate of loading of 0.5 k N/s, as shown in Figure 2. The maximum tensile force at the time of failure of the test piece is obtained, and the ITS can be calculated as follows:

$$R_i = \frac{2P}{\pi dh}\left(\sin 2\alpha - \frac{a}{d}\right) \tag{1}$$

where P is the maximum tensile force, d refers to the specimen diameter, h is the height of the test piece, a denotes the width of the batten, and α is the corresponding centre angle on the half width of the lath.

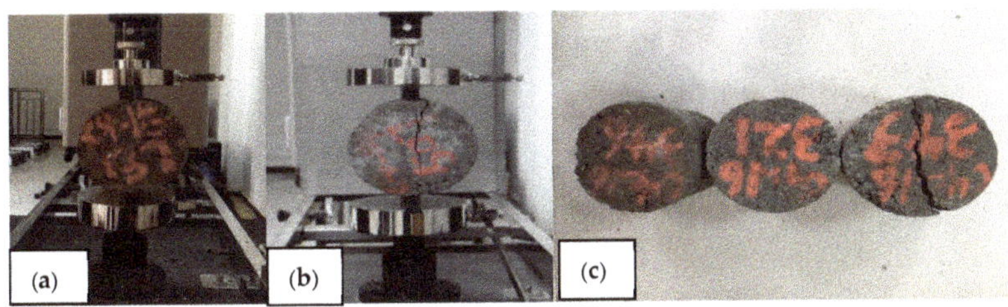

Figure 2. Indirect tensile strength test: (**a**). installation; (**b**). continuous loading; (**c**). damage.

The water stability is one of the factors considered in the study of the pavement material properties. The softening coefficient is an important indicator for characterizing a CSM. This coefficient reflects the ability of a mixture to resist water damage and has distinct influence on materials with high water absorption and porosity levels. Considering the high water absorption and porosity of the sandstone, the test pieces of each curing age were subjected to water immersion treatments, and the test pieces in the saturated and dry states were tested according to the above experimental methods. Then, the softening coefficient was calculated according to Formula (2):

$$K_p = \frac{R_w}{R_d} \tag{2}$$

where R_w refers to the ITS of a water-saturated specimen and R_d denotes the ITS of a dry specimen.

2.2.5. Freezing and Thawing

Freeze-thaw stability testing was based on JTG E51-2009 and the weight loss of the test specimens was conducted by an automatic freeze-thaw machinery. The freezing and thawing cycle experiments were carried out after the cylindrical specimens were cured for 7, 28, 60, 90, and 180 days under the specific curing conditions (20 ± 2 °C and 90 ± 5% RH). The cylindrical test-pieces were frozen at minus 20 degrees Celsius and thawed in water at 20 degrees Celsius. The freeze-thaw cycle experiments were set to 5 cycles. Usually, the ratio of the compressive strengths before and after freezing and thawing cycles is used to assessed the anti-frost property of blends, namely,

$$BDR_1 = \frac{R_{DC1}}{R_{C1}} \times 100\% \tag{3}$$

where BDR_1 represents the compressive strength loss of the specimen after freeze-thaw cycles, R_{DC1} refers to the compressive strength after freeze-thaw cycles, and R_{C1} denotes the compressive strength before freeze-thaw cycles.

In addition, the residual tensile strength ratio after a freeze-thaw cycle is used as a supplementary index. The supplementary anti-freeze index (BDR_2) can be calculated as follows:

$$BDR_2 = \frac{R_{DC2}}{R_{C2}} \times 100\% \tag{4}$$

where BDR_2 represents the ITS loss of the specimen after a freeze-thaw cycles, R_{DC2} refers to the ITS after freeze-thaw cycles, and R_{C2} denotes the ITS before freeze-thaw cycles.

3. Test Results

The following shows a series of tests results on the UCS and ITS under water-saturated and dry conditions and after freeze-thaw cycles. The results are summarized and presented in Table 4.

3.1. Unconfined Compressive Strength

The UCS is normally considered to be a major indicator for evaluating the quality of the CSM mixture. Many mixed variables affect the UCS, such as the type of aggregate, the cement content, and the curing time. The UCSs of the various mixtures are shown in Table 4 for the ages of 7, 28, 60, 90 and 180 days, and the experimental data presented are the averages of three specimens for each set of mixtures. It may be observed by comparing the experimental values of the three types of sandstone mixtures that the sandstone type has little effect on the strength. As mentioned earlier, there are few differences between the chemical compositions of the sandstones from the three producing areas. Therefore, only the data of sandstone A are used in the following analysis and comparison.

3.1.1. Influence of Cement Content

It is widely known that the cement used in CSM can effectively improve the adhesion level and mechanical properties of the mixtures. The effect of cement content on the UCS is displayed in Figure 3. The overall trend is a rise in the UCS value of the SCSM as the cement content increases, which is because the enhanced effect of cement on the strength of the material and the bonding force between the particles is enhanced by the increase in hydrated products as expected [26]. In addition, based on the slope of the curve, it can be seen that at the same age, the increase rate of the UCS is very low when the cement content adds from 3.5 to 4.0%, but the growth rate becomes significantly higher as the cement content increases from 4.0 to 4.5%. For instance, when the cement content increases from 3.5 to 4.0% at 60 days of curing, the strength of the SCSM rises by approximately 0.07 MPa, and the increase from 4.0 to 4.5% results in an approximate 0.6 MPa rise. However, the experimental results of Farhan et al. [26] show that the development rate of the UCS of a traditional CSM is almost proportional to the cement content. The reason for the above difference may be the large porosity of the sandstone. On the other hand, the strength of the SCSM with the highest cement content in the research range is still lower than those of the source rocks, which indicates that the main reason for the failure of the test piece may not be the devastation of the aggregate. Usually, the initial micro-cracks of an aggregate concrete appear in the interfacial transition zone [27]. Therefore, the failure of the CSM may be caused by the low degree of bonding between the aggregate and mortar.

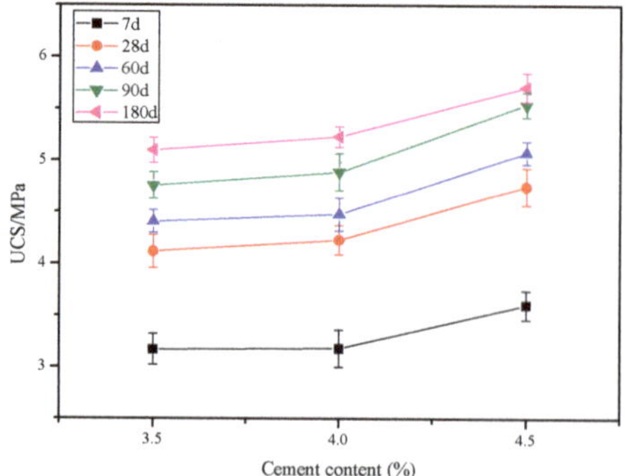

Figure 3. Relationship between the UCS and cement content.

Table 4. The experimental results of the unconfined compressive strength, indirect tensile strength.

Serial Number	UCS (MPa)										ITS (MPa)														
	Free From F-T					After 5 F-T Cycles					Dry State					Water-Saturated State					After 5 F-T Cycles				
	7 d	28 d	60 d	90 d	180 d	7 d	28 d	60 d	90 d	180 d	7 d	28 d	60 d	90 d	180 d	7 d	28 d	60 d	90 d	180 d	7 d	28 d	60 d	90 d	180 d
A1	3.17	4.12	4.41	4.76	5.10	2.79	3.63	4.03	4.34	4.59	0.32	0.42	0.43	0.46	0.48	0.26	0.35	0.37	0.40	0.42	0.25	0.31	0.34	0.37	0.39
A2	3.18	4.23	4.48	4.89	5.23	2.89	3.81	4.12	4.54	4.76	0.35	0.43	0.49	0.53	0.55	0.29	0.37	0.44	0.47	0.49	0.28	0.35	0.40	0.44	0.47
A3	3.60	4.75	5.08	5.54	5.71	3.35	4.42	4.78	5.26	5.31	0.39	0.47	0.54	0.57	0.60	0.34	0.41	0.49	0.52	0.55	0.32	0.39	0.45	0.47	0.51
B1	3.52	4.58	4.90	5.29	5.66	3.38	4.09	4.48	4.87	5.01	0.37	0.47	0.51	0.55	0.56	0.26	0.35	0.37	0.40	0.42	0.24	0.33	0.36	0.37	0.40
B2	3.61	4.80	5.09	5.55	5.94	3.47	4.09	4.47	5.19	5.40	0.39	0.49	0.57	0.60	0.63	0.34	0.41	0.49	0.52	0.55	0.32	0.38	0.45	0.47	0.53
B3	3.91	5.16	5.52	6.02	6.20	3.44	4.39	4.76	5.10	5.25	0.44	0.52	0.59	0.63	0.65	0.34	0.41	0.49	0.52	0.55	0.32	0.38	0.45	0.47	0.53
C1	3.44	4.47	4.79	5.17	5.53	3.28	4.01	4.39	4.78	4.90	0.36	0.47	0.50	0.54	0.56	0.29	0.38	0.43	0.47	0.49	0.27	0.37	0.42	0.44	0.47
C2	3.50	4.66	4.93	5.38	5.75	3.09	3.95	4.38	5.04	5.33	0.35	0.48	0.56	0.59	0.61	0.29	0.42	0.51	0.53	0.55	0.27	0.39	0.46	0.50	0.52
C3	3.88	5.12	5.48	5.97	6.15	3.44	4.36	4.77	5.08	5.32	0.45	0.52	0.59	0.63	0.64	0.38	0.45	0.54	0.57	0.59	0.36	0.41	0.51	0.53	0.57
D	5.18	7.25	8.12	8.53	9.13	4.92	6.82	7.72	8.10	8.58	0.63	0.85	0.99	1.07	1.12	0.55	0.77	0.93	1.01	1.05	0.53	0.72	0.89	0.97	1.01

3.1.2. Influence of the Aggregate Type

It is well known that the UCS of CSM is closely related to the aggregate strength in the mixture. Mixtures of sandstone A and limestone with a cement content of 4.0% were tested. It can be observed from Figure 4 that the compressive strength of the SCSM mixture is lower than the compressive strength of the CSM mixture for the same curing period. The UCS of the limestone sample is approximately twice that of the sandstone sample; correspondingly, the compressive strength of the limestone parent rock is 2.0 times that of the sandstone A parent rock. This finding means that the characteristics of the parent rock, including its chemical composition and physical mechanics, are significant factors influencing the UCS of the CSM with this rock as the aggregate. Meanwhile, the strength of the SCSM is lower than that of the CSM, which may be because the dust attached to the surface of the sandstone weakens its bond with the cement slurry.

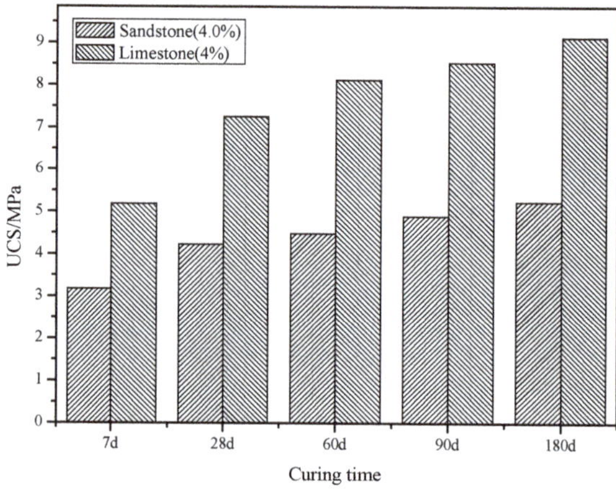

Figure 4. Relationship between the UCS and aggregate type.

3.1.3. Influence of Curing Time

In addition, a significant factor influencing the UCS of CSM is the curing period of specimens. Numerous studies have reported the curing time's influence on the UCS. The UCS development with the curing time is shown in Figure 5. It is observed from this figure that the effects of the curing period on the strengths of the SCSM blend and the CSM blend are similar in the case of the identical cement content. The longer the curing time, the greater the strength. Du carried out the same performance test on CSM with asphalt emulsion, and the growth trend of strength was similar to the test results in this paper with the increase of curing age [28]. The reason for this phenomenon is that the hydration reaction is the time-dependent action. The developing velocity of the UCS is usually proportional to the cement content, which is since the more cement is added in the blends, the more products of hydration reaction [29] and the better the strength enhancement and bonding effects. As shown in Table 5, the compressive strength increases rapidly in the first 28 days, but the increase slows down at 60 days and 180 days. In particular, from 7 to 28 days, the strength of the SCSM increased by 30%, while from 28 to 60 days, the SCSM strength only increased by 7%. This result is because the cement granules without hydration reaction are surrounded by formed cement slurry, making it difficult for water to enter the surface of the un-hydrated cement particles. This phenomenon impedes the hydration of the cement granules, thus leading to a slower increase in the UCS at late period. According to JTG D50-2017 [30], the UCS of CSM mixture bases used in medium or light traffic roads at an age of 7 days should be between 3.0 and 5.0 MPa. From the experimental results, the UCS of the 7-day SCSM successfully met the requirements.

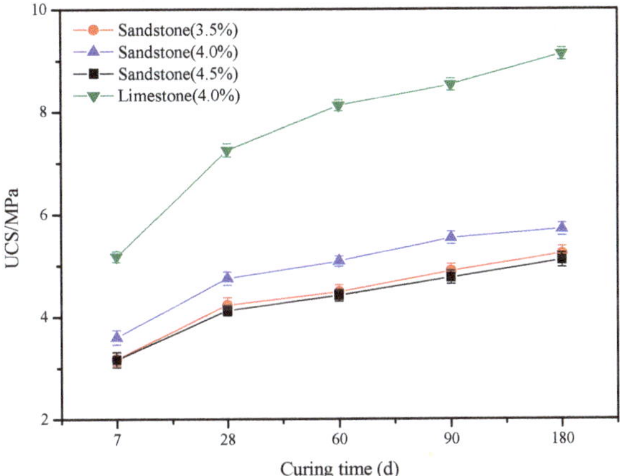

Figure 5. Relationship between the UCS and curing time.

Table 5. The growth rate (%) of UCS at different ages.

Code	7 d	28 d (Based on 7 d)	60 d (Based on 28 d)	90 d (Based on 60 d)	180 d (Based on 90 d)
A1	-	30%	7%	8%	7%
A2	-	33%	6%	9%	7%
A3	-	32%	7%	9%	3%
D	-	40%	12%	5%	7%

3.2. Indirect Tensile Strength

3.2.1. Crack Resistance

The tensile strength of the mixture was determined by the ITS test at the time of failure to evaluate the ability of the CSM to resist cracking [31]. Table 4 shows the ITS values of the SCSM with different cement contents and CSM with cement content of 4% at ages of 7, 28, 60, 90 and 180 days in water-saturated and dry situations. It was observed from Figure 6 that the values of ITS increased proportionately with the increase in the amount of cement and the curing time, whether under dry or water-saturated conditions. In fact, the mechanisms of the effects of cement content and curing age on the ITS are similar to those of the UCS. In addition, it can be observed in Figure 7 that the CSM exhibited much higher ITS than the SCSM for the same cement content at 7, 28, 60, 90 and 180 days. The reason for this phenomenon is most likely because the crushing value of the sandstone parent rock is significantly higher than that of the limestone, which can be seen in Section 2.1.2. Different from the UCS, the ITS is mainly affected by the interfacial bonding in the CSM between the cement mixture and lightweight aggregate particles [32]. This phenomenon may also be due to the high clay content of the sandstone, which is present in the form of fine aggregates or encased on the surface of the mixture, thereby significantly delaying the hydration of the Portland cement. This not only weakens the cohesion between the aggregate and the cement but also affects the ITS of the CSM. According to JTG D50-2017, the ITS of the CSM mixture base should be between 0.4 and 0.6 MPa at 90 days of age. It can be observed from the test results that if sandstone is used instead of a natural aggregate, the ITS values are higher than the criterion for cement stabilized base materials of the standard.

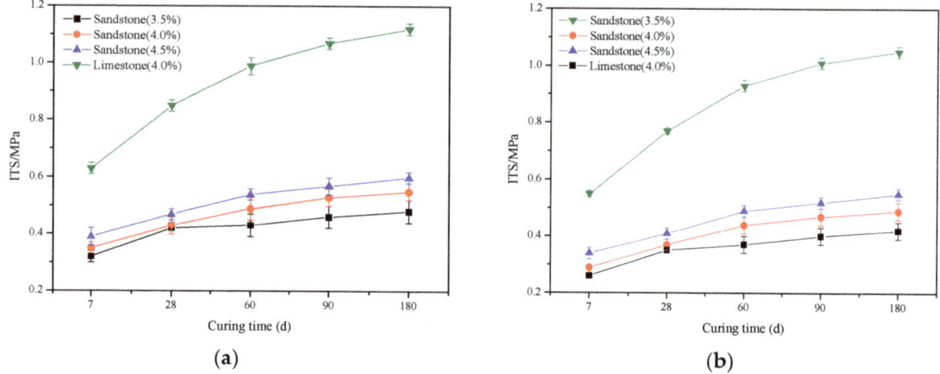

Figure 6. Relationship between the ITS and curing time. (**a**) Dry state (**b**) Water-saturated stat.

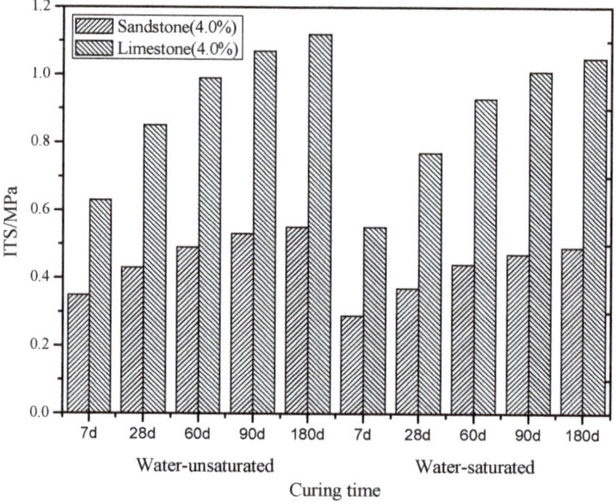

Figure 7. Relationship between the ITS and aggregate type.

3.2.2. Water Stability

Sandstone, a type of sedimentary rock, is usually affected by water action, so ITS tests were conducted in both dry and saturated states. The calculated softening coefficients are summarized in Table 6. It can be seen from the table that the softening coefficient is less than 1, which means that the water saturation has a weakening effect on the splitting tensile strengths of the SCSM and CSM. Apparently, for the same cement content, the softening coefficient of the CSM is higher than that of the SCSM. For example, the CSM value is 5% higher than the SCSM value at a curing age of 7 days. This result may be related to their different water absorption rates and porosities, as mentioned above. Meanwhile, as the curing age increased, the softening coefficient is gradually increased at a decreasing rate. With the curing age ranging from 7 to 60 days, the softening coefficients of the SCSM and CSM increased by 5% and 6%, respectively, while the coefficients were almost unchanged from 60 to 180 days. This finding might be mainly due to the increase in the curing age, which caused the transformation of the hydration products into a hydrophobic gel [33]. In addition, it is clear that the softening coefficient of the SCSM increases with the increase in the amount of cement. It is well known that the water resistance of a material is tightly related to the pore structure of the material and the method of adhesion between the particles. Therefore, effectively reducing the porosity of the sandstone is an important means to improve the performance of CSM with sandstone as the aggregate.

Table 6. The softening coefficient at different ages.

Code	7 d	28 d	60 d	90 d	180 d
A1	0.80	0.82	0.85	0.87	0.87
A2	0.83	0.86	0.90	0.89	0.90
A3	0.85	0.88	0.91	0.91	0.92
D	0.88	0.90	0.94	0.94	0.94

3.2.3. Relationship between the UCS and ITS

The fitting curve of the UCS and ITS values for SCSM and CSM blends in different curing periods is shown in Figure 8, from which it can be inferred that the ITS value is about 10% of the UCS value, which is suitable for total examined blends at different ages. The phenomenon represents there is a unique connection between the UCS and ITS, regardless of the composition of the mixture (aggregate type, cement content and curing age). In general, the tensile strength of ordinary concrete is 1/10 to 1/20 of its compressive strength. In the study of [34], for different natural aggregate mixtures with the disparate amount of cement, the results indicated that there was a linear relation, namely, UCS $=9.8 \times$ ITS. Thus, the current test results have been proven to be reasonable.

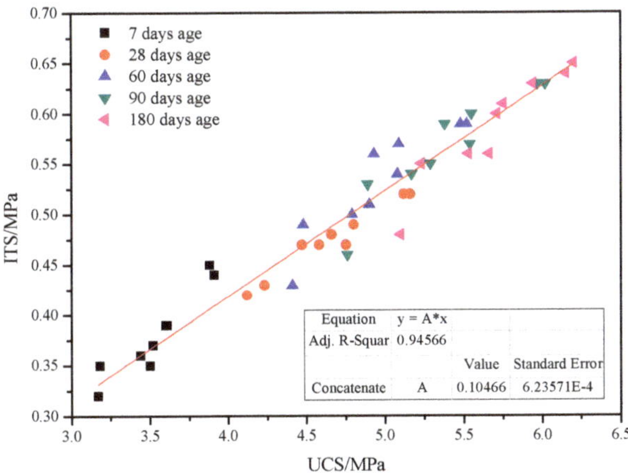

Figure 8. Relationship between the UCS and ITS.

3.3. Frost Resistance

In northern Shaanxi, the climate is characterized by cold and long winters. After repeated freeze-thaw cycles during the winter and early spring thawing, the semi-rigid base layer is susceptible to freeze-thaw failure, resulting in melt settling and frost heave. The frost heaving action of the pore water in the semi-rigid base material damages the cementing action between the particles, which is the cause of the instability of the mixture caused by freeze-thaw action. As a type of sedimentary rock, sandstone has lower mechanical properties and higher water absorption and porosity levels than those of limestone. Therefore, the frosting process (freeze-thaw cycles) that can destroy CSM is a significant problem. The anti-frost property of the material is characterized by BDR_1 and BDR_2 to determine the amplitude range of the mechanical properties of the cement stabilized substrate in cold weather conditions. The results of the different frost resistance indexes before and after freezing and thawing cycles are summarized in Figures 9 and 10. All frost resistance indexes are lower than 100%, which indicates that the freezing and thawing effect can cause the attenuation of strength. During the frosting process, the pore water of the mixture gradually freezes in the capillary chamber, creating a water pressure as frozen water bulk increases. As shown in Figure 9, the lower the cement incorporation is, the greater the intensity attenuation of the SCSM. For instance, when the cement content is 4.5%, both the UCS and ITS are attenuated by approximately 5%. However, the attenuation amplitude approaches 11% with a cement content of 3.5%. The reason for the faster deterioration of the mechanical properties of the mixture at a lower cement dosage is the cementation is the main factor of the adhesive property of the mixture. In the study of the antifreeze properties of other types of CSM bases, the experimental results showed that the cement content is an important factor affecting the antifreeze performance similarly [35,36]. Therefore, to increase the frost resistance of the SCSM mixture, increasing the cement content appropriately is an effective method. In addition, Figure 10 shows that the attenuation of the SCSM is significantly higher than that of the CSM. Compared with porous cement stabilized macadam, the strength loss of SCSM after freeze-thaw cycles is relatively smaller, which confirms the frost resistance is affected by the porosity of the material, its internal moisture and the environmental conditions. Therefore, the poor freezing resistance of the SCSM mixture is due to its high porosity and water absorption level.

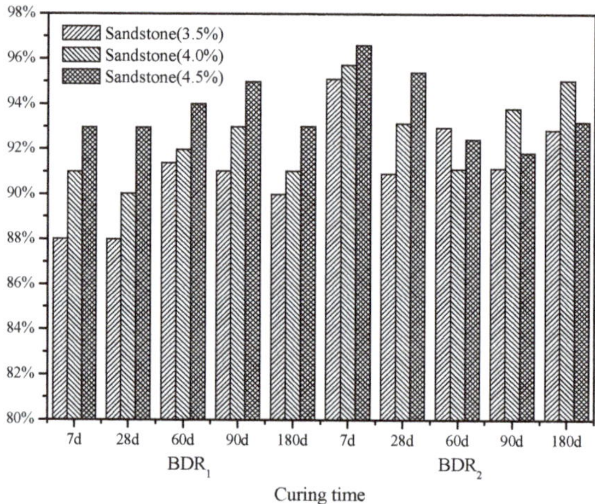

Figure 9. Relationship between the BDR and cement content.

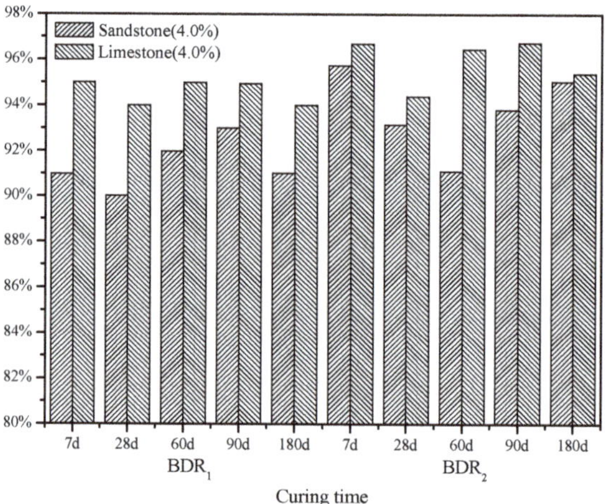

Figure 10. Relationship between the BDR and aggregate type.

4. Conclusions

In this paper, sandstone is utilized as a coarse aggregate for CSM, and the mechanical properties and influential elements of the SCSM and CSM are evaluated by laboratory tests. The results lead to the following conclusions:

1. The results show that the cement content and curing age are factors affecting the ITS and UCS. The mechanical properties of the SCSM blend increase with the cement dosage and curing period, similar to the CSM mixture.
2. The strength of the SCSM blend is significantly lower than the strength of the CSM blend. The cause of this phenomenon may be the differences in the properties of the parent rock, including the porosity, crushing value and compressive strength. It may also be due to the weak bonding at the interface between the sandstone and cement.
3. Both the UCS and ITS of the SCSM and CSM blends are affected by frost action. However, the strength degradation amplitude of the SCSM blend caused by freeze-thaw effect is larger than that of the CSM blend. The degradation amplitude increased with increasing cement content, and the curing age has little effect on the amplitude.
4. The properties of the SCSM, including the UCS, ITS, softening coefficient and frost resistance coefficient, meet the requirements of low-grade roads.

In view of the above conclusions, sandstone can be used for road base construction. Furthermore, applying sandstone to the actual construction of on-site resource utilization will bring suitable economic and environmental benefits.

Author Contributions: All authors contributed equally to this work. All authors wrote, reviewed and commended on the manuscript. All authors have read and approved the final manuscript.

Funding: This study is sponsored by the National Social Science Foundation of China (Grand No. 16CJY028), Transportation Technology Project of Shaanxi Province (Grand No. 15-06k) and the Fundamental Research Funds for the Central Universities (Grand No. 300102238303, 300102239617).

Conflicts of Interest: The authors declare no conflict of interest.

References

1. China Professional Standard. *JTJ 034, Technical Specifications for Construction of Highway Roadbase*; Ministry of Communications of the People's Republic of China; China Communication Press: Beijing, China, 2000.
2. Xuan, D.X.; Houben, L.J.M.; Molenaar, A.A.A.; Shui, Z.H. Mechanical properties of cement-treated aggregate material—A review. *Mater. Des.* **2012**, *33*, 496–502. [CrossRef]
3. Li, W.; Lang, L.; Lin, Z.; Wang, Z.; Zhang, F. Characteristics of dry shrinkage and temperature shrinkage of cement-stabilized steel slag. *Constr. Build. Mater.* **2017**, *134*, 540–548. [CrossRef]
4. Bektas, F.; Wang, K.; Ceylan, H. Effects of crushed clay brick aggregate on mortar durability. *Constr. Build. Mater.* **2007**, *23*, 1909–1914. [CrossRef]
5. Yang, C.; Zhao, K.; Liu, D.R.; Zhang, W.Q. Study on properties of cement modified weak cemented sandstone filling material. *Bull. Chin. Ceram. Soc.* **2018**, *37*, 630–634.
6. Kanji, M.A. Critical issues in soft rocks. *J. Rock Mech. Geotech. Eng.* **2014**, *6*, 186–195. [CrossRef]
7. Hu, X.M. Technical quality measurement and control of construction of red-sandstone-road-bed in freeway. *Road Mach. Constr. Mech.* **2004**, *21*, 46–48.
8. Rodgers, M.; Hayes, G.; Healy, M.G. Cyclic loading tests on sandstone and limestone shale aggregates used in unbound forest roads. *Constr. Build. Mater.* **2009**, *23*, 421–2427. [CrossRef]
9. Xuan, D.X.; Molenaar, A.A.A.; Houben, L.J.M. Compressive and Indirect Tensile Strengths of Cement Treated Mix Granulates with Recycled Masonry and Concrete Aggregates. *J. Mater. Civ. Eng.* **2012**, *21*, 577–585. [CrossRef]
10. Aliabdo, A.A.; Abd-Elmoaty, A.E.M.; Hassan, H.H. Utilization of crushed clay brick in concrete industry. *Alex. Eng. J.* **2014**, *53*, 151–168. [CrossRef]
11. Poon, C.S.; Chan, D. Feasible use of recycled concrete aggregates and crushed clay brick as unbound road sub-base. *Constr. Build. Mater.* **2006**, *20*, 578–585. [CrossRef]

12. Disfani, M.M.; Arulrajah, A.; Haghighi, H.; Mohammadinia, A.; Horpibulsuk, S. Flexural beam fatigue strength evaluation of crushed brick as a supplementary material in cement stabilized recycled concrete aggregates. *Constr. Build. Mater.* **2014**, *68*, 667–676. [CrossRef]
13. Sobhan, K. Stabilized fiber-reinforced pavement base course with recycled aggregate. *Diss. Abstr. Int.* **1997**, *58*, 2024.
14. Liu, J.G. Recycling application of construction waste in the road base. *China High Technol. Enterp.* **2012**, *10*, 46–47.
15. Wang, Y.F.; Su, H.; Wang, L.P.; Jiao, H.Z.; Li, Z. Study on the difference of deformation and strength characteristics of three kinds of sandstone. *J. China Coal Soc.* **2019**. [CrossRef]
16. Yao, H.; Jia, S.; Gan, W.; Zhang, Z.; Lu, K. Properties of crushed red-bed soft rock mixtures used in subgrade. *Adv. Mater. Sci. Eng.* **2016**, *2016*. [CrossRef]
17. Tan, S.F.; Li, L.J. Field test research for subgrade filling constructed by red sandstone in liu-li expressway. *Highw. Eng.* **2011**, *36*, 43–47.
18. Zhao, M.H.; Zou, X.J.; Zou, P.X.W. Disintegration characteristics of red sandstone and its filling methods for highway roadbed and embankment. *J. Mater. Civ. Eng.* **2007**, *19*, 404–410. [CrossRef]
19. Yang, Z.G.; Zhou, X.H.; Wang, J. Feasibility Research of Application of Sandrock in Cement Stabilized Crushed Stone Mixture of National Highway. *Munic. Eng. Technol.* **2017**, *35*, 175–177.
20. Lv, J.; Zhou, T.H.; Du, Q.; Wu, H.H. Effects of rubber particles on mechanical properties of lightweight aggregate concrete. *Constr. Build. Mater.* **2015**, *91*, 145–149. [CrossRef]
21. China Professional Standard. *JTG E41, Test Methods of Rock for Highway Engineering*; Ministry of Transport of the People's Republic of China; China Communication Press: Beijing, China, 2005.
22. China Professional Standard. *JTG E42, Test Methods of Aggregate for Highway Engineering*; Ministry of Transport of the People's Republic of China; China Communication Press: Beijing, China, 2005.
23. China Professional Standard. *JTG F30, Technical Specification for Construction of Highway Cement Concrete Pavements*; Ministry of Transport of the People's Republic of China; China Communication Press: Beijing, China, 2003.
24. China Professional Standard. *JTG E51, Test Methods of Materials Stabilized with Inorganic Binders for Highway Engineering*; Ministry of Transport of the People's Republic of China; China Communication Press: Beijing, China, 2009.
25. ASTM International. *ASTM C496, Standard Test Method for Splitting Tensile Strength of Cylindrical Concrete Specimens*; ASTM International: West Conshohocken, PA, USA, 2017.
26. Farhan, A.H.; Dawson, A.R.; Thom, N.H. Effect of cementation level on performance of rubberized cement-stabilized aggregate mixtures. *Mater. Des.* **2016**, *97*, 98–107. [CrossRef]
27. Li, W.G.; Long, C.; Luo, Z.Y.; Huang, Z.Y.; Long, B.H.; Ren, J.Q. Review of Failure Mechanism and Modification Research of Recycled Aggregate Concrete. *J. Archit. Civ. Eng.* **2016**, *33*, 60–72.
28. Du, S.W. Mechanical properties and shrinkage characteristics of cement stabilized macadam with asphalt emulsion. *Constr. Build. Mater.* **2019**, *203*, 408–416. [CrossRef]
29. Barišić, I.; Dokšanović, T.; Draganić, H. Characterization of hydraulically bound base materials through digital image correlation. *Constr. Build. Mater.* **2015**, *83*, 299–307. [CrossRef]
30. China Professional Standard. *JTG D50, Specifications for Design of Highway Asphalt Pavement*; Ministry of Transport of the People's Republic of China; China Communication Press: Beijing, China, 2017.
31. Huang, B.; Dong, Q.; Burdette, E.G. Laboratory evaluation of incorporating waste ceramic materials into Portland cement and asphaltic concrete. *Constr. Build. Mater.* **2009**, *23*, 3451–3456. [CrossRef]
32. Najjar, M.F.; Soliman, A.M.; Nehdi, M.L. Critical overview of two-stage concrete: Properties and applications. *Constr. Build. Mater.* **2014**, *62*, 47–58. [CrossRef]
33. Dong, J.L.; Wang, L.J.; Yang, D.L. Experimental Study on the Strength and Hydration Products of Alkali-Activated Pisha Sandstone. *Yellow River* **2016**, *38*, 35–38.
34. Ji, X.; Jiang, Y.; Liu, Y. Evaluation of the mechanical behaviors of cement-stabilized cold recycled mixtures produced by vertical vibration compaction method. *Mater. Struct.* **2016**, *49*, 2257–2270. [CrossRef]
35. Xu, B.; Yang, Z.H.; Zhao, L.H.; Cao, D.W.; Zhang, H.Y.; Shi, X. Study on the antifreeze performance of porous cement stabilized macadam. *Constr. Build. Mater.* **2019**, *208*, 13–22. [CrossRef]
36. Wang, T.L.; Song, H.F.; Yue, Z.R.; Hu, T.F.; Sun, T.C.; Zhang, H.B. Freeze–thaw durability of cement-stabilized macadam subgrade and its compaction quality index. *Cold Reg. Sci. Technol.* **2019**, *160*, 13–20. [CrossRef]

MDPI
St. Alban-Anlage 66
4052 Basel
Switzerland
Tel. +41 61 683 77 34
Fax +41 61 302 89 18
www.mdpi.com

Applied Sciences Editorial Office
E-mail: applsci@mdpi.com
www.mdpi.com/journal/applsci